Suresh Kumar Krishnadhas, Ananth Kumar Tamilarasan, Pramod Singh Rathore, Abhishek Kumar (Eds.)

Artificial Intelligence and Robotics

Mathematical Methods in the Digital Age

Computational Intelligence and Advancements

Edited by
Abhishek Kumar, Satya Prakash Yadav,
Ashutosh Kumar Dubey, Andrei Chtcheprov,
Sayavur Bakhtiyarov

Volume 8

Artificial Intelligence and Robotics

Transformative and Computational Algorithms for Enhanced Problem Solving

Edited by
Suresh Kumar Krishnadhas, Ananth Kumar Tamilarasan, Pramod Singh Rathore, Abhishek Kumar

DE GRUYTER

Editors
Dr Suresh Kumar Krishnadhas
Computer Science and Engineering
Sri Eshwar college of Engineering
Coimbatore 641202, Tamilnadu, India
sureshkumarkskphd@gmail.com

Dr Ananth Kumar Tamilarasan
Computer Science and Engineering
IFET College of Engineering
Villupuram 605108, Tamilnadu, India
tananthkumar@ifet.ac.in

Dr Pramod Singh Rathore
Department of Computer Science and Engineering
Manipal University Jaipur
Jaipur 303007, Rajasthan, India
pramodrathore88@gmail.com

Dr Abhishek Kumar
Department of Computer Science and Engineering
Chandigarh University
Mohali 140413, Punjab, India
Abhishekkmr812@gmail.com

ISBN 978-3-11-914520-6
e-ISBN (PDF) 978-3-11-221437-4
e-ISBN (EPUB) 978-3-11-221466-4
ISSN 2943-2731

Library of Congress Control Number: 2026930889

Bibliographic information published by the Deutsche Nationalbibliothek
The Deutsche Nationalbibliothek lists this publication in the Deutsche Nationalbibliografie; detailed bibliographic data are available on the Internet at http://dnb.dnb.de.

Typesetting: Integra Software Services Pvt. Ltd.
Printing and binding: CPI books GmbH, Leck

www.degruyterbrill.com
Questions about General Product Safety Regulation:
productsafety@degruyterbrill.com

Preface

Artificial intelligence (AI) and robotics are becoming more and more important in our lives. Hence it is important to understand the basic principles of mathematics. These fields go beyond algorithms and code, and they are based on strict mathematical rules that make their amazing abilities and ongoing growth possible. The book *Quantitative Approaches to Artificial Intelligence* is a complete guide that helps make these basic mathematical ideas clearer.

The main reason for this effort is the need for a resource that links abstract mathematical ideas to their real-world uses in AI and robotics. Many books talk about specific AI techniques, but not as many talk about all the different quantitative methods that make these new technologies possible. This book tries to fill that gap by giving readers a full picture of the logic, frameworks, and tools that make intelligent systems work.

Researchers, engineers, academics, and advanced students who want to learn more about AI and robotics from the ground up can all benefit from this book. This book is a great resource for people who want to learn more about modern AI, develop new algorithms, or design robotic systems.

This guide covers a wide range of topics, from basic ones like algorithm design, cryptography, and optimization to more advanced ones like quantum algorithms, network theory, and big data analytics. We look at the difficulties of machine learning, the interesting field of swarm robotics, and most importantly, the big moral and social issues that come up as AI gets better. The goal of each section is to explain complicated mathematical ideas while also showing how they are useful and relevant in real-world AI and robotics situations.

This collection of quantitative methods is meant to encourage more research, new breakthroughs, and a greater appreciation for the mathematical beauty behind the smart technologies that will shape our future.

https://doi.org/10.1515/9783112214374-202

Contents

List of Contributors

Dr I. Divya
Department of Artificial Intelligence and Data Science
Rajalakshmi Institute of Technology
Chennai, Tamil Nadu, India
inbarajdivya@gmail.com
Chapter 1

K. Sandhiya Rajeshwari
Department of Artificial Intelligence and Data Science
Rajalakshmi Institute of Technology
Chennai, Tamil Nadu, India
And
Department of Electronics and Communication Engineering
Saveetha College of Engineering
Chennai, Tamil Nadu, India
sandhiyarajeswari.k@ritchennai.edu.in
Chapter 1

Dr G. Revathi
Department of Electronics and Communication Engineering
Saveetha College of Engineering
Chennai, Tamil Nadu, India
revathiresearch15@gmail.com
Chapter 1

S. Sangeetha
Department of Artificial Intelligence and Data Science
Rajalakshmi Institute of Technology
Chennai, Tamil Nadu, India
sangeethasegar255@gmail.com
Chapter 1

Swetha S.
Department of Artificial Intelligence and Machine Learning
RMD Engineering College
Kavaraipettai, Tamil Nadu
India
swethasundar2611@gmail.com
Chapter 2

Joe Prathap P. M.
Department of Computer Science and Engineering
Sapthagiri NPS University
Bengaluru, Karnataka, India
joeprathap@gmail.com
Chapter 2

Dr A. Vaithiyanathan
Department of Science and Humanities (Mathematics)
IFET College of Engineering
Villupuram, Tamil Nadu, India
vaithi05@gmail.com
Chapter 3

Dr A. Mohammed Shapique
Department of Science and Humanities (Mathematics)
IFET College of Engineering
Villupuram, Tamil Nadu, India
shapique@gmail.com
Chapter 3

R. Venkatesh
Department of Physics
PSNA College of Engineering and Technology (Autonomous)
Dindigul 624622, Tamil Nadu, India
sukatesh@gmail.com
Chapter 4

V. Karthikeyan
Electronics and Communication Engineering
Mepco Schlenk Engineering College
Sivakasi, Tamil Nadu, India
velkarthi85@gmail.com
Chapter 5

Y. Palin Visu †
Electronics and Communication Engineering,
St. Mother Theresa Engineering College
Thoothukudi, India
palinvisu@gmail.com
Chapter 5

https://doi.org/10.1515/9783112214374-204

Fateh Bahadur Kunwar
Department of Computer Science and Engineering
GITS, Rajasthan Technical University
Udaipur, Rajasthan, India
fateh.kunwar@gmail.com
Chapter 6

Hitendra Singh
MSOET
Maharishi University of Information Technology
Lucknow, Uttar Pradesh, India
hitendra.singh@muit.in
Chapter 6

Rakesh Kumar Yadav
MSOET
Maharishi University of Information Technology
Lucknow, Uttar Pradesh, India
rkymuit@gmail.com
Chapter 6

Vikas Misra
Department of Mechanical Engineering
Shri Ramswaroop Memorial University
Barabanki, Uttar Pradesh, India
vikas.rml@gmail.com
Chapter 6

Nitin Tripathi
TCS, N.
Chicago, IL, USA
Tripanx.nitin@gmail.com
Chapter 6

Awaneesh Jee Srivastava
Department of Applied Science and Humanities (Mathematics)
United College of Engineering and Research
Prayagraj, Uttar Pradesh, India
awaneeshcancerian@gmail.com
Chapter 6

Senthilnayaki B.
Department of Information Technology
St. Joseph's Institute of Technology
OMR, Chennai, Tamil Nadu, India
nayakiphd@gmail.com
Chapter 7

Shanthi Thangam M.
Department of CSE
Puducherry Technological University
Puducherry, India
mshanthithangam@gmail.com
Chapter 7

Dharanyadevi P.
Department of ECE
IFET College of Engineering
Villupuram, Tamil Nadu, India
dharanyadevi@gmail.com
Chapter 7

Devi A.
Department of CSE
Sathyabama Institute of Science and Technology
Chennai, Tamil Nadu, India
deviarumugam02@gmail.com
Chapter 7

Vijayakumar K.
Department of Information Technology
St. Joseph's Institute of Technology
Chennai, Tamil Nadu, India
kvijay101983@gmail.com
Chapter 7

Deepa R.
Department of Information Technology
St. Joseph's Institute of Technology
Chennai, Tamil Nadu, India
Chapter 7

S. Saranya
IFET College of engineering
Gangarampalayam, Villupuram 605108, Tamil Nadu, India
sahaasaranya@gmail.com
Chapter 8

S. Sowmyayani
Department of Computer Science
St. Mary's College (Autonomous)
(Affiliated to Manonmaniam Sundaranar University, Tirunelveli)
Thoothukudi, Tamil Nadu, India
sowmyayani@gmail.com
Chapter 9

S. Kulandai Therese
Department of Mathematics,
St. Mary's College (Autonomous)
Thoothukudi, Tamil Nadu, India
sk.therese@gmail.com
Chapter 9

Rajeshwari Rajesh Immanuel
Department of Computer science and Engineering
Vel Tech Rangarajan Dr Sagunthala R&D Institute of Science and Technology
Chennai, Tamil Nadu, India
drrajeswari@veltech.edu.in
Chapter 9

S. Maheswari
Department of Computer science and Engineering
Vel Tech Rangarajan Dr Sagunthala R&D Institute of Science and Technology
Chennai, Tamil Nadu, India
drmaheswaris@veltech.edu.in
Chapter 9

P. Maria Joshy Roy
Department of Mathematics
St. Mary's College (Autonomous)
Thoothukudi, Tamil Nadu, India
Chapter 9

J. Kavitha
Department of Computer science and Engineering
Vel Tech Rangarajan Dr Sagunthala R&D Institute of Science and Technology
Chennai, Tamil Nadu, India
drkavithaj@veltech.edu.in
Chapters 9, 15

Soumitra Saha
Department of Computer Science and Engineering
Chandigarh University
Punjab, India
ssoumitra45@gmail.com
Chapter 10

Umesh Kumar Lilhore
Galgotias University
Greater Noida, Uttar Pradesh, India
umeshlilhore@gmail.com
Chapter 10

Sarita Simaiya
Galgotias University
Greater Noida, Uttar Pradesh, India
saritasimaiya@gmail.com
Chapter 10

R. Roselin Kiruba
Department of Computer Science and Engineering
Vel Tech Rangarajan Dr Sagunthala R&D Institute of Science and Technology
Avadi, Chennai, Tamil Nadu, India
kirubaroselin@gmail.com
Chapters 11, 15

J. P. Shritharanyaa
School of Electrical and Electronics Engineering
VIT Bhopal University
Kothrikalan, Sehore, Madhya Pradesh, India
vijayatharani@gmail.com
Chapter 11

J. Jude Moses Anto Devakanth
Department of Computer Applications
Madanapalle Institute of Technology and Science
Madanapalle, Andhra Pradesh, India
judedevaj@gmail.com
Chapter 11

Vasumathi
Department of Information Technology
Kingston Engineering College
Vellore, Tamil Nadu, India
drvasumathym@veltech.edu.in
Chapter 11

C. Saranya Jothi
Department of Computer Science and Engineering
Vel Tech Rangarajan Dr Sagunthala R&D Institute of Science and Technology
Avadi, Chennai, Tamil Nadu, India
saranyajothi22@gmail.com
Chapter 11

L. Sharmila
Department of Computer Science and Engineering
Vel Tech Rangarajan Dr Sagunthala R&D Institute of Science and Technology
Avadi, Chennai, Tamil Nadu, India
shar.hariharan@gmail.com
Chapter 11

Shijul C. V.
Department of Mathematics
Karunya Institute of Technology and Sciences
Karunya Nagar, Coimbatore 641114, Tamil Nadu
India
angithalakshmi17@gmail.com
Chapter 12

J. Catherine Grace John
Department of Mathematics
Karunya Institute of Technology and Sciences
Karunya Nagar, Coimbatore 641114, Tamil Nadu
India
catherinegracejohn@gmail.com
Chapters 12, 14

P. Xavier
Department of Mathematics
Karunya Institute of Technology and Sciences
Karunya Nagar, Coimbatore 641114, Tamil Nadu
India
xavier@karunya.edu
Chapters 12, 14

S. Usharani
Department of Artificial Intelligence and Machine Learning
IFET College of Engineering
Villupuram, Tamil Nadu, India
ushasanchu@gmail.com
Chapter 13

P. Manju Bala
Department of Artificial Intelligence and Data Science
IFET College of Engineering
Villupuram, Tamil Nadu, India
pkmanju26@gmail.com
Chapter 13

A. Balachandar
Department of Computer Science and Engineering
IFET College of Engineering
Villupuram, Tamil Nadu, India
balaifet20@gmail.com
Chapter 13

K. Rajkumar
Department of Data Science and Business Systems
SRM Institute of Science and Technology
Chennai, Tamil Nadu, India
rajkumak5@srmist.edu.in
Chapter 13

Morenike Eunice Coker
Department of Pharmaceutical Microbiology and Biotechnology
University of Ibadan
Ibadan, Nigeria
morencoker2002@yahoo.com
Chapter 13

G. B. Priyanka
Department of Mathematics
Karunya Institute of Technology and Sciences
Karunya Nagar, Coimbatore 641114, Tamil Nadu
India
priyankagb03@gmail.com
Chapter 14

M. Babila Revathy
Computer Applications
Govindammal Aditanar College for Women
Tiruchendur
Tiruchendur, Tamil Nadu, India
bilarev@gmail.com
Chapter 15

P. Arockia Jansi Rani
Department of Computer Science and Engineering
Manonmaniam Sundaranar University
Tirunelveli, Tamil Nadu, India
jansi_cse@msuniv.ac.in
Chapter 15

Lavanya Devi N.
Department of Electronics and Communication Engineering
PSNA College of Engineering and Technology
Dindigul, Tamil Nadu, India
lavanyadevinadarajan@psnacet.edu.in
Chapters 16, 18

C. N. Savithri
Department of Electronics and Communication Engineering
Sri Sairam Engineering College
Chennai, Tamil Nadu, India
savithri.ece@sairam.edu.in
Chapters 16, 18

Femila L.
Department of Electronics and Communication Engineering
St. Xavier's Catholic College of Engineering
Chunkankadai 629003, Tamil Nadu, India
femilajenin@sxcce.edu.in
Chapters 16, 18

R. Karthick Manoj
Department of Electrical and Electronics Engineering
AMET Deemed to be University
Kanathur, Tamil Nadu, India
Karthickmanoj.r@gmail.com
Chapters 16, 24

B. Karthick
Department of Mathematics
College of Engineering and Technology
SRM Institute of Science and Technology
Kattankulathur
Chennai 603203, Tamil Nadu, India
karthickbalasubramani96@gmail.com
karthicb@srmist.edu.in
Chapter 17

Chitra R.
Department of Electronics and Communication Engineering
Sri Sairam Engineering College
Chennai, Tamil Nadu, India
chitra.ece@sairam.edu.in
Chapter 18

G. Prema
Department of Electronics and Communication Engineering
Mepco Schlenk Engineering College
Sivakasi 626005, Tamil Nadu, India
prema@mepcoeng.ac.in
Chapter 19

G. Shobana
Department of Electronics and Communication Engineering
Mepco Schlenk Engineering College
Sivakasi 626005, Tamil Nadu, India
gshobana@mepcoeng.ac.in
Chapter 19

V. Gnanalakshmi
Department of Electronics and Communication Engineering
Mepco Schlenk Engineering College
Sivakasi 626005, Tamil Nadu, India
v.gnanalakshmi@mepcoeng.ac.in
Chapter 19

M. Parthiban
Department of Computer Science and Engineering
Sasi Institute of Technology and Engineering
Tadepalligudem, Andhra Pradesh, India
parthimitit@sasi.ac.in
Chapter 20

D. Amaravathi
Department of Computer Science and Engineering
Sasi Institute of Technology and Engineering
Tadepalligudem, Andhra Pradesh, India
amaravathi@sasi.ac.in
Chapter 20

Deepika Attavar
Department of Computer Science and Engineering
Sasi Institute of Technology and Engineering
Tadepalligudem, Andhra Pradesh, India
deepika.a@sasi.ac.in
Chapter 20

P. Krishnamoorthy
Department of Computer Science and Engineering
Sasi Institute of Technology and Engineering
Tadepalligudem, Andhra Pradesh, India
krishnancse0206@sasi.ac.in
Chapter 20

R. Srinivasan
Department of Electronics and Communication Technology
Sasi Institute of Technology and Engineering
Tadepalligudem, Andhra Pradesh, India
srinivasan.researcher@gmail.com
Chapter 20

K. Krishnakumar
Department of Computer Science and Engineering
Sasi Institute of Technology and Engineering
Tadepalligudem, Andhra Pradesh, India
krishnakumar.phd16@sasi.ac.in
Chapter 20

Palanikumar R.
Department of Computer Science and Engineering
PSR Engineering College
Sivakasi, Tamil Nadu, India
palanikumar@psr.edu.in
Chapter 21

Kirthika R.
Department of Artificial Intelligence and Data Science
PSR Engineering College
Sivakasi, Tamil Nadu, India
kirthikar219@gmail.com
Chapter 21

Arunkumar R.
Department of Artificial Intelligence and Data Science
Sri Krishna College of Technology
Coimbatore, Tamil Nadu, India
arunkumar.r@skct.edu.in
Chapter 21

Ramathilagam A.
Department of Computer Science and Engineering
PSR Engineering College
Sivakasi, Tamil Nadu, India
thilla2012@gmail.com
Chapter 21

Anupam Tiwari
Department of Computer Science and Engineering
The Institution of Electronics and Telecommunication Engineers
New Delhi, India
anupamtiwari@proton.me
Chapter 22

Neha Sharma
Chitkara University Institute of Engineering and Technology
Rajpura, Punjab, India
Chapter 22

R. Manimegalai
PSG Institute of Technology and Applied Research
Coimbatore 641402, Tamil Nadu, India
drrm@psgitech.ac.in
Chapter 23

S. Kavisri
PSG Institute of Technology and Applied Research
Coimbatore 641402, Tamil Nadu, India
Chapter 23

S. Lakshmi Priya Varshini
PSG Institute of Technology and Applied Research
Coimbatore 641402, Tamil Nadu, India
Chapter 23

C. S. Shruti
PSG Institute of Technology and Applied Research
Coimbatore 641402, Tamil Nadu, India
Chapter 23

S. K. Ayesha
Vellore Institute of Technology
Chennai 641402, Tamil Nadu, India
Chapter 23

Sankarsan Panda
ASMIE
Asind, Bhilwara, Rajasthan, India
pandasankarsanasind@gmail.com
Chapter 24

Avanindra Vasuki Vyas
ASMIE
Asind, Bhilwara, Rajasthan, India
avanindravasukivyas@gmail.com
Chapter 24

Desh Bandhu Bhatt
VPTC
Bhilwara, Rajasthan, India
Deshbandhubhatt99@gmail.com
Chapter 24

Shweta Solanki
JIET Universe
Jodhpur, Rajasthan, India
shweta.solanki@jietjodhpur.ac.in
Chapter 24

T. Grace Shalini
Department of Computational Intelligence
School of Computing
SRM Institute of Science and Technology
Kattankulathur, Chennai, Tamil Nadu, India
gracesht@srmist.edu.in
Chapter 25

P. Geetha
Department of Computational Intelligence
School of Computing
SRM Institute of Science and Technology
Kattankulathur, Chennai, Tamil Nadu, India
geethap4@srmist.edu.in
Chapter 25

B. Anandasabarirajan
Department of Artificial Intelligence and Data Science
KLN College of Engineering
Pottapalaiyam, Tamil Nadu, India
er.anandbalakrishnan@gmail.com
Chapter 25

I. Divya, K. Sandhiya Rajeshwari, G. Revathi, S. Sangeetha

1 Mathematical and Statistical Techniques in Artificial Intelligence Designing AI Systems Grounded in Math and Statistics

Abstract: The foundation of modern artificial intelligence (AI) is deeply rooted in mathematical and statistical principles that enable systems to learn from data, reason under uncertainty, and make intelligent decisions. This chapter presents a comprehensive exploration of the mathematical and statistical techniques that underpin AI system design. Beginning with the fundamentals of probability, random variables, and statistical distributions, it delves into their role in developing probabilistic models such as Bayesian networks, Markov models, and Gaussian mixtures, which are pivotal for reasoning and learning in uncertain environments. It further outlines essential optimization techniques, including gradient descent variants and heuristic approaches like genetic algorithms and simulated annealing, that facilitate efficient learning and model tuning in complex spaces. Advanced mathematical constructs such as linear algebra, calculus, and graph theory are highlighted for their contributions to neural networks, deep learning, and graph-based models. Statistical methodologies, including regression analysis, classification, clustering, and regularization, are detailed alongside their machine learning applications in supervised, unsupervised, and reinforcement learning paradigms. Reinforcement learning is explored in depth through Markov decision processes and policy optimization strategies, showcasing its relevance to robotics, healthcare, and financial decision-making.

Furthermore, the chapter elaborates on statistical learning, inference, hypothesis testing, confidence intervals, and Bayesian inference, emphasizing their significance in uncertainty quantification and decision support systems. Real-world case studies – from predictive maintenance and fraud detection to personalized healthcare and smart grids – illustrate the transformative impact of integrating mathematical and statistical techniques into AI. Concluding with a forward-looking perspective, the chapter anticipates future advances in interpretable AI, quantum optimization, and causal inference, affirming the indispensable role of quantitative methods in driving robust, ethical, and high-performing AI systems.

Keywords: Artificial intelligence, probability theory, statistical learning, optimization techniques, neural networks, reinforcement learning, Bayesian inference

https://doi.org/10.1515/9783112214374-001

1.1 Introduction to Mathematical and Statistical Techniques

Artificial intelligence (AI) has developed from a concept to a transformative technology that has a lot of influence on industries like medicine, financial services, and transport. AI technology is based on well-defined mathematical and statistical concepts – machines are capable of processing information, drawing conclusions, and even making predictions using data. Such instruments form the basis of AI development and this chapter aims at a detailed study of some of them as mathematical and statistical method [1].

Mathematics and statistics are fundamental to the AI system as they sidestep complexities in data, manage uncertainties, assist in algorithm automation, and enhance decision-making. Basic concepts such as machine learning (ML), probability, optimization, and statistical inference define the architecture for designing, training, and evaluating intelligent systems. For example, Bayesian networks adjust beliefs based on new information, demonstrating how probability gives AI the ability to handle ambiguity and make decisions even when evidence is inadequate. To find the best answers in large datasets or challenging problems, AI relies on optimization techniques like gradient descent (GD) and linear programming (LP). These techniques are crucial for a wide range of AI applications, such as natural language processing (NLP), where probabilistic models help understand and produce human language, and ML, where statistical models are used for pattern identification and forecasting. Together, these statistical and mathematical techniques give AI systems the ability to address complex real-world data and decision-making challenges by making them accurate, efficient, and situation-aware.

1.2 Fundamentals of Probability and Statistics

Modern data science, AI, and cross-disciplinary decision-making all rely on probability and statistics as fundamental building blocks. Additionally, they aid in the development of prediction models and the extraction of useful information from intricate data sets by offering methodical approaches to data interpretation, random event measurement, and uncertainty quantification. The principles, significance, and relevance of probability and statistics in data-centric applications are covered in this chapter [2].

1.2.1 Probability: The Foundation of Uncertainty

In order to assess uncertainty, probability theory assigns events numerical probabilities. It enables forecasting of results in circumstances including randomness.

1.2.1.1 Probability of Events

$P(U)$ = Number of favorable outcomes in U/total number of outcomes in S is the probability of U if S is the sample space that represents all potential outcomes and $U \subseteq SA$ is an event. When all of the possible possibilities in S have the same probability, this is referred to as the classical definition of probability.

1.2.1.2 Axioms of Probability

The foundations of probability theory are based on three principles established by Andrey Kolmogorov:

- **Non-negativity**: $P(U) \geq 0$ $P(U)$ for any event U.
- **Normalization**: $P(U) = 1$ meaning the probability of the sample space is always 1.
- **Additivity**: If G and H are mutually exclusive events (they cannot both happen simultaneously), then:

$$P(G \cup H) = P(G) + P(H) \tag{1.1}$$

1.2.1.3 Joint, Conditional Probability, and Independence

- **Joint Probability**: The probability of two events G and H both occurring, denoted $P(G \cap H)$ is:

$$P(G \cap H) = P(GH) \cdot P(H) \tag{1.2}$$

- **Conditional Probability**: The probability of G given that H has occurred, represented as is:

$$P(G \cap H) = \frac{G \cap H}{P(H)} P(H) \neq 0 \tag{1.3}$$

- **Independence**: Two events G and H are independent if the occurrence of one does not affect the probability of the other:

$$P(G \cap H) = P(G) \cdot P(H) \tag{1.4}$$

1.2.2 Random Variables and Probability Distributions

A **random variable** is a variable that takes on different values depending on the outcome of a random process. Random variables can be **discrete** (finite or countable outcomes) or **continuous** (infinite and uncountable outcomes).

1.2.2.1 Discrete Random Variables

A random variable is **discrete** if it can take on a finite or countable number of values.

- **Example**: The number of heads in 10 coin tosses.
- **Probability Mass Function (PMF)**: The PMF $P(Y=y)$ specifies the probability of each possible value y:

$$P(Y=y)=p_y, \text{ where } \sum_y P(Y=y)=1 \tag{1.5}$$

- **Example PMF**: A fair six-sided die:

$$P(Y=y)=\begin{cases} \frac{1}{6}, \text{ if } y \ \varepsilon \ \{1,2,3,4,5,6\} \\ \\ 0, \text{ otherwise} \end{cases} \tag{1.6}$$

1.2.2.2 Continuous Random Variables

A random variable is **continuous** if it can take on an infinite number of values within an interval:

- **Example**: The height of people in a population.
- **Probability density function(PDF)**: Let X represent the height of people in a population. Suppose it follows a normal distribution with mean μ and standard deviation σ. The PDF is given by:

$$f_X(x)=\frac{1}{\sigma\sqrt{2\pi}}e^{-\frac{1}{2}\left(\frac{X-\mu}{\sigma}\right)^2}, \quad -\infty < X < \infty \tag{1.7}$$

As shown in Table 1.1, a discrete random variable takes distinct, countable values with probabilities given by a PMF, where $P(X=x)$ can be non-zero. In contrast, a continuous random variable takes values from an interval of real numbers, described by a PDF. For continuous cases, $P(X=x)=0$, and probabilities are calculated over intervals using integration.

Table 1.1: Comparison of discrete and random variable.

Property	Discrete random variable	Continuous random variable
Definition	Takes specific, countable values	Takes any value in a continuous range
Probability function	PMF: $P(X = x)$	PDF: $f_X(x)$
Probability of a single value	$P(X = x) \neq 0$valid x	$P(X = x) = 0$
Probability of interval/set	Summation: $\sum P(X = x)$	Integration: $\int f_X(x)\, dx$

1.2.2.3 Probabilistic Models in AI

AI relies heavily on probabilistic models, especially for jobs requiring uncertainty and imperfect data. These models represent, analyze, and forecast uncertain events by utilizing statistics and probability theory. The main probabilistic models in AI are examined in Table 1.2, along with the statistical and mathematical foundations for each [3].

Table 1.2: Comparison of the AI mathematical models.

Models	Formulas	Applications
Bayesian networks	$P(U_1, U_2, \ldots, U_n) = \prod_{i=1}^{n} P(U_i \mid \text{Parents}(U_i))$	Diagnostic systems (e.g., medical diagnosis)
Markov models	$P(U_{t+1} = j \mid U_t = i) = P_{ij}$	Speech recognition
Probabilistic graphical models (PGMs)	$P(U) \propto \prod_C \emptyset_C(U_c)$	Image processing
Gaussian mixture models (GMMs)	$P(U) = \sum_{k=1}^{K} \pi_k \aleph(U \mid \mu_k, \sum_k$ where π_k represent mixing coefficients	Clustering
Probabilistic latent semantic analysis (PLSA)	$P(d, w) = \sum_z P(z)P(d \mid z)P(w \mid z)$ where z is the latent topic	Text mining
Naive Bayes classifier	$P(C \mid X) = \frac{P(C) \prod_{i=1}^{n} P(x_i \mid C)}{P(X)}$	Sentiment analysis

1.2.2.4 AI Optimization Techniques

Within the domain of AI, many mathematical and statistical methods are heavily reliant on optimization techniques. Fundamentals of Algorithm Design- if you want your algorithms to work optimally, if you want your models to learn properly, and most importantly, if you want to have an approach to problem-solving that is computationally tractable. Based on implementation usages and areas of focus, below are the most commonly utilized optimization techniques [4].

1.3 Optimization by the Gradient

These are typically used to train deep learning, or other ML models. GD: To minimize the loss function, iteratively modifies parameters by traveling in the direction of negative gradients.

1.3.1 Variants

- Batch GD: This is expensive in terms of time and resources as it needs the entire dataset to run.
- Stochastic GD: Stochastic GD is fast, but messier as it uses only one data point to work with.
- Mini-batch GD: This is a hybrid of both the advantages of the batch and the advantages of the SGD.

1.3.2 Extensions

Momentum is the process of taking a fraction of the previous update and incorporating it into the present step. These adaptive algorithms where the learning rates are automatically adjusted are Adam and RMS prop, and finally multiplying algorithms to enhance convergence speed while increasing the overhead cost, uses second derivatives (Hessian matrix).

1.4 Optimization Through Heuristics

- These approaches can be applied when gradients are hard to obtain or when the search area is too large.
- Genetic algorithms (GAs) optimally crossover, mutate, and select semi-randomly inspired by the thoughts of evolution.
- Simulated annealing optimally reduces the range of search based on strategies of annealing in metallurgy.
- Like particle collaboration which is used for locating optimal solutions in a bird swarm.
- Ant colony optimization locates the optimal paths by imitating the way ants search for food.

1.4.1 Convex Optimization

We have focused on problems where the objective and constraints are convex, ensuring global optima. Optimization of a linear objective function under linear constraints is known as linear programming (LP) and LP is extended with quadratic objective functions in quadratic programming. Then the semi-definite programming for complex scenarios is a component of conic programming.

1.4.2 Optimization Using Combinatorial Methods

It is useful in vast, discrete solution spaces; it is often used in AI scheduling and planning.

- Branch and bound: Lists potential solutions in an organized manner.
  Using cutting-plane methods, the feasible regions are iteratively refined to obtain optimal solutions.
- Dynamic programming breaks down problems into smaller ones and solves them iteratively.

1.4.3 Optimization via Bayesian

- It was used in expensive-to-evaluate functions and hyper-parameter optimization.
- Gaussian processes are modeled as a probabilistic distribution of the objective function.
- Acquisition functions: Strike the balance between exploitation, where proven, effective solutions are improved, and exploration, where new areas are sought.

1.4.4 Heuristic Search Methods

Heuristic search techniques are important in AI, especially when it can be difficult to find highly useful or nearly optimal solutions due to large search areas or implementation restrictions used the definition of the heuristic seek to be applied to this domain name is given below [5].

1.5 Heuristic Functions

- By calculating the cost or distance to a goal from a specific condition, heuristic functions play a crucial role in directing searches towards the best answers.
- Heuristic functions in AI are intended to increase search algorithm efficiency, especially in cases where the search space is large.
- In order to increase the precision and speed of solution discovery for mathematical and statistical issues, heuristics frequently integrate domain-specific knowledge (such as route distance in logistics) or approximations (such as Euclidean distance in clustering).

1.5.1 Popular Heuristic Search Algorithms

- **A***: Directs the search by a heuristic estimate of the remaining value in addition to the cost of node acquisition. It is widely used in combinatorial optimization and path finding cases, especially where efficiency is required but solution accuracy is desired.
- **Greedy Best-First Search**: This fastest search is usually local optima because it gives priority to the nodes with the lowest guess cost.
- **Simulated Annealing**: This method adds unpredictability to optimization issues such as statistical model fitting and scheduling to prevent local minima.
- **GAs**: This algorithm uses natural selection to identify optimal solutions in sustainable fields; they help in cases of aggregation and high-performance calculations.
- **Hill Climbing**: This iterative process, which is frequently used in conjunction with restart techniques for improved outcomes in non-deterministic contexts, advances towards better solutions depending on local information.

1.5.2 Application in Optimization Problems

- **Route Optimization**: The logistics industry often uses heuristic search (such as A* or GAs) to find the best routes and optimize logistics based on cost, time, and distance, multi-objective balancing is often performed.
- **Statistical Model Tuning**: Statistical techniques such as particle swarm optimization and GAs can be used to tune parameters in complex statistical models to reduce error rates or increase probabilities.
- **Combinatorial Optimization**: Heuristic methods are often used to generate nearly optimal solutions to issues such as passenger vendor and traffic routing problems when the exact methods are not computationally practical.

1.5.3 Handling Uncertainty and Probabilistic Heuristics

- **Probabilistic Heuristics:** These are applied to searches when there are non-deterministic activities or partially observable settings.
- **Bayesian Heuristics**: Bayesian methods include probability distributions to direct search decisions and model uncertainty. They are useful for problems such as probabilistic belief-state planning.
- **Monte Carlo Methods**: Heuristic Monte Carlo simulations aid in assessing the possible outcomes of complex decision-making under uncertainty and statistical issues.

1.5.4 Continuous and High-Dimensional Spaces

- Gradient-based heuristics or hybrid approaches that combine search and optimization (such as GD) are employed for problems in continuous domains (such as parameter tuning in ML).
- Particle swarm optimization and evolutionary algorithms manage high-dimensional spaces where conventional search methods fall short. The comparison of the specific AI applications for the linear algebra, calculus, and the graphs are shown in Table 1.3.

Table 1.3: Comparison-specific AI applications for linear algebra, calculus, and graph.

Mathematical tool	Key concepts	Formulas/operations	AI applications
Linear algebra	– Vectors – Matrices – Eigenvalues and eigenvectors	– Matrix multiplication: $X\beta = Y$ Eigendecomposition: $A = Q\Lambda Q - 1$	– **Data representation**: Representing data points as vectors or matrices. – **Dimensionality reduction**: Techniques like PCA to reduce feature space. – **Linear regression** and **neural networks**: Calculating weights and activations.
Calculus	– Derivatives – Gradients – Optimization	– Gradient descent update: $\theta = \theta - \eta \nabla L(\theta)$ – Loss function: $L(\theta) = 1/n\sum(y - \hat{y})^2$	– **Optimization**: Gradient-based optimization in training models (e.g., neural networks and SVMs). – **Model tuning**: Minimize error/loss functions. – **Gradient-based learning**: Backpropagation in neural networks.

Table 1.3 (continued)

Mathematical tool	Key concepts	Formulas/operations	AI applications
Graph theory	– Nodes – Edges – Adjacency matrix – Graph traversal	– Adjacency matrix: $A_{ij} = 1$ if there is an edge from node i to j – Markov transition matrix: P_{ij} for state transitions	– **Markov models**: Represent sequential decision-making, e.g., hidden Markov models. – **Graph neural networks**: Learning node embeddings in a graph. – **Reinforcement learning**: Markov decision processes for modeling states and actions.

1.6 Statistical Techniques in Machine Learning

Statistics offers ways to characterize uncertainty, quantify correlations between variables, and draw conclusions from data, it is essential to ML. Many ML methods are based on probability theory, particularly those that simulate uncertainty, including Gaussian mixed models and Bayesian networks. The comparison of the statistical techniques in ML is shown in Table 1.4. Based on sample data, ML algorithms can forecast the population – thanks to statistical inference. One of the most used statistical methods is regression analysis, which makes it possible to describe the connections between input data and output predictions. Discrete outcomes are predicted using classification techniques like logistic regression and decision trees, while comparable data points without labels can be grouped using clustering techniques like *k*-means [6].

Table 1.4: Comparison of the statistical techniques in machine learning.

Key concept	Description	Machine learning techniques
Characterizing uncertainty	Quantifies uncertainty and variability in data.	– Gaussian mixture models (GMM) – Bayesian networks
Quantifying correlations	Identifies and measures relationships between variables.	– Regression analysis: Describes relationships between inputs and outputs (e.g., linear regression)
Drawing conclusions	Using data to make predictions and decisions.	– Regression: Predict continuous outcomes (e.g., linear regression) – Classification: Predict discrete outcomes (e.g., logistic regression and decision trees) – Clustering: Grouping unlabeled data (e.g., *K*-means)

1.6.1 Machine Learning Techniques

Various ML algorithms, which can be broadly divided as supervised learning, unsupervised learning, and reinforcement learning, use the mathematical and statistical methods previously discussed and it has been show in Table 1.5.

Table 1.5: Machine learning techniques.

Machine learning type	Description	Tasks
Supervised learning	Learning from labeled data where the correct output is provided.	– **Classification**: Predicting categorical outcomes (e.g., spam detection and image classification). – **Regression**: Predicting continuous outcomes (e.g., price prediction and stock market forecasting).
Unsupervised learning	Learning from unlabeled data to find hidden patterns.	– **Dimensionality reduction**: Reducing feature space while retaining important information (e.g., PCA and t-SNE). – **Clustering**: Grouping similar data points (e.g., K-means and hierarchical clustering).
Reinforcement learning	An agent learns by interacting with its environment to maximize rewards.	– **Agent-environment interaction**: The agent takes actions based on the current state. – **Reward maximization**: The agent adjusts its actions to maximize cumulative reward (e.g., in robotics and game playing).

1.6.2 Optimization and Regularization

The core of ML is optimization. The loss function, which quantifies the mistake or discrepancy between expected and actual results, is minimized by making sure the model parameters are adjusted accordingly. Iterative parameter adjustments are made using methods such as GD. Additionally, by include penalty terms in the loss function, regularization approaches like L_1 and L_2 regularization assist avoid over fitting by discouraging overly complex models that might perform well on training data but not generalize to new, unknown data.

1.6.3 Applications of Mathematical and Statistical Techniques in AI

Because of the use of these statistical and mathematical approaches, significant advancements have been made in several AI fields. While algorithms in NLP use linear algebra and optimization approaches to understand images, convolutional neural networks (CNNs) in computer vision use statistical models for human language production and comprehension. Reinforcement learning uses statistical methods to simulate sequential decision-making in dynamic scenarios for robotic control or AI that plays games. The widespread use of ML in industries including healthcare, banking, autonomous vehicles, and entertainment demonstrates the transformative potential of these mathematical and statistical foundations. Understanding and using these strategies is essential to creating AI systems that work well and can learn from data, which will progress a variety of industries, including robotics and healthcare.

1.7 Advanced ML Techniques

Building upon the fundamental ideas of mathematics and statistics, advanced ML techniques use increasingly intricate algorithms and methodologies to address ever-more-complex AI problems. These methods let machines to process enormous, high-dimensional datasets, learn from data in previously unattainable ways, and produce predictions that are more accurate. Here, we will examine a few of the more sophisticated ML algorithms that are based on statistical and mathematical concepts.

1.7.1 Neural Networks and Deep Learning

Neural networks with more than one layer – subsequently the term "deep" networks – are the inspiration of deep gaining knowledge of, a subset of machine studying, which allows the version to apprehend extremely complicated patterns in records. Calculus (for GD optimization), hazard concept (for stochastic strategies and activation capabilities), and linear algebra (for records illustration) are the mathematical underpinnings of deep analyzing [6]:

- Backpropagation is a vital deep analyzing approach that computes gradients and updates the community's weights with the useful resource of making use of the chain rule from calculus.
- While recurrent neural networks and long short-term reminiscence (LSTM) networks are made to deal with sequential enter, such time collection or herbal languages, CNNs are regularly hired for laptop imaginative and prescient applications.

- Deep community schooling is stepped forward via optimization strategies like Adam and RMS prop.

1.7.2 Kernel Methods and Support Vector Machines (SVMs)

Strong supervised learning methods for classification and regression applications are SVM. SVMs are based on the idea of kernels, which are mathematical operations that convert input data into a higher-dimensional space where classes can be distinguished using a linear decision boundary [7]:
- Foundation in Mathematics: SVMs are based on convex optimization and mainly use ideas from calculus and linear algebra to determine the best hyper plane for maximizing the margin between classes.
- Nonlinear data is handled via kernel methods, which map it to higher dimensions. Gaussian (RBF) and polynomial kernels are common kernels.

1.7.3 Ensemble Methods

- By decreasing variance, bias, or both, ensemble getting to know tactics integrate numerous models to enhance average performance [8].
- Random forests and gradient boosting machines (GBMs) are the two maximum extensively used ensemble techniques.
- Decision trees, the muse of random forests, are skilled on a random subset of the information. A final prediction is determined by voting or by averaging over all trees.
- GBM create images iteratively, correcting errors in previous images. Advanced GBM implementations such as XG-Boost and light GBM increase performance by implementing optimization techniques such as regularization and early closure.
- Statistical considerations and optimization.

1.7.4 Graphical Models Based on Probabilities

Using a graph structure, probabilistic graphical models (PGMs) depict the conditional interdependence between random variables. They are extensively employed in fields like computer vision and NLP, as well as in situations with ambiguous or insufficient data [9]. Bayesian networks, which depict probabilistic dependencies among variables, are directed acyclic graphs. As fresh data is observed, they update the probabilities using Bayesian reasoning:

- Undirected graphs called Markov networks, also known as Markov random fields, are used to simulate joint distributions across a collection of random variables.
- One kind of PGM used in sequence modeling is called a hidden Markov model (HMM), in which the system alternates between hidden states that can only be seen by signals that are released

1.7.5 Reinforcement Learning (RL)

A subfield of ML called RL is concerned with discovering the best course of action for making decisions by interacting with the environment. An agent in RL maximizes cumulative rewards over time by learning by trial and error [10]. RL challenges are modeled using a mathematical framework called Markov decision processes. The objective is to develop a policy that maximizes long-term rewards, with probabilistic transitions governing the agent's actions and the states that ensue. To determine the best action-value function that approximates the predicted reward for a particular state-action pair, Q-learning and deep Q-networks (DQNs) employ dynamic programming and value iteration. Actor-critic and REINFORCE are two examples of policy gradient approaches that use gradients to directly optimize the policy. These techniques are essential for deep learning models of RL for activities such as gaming and controlling robots.

1.8 Reinforcement Learning: Concepts and Applications

A key concept in AI is RL, in which an agent interacts with its surroundings to maximize a cumulative reward in order to learn how to make decisions. Because it provides a formal framework for resolving sequential decision-making issues under uncertainty, reinforcement learning is especially well-suited for applications in the mathematical and statistical areas. Here's a look at core RL concepts and their applications in mathematical and statistical techniques within AI [11].

1.8.1 Core Concepts of RL

1.8.1.1 The Environment and the Agent

- Agent: The RL setup's decision-maker, which engages with the surroundings to discover the best course of action.

- Environment: The system that responds to actions and gives feedback (in the form of rewards or penalties) to the agent.

1.8.1.2 State, Action, and Rewards

- State: An illustration of the environment's current state. It gives the agent the data that they need to make choices.
- Action: An agent-made choice that modifies the surroundings and creates new states.

Reward: The environment's reaction to every action. Incentives assist the agent learn the best policies and steer it towards positive activities.

1.8.1.3 Policy (π)

- By connecting states to actions, the policy establishes the agent's approach. It can be stochastic (probabilistic action selection) or deterministic (one action per state).
- Iterative policy improvement is possible, since the agent gradually learns to optimize cumulative reward

1.8.2 Q-Function (Q) and Value Function (V)

- Value Function (V): Indicates the attractiveness of states by estimating the predicted cumulative reward from a certain state.
- Q-Function (Q): Helps determine the optimal course of action in each state by estimating the expected cumulative reward from a given state-activity pair.

1.8.3 Investigating as Opposed to Exploiting

- The agent must choose between using known actions to maximize rewards and investigating novel actions to obtain information.
- To strike a balance between exploration and exploitation, strategies like upper confidence bound and ε-greedy policies are employed.

1.8.4 Algorithms for Learning

- Q-Learning: A well-liked off-policy technique in which the agent discovers the best course of action by learning an ideal Q-function.
- SARSA (State-Action-Reward-State-Action): It is an on-policy technique that uses the agent's current policy to learn Q-values.
- Policy Gradient Methods: These techniques directly optimize the policy and are used for continuous environments or complex action spaces.
- Deep RL (DRL): Expands reinforcement learning into high-dimensional, complex domains by combining RL with deep neural networks.

1.8.5 Applications of Reinforcement Learning in AI and Statistics

- **Robotics and Logistics Control and Optimization**
  - RL is frequently used in robotics for problems where the agent must maximize actions under restrictions, such as path planning, motion control, and manipulation.
  - RL facilitates supply chain scheduling, inventory control, and route optimization in logistics, where it continuously adjusts to changes and optimizes choices to cut expenses.
- **Optimization of Strategies and Game Theory**
  - RL techniques that handle complicated, multi-agent settings, such as policy gradients and Q-learning, are used to optimize strategies in games.
  - RL is used in conjunction with game theory techniques to address issues in adversarial environments, including as cyber security and financial trading.
- **Statistical Model Tuning and Adaptive Decision-Making:**
  - In ML, RL iteratively adjusts model configurations and hyper parameters.
  - RL aids in the selection of data points or activities that optimize information gain or predicted accuracy in adaptive sampling and experimental design.
- **Medical Care and Tailored Treatment Strategies**
  - RL is used in adaptive therapy design, where responses to prior treatments are used to learn the best policies for patient care.
  - RL optimizes sequential decision-making in illness management and diagnostics to enhance patient outcomes in the face of ambiguity.

1.8.6 Risk and Finance Management

By adjusting to shifting markets, forecasting future conditions, and striking a balance between risk and reward, RL makes portfolio management and dynamic pricing possible. RL optimizes decision rules based on probabilistic models of risk and behavior in fraud detection and credit scoring.

1.9 Statistical Learning and Inference in AI

In the field of AI, mathematical and statistical techniques are fundamentally based on statistical learning and inference. These methodologies provide the necessary tools and theoretical foundations to identify patterns, develop predictive models, and draw conclusions from data. An in-depth examination of these methods and their applications in AI is given below. Statistical models illustrate the connections between variables within the data [12].

1.9.1 Core Concepts in Statistical Learning

1.9.1.1 Statistical Models and Probabilistic Assumptions

- Statistical models translate the relationships that exist between variables in recorded data. The models operate under probabilistic systems by assuming data points follow distribution patterns with hidden parameters.
- Several common statistical approaches include linear regression and logistic regression, as well as generalized linear models since these models permit the estimation of uncertainty and deliver robust results through probabilistic assumptions.

1.9.1.2 Supervised and Unsupervised Learning

- Models within supervised learning receive labeled data for training functions to predict outcomes from given input variables. The classification grouping contains support vector machines and decision trees while the regression grouping has linear regression as an example.
- Models operating under this realm discover essential patterns within datasets which lack output labels. The methods for clustering and dimensionality reduction include *K*-means and hierarchical clustering and PCA, respectively.

1.9.1.3 Parameter Estimation

- The procedure of statistical learning finds model parameters, which properly match the selected framework through data estimation. Methods include the following:

- MLE operates through a process that selects parameters for maximum data likelihood.
- Bayesian estimation takes prior distribution beliefs and observed data to produce a posterior distribution.

1.9.1.4 Regularization and Complexity Control

- Model parameters receive added penalties through regularization techniques such as Lasso and Ridge and Elastic Net so that overfitting is prevented by maintaining a fit with simple models.
- Regularization remains necessary to maintain stability in spaces with a large number of features compared to the available observations.

1.9.1.5 Model Evaluation and Selection

Statistical learning relies on cross-validation together with bootstrapping methods to test generalization capabilities of models during evaluation processes. The choice among statistical models becomes possible through the use of model selection criteria including Akaike information criterion (AIC) and Bayesian information criterion (BIC).

1.9.2 AI Statistical Inference

1.9.2.1 Hypothesis Testing and Significance Testing

By separating genuine patterns from chance fluctuations, hypothesis testing enables AI models to ascertain if observed effects are statistically significant. To determine the correlations between variables in data, methods such as ANOVA, chi-squared tests, and *t*-tests are employed [13].

1.9.2.2 Quantification of Uncertainty and Confidence Intervals

Confidence intervals, which represent uncertainty in model parameters or forecasts, offer a range around estimations. Uncertainty quantification is crucial for probabilistic models in decision-making applications where AI models must convey the degree of prediction confidence (e.g., medical diagnostics).

1.9.2.3 Bayesian Inference

When data is scarce or the model must take into account prior assumptions, Bayesian inference updates the distribution of parameters by combining evidence from data with prior beliefs. Bayesian techniques in AI facilitate adaptive learning, which lets models improve their predictions as new information becomes available.

1.9.2.4 Resampling and Bootstrapping

Bootstrapping is a non-parametric technique that uses replacement sampling to estimate distributions and quantify uncertainty. It allows for strong inference regarding model stability and parameter estimates, which is especially helpful when analytical solutions are not feasible.

1.9.3 Applications of Statistical Learning in AI

Forecasting and Predictive Modeling
- From projecting stock prices to anticipating customer behavior in marketing, statistical learning techniques are extensively employed in predictive modeling [14].
- Time series analysis, regression models, and Bayesian models are commonly used to forecast the future using historical data.

NLP
- Many NLP models, such as sentiment analysis, language modeling, and text categorization, are based on statistical techniques.
- For tasks like named entity recognition and part-of-speech tagging, probabilistic models such as Conditional Random Fields (CRFs), HMMs, and Naïve Bayes classifiers rely on statistical learning.

Computer Vision
- Image processing tasks like facial recognition and object detection use statistical learning.
- Gaussian models assist in managing uncertainty in identifying patterns and features, while methods such as principal component analysis (PCA) minimize dimensionality in high-dimensional image data.

Identifying Anomalies
- Statistical inference and learning aid in the detection of abnormalities that departs from typical patterns in domains like fraud detection and cyber security.
- Methods such as statistical outlier identification and Gaussian Mixture Models (GMMs) offer probabilistic frameworks for spotting anomalous occurrences, such fraudulent transactions.

Recommendation Systems
- Statistical learning is the base of the matrix factorization and collaborative filtering used in recommendation engines.
- The methods of probabilistic managing of individual preference and uncertainty in large datasets are Bayesian Personalized Ranking and Gaussian process models.

1.9.4 High-Level Statistical Techniques of AI

Models of Hierarchies
- Multi-level models, also called hierarchical models, deal with data that has considerable variability such as user-specific behavior in recommendation systems.
- Bayesian hierarchical models are very often applied for effect estimation at various levels of abstraction, such as individual and group levels, when aggregating data from several groups to obtain better precision.

Bayesian Networks and Graphical Models with Probabilities
- Bayesian networks use a directed acyclic graph (DAG) to model the dependencies between variables, enabling probabilistic inference in complicated systems.
- The network structure is used to depict causal links in decision-support systems and diagnostic systems (such as medical diagnostics).

State-Space Models and HMMs
- HMMs are helpful in time series analysis, speech recognition, and biological sequence analysis because they can represent observational sequences with hidden underlying states.
- State-space models extend HMMs to continuous state spaces, which are utilized for real-time tracking and prediction in robotics and control systems.

Gaussian Processes (GPs)
- GPs are non-parametric Bayesian models that are used for classification and regression; instead of using point estimates for predictions, GPs employ distributions.

- They are employed in applications like active learning and spatial modeling, when data is scarce or costly to acquire, and they are effective at managing uncertainty

Causal Inference
- For tasks like treatment effect estimates in healthcare and policy-making, causal inference techniques determine cause-and-effect links beyond correlations.
- By taking into consideration the confounding variables, some techniques such as structural equation modeling, instrumental variable analysis, and propensity score matching support more effective decisions.

1.10 Case Studies of Quantitative AI Applications

Quantitative applications of AI using statistical and mathematical methods have resolved real-world issues in the space of optimization, predictive modeling, and data-driven decision-making. Case studies show comprehensive details about the various quantitative methods increasing AI utility in various applications across multiple domains.

1.10.1 Predictive Maintenance in Manufacturing

The failure of any piece of equipment results in expensive downtime, thus increasing maintenance costs.

Solution:
Data Collection: Sensor data, machine usage logs, and maintenance records are captured from the shop floor equipment.
- Statistical Models Survival analysis: Survival time-to-event, where modeling uses these Techniques the Cox Proportional Hazards and also Kaplan Maier estimates provided that ensure for them they provide the actual end usage period the machine is left.
- ML and deep learning
- Optimization and decision-making
- RL techniques are applied to optimize the maintenance schedule, thereby achieving a balance between cost and uptime of the machinery.

1.10.2 Fraud Detection in Financial Services

However the issue remains to be: How to stop or rather detect fraudulent transactions in real-time which is necessary as it helps cutting down on financial losses and improving trust of customers.

Solution:
Since anomaly detection primarily uses statistical methods like z-scores and multivariate Gaussian distribution to identify abnormal patterns in transaction data, it is easy to learn.

- **Supervised Learning:** These use the training data of labeled fraudulent and legitimate transactions: Logistic regression, Decision Trees, and Support Vector Machines. More complex models are used to capture deeper, more nonlinear relationships in the transactional data, including ensemble methods like XG-Boost and neural networks.

Bayesian Networks in Real Time: These PGMs and Bayesian Networks update the probability of fraud in real-time as new data comes in, you can think of it an analogy to focusing a camera.

1.10.3 Optimizing Supply Chain and Logistics with Predictive Analytics

The problem addressed in this chapter is related to operational costs within the supply chain and requires improved route planning, demand forecasting, and inventory management.

Solution:
- ARIMA, Exponential Smoothing, and Prophet Models: The time series forecasting techniques also forecast the demand by using the historical sales and seasonality data.
- Dynamic Routing Through Reinforcement Learning: The algorithms used in reinforcement learning, such as Q-learning and deep Q-networks, optimize the paths of delivery in real time, considering traffic conditions, weather, and other variable parameters.
- Inventory Optimization Using Probabilistic Models: Bayesian models enable the estimation of lead times and demand, and thus optimal reorder points and stock levels can be computed.
- Network Optimization and Simulation: Using mathematical optimization methods, such as LP, and Monte Carlo simulations, one can determine the optimal locations of inventories so that the total transportation cost is minimized.

1.10.4 Personalized Healthcare and Treatment Optimization

Problem: Design individually tailored treatment approaches using past as well as existing data that optimizes outcome in each case.

This is achieved through predictive modeling, which relies on genetic information, medical history, and demography in the estimation of patient risk variables and survival probabilities through logistic regression and Cox proportional hazard models:
- Bayesian Inference and Probabilistic Models: Bayesian networks enable doctors to balance the potential benefits and drawbacks of certain treatments by modeling the probability of specific outcomes.
- RL for Personalized Treatment: RL is applied to the timing and dosing of cancer treatments based on patient response, and the treatment is adjusted accordingly to minimize side effects and maximize efficacy.
- Genetic and Genomic Analysis with ML: Deep learning algorithms scan genetic data to find biomarkers and predict disease risk, enabling tailored treatment plans.

1.10.5 Credit Scoring and Risk Assessment in Banking

Problems: Problems in determining applicants' creditworthiness to minimize defaults and financial risk. This solution applies supervised learning models, using the decision tree along with logistic regression to classify a candidate based on his default risk. Features such as employment stability, credit history, and income are utilized:
- Ensemble Methods: Random forests and gradient boosting models improve prediction accuracy by taking the best features from multiple models to find different patterns in data.
- Bayesian Networks for Probabilistic Risk Assessment: Bayesian inference provides a degree of confidence with each decision and measures the uncertainty associated with credit risk.
- Explainable AI and Model Interpretability: LIME and SHAP are applied to interpret complex models for the purpose of ensuring transparency and regulatory compliance in credit scoring.

1.10.6 E-Commerce Dynamic Pricing Optimization

Problem: Quantity of supply relative to a competing product must be determined in real time in terms of inventory levels, competing prices, and demand.

Answer:
- Demand forecasting to past sales data, seasons, and external variables, using ML techniques such as time series modeling and ARIMA and XG-Boost forecast demand.

- RL for Pricing Strategies: The RL models may identify the appropriate pricing strategies in place by formulating various types of pricing action and monitoring instantaneous client responses.
- Elasticity Estimation via Regression Models: Statistical techniques, in the form of multiple regression methods, are implemented to estimate the price elasticity of demand in evaluating the client sensitivity towards prices.
- Optimization and Simulation: Monte Carlo simulations and LP are applied to analyze the effects of various pricing methods under ambiguous demand.

1.10.7 Predicting and Improving Energy Use in Smart Grids

Predict energy consumption and optimize the grid to reduce costs and promote the use of renewable energy sources:
- Time Series Forecasting: Considering weather and seasonality, ML models like ARIMA, LSTM, and Prophet predict short-term and long-term energy consumption.
- Load Balancing Optimization Algorithms: Integer and LP balance supply and demand across the grid, reducing costs and preventing overloads.
- Demand Response Using Reinforcement Learning: The RL-based algorithms encourage the users to alter their consumption patterns in response to real-time energy pricing adjustments, which leads to lower peak loads and a balance of renewable energy sources.
- Anomaly Detection: Statistical anomaly detection techniques can help identify unusual patterns of use to assist in the detection of equipment failures or energy theft.

1.11 Future Directions in Mathematical and Statistical AI

The future of AI is inseparable from advances in certain mathematical and statistical approaches, through focus on areas such as interpretable and explainable models; optimization; and quantum machine learning. Research efforts underway are focused on increasing the robustness, generalization, and privacy of AI (e.g., adversarial learning, distributed optimization, and differential privacy). The combination of symbolic and statistical methods applied to AI will improve the reasoning capabilities; reinforcement learning will develop within multi-agent and evolving environments. Causal inference will help in understanding the cause-and-effect relationships and this is very critical in fields like healthcare and economics.

It is further augmented for automated ML, neural architecture search, with improvements to capacity to think; therefore, RL develops in dynamics, as well as in multi-agent conditions. Understanding links for cause-and-effect links will play an important role for economics, health, among disciplines. It has enabled the practice of causal inference with a proper outcome of it all. AutoML, as well as NAS improves in development models to enable higher efficiency with scalabilities provided with high-dimensional data approaches. Advancements in low-energy AI, fairness algorithms, and sparse representations will increase the contribution of AI towards scientific research, sustainability, and ethical issues. These developments prognosticate the development of future AI systems that are more opening, effective, and socially conscious.

1.12 Conclusion

In conclusion, developments in statistical and mathematical techniques are closely linked to the future of AI. Making models easier to understand and interpret will be a major priority as AI develops, allowing for increased openness and confidence. The investigation of quantum ML and optimization methods will improve the scalability and effectiveness of AI systems. Moreover, integration of statistical and symbolic methods should significantly strengthen the reasoning capability of AI, and reinforcement learning will progress in dynamic, multi-agent environments. It will be possible to better understand cause-and-effect interactions due to the emphasis on causal inference, especially in such critical fields as economics and healthcare. With advancements such as adversarial learning, differential privacy, and fairness algorithms propelling AI ahead, the pace of incorporating privacy, fairness, and ethical considerations will accelerate as AI speeds up. AI will continue to move ahead with neural architecture search and automated ML, accelerating model building to allow for fast, more aggressive and efficient AI system development.

Investigating low-energy AI systems will further improve sustainability and reduce the impact of AI on the environment. At the end of this journey, there will be far more capable and socially aware AI systems; and they would be making major contributions to sustainability, scientific research, and solutions to urgent ethical problems. This may eventually make future AI develop open, efficient, and morally sound systems which are beneficial for society in different aspects.

References

[1] Jiang Y, Li X, Luo H, Yin S, Kaynak O. Quo vadis artificial intelligence? Discover Artificial Intelligence. 2022 Mar 7;2(1):4.

[2] Koller D, Friedman N. Probabilistic Graphical Models: Principles and Techniques. MIT press; 2009 Jul 31.

[3] Samiullah M, Albrecht D, Nicholson AE, Ahmed CF. A review on probabilistic graphical models and tools. Dhaka University Journal of Applied Science and Engineering. 2021;6(2):82–93.

[4] Kumar SA, Kumar A, Dutt V, Agrawal R. Multi Model Implementation on General Medicine Prediction with Quantum Neural Networks. In 2021 Third International Conference on Intelligent Communication Technologies and Virtual Mobile Networks (ICICV) 2021 Feb 4 (pp. 1391–1395). IEEE.
[5] Hart PE, Nilsson NJ, Raphael B. A formal basis for the heuristic determination of minimum cost paths. IEEE Transactions on Systems Science and Cybernetics. 1968 Jul 31;4(2):100–107.
[6] Van Doremaele ER, Stevens T, Ringeling S, Spolaor S, Fattori M, Van de Burgt Y. Hardware implementation of backpropagation using progressive gradient descent for in situ training of multilayer neural networks. Science Advances. 2024 Jul 12;10(28):eado8999.
[7] Payal M, Kumar TA, Kumar KS. Integrating Natural language processing (NLP) and Machine learning Techniques for Healthcare Industries.
[8] Natekin A, Knoll A. Gradient boosting machines, a tutorial. Frontiers in Neurorobotics. 2013 Dec 4;7:21.
[9] Williams RJ. Reinforcement learning and Markov decision processes. CSG220, Spring. 2007.
[10] Blei DM, Kucukelbir A, McAuliffe JD. Variational inference: A review for statisticians. Journal of the American Statistical Association. 2017 Apr 3;112(518):859–877.
[11] Ramu V, Naresh U, Prakash PA, Shyamala G, Saritha P, Atheeswaran A. Real-time Air Quality Monitoring with Edge AI and Machine Learning Algorithm. In 2024 8th International Conference on Electronics, Communication and Aerospace Technology (ICECA) 2024 Nov 6 (pp. 1204–1209). IEEE.
[12] Angelopoulos AN, Bates S, Fannjiang C, Jordan MI, Zrnic T. Prediction-powered inference. Science. 2023 Nov 10;382(6671).
[13] Oliveira TA, Teodoro MF. Mathematical and Statistical Frameworks Fostering Advances in AI Systems and Computing. In International Conference on Computational Science and Its Applications 2024 Jul 1 (pp. 235–246). Cham: Springer Nature Switzerland.
[14] Abolade YA. Bridging mathematical foundations and intelligent systems: A statistical and machine learning approach. Communication In Physical Sciences. 2023 Sep 19;9(4):774–784.

S. Swetha, P. M. Joe Prathap

2 Linear Algebra in AI and Robotics

Abstract: Linear algebra renders an intrinsic role in artificial intelligence (AI) and robotics, which gives a mathematical foundation for representing and manipulating the data, modeling systems, and optimizing algorithms. Data is often appeared as vectors and matrices. Pretend that, an image is portrayed with a matrix of pixel values, and a document as a vector in high-dimensional space. It is widely used for feature extraction, dimensionality reduction, and helps to simplify large datasets while retaining important information. Optimization techniques like gradient descent rely on linear algebra to minimize loss functions. Linear algebra is explicative for figuring the positions, orientations, and velocities of robotic arms and legs. These estimations involve matrix transformations to unearth the joint angles and position in 3D space and to pretend dynamic systems in robotics through state space representation. This implicates matrices that chronicle the systems' evolution over time, which is rudimentary for control and stability analysis. It provides the mathematical chief support for many algorithms and models in AI and robotics, facilitates machines to learn from data, and make decisions and interact with their environment in precise and controlled ways.

Keywords: Artificial intelligence, robotics, linear algebra, matrix, vector

2.1 Introduction

Linear algebra is the foundation for mathematical discipline, which plays a pivot role in artificial intelligence (AI) and robotics and it defines the study of planes, vectors, matrices, lines, and mappings required for linear transformations. Linear algebra stores your data in vectors, matrices, and tensors and then operates as per your instructions. Linear algebra is a storage space from representing data as matrices and vectors to performing complex optimizations and transformations [1–3]. Data is an important factor in the outcome of any event. The four pillars aforementioned function as a repository to accumulate the data and preform operations as per the instructions given. It provides the required components to AI and robotics. In AI, linear algebra facilitates the handling of high dimensional data in data image, text, and audio processing through algorithm's machine learning. It supports the very foundation of neural networks since the flow of information across layers and its optimization are done through matrix operations and so is the accuracy improvement through information flow within layers. In robotics, linear algebra covers spatial transformations enabling robot navigation and understanding of environments. Robots are able to move, perceive objects, and interact with the world through vision using matrices to define rotations, translations, and scaling [4–6]. Concepts such as kinematic govern the feedback control systems and robotic

https://doi.org/10.1515/9783112214374-002

motion that enable precise movements have strong dependence on linear algebra. To put it simply, linear algebra connects the practicality of the real world with the world of theory, thus serving as a crucial asset in the fields of AI and robotics for processes involving data management, decision-making, and advanced reasoning spatially. Data storage is a prerequisite for computing, as is the basic framework of unsupervised learning in which linear algebra is utilized (for instance, k-means clustering), principal component analysis (PCA), non-negative matrix factorization, and more advanced clustering methods. Advanced algebra can be blended with regression and classification techniques, including but not limited to, logistic regression, support vector machines, and scalable parameter estimation in model computation for very large datasets [6–9]. The linear algebra with mathematical effects is shown in Figure 2.1.

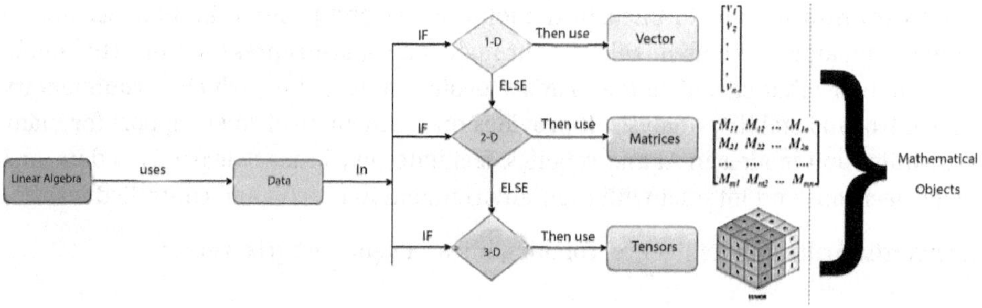

Figure 2.1: Linear algebra with mathematical objects.

2.2 Data Representation Technique Using Linear Algebra

Data representation and manipulation are judgmental in AI and robotics, as data is habitually composite, multidimensional, and essential systematic management. Linear algebra extends an efficacious representation for preside over this data, from articulating underdone data to remodel it for decision-making, model training, and environmental interaction [11]. Data points are recurrently represented as vectors. A vector marshals data into a list of numbers, creating it as uncomplicated to do transformations, comparisons, and calculations. Matrixes are frequently worn to clasp enormous data sets such as images (pixel grids) or time series measurements. This systemized method makes intricate jobs easier, such as image processing in AI or mapping in robots [12]. Tensors are multidimensional arrays that broaden matrices into more advanced dimensions. Tensors enable us to proficiently preside over and commute tangled data structures, making performance such as convolutional filtering or sequence processing significantly more coherent [13]. Normalization or scaling of characteristics is crucially stand in need of prior to data processing in AI and robotics. Transformations of a linear nature help to put off ranges adjusting biases and increasing model performance. PCA is a method of reducing dimensionality by conserving important features and eliminating redundant or trivial details. The simplification improves computational efficiency that is highly crucial in real-time decision-making processes in robotics [14]. Categorical data again and again has to be represented numerically. One-hot encoding and embeddings transfigure these classifications into numerical vectors, which models may recognize pertinently. Word embeddings in natural language processing are by way of illustration of how words are landed out to vectors to apprehend semantic information. In robotics, sensors generate data streams such as locations, velocities, and orientations. This information is again and again hoarded in vectors and processed in matrices, give authorization to the robot to build a structured representation of its state and surroundings [15]. Synthesis calculations, such as those found in neural networks, rely on matrix multiplications to move around data between model layers. Each and every multiplication mixes inputs, weights, and biases to aid pattern recognition in data. Matrix inversion and transposition are commonly used in robotics for tasks such as coordinate transformations and solving systems of equations, as well as for mapping and localization [16]. The data representation and manipulation are shown in Figure 2.2.

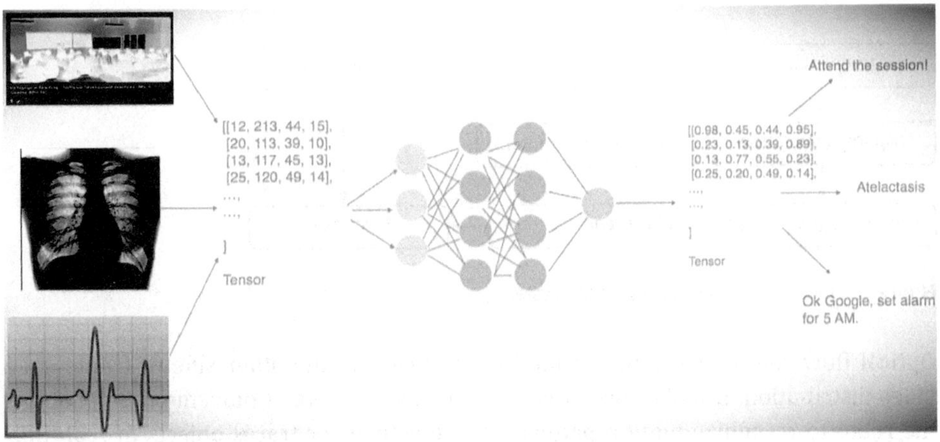

Figure 2.2: Data representation and manipulation.

Transformations are one of the core characteristics of computer vision robotics in which transformations allow robots to exhibit, represent, and work together with the environment in an effective manner. Linear algebra provides the mathematical framework mandatory for representing the transformations, including rotation, translation, and scaling, that are ruthless for work such as localization, 3D reconstruction, and object detection [17]. Linear transformations are second to manage an object's position, orientation, and size in a scene. Robots use translation vectors to circulate an object's position and rotation matrices to fit orientation. Scaling matrices swap the object's size. By combining these transformations, robots can truthfully map and utilize objects in their environment. To make transformations trouble free to estimate and coalesce, robotics in many cases uses homogeneous coordinates [18]. Camera calibration is pivotal in computer vision to be in cognizant of how a camera captures images. Intrinsic parameters depict internal camera properties while extrinsic describe the camera's relationship to the environment. Perspective projection needs to be done in order to depict a 3D scene onto a 2D image plane. This transformation demonstrates an example of a matrix that monitors depth cues while converting 3D coordinates into 2D coordinates. In applications like augmented reality, where virtual objects must be placed realistically in real-world settings, perspective projection is crucial [19]. Linear algebra empowers robots to ascertain the distance downwards of objects in the scene by scrutinizing images from different viewpoints (stereo vision). This concerns calculating disparities between two images and exploiting the difference of opinion to estimate distances [20]. Structure force motion is a methodology where the robot moves through its natural world and ensnare images at different angles. By examining how objects repositioned in the images, the robot can remodel the 3D structure of its environment. This is serviceable for mapping and navigating undetermined environments. The block diagram of the synthesis system is shown in Figure 2.3.

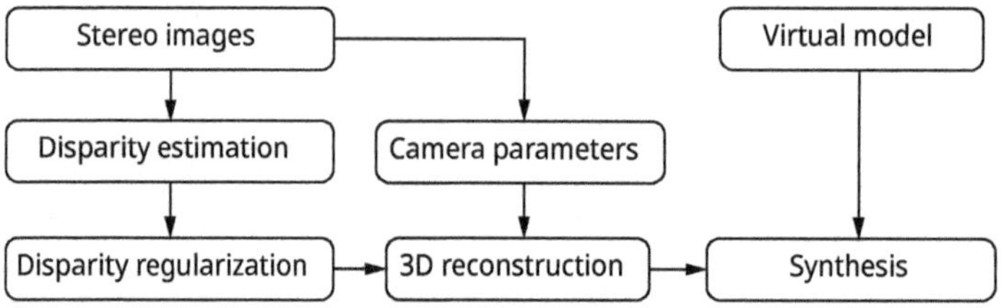

Figure 2.3: Block diagram of the synthesis system.

Optical flow counts the manifest motion of objects to the other side of frames in a video distribution. It makes use of vectors to have the vote of movement, so enabling the robot to recognize motion patterns, value actions, or traces objects in real-time. Visual odometry consent robots use changes in camera images across time to appraise their own motion. By keeping an eye on key points to the others side of frames, robots can calculate their development in 3D space, which is castigatory for navigation and obstacle avoidance. In object detection, transformations assist in resizing, cropping, and accommodating bounding boxes close to objects. As objects move or as the camera barter angles, affine transformations – which combine rotation, scaling, and translation – trace and control these boxes. Pose estimation relates to object orientation and position judgment. Key point estimation and matrix transformations enable robots to detect the 3D posture of an object, which is absolutely crucial for grasping or working with the object. Image rectification coincides images from poles apart viewpoints so they appear from a familiar perspective. This transformation is often put in an application in stereo vision to make vertical drop assessment more accurate. Image warping guilt trips images by putting out transformations like shearing or twisting. In robotics, warping can be nearly new to turn out panoramic images, correct distortions, or simulate different viewpoints, make preparations for the robot with a better understanding of its surroundings.

2.3 Optimization and Machine Learning

A rudimentary machine learning representation utilizes linear algebra, where the interconnection among variables is manifested as linear equations. In a simple linear regression model, we aim to replicate, via a linear equation, the interrelation between a single input feature x and an output y. Multiple linear regression allows us to handle several features. We show the relationship between the input features and the target variable y for a dataset comprising N samples and z features as follows:

$$y = a_0 + a_1x_1 + a_2x_2 + \cdots + a_zx_z + \epsilon \tag{2.1}$$

Because of its ease of use, interpretability, and suitability for a range of estimation, calibration, and prediction tasks, linear regression is a flexible tool in AI and robotics. In order to iteratively update model parameters for reducing errors or increasing likelihood, optimization techniques such as gradient descent rely on linear algebra:

$$ai{:} = aj - \eta \frac{\delta L}{\delta aj} \tag{2.2}$$

where η is the learning rate, controlling the step size of each update. $\frac{\delta L}{\delta aj}$ is the partial derivative of the loss function, which tells us the direction and rate of change.

PCA is a dimensionality dwindle strategy that employs matrix factorization, which has eigen values and eigen vectors to cut down data trademark while sustaining key information, accommodating in following components:

2.3.1 Data Preprocessing and Noise Reduction

- In innumerable AI applications, PCA promotes preprocess and noise cancellation of data by lessen irrelevant dimensions, removes the complexities from the dataset without loss track of supreme information. This is serviceable for sensor data, image data, and other types of high-dimensional inputs that may have redundant or noisy features.

2.3.2 Feature Extraction

- New, more insightful aspects that capture the essence of the data can be created through PCA. This enables AI algorithms to eliminate redundant or unnecessary data and concentrate on the most important patterns.

2.3.3 Visualization

- Visualizing high-dimensional data, such as text embeddings and images, can be difficult. By reducing it to two or three dimensions for easy visualization, PCA enables researchers to observe separations, clustering patterns, and other insights.

2.3.4 Improving Model Efficiency

- Using PCA to reduce the number of features can speed up model training and inference because smaller datasets require less processing power. This is especially helpful when working with large datasets or plotting with limited processing power.

2.3.5 Pattern Recognition and Anomaly Detection

- PCA highlights the key trends in data and can be used in domains such as cyber security, finance, and computer vision. Variations from typical patterns in the PCA, which may indicate outliers or unusual occurrences, qualify anomaly detection.

2.3.6 Compression in Image and Signal Processing

- PCA is frequently used to compress signals and images by preserving only the principal components, enabling effective transmission and storage. As an example, PCA allows for the reduction of an image's pixel count while maintaining visual quality, which is helpful for medical imaging and facial recognition.

2.4 Neural Networks and Deep Learning

Neural networks confide with leaden steps on matrix multiplications, primarily for weight updates and forward passes through layers of neurons. Each layer in a neural network carries out linear transformations followed by non-linear activations. These matrices lay hold off the parameters that the network comprehends during training, with each entry in the matrix be the representative of connection strength between two neurons. The weight matrix correlates neurons in the middle of adjacent layers, determining the influence of each input on the resulting output. When data flows through a neural network, it passes through each layer, where it is transfigured by a fusing of matrix multiplications (as shown in Figure 2.4) and activation functions. Matrix multiplication is a core operation here, where the input vector or matrix is multiplied by the weight matrix, followed by adding the bias. These operations allow the network to produce the next layer's outputs based on the learned parameters.

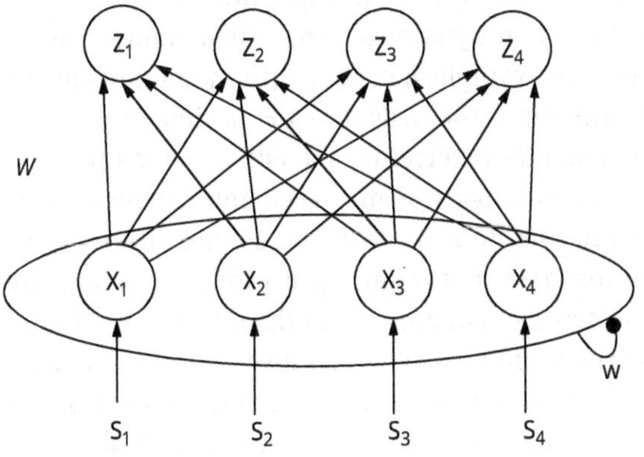

Figure 2.4: Matrix multiplication in neural networks.

Given an input vector x and a weight matrix W for a layer, the output z is computed as:

$$z = W \times x + b \tag{2.3}$$

where b is a bias vector that is appended to the output of every neuron. In order to optimize model weights, backpropagation – a crucial step in neural network training – calculates gradients using linear algebra. Backpropagation determines the gradient of the loss function for each network weight during training. These gradients are efficiently computed for the entire network using linear algebra, more especially matrix calculus, which guarantees the best possible weight adjustment. Neural networks learn to minimize prediction errors by updating weights through linear transformations (e.g., gradient descent). Neural networks can process data efficiently through matrix multiplication, vector operations, and other linear transformations, and the basic learning processes in these AI models are supported by the concepts of linear algebra.

2.5 Control Systems in Robotics

In feedback and state estimation feedback control systems, robot actions are controlled based on current and predicted states, and in control, state estimation techniques, such as the Kalman filter, are supported by linear algebra. Matrix multiplication is frequently used in control systems to depict system dynamics. The new state can be calculated given a control input u and a state vector x by:

$$x' = Ax + Bu \tag{2.4}$$

where A represents the system matrix and B represents the input matrix.

Robot kinematics (describing motion) and dynamics (force calculations) are described by systems of linear equations. For example, a robotic arm's joint movements are represented as a system of linear equations that calculate the position and orientation of the end effector. Forward kinematics determines the end-effector's position and orientation given the joint parameters (angles or displacements). Inverse kinematics solves for the joint parameters needed to achieve a desired end-effector position and orientation. This is essential for tasks that require precision, like picking up an object at a specific location. Where forward dynamics gives the forces and torques, it predicts the resulting motion of the robot. Inverse dynamics gives a desired motion, it calculates the necessary forces and torques. This is especially useful for control systems that need to know exactly how much force each motor should apply to achieve a specific movement.

2.6 Conclusion

In robotics, linear algebra provides the necessary tools for spatial analysis, motion control, and system stability, while in AI, it enables the creation of models that are capable of learning patterns and making predictions. Proficiency in linear algebra is a prerequisite for anyone interested in advancements in AI or robotics, as it is crucial for both developing fast algorithms and understanding intricate robotic systems. Intricate data can be simplified into numerical formats that algorithms can understand and work with the help of linear algebra. It provides tools to acknowledge the associations within data, make predictions, and categorize information. Eventually, without linear algebra, the computation and implementation of AI algorithms would be highly counterproductive, if not impossible. Linear algebra empowers the precise calculation, manipulation, and control of robotic systems, making it a rudimentary tool in robotics. A solid understanding of linear algebra allows engineers and researchers to design systematic algorithms for navigation, manipulation, and interaction, all of which are vitally important for the furtherance of robotic technologies.

References

[1] Tsujiuchi N, Takayuki K, Yoneda M Manipulation of a Robot by EMG Signals Using Linear Multiple Regression Model. In 2004 IEEE/RSJ International Conference on Intelligent Robots and Systems (IROS)(IEEE Cat. No. 04CH37566) 2004 Sep 28 2 (pp. 1991–6). IEEE.

[2] Hölgyesi Á, Zrubka Z, Gulácsi L, Baji P, Haidegger T, Kozlovszky M, Weszl M, Kovács L, Péntek M Robot-assisted surgery and artificial intelligence-based tumour diagnostics: Social preferences with a representative cross-sectional survey. BMC Medical Informatics and Decision Making. 2024 Mar 27;24(1):87.

[3] Hines L, Petersen K, Lum GZ, Sitti M. Soft actuators for small-scale robotics. Advanced Materials. 2017 Apr;29(13):1603483.

[4] Gunawardane PD, Budiardjo N, Alici G, De Silva CW, Chiao M Thermoelastic Strain-limiting Layers to Actively-control Soft Actuator Trajectories. In 2022 IEEE 5th International Conference on Soft Robotics (Robosoft) 2022 Apr 4 (pp. 48–53). IEEE.

[5] Hu W, Alici G Bioinspired three-dimensional-printed helical soft pneumatic actuators and their characterization. Soft Robotics. 2020 Jun 1;7(3):267–82.

[6] Adams R, Soe SP, Santiago R, Robinson M, Hanna B, McShane G, Alves M, Burek R, Theobald P A novel pathway for efficient characterisation of additively manufactured thermoplastic elastomers. Materials & Design. 2019 Oct 15;180:107917.

[7] Gunawardane PD, Cheung P, Zhou H, Alici G, De Silva Cw, Chiao M A versatile 3d-printable soft pneumatic actuator design for multi-functional applications in soft robotics. Soft Robotics. 2024 Aug 1;11(4):709–23.

[8] AK D, Kumar A, Sr K, Gayathri N, Das P, edited by. AI and IoT-based Intelligent Automation in Robotics. John Wiley & Sons; 2021 Apr 30.

[9] Eppenberger T, Cesari G, Dymczyk M, Siegwart R, Dubé R Leveraging Stereo-camera Data for Real-time Dynamic Obstacle Detection and Tracking. In 2020 IEEE/RSJ International Conference on Intelligent Robots and Systems (IROS) 2020 Oct 24 (pp. 10528–35). IEEE.

[10] Mir-Nasiri N Camera-based 3D Object Tracking and following Mobile Robot. In 2006 IEEE Conference on Robotics, Automation and Mechatronics 2006 Jun 1 (pp. 1–6). IEEE.

[11] Joseph AJ, Sani N, KS K, TA K, Nishanth R Towards a Novel and Efficient Public Key Management for Peer-peer Security in Wireless Ad-hoc/sensor networks. In 2022 International Conference on Smart Technologies and Systems for Next Generation Computing (ICSTSN) 2022 Mar 25 (pp. 1–4). IEEE.

[12] Li G, Liao X, Huang H, Song S, Liu B, Zeng Y Robust stereo visual slam for dynamic environments with moving object. IEEE Access. 2021 Feb 16;9:32310–20.

[13] Lou Q, González F, Kövecses J Kinematic modeling and state estimation of exploration rovers. IEEE Robotics and Automation Letters. 2019 Jan 25;4(2):1311–18.

[14] Ji Y, Liu D, Guo Y Adaptive neural network based position tracking control for dual-master/single-slave teleoperation system under communication constant time delays. ISA Transactions. 2019 Oct 1;93:80–92.

[15] Ma Z, Liu Z, Huang P Discrete-time practical robotic control for human–robot interaction with state constraint and sensorless force estimation. ISA Transactions. 2022 Oct 1;129:659–74.

[16] Hua CC, Yang Y, Guan X Neural network-based adaptive position tracking control for bilateral teleoperation under constant time delay. Neurocomputing. 2013 Aug 3;113:204–12.

[17] Ji Y, Gong Y Adaptive control for dual-master/single-slave nonlinear teleoperation systems with time-varying communication delays. IEEE Transactions on Instrumentation and Measurement. 2021 Jun 8;70:1–5.

[18] Swetha S, JoePrathap PM. "A Study on a Decentralized Network Secured Data Sharing Using Blockchain,". In 2022 1st International Conference on Computational Science and Technology (ICCST) 2022 (pp. 620–4). CHENNAI, India.
[19] Mahadevaswamy CL, Baltes J, Chang HT Humanoid Pose Estimation through Synergistic Integration of Computer Vision and Deep Learning Techniques. In 2024 International Conference on Advanced Robotics and Mechatronics (ICARM) 2024 Jul 8 (pp. 771–6). IEEE.
[20] Panigrahi PK, Bisoy SK Localization strategies for autonomous mobile robots: A review. Journal of King Saud University-Computer and Information Sciences. 2022 Sep 1;34(8):6019–39.

A. Vaithiyanathan, A. Mohammed Shapique

3 Advanced Statistical Tools for Data Science

Abstract: This chapter explores advanced statistical tools that form the backbone of modern data science, providing the mathematical and computational foundation for extracting insights from complex, high-dimensional, and structured data. Beginning with an overview of their relevance across disciplines such as robotics, healthcare, finance, and manufacturing, the chapter details the theoretical and applied aspects of cutting-edge methods. Techniques like Markov chain Monte Carlo, canonical correlation analysis, discriminant analysis, and multivariate analysis of variance (MANOVA) enable rigorous modeling, classification, and inference across multivariate datasets. The utility of hierarchical and mixed-effects models is demonstrated in structured data environments, facilitating robust parameter estimation and variance partitioning. Causal inference methodologies, including structural causal models and do-calculus, extend traditional statistical reasoning to uncover underlying cause-effect relationships. High-dimensional statistical techniques such as Lasso, Elastic Net, random forests, and gradient boosting machines enhance model accuracy, feature selection, and interpretability in settings with many predictors. Time series analysis using autoregressive integrated moving average and seasonal autoregressive integrated moving average, and long short-term memory networks underscores the chapter's commitment to temporal forecasting, while nonparametric methods like kernel density estimation and spline regression offer flexibility in modeling nonlinear patterns.

By integrating theory, real-world examples, and visualization, this chapter serves as a comprehensive guide for data scientists and researchers aiming to leverage statistical sophistication in decision-making, prediction, and exploratory analysis. The methodologies discussed bridge classical statistical foundations with contemporary data-driven applications, empowering practitioners to address uncertainty, variability, and complexity in diverse analytical environments.

Keywords: Statistical tools, data science, multivariate analysis, causal inference, high-dimensional statistics, time series analysis, nonparametric methods

3.1 Introduction

Data science is an interdisciplinary domain that applies methodologies from statistics, computer science, mathematics, and domain-specific expertise to extract actionable insights from extensive and commonly unstructured datasets [1]. As the modern world produces immense volumes of information daily from social media interactions to online purchases data science serves the crucial role of transforming this raw, disorganized data into actionable intelligence. Its applications are broad and impactful [2]. In

https://doi.org/10.1515/9783112214374-003

public health, it helps predict disease outbreaks through demographic analysis; in business, it enables personalized marketing by understanding consumer behavior. At its core, data science empowers better decision-making by uncovering patterns, trends, and relationships that would otherwise remain hidden in the noise of data [3].

The data science life cycle is a structured yet flexible process that transforms raw data into meaningful insights for informed decision-making. It begins with obtaining relevant data from various sources like databases, sensors, APIs, or online platforms. Once collected, the data undergoes cleaning and preprocessing to correct errors, handle missing values, and ensure consistency, an essential step that makes the data reliable for analysis. Next, through exploratory data analysis (EDA), data scientists examine patterns and relationships using univariate, bivariate, or multivariate techniques to better understand the data and form hypotheses. This is followed by modeling, where statistical methods or machine learning algorithms are applied to generate predictions or explain behaviors within the data. Finally, the results are interpreted and communicated clearly to stakeholders, enabling effective decision-making and real-world impact [4].

In data science, four core techniques of analysis serve distinct but complementary purposes depending on the objective of the study. Descriptive analysis summarizes and highlights the main features of a dataset using measures like averages, percentages, and frequencies such as evaluating the average spending of customers in a retail setting. Diagnostic analysis seeks to uncover the reasons behind particular outcomes; for instance, healthcare professionals might use it to identify factors contributing to high patient readmission rates. Predictive analysis employs statistical models, such as regression or classification, to forecast future events based on historical data like projecting sales trends or predicting customer behavior. Finally, prescriptive analysis takes prediction a step further by suggesting actionable strategies, enabling organizations to make informed decisions regarding resource management, planning, or targeted recommendations. Together, these techniques form the analytical backbone of data-driven decision-making

3.1.1 Advanced Statistical Tools and its Application in Data Science

Advanced statistical tools play a crucial role by enhancing the depth, precision, and reliability of analyses. While descriptive and diagnostic analyses benefit from tools like hypothesis testing and correlation analysis to validate observations and investigate causal relationships, predictive and prescriptive analyses rely heavily on statistical models such as regression, Bayesian methods, and time series forecasting to generate accurate forecasts and actionable insights. Techniques like multivariate analysis and dimensionality reduction further enable data scientists to manage and interpret large, complex datasets. Together, these statistical tools enrich each stage of the analytical process, making it possible to extract meaningful knowledge from data and guide sound [5].

3.1.2 Advanced Statistical Tools in Robotics

Modern robotics heavily relies on sophisticated statistical methods to empower robots to function intelligently within unpredictable and ever-changing environments. By employing probabilistic approaches, robots can decipher ambiguous sensor readings and determine their location and surroundings, exemplified by techniques like simultaneous localization and mapping (SLAM) which utilizes filters such as the extended Kalman filter and particle filter. Bayesian inference further allows robots to learn and make informed choices as they gather new data, with Bayesian networks and hidden Markov models being key in applications like speech recognition and activity monitoring. Machine learning algorithms, including support vector machines, Gaussian mixture models, and neural networks, leverage statistical principles to enable robots to identify patterns, forecast results, and enhance their capabilities through experience. Stochastic models like Markov decision processes (MDPs) and partially observable MDPs (POMDPs) guide robots in making optimal decisions despite uncertainty, crucial for autonomous driving and robotic manipulation. In manufacturing, statistical process control ensures consistent quality and identifies deviations. Combining data from various sensors like cameras, LIDAR, and inertial measurement units through sensor fusion depends on statistical techniques like Kalman filters and expectation-maximization algorithms for accurate and dependable information crucial for navigation and interaction. Moreover, the emergence of deep reinforcement learning, rooted in statistical theory, allows robots to learn optimal behaviors through experimentation in complex settings. Collectively, these advanced statistical tools are revolutionizing robots from inflexible, programmed devices into adaptable, intelligent agents capable of operating effectively in real-world uncertainties.

3.2 Markov Chain Monte Carlo methods

Markov chain Monte Carlo (MCMC) is a powerful computational technique in Data Science, particularly vital for Bayesian inference, as it enables sampling from complex probability distributions. Unlike traditional Monte Carlo methods [6] that rely on independent samples, MCMC generates dependent samples by constructing a Markov chain designed to eventually converge to the desired target distribution. The theoretical foundation of MCMC lies in Markov chains, which are sequences of random variables where each subsequent state depends solely on the current state (the Markov property).

A sufficient condition for a Markov chain to converge to the target distribution $\pi(x)$ is

$$\pi(x)T(x \to x') = \pi(x')T(x' \to x) \tag{3.1}$$

Example: Bayesian inference for normal distribution.

Suppose we observe data $D = \{x_1, x_2, \ldots x_n\}$ drawn from a normal population $N(\mu, \sigma^2)$, we want to estimate μ and σ using MCMC. Bayesian inference for normal distribution is presented in Figure 3.1:

Likelihood

$$p(D|\mu, \sigma^2) = \prod_{i=1}^{n} N(x_i|\mu, \sigma^2) \tag{3.2}$$

where

$\mu \sim N(\mu, 10)$, σ Half-Cauchy (0,5)

Posterior distribution $p(\mu, \sigma|D) a p(D|\mu, \sigma)p(\mu)p(\sigma)$

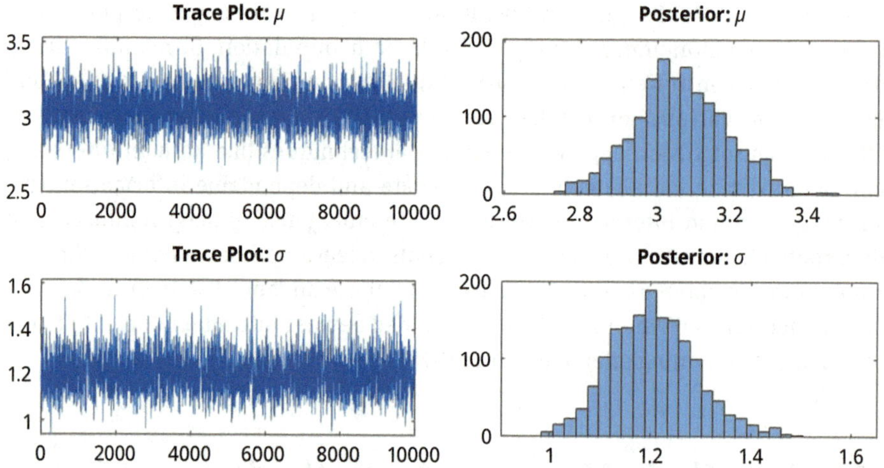

Figure 3.1: Bayesian inference for normal distribution.

The MATLAB-generated plots from the Metropolis-Hastings MCMC simulation provide key insights into the Bayesian estimation of the normal distribution parameters. Here's the interpretation:

The trace plots for both the mean (μ) and standard deviation (σ) provide strong evidence of proper mixing and convergence in the Markov Chain. The trace plot for μ fluctuates randomly around the true mean value ($\mu = 3$), with no visible long-term trends or signs of "stickiness," which indicates that the chain has successfully converged to the stationary distribution. Similarly, the trace plot for σ shows rapid and consistent exploration of the posterior distribution, suggesting efficient sampling behavior. These patterns in both plots lead to the conclusion that the Markov Chain has effectively converged to the target posterior distribution.

The posterior histograms for the parameters μ and σ effectively illustrate the uncertainty associated with their estimates. The posterior distribution for μ is centered near the true value of 3 and exhibits a narrow spread, indicating a high degree of confidence in the estimate. Its approximately Gaussian shape aligns with expectations given a normal likelihood and a conjugate prior. In contrast, the posterior for σ peaks around the true value of 1.5 but displays a broader spread, reflecting greater uncertainty, an expected outcome, as standard deviation is inherently more difficult to estimate than the mean. Additionally, a slight right skew is observed in the σ histogram, a result of the half-Cauchy prior, which enforces positivity. Overall, the Bayesian framework offers full posterior distributions, allowing for a comprehensive quantification of uncertainty in both μ and σ.

The MCMC method successfully estimated the parameters μ and σ, with the posterior means aligning closely with the true generative values. This demonstrates the effectiveness of MCMC in parameter estimation. From a data science perspective, the practical implications are significant. MCMC not only provides point estimates but also facilitates uncertainty quantification through credible intervals. The trace plots confirm that the algorithm thoroughly explored the parameter space, ensuring reliable convergence, while the posterior histograms offer insights into the distribution of estimates – serving as Bayesian analogs to confidence intervals. For instance, the 95% credible interval for σ might fall within [1.3, 1.7], clearly illustrating the range of plausible values.

3.3 Canonical Correlation Analysis (CCA)

Canonical correlation analysis (CCA) is a foundational multivariate statistical technique designed to uncover and quantify the relationships between two sets of variables. Introduced by Harold Hotelling in 1936, CCA extends the concept of simple correlation (between two variables) to multidimensional spaces. It identifies pairs of linear combinations – termed canonical variates – derived from each variable set, such that their correlation is maximized. This maximum correlation, the first canonical correlation, captures the dominant shared dimension between the datasets. Subsequent orthogonal pairs reveal additional layers of association, each uncorrelated with the prior.

Unlike methods focused on prediction (e.g., regression) or dimensionality reduction within a single set (e.g., PCA), CCA specializes in exploring cross-domain dependencies. By revealing latent structures that link two complex observational domains such as genetic markers and clinical symptoms, or environmental factors and ecological responses CCA provides a powerful "bridge" for interdisciplinary analysis. Its versatility makes it indispensable in fields including neuroscience, genomics, psychometrics, and machine learning, where understanding joint dynamics between complementary data modalities is critical. CCA finds relationships between two sets of variables. It is used for multi-view learning, linking different data sources (text and images), recommendation systems (user preferences and item features), and identifying complex dependencies [7].

Example: Suppose we have data on students' academic performance and their psychological traits:

Set 1 (Academic Variables): Math score and English score

Set 2 (Psychological Variables): Motivation score and anxiety score

We want to investigate how the academic performance variables relate to the psychological traits using CCA.

X = [88 75; 92 82; 85 78; 90 80; 87 76; 93 85; 91 79; 86 77; 89 81; 84 74];

Y = [80 45; 85 40; 78 48; 82 42; 79 44; 87 39; 84 41; 77 46; 83 43; 76 47];

The CCA of the data is presented in Figure 3.2.

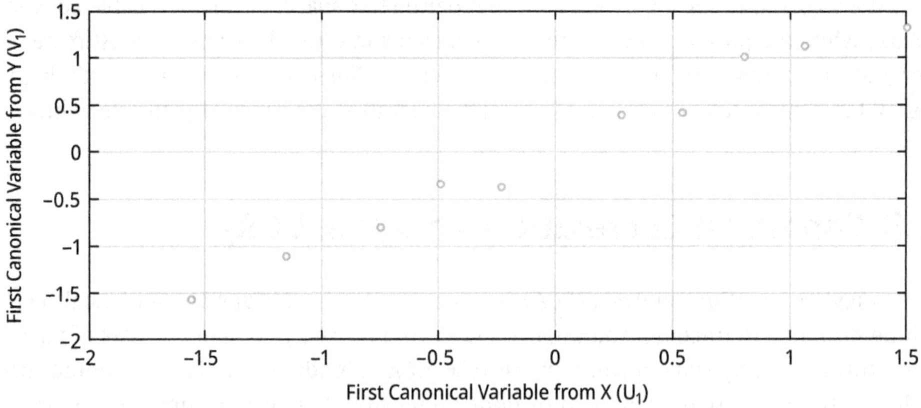

Figure 3.2: Canonical correlation analysis (first pair).

If the first canonical correlation (first value of r) is high (e.g., >0.8), it indicates a strong linear relationship between some combination of academic performance and psychological traits.

3.4 Discriminant Analysis (LDA and QDA)

Discriminant analysis (DA) encompasses a set of supervised classification methods aimed at distinguishing between two or more predefined groups based on observed characteristics. In contrast to opaque "black-box" models, DA offers interpretable probabilistic classification grounded in assumptions of multivariate normality. Among its most widely used forms are linear DA (LDA) and quadratic DA (QDA) , both of which utilize Gaussian distributions to derive decision boundaries. LDA emphasizes simplicity by assuming equal covariance among groups, while QDA provides greater flexibility by allowing distinct covariance structures, thus achieving a balance between interpretability and adaptability [8].

Example: In a semiconductor manufacturing plant, sensors monitor two features from each produced chip: Temperature (°C) of the chip during testing and voltage stability (V) of the chip output. Based on these measurements, chips are classified as Class 1 (good chip) and Class 2 (faulty chip).

We aim to use LDA and QDA to build a model that classifies new chips and to understand which method performs better given potential differences in group variance:

X = [72 1.02; 75 1.01; 73 1.03; 74 1.00; 76 1.02; 82 0.96; 85 0.95; 81 0.94; 83 0.97; 84 0.93]

Y = [1; 1; 1; 1; 1; 2; 2; 2; 2; 2];

LDA Classification: Class 2 (1 = good, 2 = faulty)

QDA Classification: Class 2 (1 = good, 2 = faulty)

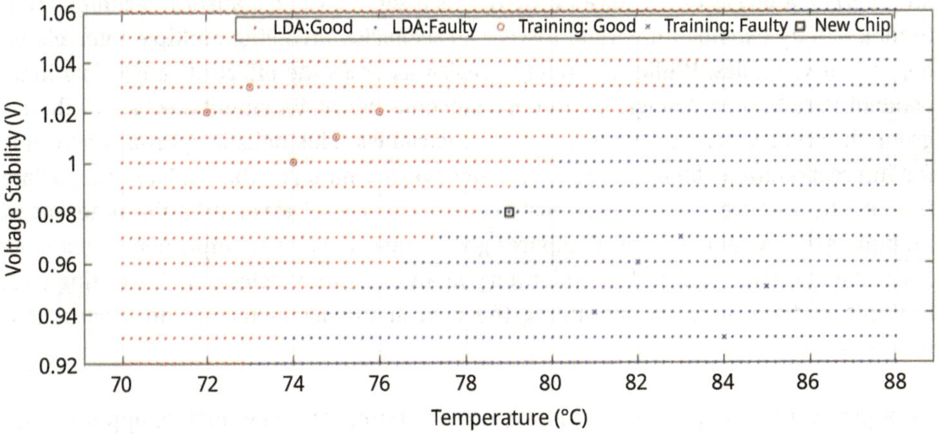

Figure 3.3: LDA decision boundary.

Figures 3.3 and 3.4 indicate that the new chip is classified as faulty by both LDA and QDA, based on learned patterns.

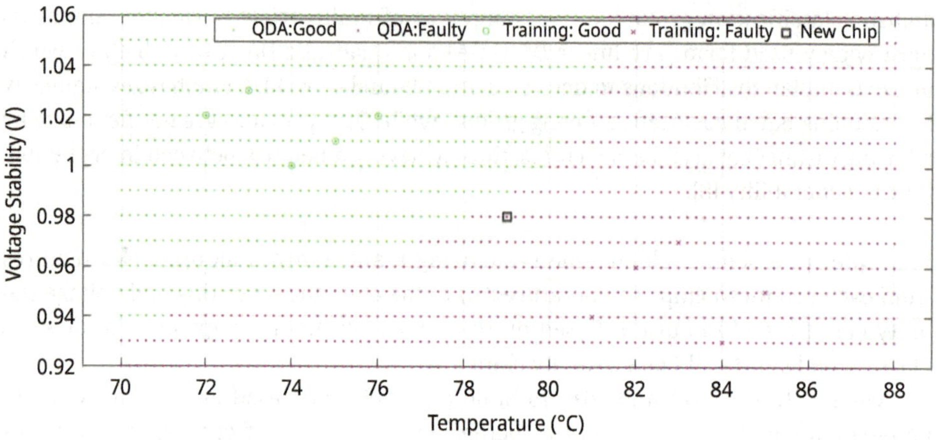

Figure 3.4: QDA decision boundary.

3.5 Multivariate ANOVA (MANOVA)

Multivariate analysis of variance (MANOVA) is a sophisticated statistical technique that fundamentally extends univariate ANOVA to scenarios involving multiple interrelated dependent variables. While traditional ANOVA assesses the effect of categorical independent variables on a single continuous outcome, MANOVA simultaneously evaluates group differences across several dependent variables. This holistic approach accounts for intercorrelations among outcomes – uncovering patterns that isolated univariate tests might obscure – while rigorously controlling Type I error inflation inherent in multiple ANOVA comparisons. Compares group means across multiple dependent variables simultaneously. Used for exploratory analysis, feature selection, A/B testing with multiple metrics, and understanding the overall impact of factors on several outcomes [9].

Example: A tech company is evaluating three versions of a new mobile app interface (A, B, and C). They test each version on a group of users and collect two usability scores: Task completion time (in seconds) and user satisfaction rating (scale 1–10):

Group A = [30 8.5; 32 9.0; 28 7.5; 35 9.2]
Group B = [40 6.5; 42 6.8; 38 6.2; 41 7.0]
Group C = [33 8.0; 36 7.8; 34 7.9; 37 8.1]

We test whether the app version has a statistically significant effect on either or both variables. The task completion time by app version and user satisfaction is graphically presented in Figure 3.5. The MANOVA results are follows:

MANOVA RESULTS
Wilks' Lambda *p*-value: 0.0000
Wilks' Lambda *p*-value: 0.7315
Group Means (Centroids):
Group 1: Task time: 31.25; satisfaction: 8.55
Group 2: Task time: 40.25; satisfaction: 6.63
Group 3: Task time: 35.00; satisfaction: 7.95

Since $p < 0.05$, we conclude that the app version has a significant multivariate effect on user experience. The MANOVA results are presented in Figure 3.6.

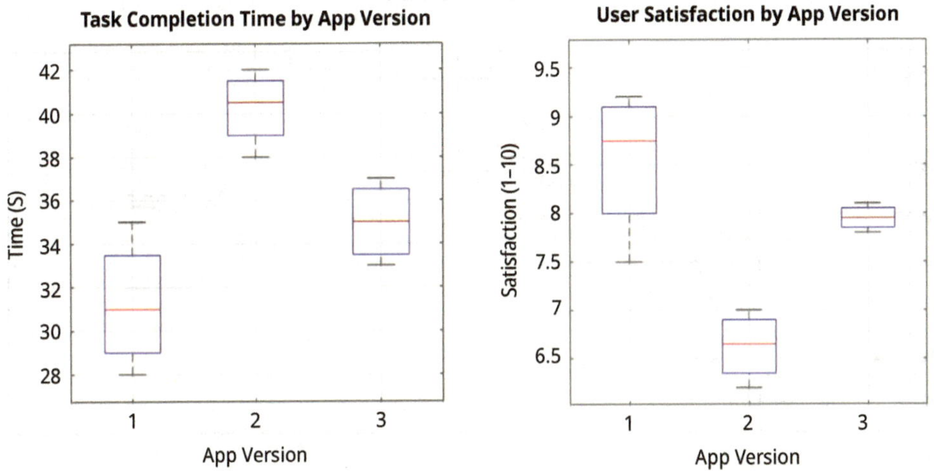

Figure 3.5: Task completions versus user satisfaction.

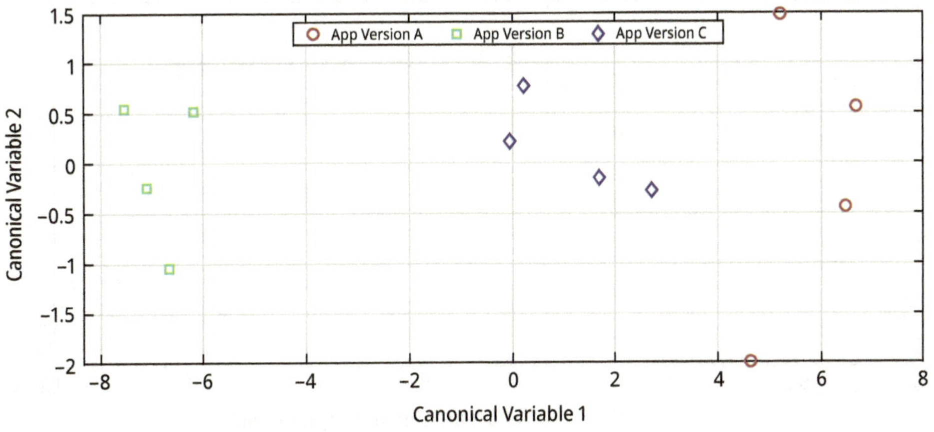

Figure 3.6: MANOVA: canonical discriminant variables.

To understand which dependent variables are contributing to the overall MANOVA significance. We will use ANOVA in MATLAB for task completion time and user satisfaction rating and then apply Tukey's honestly significant difference (HSD) test using Mult compare. See Tables 3.1 and 3.2, and the results are graphically illustrated in Figures 3.7–3.10.

Table 3.1: ANOVA table 1.

Source	SS	df	MS	*F*	Prob > *F*
Groups	163.5	2	81.75	16.17	0.001
Error	45.5	9	5.0556	–	–
Total	209	11	–	–	–

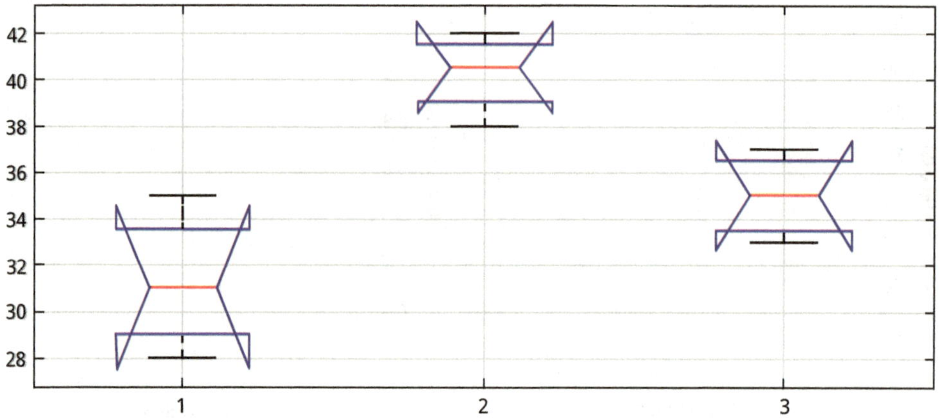

Figure 3.7: Post hoc: task completion time.

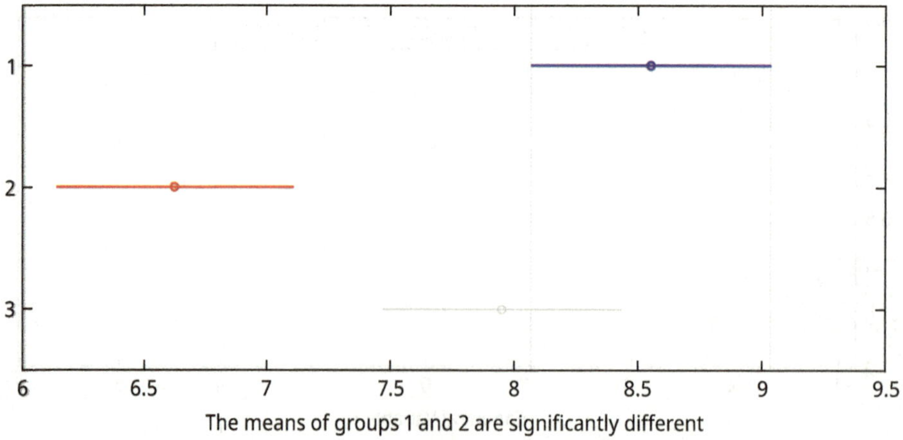

Figure 3.8: Post hoc: user satisfaction.

Table 3.2: ANOVA table 2.

Source	SS	df	MS	*F*	Prob > *F*
Groups	7.76167	2	3.88083	16.26	0.001
error	2.1475	9	0.23861	–	–
total	5.90917	11	–	–	–

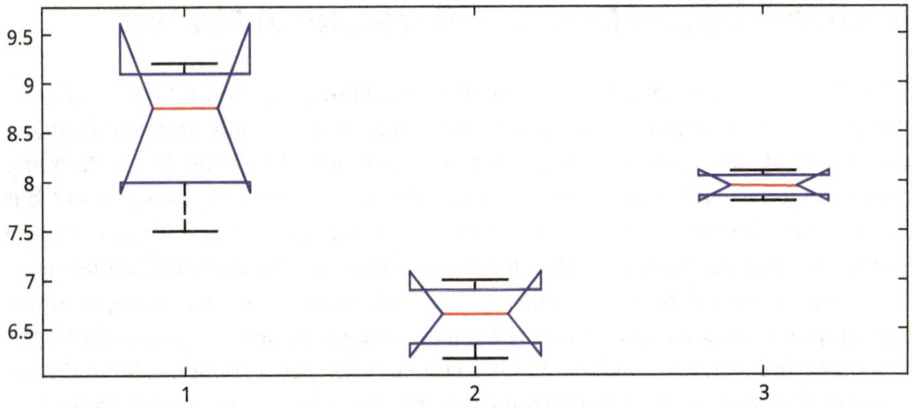

Figure 3.9: Post hoc: task completion time.

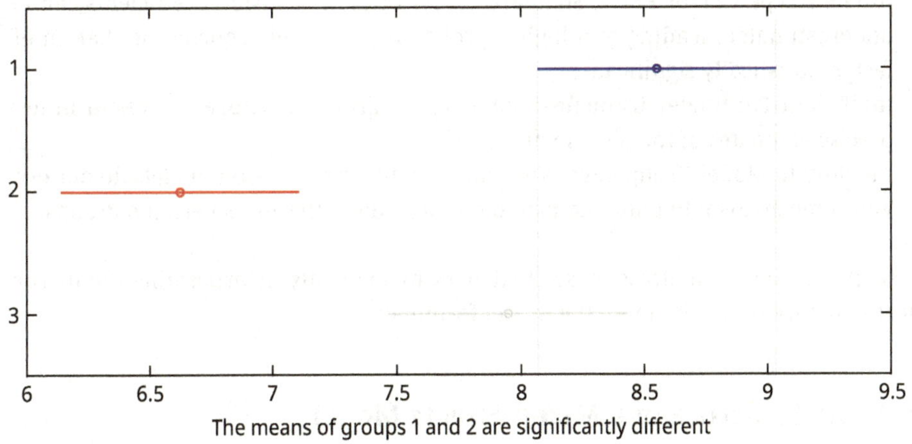

Figure 3.10: Post hoc: user satisfaction.

Analysis	Interpretation
MANOVA *p*-value	Shows if there's a significant multivariate difference.
ANOVA for task time	Whether task times differ by app version.
ANOVA for satisfaction	Whether user satisfaction differs by version.
Tukey's post hoc	Shows which pairs (A vs. B, B vs. C, etc.) differ.

3.6 Mixed-Effects Models (Hierarchical Models)

This section introduces mixed-effects models, also known as hierarchical models or multilevel models, a powerful class of statistical models specifically designed to analyses such data by simultaneously accounting for variation at multiple levels. In many real-world data science applications, observations are not entirely independent but rather exhibit a hierarchical or clustered structure. This inherent grouping can arise from various sources, such as repeated measurements on the same individuals over time, students nested within classrooms and schools, or customers belonging to different geographical regions. Traditional regression models, as discussed in earlier, typically assume independence among all data points. A key property of clustered data is the increased homogeneity of observations within the same cluster relative to the heterogeneity of observations across different clusters [10]. This intra-group correlation violates the independence assumption and can have significant consequences:

- Inflated Type I Error Rates: Standard errors of the regression coefficients can be underestimated, leading to a higher probability of falsely concluding that an effect is statistically significant.
- Inefficient Parameter Estimates: Ignoring the group structure can result in less precise estimates of the model parameters.
- Inability to Model Group-Level Variation: Standard regression models do not provide a mechanism to understand and quantify the differences between groups.

Mixed-effects models address these challenges by explicitly incorporating the hierarchical structure of the data into the model formulation.

3.6.1 The Structure of a Mixed-Effects Model

A basic linear mixed-effects model (LMM) can be represented as:

$$y_{ij} = (\beta_0 + u_{0j}) + (\beta_1 + u_{1j})x_{ij} + \xi_{ij} \tag{3.3}$$

Example

It is advisable to gather reaction time data from several participants, each undertaking multiple trials under two different treatment conditions, labeled A and B. The objective is to analyze the effects of these treatments while considering the variability between individuals.

Sample Data (created manually for illustration)

3.7 Linear Mixed-Effects Model Fit by ML

LMMs offer a robust statistical approach for analyzing data characterized by hierarchical or nested structures, where observations are grouped within broader units such as students within schools, repeated measures within individuals, or plants within ecological plots. By incorporating both fixed effects (representing overall population-level trends) and random effects (capturing variability across groups), LMMs effectively account for the correlation among observations that violates the independence assumption of standard linear regression. When fitted using maximum likelihood (ML) estimation, LMMs facilitate coherent statistical inference, hypothesis testing, and model selection in complex and structured datasets.

The model uses 36 observations, includes 2 fixed effects, 6 random effects, and estimates 2 covariance parameters (Tables 3.3–3.6). The results are graphically illustrated in Figures 3.11 and 3.12.

Formula: Reaction time ~1 + treatment + (1 | subject)
Model fit statistics:

Table 3.3: Fixed effects coefficients (95% CIs).

AIC	BIC	Log likelihood	Deviance
194.57	200.91	−93.286	186.57

Table 3.4: Coefficient estimates.

Name	Estimate	SE	*t*-Stat	DF	*p*-Value	Lower	Upper
Intercept	94.131	3.215	29.279	34	1.06E-25	87.598	100.66
Treatment	10.11	0.75402	13.408	34	3.93E-15	8.5776	11.642

Table 3.5: Random effects estimates.

Group	Levels	Name 1	Name 2	Type	Estimate	Lower	Upper
Subject	6	Intercept	Intercept	std	7.7661	4.3752	13.785

Table 3.6: Residual variance estimates.

Name	Estimate	Lower	Upper
Res Std	2.2621	1.7564	2.9134

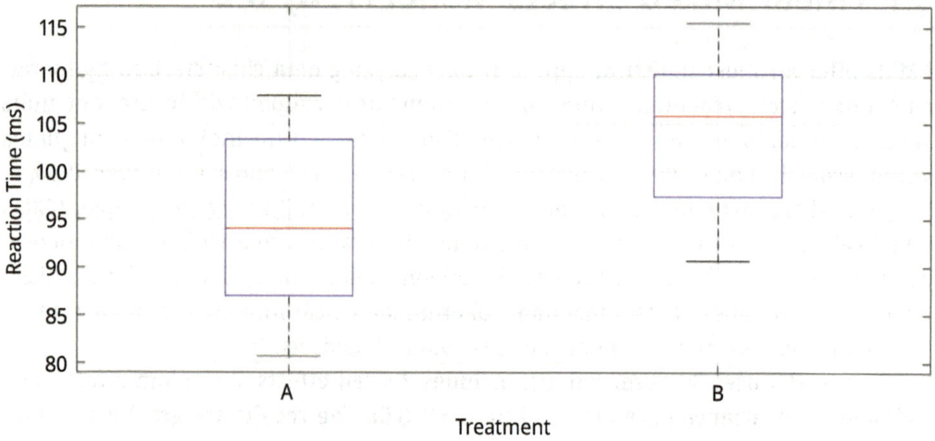

Figure 3.11: Reaction time by treatment.

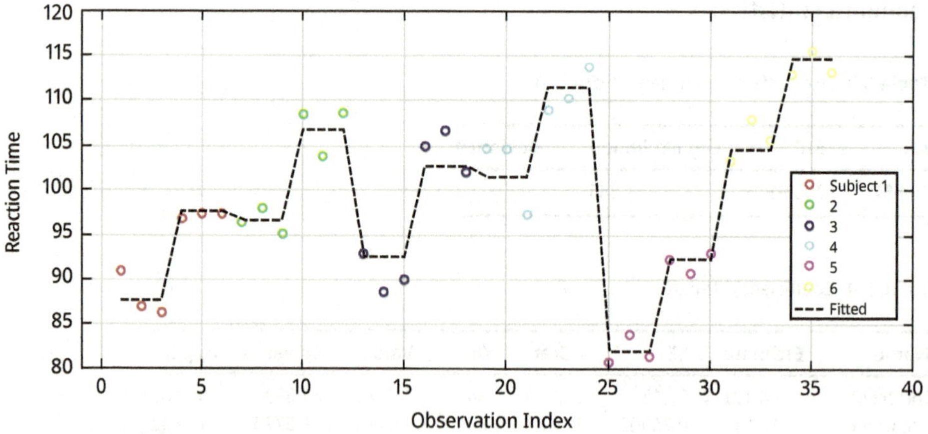

Figure 3.12: Observed versus predicted reaction times.

In this section, we explore a structural equation model assessing technology user satisfaction. Employing MATLAB for data analysis, we validate latent constructs using factor analysis and use regression-based path modeling to explore causal relationships.

Example: To uncover latent constructs that explain the variability in a set of observed technological indicators gathered from a sample of 50 tech products (Table 3.7). Consider a dataset consisting of 50 observations, each representing a distinct technological product.

Table 3.7: Performance metrics scores.

Q1	*Q2*	*Q3*	*Q4*	*Q5*	*Q6*	*Q7*	*Q8*	*Q9*	*Q10*	*Q11*	*Q12*
4	5	1	2	4	1	1	2	5	5	2	3
5	3	5	4	5	3	5	3	5	2	2	1
2	2	2	2	1	5	2	1	1	4	3	2
1	4	2	2	4	5	1	2	4	5	3	5
1	4	2	3	1	5	2	3	3	2	1	2
4	5	4	2	4	1	1	3	3	2	4	5
3	1	1	2	2	4	1	1	5	3	5	4
4	1	4	5	4	5	5	5	2	1	5	3
1	3	5	2	2	1	3	5	5	1	5	2
5	2	1	3	4	2	3	5	5	3	3	1
2	3	3	1	2	3	5	1	2	3	2	5
3	1	1	2	1	2	4	2	2	5	5	4
1	4	1	4	2	3	5	1	1	4	2	2
1	1	5	3	4	5	3	1	4	4	2	3
5	4	1	1	5	3	5	3	1	4	1	1
5	3	2	1	5	4	1	2	3	5	5	4
4	4	4	3	4	4	5	4	3	5	5	1
4	4	1	4	2	1	3	3	3	1	3	2
5	1	5	5	2	4	2	5	2	3	2	1
1	3	5	2	1	1	4	2	1	5	4	3
4	5	5	4	1	1	1	5	2	5	2	3
3	5	4	5	5	1	4	3	5	4	5	3
4	1	3	2	4	3	1	2	5	2	4	4
2	3	1	3	5	1	3	5	4	5	4	1
5	3	3	5	2	3	2	2	2	1	5	5
3	3	4	2	5	1	1	4	3	5	2	4
2	2	3	5	1	4	1	5	3	4	2	2
5	5	1	3	2	4	1	2	1	3	3	5
4	3	3	3	1	3	1	5	2	1	3	4
1	5	4	4	4	1	4	2	3	1	2	5
3	4	5	5	3	2	3	1	4	4	3	1
1	1	1	3	5	1	5	2	3	2	3	5
2	4	2	2	3	5	2	1	3	2	3	1
1	4	5	2	1	5	1	4	3	5	4	3
5	3	5	1	1	5	3	3	5	3	4	1
1	5	4	5	4	4	5	1	4	2	5	5
4	3	3	3	3	3	1	3	2	3	4	1
1	2	2	4	4	4	2	4	4	4	2	4
1	5	1	3	5	5	3	1	4	3	5	1

Table 3.7 (continued)

Q1	*Q2*	*Q3*	*Q4*	*Q5*	*Q6*	*Q7*	*Q8*	*Q9*	*Q10*	*Q11*	*Q12*
5	3	4	5	3	5	2	4	5	4	1	4
1	5	4	1	4	2	5	5	3	3	5	3
2	3	4	2	4	4	1	5	4	4	4	4
1	3	4	2	4	3	4	3	3	1	5	1
2	4	1	1	1	2	3	5	5	3	1	2
1	1	3	5	4	1	3	2	4	4	2	5
5	5	1	2	1	2	3	3	5	3	5	1
3	5	5	2	2	2	2	5	4	5	2	2
2	3	5	4	3	4	4	3	4	1	3	1
2	1	1	4	5	1	1	1	1	2	1	2
4	4	5	1	2	1	4	3	2	2	1	1

where ***Q*1**, processing speed; ***Q*2**, power consumption; **Q3**, memory usage; ***Q*4**, thermal efficiency; ***Q*5**, network latency; **Q6**, battery backup; **Q7**, data throughput; **Q8**, cost-efficiency; **Q9**, user experience score; **Q10**, error rate; **Q11**, build quality; **Q12**, reliability index.

The factor model is

$$x = \mu + \Lambda f + \xi \tag{3.4}$$

where x is $p \times 1$ vector of observed variables; μ is $p \times 1$ vector of means of the observed variables; Λ is $p \times k$ matrix of factor loadings; f is $k \times 1$ vector of latent factors, assumed to be standard normal: $E[f] = 0$, $\text{cov}[f] = I_k$; ξ is the **error term** uncorrelated with factors, $E[\xi] = 0$ $\text{cov}[\xi] = D$, D is the diagonal matrix.

To derive the covariance matrix, we compute $\text{cov}[x] = \Lambda\Lambda^T + \Psi$, where Ψ is the error (unique variance).

The aim is to explore whether a smaller set of unobserved (latent) factors can explain the correlations among these 12 observed variables. For example, latent factors such as system performance, power efficiency, and build reliability may be responsible for the patterns observed in the data. The aim is to explore whether a smaller set of unobserved (latent) factors can explain the correlations among these 12 observed variables. For example, latent factors such as system performance, power efficiency, and build reliability may be responsible for the patterns observed in the data. Exploratory factor analysis is presented in Figure 3.13.

X_5 shows a very strong loading on F_3 (0.9306) – it's a key indicator for this factor.

X_7 has a strong negative loading on F_1 (−0.8978), and moderately positive on F_2 (0.3784).

X_8 loads highly on both F_1 and F_2, which might indicate that it is influenced by more than one latent construct.

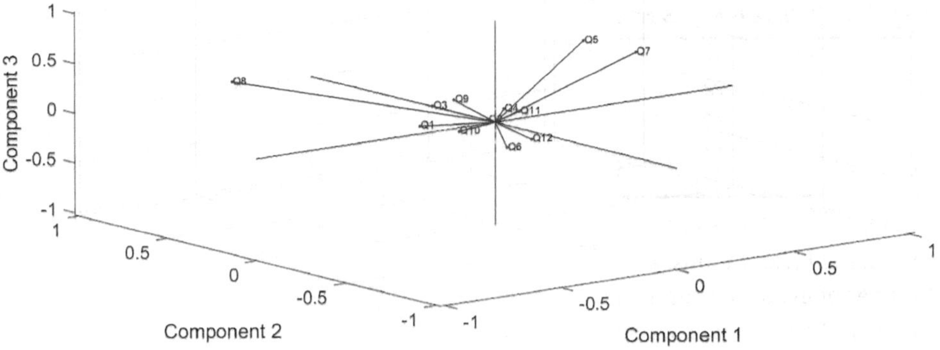

Figure 3.13: Exploratory factor analysis (proxy for SEM).

3.8 Survival and Hazard Models

Consider an example of device failure analysis in smart manufacturing systems. To analyze the time until failure of devices and estimate hazard rates, helping predict reliability and schedule preventive maintenance. To monitor 20 smart devices (Table 3.8) over time, each device has a recorded failure time (in days) or is right-censored (i.e., still working at last observation). The results are presented in Figure 3.14 and Table 3.9.

Table 3.8: Device reliability data.

Device ID	Failure time	Status
1	12	1
2	18	1
3	20	0
4	10	1
5	25	0
6	14	1
7	30	0
8	9	1
9	21	1
10	19	1
11	28	0
12	11	1
13	24	0
14	16	1
15	17	1

Table 3.8 (continued)

Device ID	Failure time	Status
16	13	1
17	22	1
18	26	0
19	29	0
20	15	1

Number of observations: 20
Number of events (failures): 13
Number of censored observations: 7
Status = 1 ⇒ Device failed
Status = 0 ⇒ Device still working (censored)
Analysis Objectives

- Estimate device reliability over time
- Compare failure rates between device types or usage conditions
- Predict mean time to failure

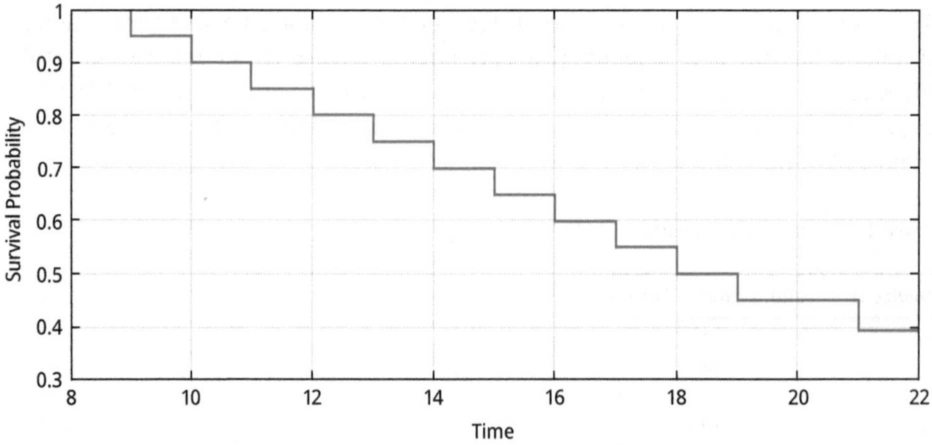

Figure 3.14: Kaplan-Meier survival curve.

Table 3.9: Log-likelihood and stats.

Variable	F_1	F_2	F_3	ψ
X_1	0.2444	0.1009	0.0026	0.9301
X_2	−0.0347	0.0893	−0.029	0.9900
X_3	0.0216	0.3215	0.0241	0 .8956
X_4	−0.0391	0.0021	0.1196	0.9842

Table 3.9 (continued)

Variable	F_1	F_2	F_3	ψ
X_5	−0.1225	−0.3376	0.9306	0 .0050
X_6	0.048	−0.1286	−0.1777	0 .9496
X_7	−0.8978	0.3784	0.2138	0.0050
X_8	0.5052	0.8081	0.2263	0 .0405
X_9	0.1468	0.0405	0.2585	0 .9100
X_{10}	0.2055	−0.0667	0.0117	0.9532
X_{11}	−0.1067	0.0022	0.0708	0.9836
X_{12}	0.0833	−0.3117	−0.0002	0 .8959

Log-likelihood and stats:
Log-likelihood: −0.2887
dfe: 33

Methodology
An exploratory factor analysis is applied to identify and estimate a small number of common factors. The process involves:
1. Factor Extraction: Estimating how many latent variables adequately explain the observed data structure.
2. Factor Loadings: Determining how strongly each observed variable is associated with each latent factor.
3. Unique Variance Estimation: Assessing how much of each observed variable's variability is unique and not explained by common factors.
4. Visualization: Generating a factor plot to observe how variables cluster and align with the latent constructs.

Interpretation of Results
- Factor loadings show the correlation between observed variables and latent constructs. For instance, high loadings of "processing speed," "memory usage," and "data throughput" on one factor would suggest the presence of a Performance dimension.
- Unique variances help identify which variables are mostly independent or poorly explained by the model, aiding in refinement or exclusion.
- Factor structure visualization (e.g., via a biplot) reveals natural groupings and the relationship of variables with each dimension.

Applications
This analysis assists in:
- Reducing dimensionality for downstream predictive models
- Defining meaningful constructs from empirical data
- Informing the development of full structural models, where relationships among latent variables themselves are theorized and tested

3.9 Causal Inference

Statistical techniques are crucial for data scientists aiming to move beyond basic correlations and uncover the true causal relationships between variables. They offer a robust set of tools for making well-informed decisions and deriving meaningful insights from data [11].

Consider a binary variable $Z \in \{0, 1\}$, representing control (0) and treatment (1) groups. We aim to understand the impact of Z on an outcome variable Y. The potential outcomes framework, also known as the Rubin causal model (RCM), extends the joint distribution of (Z,Y) by introducing two dimensional random variables: $(Y(1),Y(0))$. These represent the potential outcomes of Y under treatment and control conditions, respectively. Since Y is the observed outcome, and by definition we have:

$$Y = \begin{cases} Y(1), & \text{if } Z = 1 \\ Y(0), & \text{if } Z = 0 \end{cases} \tag{3.5}$$

When $Z = 1$, $Y(0)$ is not observed and is the counterfactual, and when $Z = 0$, the unobserved outcome $Y(1)$ becomes the counterfactual.

3.9.1 Fundamental Problem of Causal Inference, Confounding Variables, and Selection Bias, Along with Expanded Examples and Implications

Fundamental problem of causal inference:

At the core of causal inference lies a critical challenge:

We can never observe both potential outcomes $Y(1)$ and $Y(0)$ for the same individual.

- If a person receives the treatment ($Z = 1$), we observe only their outcome under treatment $Y(1)$, but we do not know what would have happened if they not received the treatment $Y(0)$.
- Conversely, if they do not receive the treatment ($Z = 0$), we observe $Y(0)$, but never $Y(1)$.

This is known as the **"missing counterfactual"** problem. In this situation, we focus on **population-level causal effects**, such as:

- Average treatment effect:

$$A_{\mathrm{TE}} = E[Y(1) - Y(0)] \tag{3.6}$$

- Average treatment effect on the treated:

$$A_{\mathrm{TT}} = E[Y(1) - Y(0) / Z = 1] \tag{3.7}$$

These quantities estimate causal impact **on average**, which is often sufficient for decision-making in policy, business, or medicine.

3.9.2 Structural Causal Models (SCMs)

Structural causal models (SCMs) are a foundational framework for causal reasoning in data science, developed and popularized by Judea Pearl. While traditional data science focuses on correlations and predictions, SCMs go further by enabling us to ask and answer causal questions [12].

Consider Table 3.10, which represents our simulated observations for hours studied (X), prior knowledge (Z), and test score (Y).

Table 3.10: Structural causal models.

Hours studied	Prior knowledge	Test score
2	3	60
5	2	75
3	4	70
7	1	85
4	3	72
6	4	88
1	2	55
8	5	92
3	1	65
5	3	80

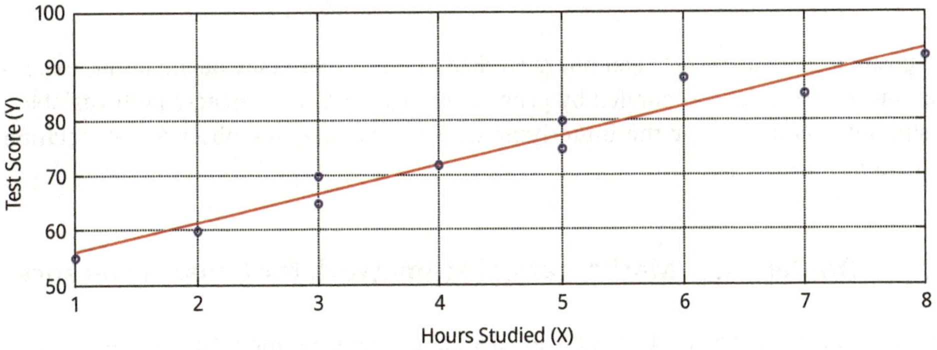

Figure 3.15: Observed relationship: hours studied versus test score.

Scatter plots (Figures 3.15–3.17) display the observed relationships between each pair of variables, with the least-squares regression line indicating the general trend. While

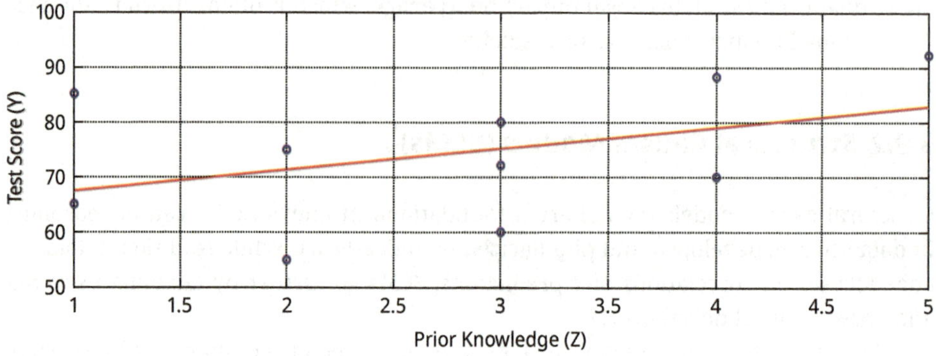

Figure 3.16: Observed relationship: prior knowledge versus test score.

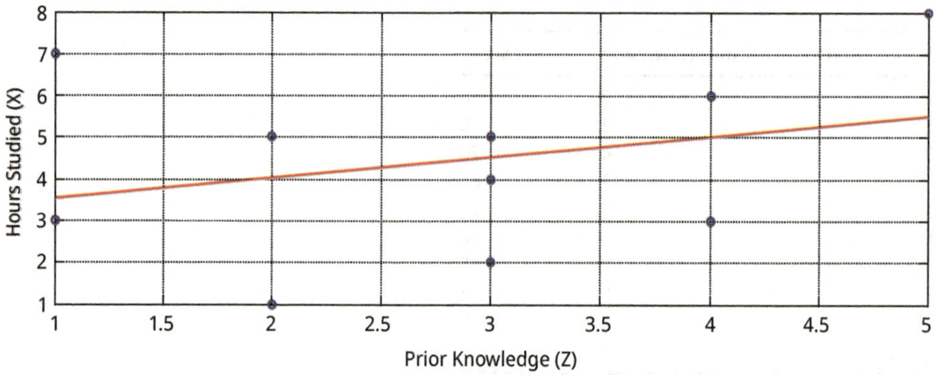

Figure 3.17: Observed relationship: prior knowledge versus hours studied.

a positive correlation between hours studied and test score may be apparent, this relationship could be confounded by prior knowledge, which influences both variables. Without accounting for the underlying causal structure, the observed association might be misleading.

3.9.3 Do-Calculus: Mathematical Framework for Causal Inference

Do-calculus provides a set of rules for deducing causal relationships and estimating the effects of interventions – symbolized as do($X = x$) – from observational data. It operates on causal diagrams, which are also known as directed acyclic graphs (DAGs). In simple terms:

- Standard Probability: Describes associations (e.g., $P(Y \mid X)$)
- Causal Inference: Describes interventions (e.g., $P(Y \mid (X))$)

The key challenge is that $P(Y \mid X)$ is NOT always equal to $P(Y \mid (X))$, because of confounding variables or indirect effects. Consider Table 3.11, which shows the dataset of persons' exercise, weight, and blood pressure.

Table 3.11: Dataset.

Person	Exercise (*X*)	Weight (*Z*)	Blood pressure (*Y*)
1	1	1	1
2	1	1	0
3	1	0	0
4	0	0	0
5	0	0	1
6	0	1	1
7	1	1	1
8	1	1	1
9	0	0	0
10	1	1	1

From the above, X affects Z (exercise helps normalize weight) and Z affects Y (body weight impacts blood pressure). We want to calculate: $P(Y = 1 \mid do(X = 1))$, i.e., the effect of exercise on normal blood pressure. Also, we visualize through the following graphs.

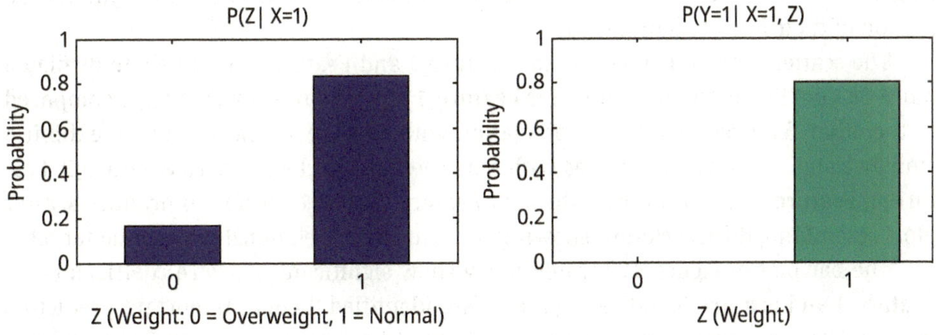

Figure 3.18: Conditional probabilities.

Using do-calculus, vi back door adjustment we conclude the value $P(Y = 1 \mid X = 1) = 4/5 = 0.8$. Figure 3.18 shows the distribution $P(Z \mid X = 1)$ that is how often each weight category occurs when people exercise. The **second plot** shows $P(Y = 1 \mid X = 1, Z)$ that is the chance of normal blood pressure given weight and exercise. Together, these components are multiplied and summed to compute $P(Y = 1 \mid (X = 1))$.

3.10 High-Dimensional Statistics

3.10.1 Lasso and Elastic Net Regression

Lasso is effective for feature selection and building sparse models, particularly when many features may be irrelevant. Elastic Net offers a balanced approach that better handles multicollinearity and often yields more stable and accurate predictions, especially when features are highly correlated. The choice between the two depends on the specific dataset and modeling objectives, with cross-validation commonly used to determine the optimal method and tuning parameters [13]. Lasso and Elastic Net are powerful techniques for developing robust, interpretable, and accurate regression models, especially in situations involving high dimensionality and multicollinearity. These methods are extensively applied in a wide range of data science domains. Consider the dataset presented in Table 3.12.

Figure 3.19 shows the pairwise relationships between a subset of the features and the target variable. It is useful for gaining a general understanding of correlations and distributions. Figure 3.20 presents the correlation matrix of all features and the target variable, where the color intensity and annotations reflect the strength and direction of their linear relationships.

The scatter plots in Figure 3.21 for Feature_1 and Feature_3 are likely to display a more distinct linear trend positive for Feature_1 and negative for Feature_3 compared to the other features. Due to its correlation with Feature_1, Feature_2 may exhibit a similar trend. In contrast, features with true coefficients close to zero (Feature_2, Feature_4, Feature_5, and Feature_7 through Feature_10) are expected to produce scatter plots resembling diffuse clouds, showing no clear linear relationship with the target.

The bar plot in Figure 3.22 would likely show significant non-zero coefficients for Feature_1 and Feature_3, indicating that Lasso identified them as important predictors (Figure 3.23). The coefficient for Feature_2 might be very close to zero or even zero due to its high correlation with Feature_1 and the L1 penalty's tendency to select one feature from a group of correlated ones. The coefficients for the other less relevant features (Feature_4, Feature_5, Feature_7 through Feature_10) would likely be shrunk towards or exactly to zero, demonstrating Lasso's feature selection capability. Feature_6's small true coefficient might result in a small non-zero coefficient in the Lasso model.

Table 3.12: Machine learning dataset or feature and target data.

Feature_1	Feature_2	Feature_3	Feature_4	Feature_5	Feature_6	Feature_7	Feature_8	Feature_9	Feature_10	Target
0.549	0.436	0.792	0.899	0.265	0.774	0.122	0.671	0.827	0.970	0.728
0.603	0.273	0.276	0.391	0.946	0.791	0.568	0.019	0.618	0.612	-0.388
0.646	0.894	0.083	0.821	0.097	0.958	0.531	0.164	0.124	0.314	-1.085
0.926	0.792	0.759	0.489	0.723	0.430	0.208	0.494	0.053	0.378	1.373
0.087	0.130	0.676	0.712	0.016	0.291	0.894	0.722	0.418	0.950	0.442
0.778	0.871	0.979	0.127	0.059	0.093	0.135	0.392	0.266	0.568	2.119
0.940	0.122	0.143	0.945	0.522	0.415	0.265	0.774	0.122	0.671	-0.568
0.029	0.857	0.291	0.021	0.769	0.234	0.646	0.894	0.083	0.821	1.306
0.208	0.949	0.665	0.193	0.685	0.871	0.979	0.127	0.059	0.093	0.153
0.569	0.302	0.332	0.288	0.761	0.122	0.143	0.945	0.522	0.415	-0.814

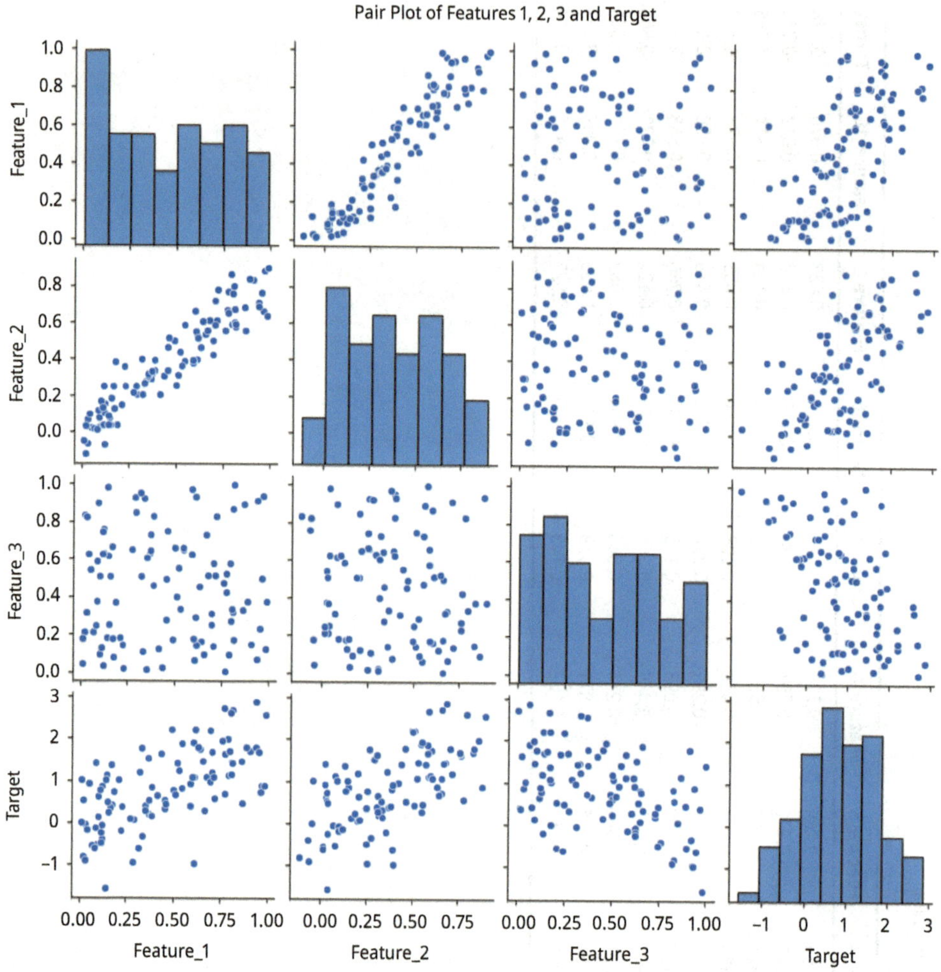

Figure 3.19: Pair plot of features 1, 2, 3, and target.

The generated graphs collectively highlight:

- The underlying relationships and correlations within the synthetic dataset.
- Lasso's tendency to perform feature selection by driving the coefficients of less important or redundant (i.e., correlated) features toward zero, resulting in a sparser model.
- How Elastic Net also applies regularization and feature shrinkage but manages correlated features by shrinking their coefficients together, rather than eliminating them outright.
- A visual evaluation of the Lasso model's predictive performance on unseen data.

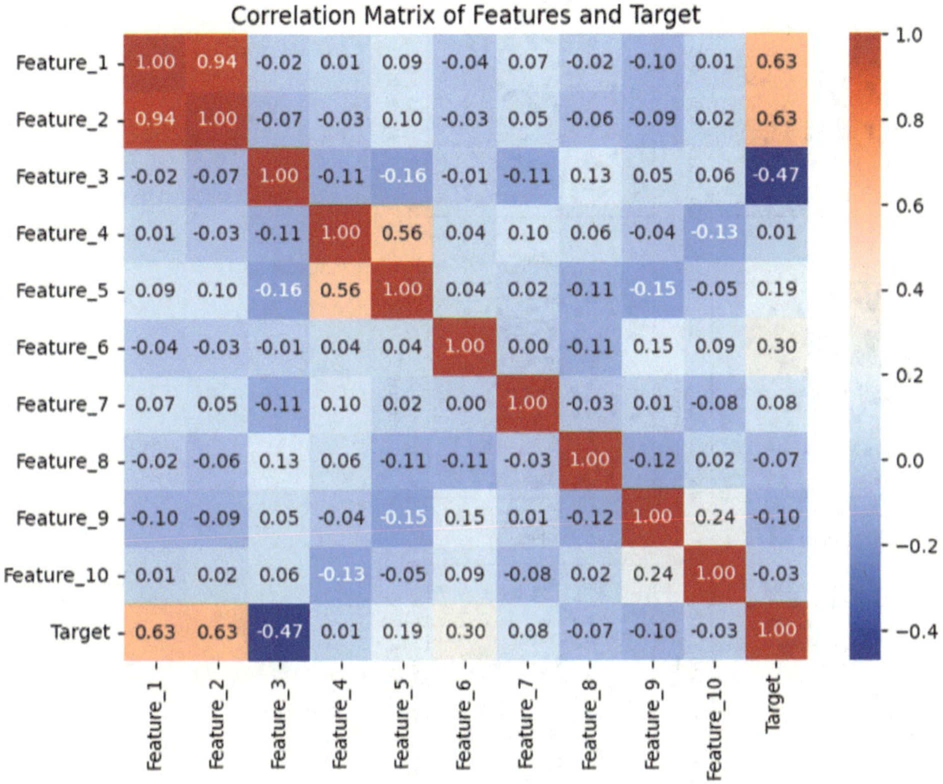

Figure 3.20: Correlation matrix of features and target.

Together, these visualizations offer a clearer understanding of how Lasso and Elastic Net operate in practice, especially when dealing with datasets that include correlated or irrelevant features.

3.10.2 Random Forests

Random forests construct multiple slightly varied decision trees and aggregate their results to produce a model that is more accurate, stable, and resistant to overfitting. This "wisdom of the crowd" strategy is what makes random forests highly effective across diverse applications. The concept "random forest" is so popular in data science due to the following reasons [14]:

- High Accuracy: Random forests typically deliver superior predictive performance compared to many other machine learning models.

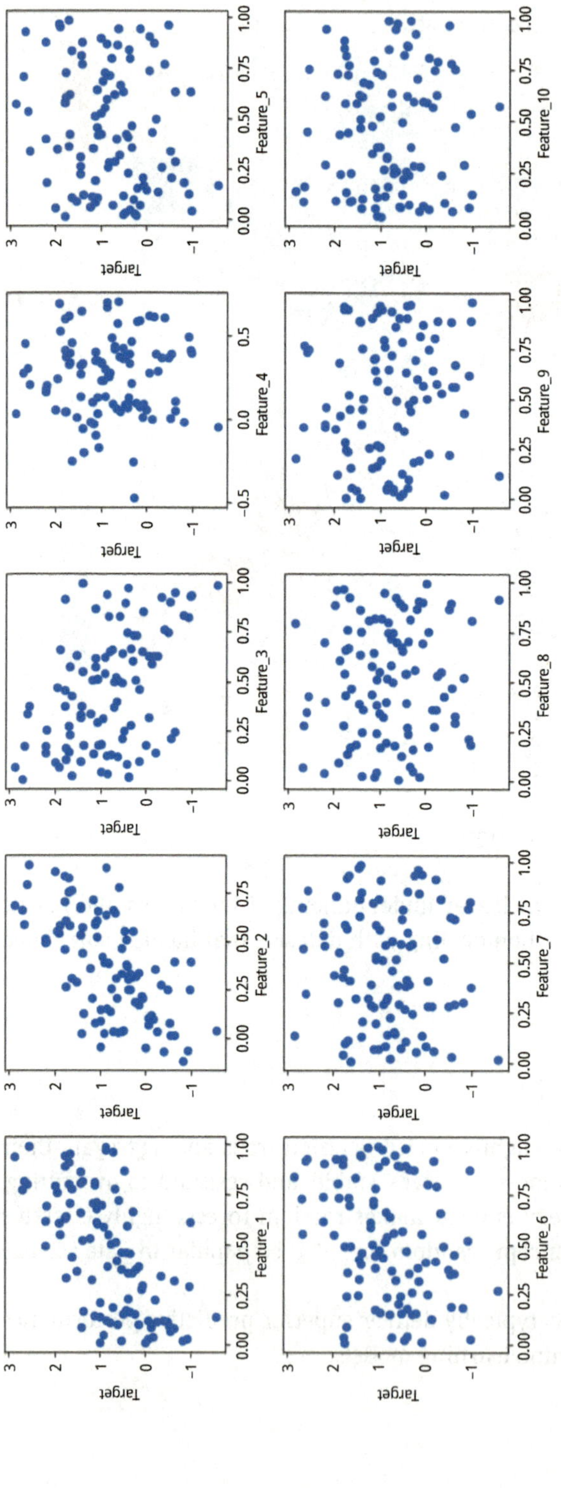

Figure 3.21: Scatter plots of features versus target.

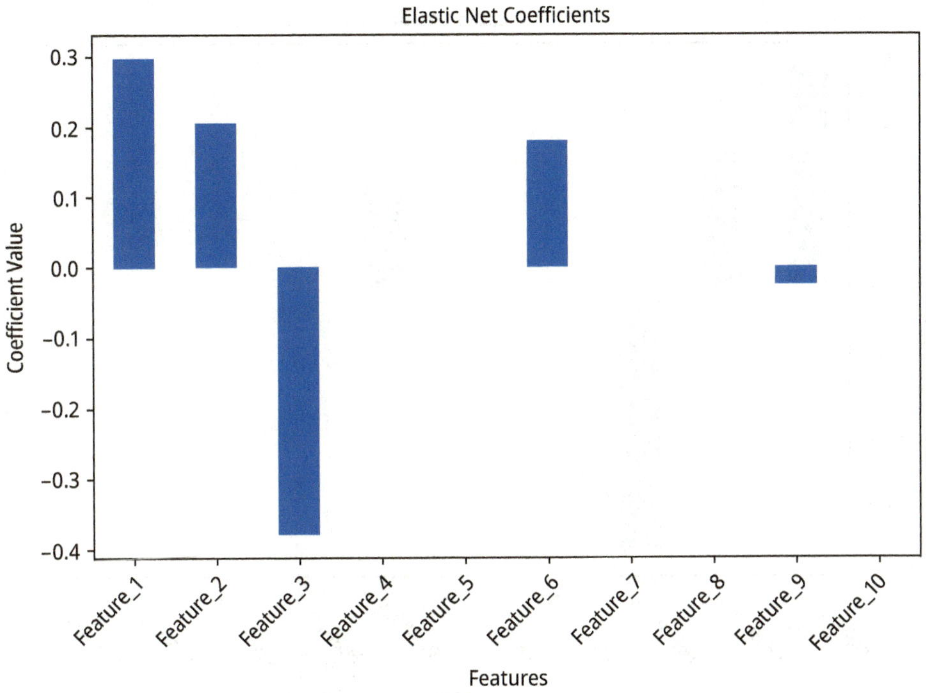

Figure 3.22: Elastic net coefficients.

- Resistant to Overfitting: Their ensemble structure and use of random sampling reduce the risk of overfitting, which is often a challenge with individual decision trees.
- Effective with Complex Data: They perform well with high-dimensional data and can implicitly handle feature selection during training.
- Versatile Applications: Suitable for both classification and regression problems.
- Robust to Missing Data: They include mechanisms for managing missing values without requiring imputation.
- Feature Importance Insights: They offer built-in methods to estimate the significance of each feature in making predictions.
- User-Friendly Tuning: Although hyperparameters exist, random forests are generally more forgiving and easier to tune than many other algorithms.

3.10.3 Gradient Boosting Machines

Gradient boosting machines are a highly effective and flexible ensemble learning method applicable to both classification and regression problems. As part of the boosting family, they work by sequentially building a series of weak learners usually

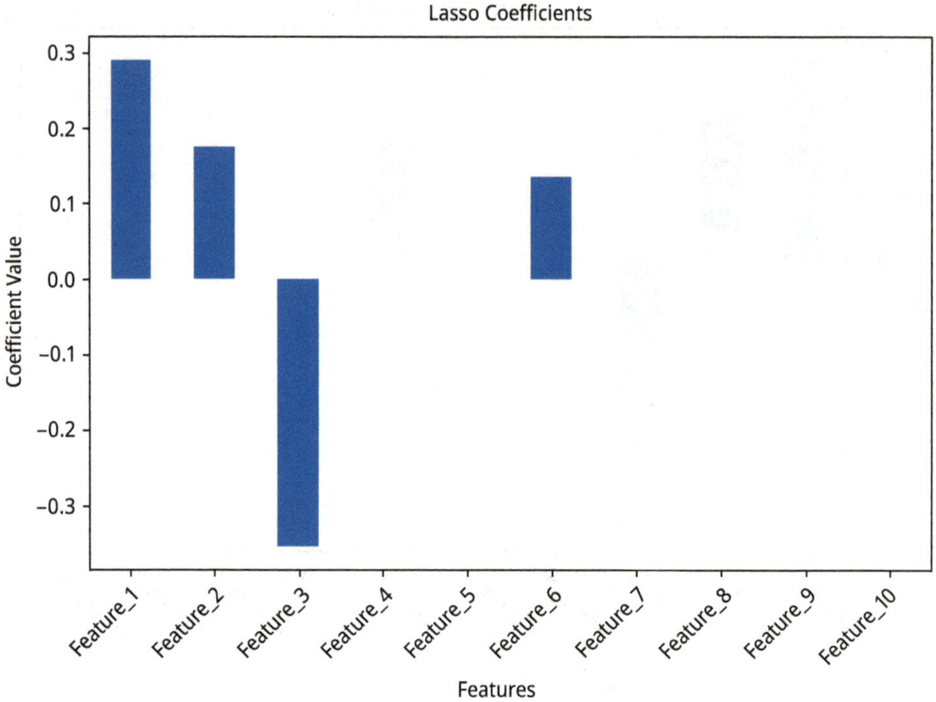

Figure 3.23: Lasso coefficients.

decision trees and combining them to form a strong, accurate predictive model. Consider the dataset presented in Table 3.13.

Table 3.13: Iris dataset sample.

Sepal length (cm)	Sepal width (cm)	Petal length (cm)	Petal width (cm)	Species
5.1	3.5	1.4	0.2	Setosa
4.9	3	1.4	0.2	Setosa
4.7	3.2	1.3	0.2	Setosa
5	3.6	1.4	0.2	Setosa
5.4	3.9	1.7	0.4	Setosa
5.8	2.7	5.1	1.9	Versicolor
6	2.9	4.5	1.5	Versicolor
5.7	2.6	3.5	1	Versicolor
6.3	3.3	4.7	1.6	Versicolor
4.9	2.4	3.3	1	Versicolor

The above Iris dataset is a well-known benchmark for classification tasks. A random forest model can be used to classify flowers into their respective species based on their features. It does so by constructing multiple decision trees, each trained on a random subset of the data and considering a random subset of features at each split. The final prediction for a new flower is determined by the majority vote across all the trees in the forest.

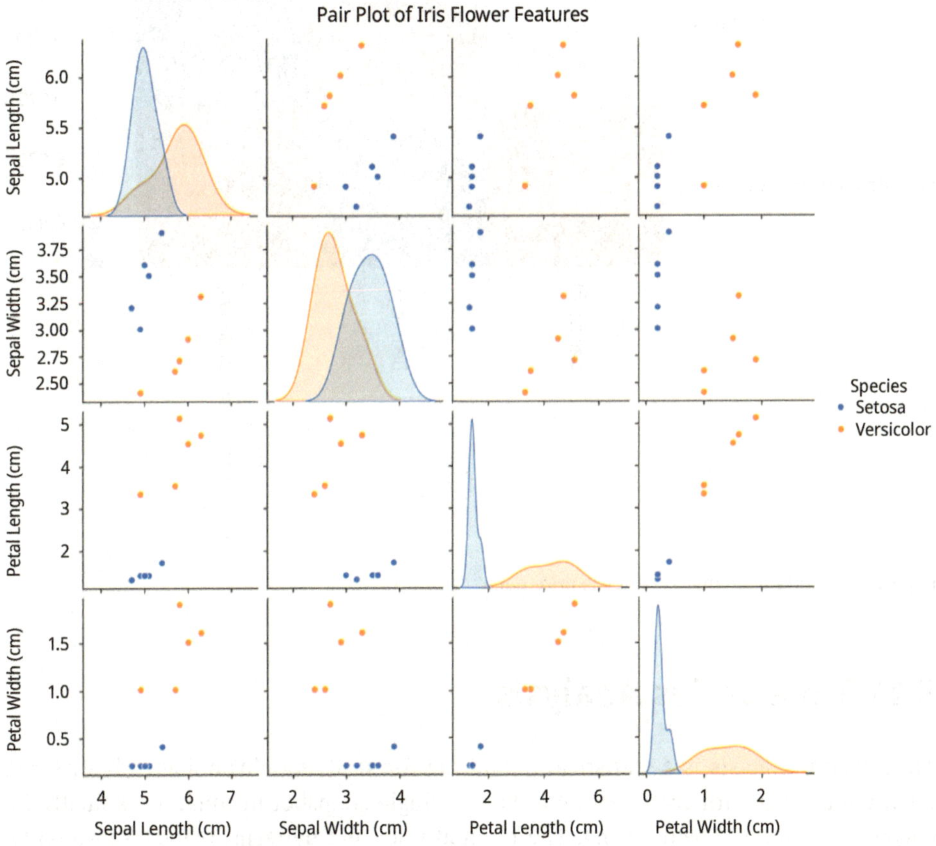

Figure 3.24: Pair plot of *Iris* flower features.

Pair plot (Figure 3.24) is a matrix of scatter plots displaying pairwise comparisons of the *Iris* flower features, with data points color-coded to differentiate between the "setosa" and "versicolor" species. This visualization helps illustrate how effectively the species can be distinguished based on their feature measurements. Figure 3.25 displays the linear correlations among size, bedrooms, and price. It compares the house prices predicted by a trained random forest regressor on a test set with the actual prices, indicating the model's performance.

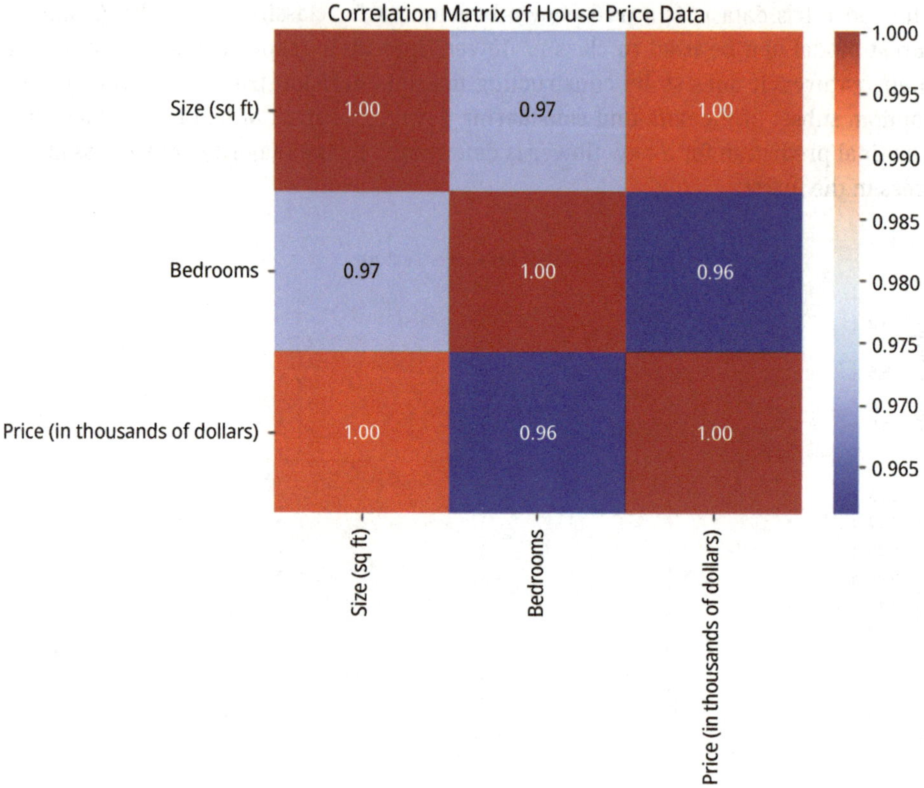

Figure 3.25: Correlation matrix of house price data.

3.11 Time Series Analysis

Time series analysis is a statistical technique for studying chronologically ordered data. Contrasted with cross-sectional data (a single snapshot in time), it explicitly incorporates temporal sequencing. The method uncovers patterns, trends, seasonality, and dependencies within the data to enable forecasting, support decisions, and reveal insights into dynamic processes.

3.11.1 Autoregressive Integrated Moving Average (ARIMA) and Seasonal Autoregressive Integrated Moving Average (SARIMA)

ARIMA and its extended form, SARIMA, are robust and widely adopted statistical techniques for time series analysis and forecasting. These models seek to capture the underlying structure of a time series by leveraging its historical values, allowing for the prediction of future data points based on identified trends and seasonal patterns [15]. Consider the real dataset in Table 3.14.

Table 3.14: Sales performance data.

Month	Sales	Month	Sales	Month	Sales	Month	Sales
Jan-20	200	Jul-21	330	Oct-20	390	Apr-22	290
Feb-20	220	Aug-21	340	Nov-20	400	May-22	310
Mar-20	250	Sep-21	360	Dec-20	420	Jun-22	330
Apr-20	270	Oct-21	380	Jan-21	210	Jul-22	350
May-20	300	Nov-21	400	Feb-21	230	Aug-22	370
Jun-20	310	Dec-21	420	Mar-21	250	Sep-22	390
Jul-20	330	Jan-22	230	Apr-21	270	Oct-22	410
Aug-20	340	Feb-22	250	May-21	290	Nov-22	440
Sep-20	360	Mar-22	270	Jun-21	310	Dec-22	460

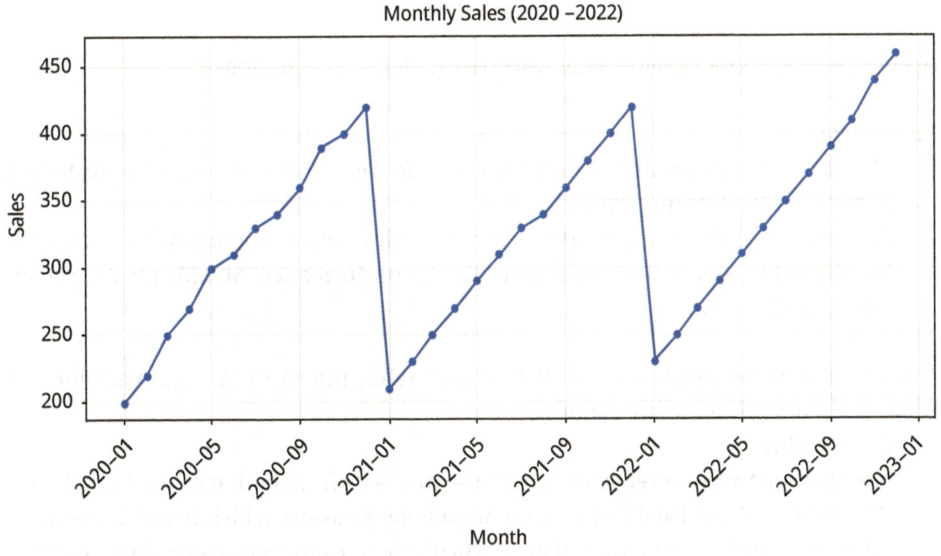

Figure 3.26: Monthly sales with SARIMA forecast (2020–2022).

From the line graphs (Figures 3.26 and 3.27), it is observed that the original "Observed" sales data alongside the "ARIMA forecast" (usually in a distinct color) extends into the test period (2022).

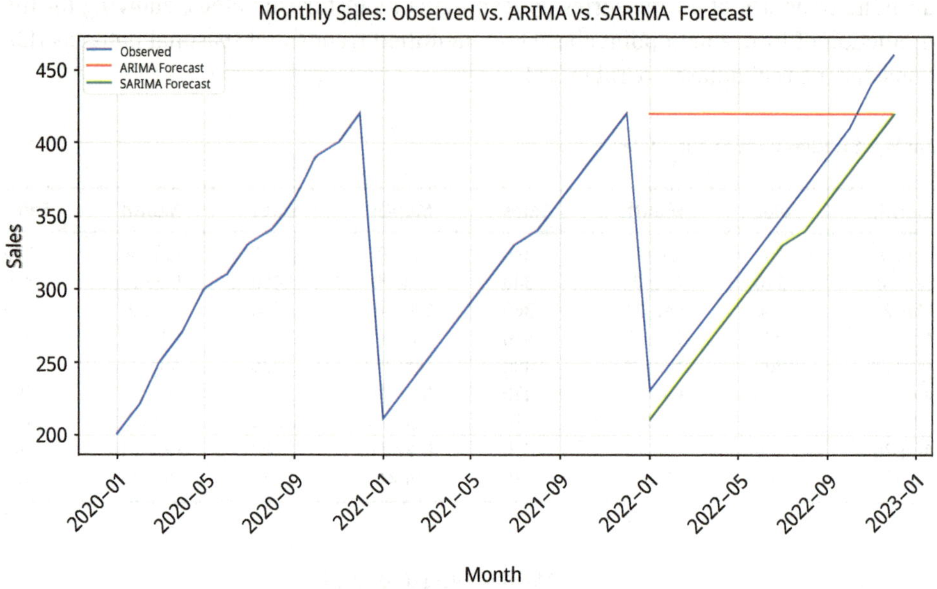

Figure 3.27: Monthly sales observed versus ARIMA versus SARIMA forecast (2020–2022).

Key Insights:

- Model Fit: Visually evaluate how well the ARIMA model represents the historical patterns in the training data.
- Forecast Accuracy: Compare the forecasted sales with the actual sales data in the test period to assess the model's prediction performance. Significant discrepancies suggest lower accuracy.

A single line graph displaying the "Observed" sales, the "ARIMA Forecast," and the ""SARIMA Forecast" all on the same plot.

Key Insights:

- Model Comparison: Directly compares the forecasts from both the ARIMA and SARIMA models. This visual comparison allows you to assess which model more accurately follows the actual sales data, particularly in capturing seasonal fluctuations. Along with the MSE and RMSE values displayed in the console, this comparison aids in determining which model is better suited for forecasting this time series.

3.11.2 Long Short-Term Memory (LSTM) Networks

Long short-term memory (LSTM) networks, an RNN derivative, mitigate vanishing/exploding gradient limitations through memory cells that preserve state over prolonged intervals. Consequently, they adeptly capture long-range dependencies essential for natural language processing, speech recognition, and time series forecasting [16].

The core mechanisms that empower LSTMs include:

- Cell State: The LSTM's persistent memory vector. It propagates information along the sequence, undergoing selective updates at each time step based on new inputs.
- Gates: Parameterized functions (input, forget, output) that gate information flow. Their combined action controls the addition of new information to the cell state, the retention of existing information, and the emission of the current hidden state/output.

LSTM networks are particularly effective for forecasting complex temporal dynamics in domains such as stock markets, weather, and sales, leveraging their core capability to model long-range dependencies and adapt to non-stationary time series patterns.

3.12 Nonparametric Methods

In data science, nonparametric methods comprise a wide range of techniques that analyze data without relying on strict, predefined assumptions about the data's underlying distribution. Unlike parametric methods, which depend on specific distributional assumptions – such as normality – and involve a limited set of parameters, nonparametric approaches are more adaptable. They can conform to the natural patterns and structures within the data, offering a flexible and effective way to analyze complex datasets [17].

3.12.1 Kernel Density Estimation

Kernel density estimation (KDE) is a nonparametric method that estimates the probability density function of a random variable. It works by smoothing discrete data points into a continuous curve, visualizing the probability distribution underlying the data. This smooth representation contrasts with histograms' discrete bars, offering a fluid depiction of relative likelihoods [18]. KDE offers several advantages that make it a powerful tool in data science, particularly for understanding and visualizing data distributions:

- Nonparametric Nature: KDE does not assume any predefined distribution for the data, making it highly adaptable for analyzing data with unknown or complex underlying patterns.
- Smooth and Continuous Output: Unlike histograms, which can be blocky and sensitive to bin choices, KDE produces a smooth and continuous curve that more ac-

curately reflects the distribution, resulting in clearer and more aesthetically pleasing visualizations.
- Reveals Distributional Features: KDE can uncover key characteristics of a dataset – such as multiple peaks (modes), skewness, and the overall shape – that might be hidden when using histograms.
- Ideal for Data Exploration and Visualization: It is an excellent tool for examining the shape of a dataset and for comparing distributions across different groups or conditions.
- Useful in Feature Engineering: In some machine learning workflows, the estimated density from KDE can be used as a feature to enrich models.

In summary, KDE is a versatile and insightful method for estimating and visualizing the distribution of data without relying on strict parametric assumptions. With careful selection of the kernel function and bandwidth, KDE can produce meaningful and informative representations that enhance data exploration and analysis.

3.12.2 Spline Regression

Spline regression is a versatile nonparametric regression method widely used in data science and statistics to model complex, nonlinear relationships between an independent (predictor) variable and a dependent (response) variable. Unlike standard linear regression, which fits a single straight line through the data, spline regression allows for more flexibility by accommodating curves and changes in direction within the data [19]. The key concept behind spline regression is to break the predictor variable's range into multiple intervals and fit a separate low-degree polynomial most commonly cubic within each segment. These individual polynomial pieces are then smoothly connected at specific points called knots. The result is a continuous and smooth curve that can closely follow the shape of the data, providing a more accurate representation of nonlinear trends.

Suppose we're examining how crop yield responds to varying amounts of fertilizer. We suspect the relationship isn't linear – while applying more fertilizer may initially lead to a substantial increase in yield, there may come a point where the benefit levels off, or excessive fertilizer could even reduce the yield [20]. Consider the dataset in Table 3.15.

The first plot (Figure 3.28) will be a scatter plot displaying individual data points for "fertilizer" versus "yield." This visualization will help confirm the suspected nonlinear trend – an initial rise in yield with increased fertilizer, followed by a plateau, and potentially a decline.

The second plot (Figure 3.29) will overlay a linear regression line on the same scatter plot. You'll notice that the straight-line fit struggles to capture the curvature in the data, highlighting the limitations of a simple linear model in this context.

Table 3.15: Fertilizer and crop yield data.

Fertilizer (units)	Yield (kg/hectare)
10	55
20	78
30	95
40	105
50	110
60	112
70	110
80	105
90	98
100	90

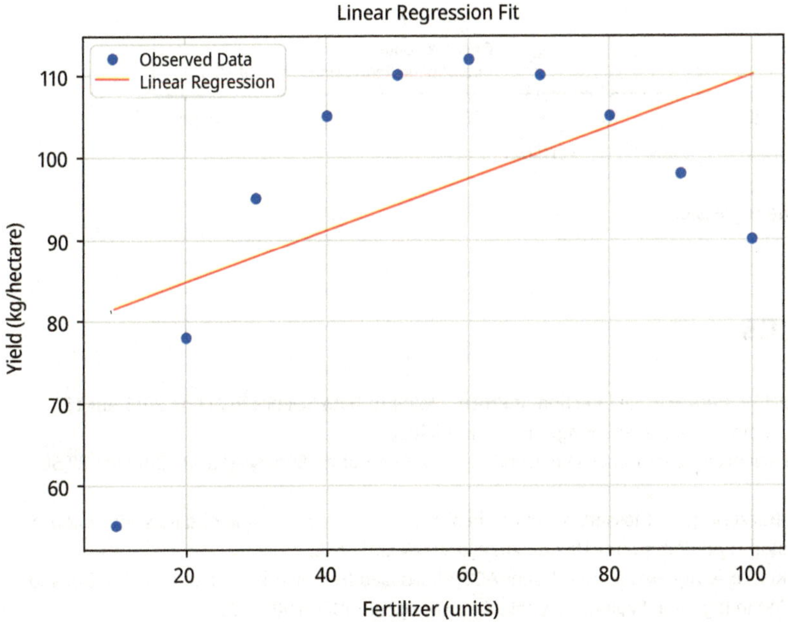

Figure 3.28: Linear regression fit.

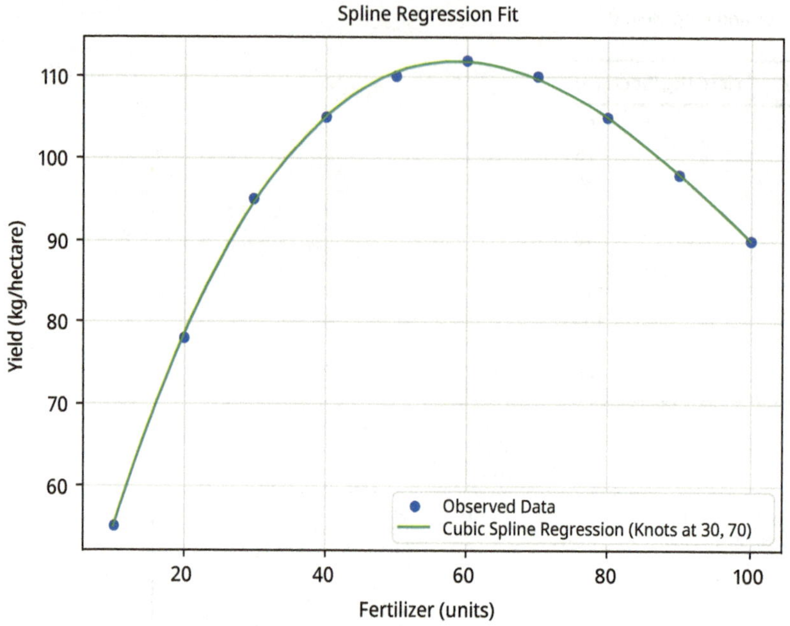

Figure 3.29: Spline regression fit.

References

[1] Van Der Aalst W. Data Science in Action. InProcess Mining: Data Science in Action 2016 Apr 16. (pp. 3–23). Berlin, Heidelberg: Springer Berlin Heidelberg.

[2] Cao L. Data science: A comprehensive overview. ACM Computing Surveys (CSUR). 2017 Jun 29;50 (3):1–42.

[3] Chiolero A, Buckeridge D. Glossary for public health surveillance in the age of data science. Journal of Epidemiology and Community Health. 2020 Jul 1;74(7):612–616.

[4] Dubey AK, Kumar A, Agrawal R. An efficient ACO-PSO-based framework for data classification and preprocessing in big data. Evolutionary Intelligence. 2021 Jun;14(2):909–922.

[5] Chen DG, Jin Z, Li G, Li Y, Liu A, Zhao Y. New advances in statistics and data science. Statistics. 2015 Aug 1.

[6] Bardenet R, Doucet A, Holmes C. On Markov chain Monte Carlo methods for tall data. Journal of Machine Learning Research. 2017;18(47):1–43.

[7] Yang X, Liu W, Liu W, Tao D. A survey on canonical correlation analysis. IEEE Transactions on Knowledge and Data Engineering. 2019 Dec 8;33(6):2349–2368.

[8] Ghojogh B, Crowley M. Linear and quadratic discriminant analysis: Tutorial. arXiv preprint arXiv:1906.02590. 2019 Jun 1.

[9] Warne R. A primer on multivariate analysis of variance (MANOVA) for behavioral scientists. Practical Assessment, Research, and Evaluation. 2014 Jan 2;19(1).

[10] Si S, Gu JW, Tian M. Leveraging single-case results to Bayesian hierarchical modelling. Computational Statistics. 2025 Feb;40(2):795–819.

[11] Rubin DB. Estimating causal effects of treatments in randomized and nonrandomized studies. Journal of Educational Psychology. 1974 Oct;66(5):688.

[12] Maathuis MH, Colombo D, Kalisch M, Bühlmann P. Predicting causal effects in large-scale systems from observational data. Nature Methods. 2010 Apr;7(4):247–248.

[13] Zou H, Hastie T. Regularization and variable selection via the elastic net. Journal of the Royal Statistical Society Series B: Statistical Methodology. 2005 Apr;67(2):301–320.

[14] Belgiu M, Drăguţ L. Random forest in remote sensing: A review of applications and future directions. ISPRS Journal of Photogrammetry and Remote Sensing. 2016 Apr 1;114:24–31.

[15] Kumar KS, Kumar TA, Radhamani AS, Sundaresan S. Blockchain Technology: An Insight into Architecture, Use Cases, and Its Application with Industrial IoT and Big Data. In Blockchain Technology 2020 Nov 9 (pp. 23–42). CRC Press.

[16] Hochreiter S, Schmidhuber J. Long short-term memory. Neural Computation. 1997 Nov 15;9 (8):1735–1780.

[17] Smeeton N, Spencer NH, Sprent P. Applied Nonparametric Statistical Methods. CRC press; 2025 Mar 31.

[18] Vani SS, Ramya P, Nagendar Y, Prakash PA, Chanti M, Atheeswaran A An Efficient Approach to Data Clustering Using the K-Means Algorithm in Big Data Analytics. In 2025 International Conference on Intelligent and Innovative Technologies in Computing, Electrical and Electronics (IITCEE) 2025 Jan 16 (pp. 1–6). IEEE.

[19] Chamidah N, Lestari B, Susilo H, Alsagaff MY, Budiantara IN, Aydin D. Spline estimator in nonparametric ordinal logistic regression model for predicting heart attack risk. Symmetry. 2024 Oct 30;16(11):1440.

[20] Hamad F, Younus N, Jaber M. Discovering the best choice for spline's knots and intervals using order of polynomial regression model. Open Journal of Statistics. 2024 Nov 5;14(6):743–756.

R. Venkatesh, J. Femila Mercy Rani, S. Devi, G. Santha Meena

4 The Heartbeat of Innovation: Investigating the Essential Role of Differential Equations in the Evolution of Digital Technologies

Abstract: This chapter addresses the underpinning role differential equations play in the development of digital technologies; it brings into perspective differential equations' role in the modeling, analysis, and simulation within most domains. Differential equations are basically vital to explaining dynamic systems hence revealing knowledge relevant to computer science, engineering, and applied mathematics. It has enabled the description of how systems change in time making it very important for the prediction of behaviors and optimum performance of system performance. Historically, it was important to the development of differential equations because it contributed to the progress of technology. Differential equations were used initially in physics for early applications that paved the way for innovation in engineering. Nowadays, differential equations aid the computation of complex simulations that are essential in developing algorithms for signal processing, crucial for telecommunications and multimedia technologies. Further, the frames of patterns and behaviors within large datasets are further provided in order to understand the data and thus assist in making decisions in the right direction. Applications have evolved into machine learning and control systems, where differential equations are used in designing adaptive and real-time optimizing algorithms. Engineers and scientists apply differential equations to make the models learn from the data, which in turn makes them more intelligent and more responsive technologies. All the more so in the sense that it is in face of ongoing integrations into emerging fields, such as quantum computing and advanced robotics, that the potential for transformatory effects will simply grow. By knotting together theoretical insights and practical examples, this chapter makes vivid how these mathematical constructs are tools and also essential foundational elements, shaping and redefining the digital landscape.

Keywords: Differential equations, digital technologies, dynamic systems, mathematical modeling, signal processing, machine learning, control systems

https://doi.org/10.1515/9783112214374-004

4.1 Introduction

4.1.1 Historical Context of Differential Equations

The history of differential equations can be traced back to the seventeenth century, when mathematical and physical development witnessed a great surge (Figure 4.1). The ideas of Isaac Newton and Gottfried Wilhelm Leibniz, who initiated calculus, were a good background for differential equations to present the actual concept of change and motion in mathematics. Newton's laws of motion, relating the motion of an object and the forces acting upon it, fundamentally depend on these mathematical constructs.

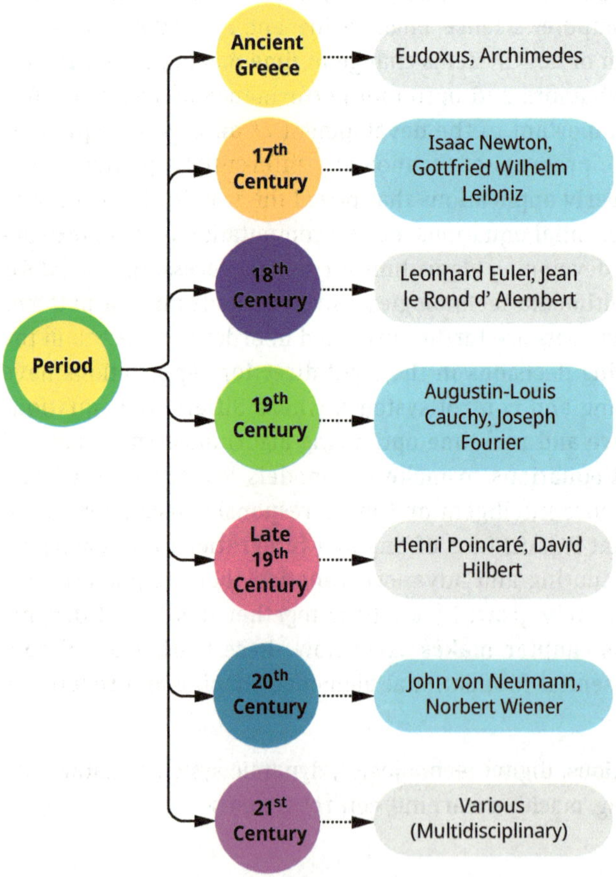

Figure 4.1: A journey through the history of differential equations.

In the eighteenth century, mathematicians such as Leonhard Euler and Joseph-Louis Lagrange opened up the field by developing methods to solve ordinary differential equations (ODEs) and paved their way into higher-order equations. Ultimately, work by Euler into the modeling of some physical phenomena, like fluid dynamics and celestial mechanics, proved that differential equations would have many uses in real life.

Partial differential equations (PDEs) were also introduced during the nineteenth century, an extremely important area of mathematics for modeling systems with multiple variables, such as heat conduction and wave propagation. Cauchy and Poincaré have been outstanding mathematicians who contributed to the theory of PDEs in ensuring proper understanding of stability and dynamic systems.

The importance of differential equations grew in the twentieth century with the development of modern physics and engineering. Differential equations became the finest tools in the formulation of theories about electromagnetism, quantum mechanics, and general relativity. Computational methods that began to appear in the mid-twentieth century transformed the application of differential equations, as analyses that were impossible up to then became possible.

Table 4.1: The history and impact of differential equations.

Period	Key figures	Contributions	Statistical impact
Ancient Greece	Eudoxus and Archimedes	Early concepts of rates of change	Limited formalism and foundational ideas on motion
Seventeenth century	Isaac Newton and Gottfried Wilhelm Leibniz	Development of calculus and formulation of differential equations	Newton's *Principia* (1687) laid groundwork for physics and Leibniz's notation established standards still in use
Eighteenth century	Leonhard Euler and Jean le Rond d'Alembert	Introduction of ordinary differential equations (ODEs) and study of vibrations	Euler published over 800 papers, significantly advancing mathematical physics
Nineteenth century	Augustin-Louis Cauchy and Joseph Fourier	Rigorous foundations of analysis and Fourier series for heat equation	Cauchy's work led to modern definitions of limits and continuity. Fourier's work influenced engineering and physics significantly
Late nineteenth century	Henri Poincaré and David Hilbert	Development of qualitative theory of differential equations and existence theorems	Poincaré's work on dynamical systems influenced modern chaos theory and topology

Table 4.1 (continued)

Period	Key figures	Contributions	Statistical impact
Twentieth century	John von Neumann and Norbert Wiener	Application of differential equations to fields like economics and biology	Growth of computational methods and differential equations in modeling population dynamics and financial markets
Twenty-first century	Others (multidisciplinary)	Advances in numerical methods, applications in AI, and complex systems	Increasing use in machine learning algorithms and simulations in climate science show complex interactions modeled by differential equations

Differential equations form a central part of most contemporary scientific and engineering endeavors. They are the mathematical technology that makes possible the understanding and prediction of behavior in dynamic systems and are at the core of many scientific and engineering disciplines. Their historical development reflects a trajectory of increasing sophistication in tandem with that of technology itself and the growth of our ability to model the natural world [1]. In this regard, differential equations continue to hold relevant values with the growth of digital technologies toward innovations in artificial intelligence (AI), robotics, or data science. Table 4.1 captures the evolution and significance of differential equations through differing historical contexts.

4.1.2 Evolution of Digital Technologies

Indeed, the development of digital technologies forms a marvelous journey that has transformed almost every aspect of modern life (Figure 4.2). It started in the mid-twentieth century with innovations in the first electronic computers reviving mass development. The early machines, ENIAC and UNIVAC, indeed preceded digital computation and paved the way for complex calculations that were once unimaginable by humans.

This phase of development reached its watershed with the creation of the microprocessor in the 1960s and 1970s. Computers became smaller, more powerful, and more accessible as new integrated circuits came into being. Therefore, it also saw the arrival of programming languages that helped the programmers come up with applications for software that could make these machines functional.

The widespread penetration of PCs in the 1980s and 1990s heralded a new generation of interaction with technology in numerous aspects of people's lives. Graphical user interfaces allowed better interactivity with the use of computers, and the World Wide Web has made it possible to connect millions of people around the globe to new information levels and communicating through new forms of channels.

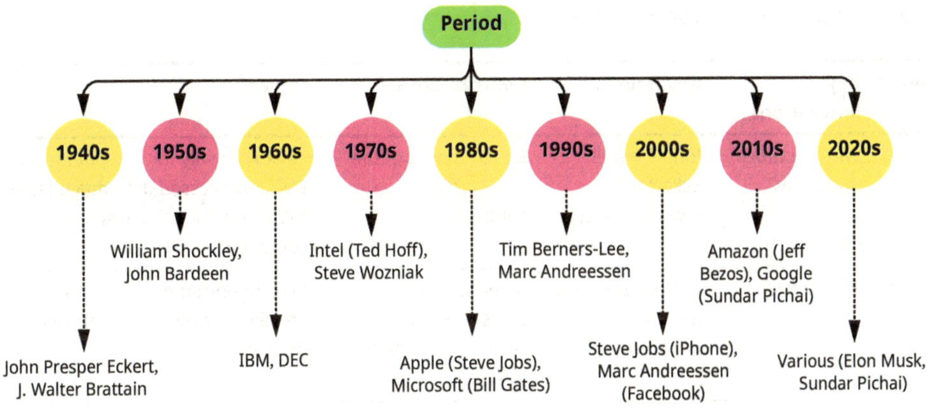

Figure 4.2: The growth of digital technologies.

The intensifying growth of the internet was accompanied by the ever-developing digital technologies. This led to innovations in mobile computing as well as broadband connectivity, which further completely changed the access and consumption of information. At the same time, social media platforms started changing the whole landscape of communication and interaction.

Recent years have taken digital technologies to unprecedented heights with big data, cloud computing, and AI [2]. This has propelled them to support vast amounts of data, stored and analyzed for all-time informative decision-making functions across industrial lines. And among these, the popularity of one sub-area of AI is machine learning, since algorithms based on differential equations, which model complex patterns and behaviors are used quite extensively.

But new industries, for example, in quantum computing and advanced robotics, are expected to reshape the digital space even further. Quantum computing uses the principles of quantum mechanics to process information at hitherto unknown speeds, while advanced robotics integrates AI and machine learning to construct adaptive, intelligent systems that can learn from their environment.

Throughout this process, differential equations have played the most important role, providing the mathematical basis for modeling dynamic systems within these technologies. And with the ongoing advances in digital technology, the integration of differential equations remains necessary, allowing for a deeper understanding and for innovation at higher level in this ever-changing landscape [3]. Table 4.2 captures major milestones and impactful changes made by digital technologies throughout the years.

Table 4.2: Milestones in digital technology.

Period	Key innovations	Key figures	Contributions	Impact
1940s	Vacuum tubes and ENIAC	John Presper Eckert and J. Walter Brattain	Development of the first electronic computers	Paved the way for computing and ENIAC processed data much faster than earlier mechanical computers
1950s	Transistors	William Shockley and John Bardeen	Invention of the transistor and replacing vacuum tubes	Revolutionized electronics and enabling smaller, more efficient devices
1960s	Mainframe computers	IBM and DEC	Introduction of time-sharing and batch processing	Allowed multiple users to access a single computer and increasing efficiency in organizations
1970s	Microprocessors	Intel (Ted Hoff) and Steve Wozniak	Birth of personal computing and introduction of the microprocessor	Sparked the personal computer revolution and affordable computing for homes and small businesses
1980s	Graphical user interfaces (GUIs)	Apple (Steve Jobs) and Microsoft (Bill Gates)	Popularization of GUIs with Macintosh and Windows	Made computers user-friendly, and boosted software development and accessibility
1990s	The Internet and World Wide Web	Tim Berners-Lee and Marc Andreessen	Creation of the first web browser and introduction of web pages	Revolutionized communication, information sharing, and e-commerce, and global connectivity
2000s	Mobile devices and Wi-Fi	Steve Jobs (iPhone) and Marc Andreessen (Facebook)	Launch of smartphones and social media platforms	Transformed how people communicate and access information, and social media reshaped society and marketing
2010s	Cloud computing and IoT	Amazon (Jeff Bezos) and Google (Sundar Pichai)	Rise of cloud services and interconnected devices	Enabled scalable storage and processing and facilitated the rise of data analytics and smart devices
2020s	Artificial intelligence (AI) and 5G	Others (Elon Musk and Sundar Pichai)	Advances in machine learning and connectivity	Enhanced automation and real-time data processing and transformative potential in numerous sectors, including healthcare, finance, and transportation

4.1.3 Literature Survey

Muslim examines how technology, in the form of computer programs, simulations, online sources, and interactive applications, has the potential to lessen the difficulty of learning differential equations and make it more fun. The authors suggests that students develop their understanding of conceptually difficult contexts better, play with different scenarios, and observe real-life applications of differential equations. The authors stress that technology should supplement traditional instruction, not as a replacement for direct contact with teachers who can provide an effective mix of technology and traditional instruction [27].

Knoke and Wick investigate case studies looking at the challenges faced using neural networks to solve ODEs and describe challenges such as sensitivity to initial conditions and also that may lead to misconvergence. The authors highlight that training parameters are important to consider in order to improve the accuracy of neural network techniques [28].

Guo et al. present a way to improve upon methods for solving PDEs in a class of problem phrased as physics-informed neural networks (PINNs), which use the laws of physics for regularization to improve performance. The study calls light upon the issues with obtaining analytical solutions for PDEs and the possibilities that deep learning techniques present for these problems. They demonstrated the power of using physics with neural networks using experimental analysis with particular equations while suggesting more work in this area [29].

Kurniadi et al. carry out a systematic literature review to investigate teaching and learning ODEs in higher education and reveals four major learning methods, which include active learning, mathematical modeling, information and technology communication, and geometric and qualitative solutions. It considers some important topics discussed in the literature, such as first-order ODEs and the Euler method, and gaps in research, particularly in advanced methods and qualitative behavior of solutions. The findings are expected to provide a general overview and foreshadow future research in this area [30].

4.2 The Mathematical Foundations of Differential Equations

4.2.1 Types of Differential Equations

Differential equations can be classified according to several criteria: type of derivatives involved, order of the equation, and linearity of the equation (Figure 4.3). Knowing these classifications is important in finding and solving differential equations with their applications in various fields.

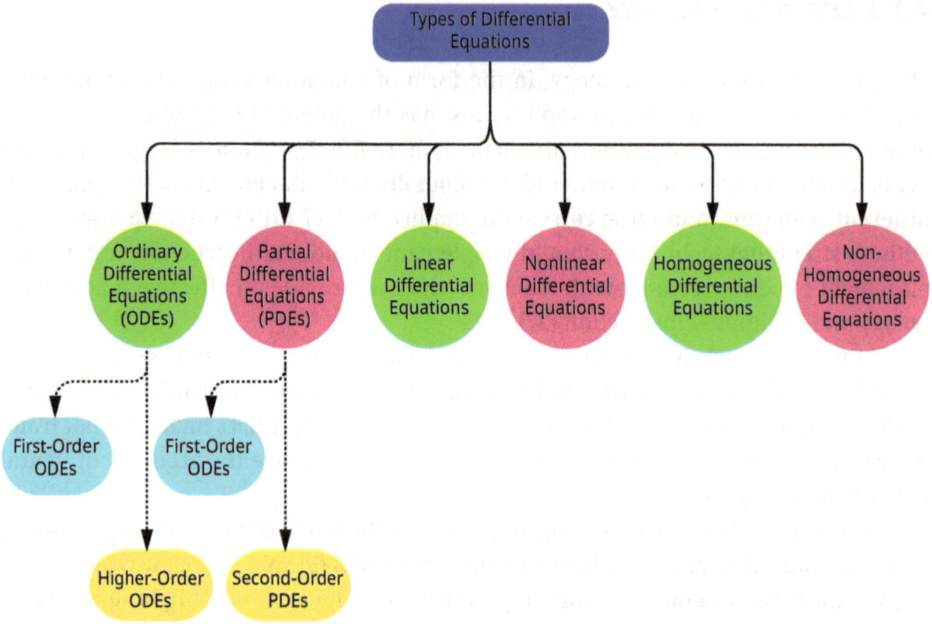

Figure 4.3: Exploring the different types of differential equations.

4.2.1.1 Ordinary Differential Equations (ODEs)

ODEs contain functions of a single variable and their derivatives. Based on order, they can be divided based on the presence of any type of highest-order derivative in that equation.

First-Order ODEs: Involving the first derivative of a function; the most general form is as follows:

$$\frac{dy}{dx} = f(x,y) \tag{4.1}$$

Some of them are exponential growth models and Newton's law of cooling.

Higher-Order ODEs: These contain derivatives of order 2 or greater. A typical form for an ODE of order 2 is as follows:

$$\frac{d^2y}{dx^2} + p(x)\frac{dy}{ds} + q(x)y = g(x) \tag{4.2}$$

Higher-order ODEs are used in virtually every study on mechanical vibrations and circuit analysis.

4.2.1.2 Partial Differential Equations (PDEs)

PDEs describe functions of multiple variables and partial derivatives. They are really essential to models of physical and engineering phenomena involving lots of factors. PDEs can also be classified according to order as follows:

First-Order PDEs: These ones include the partial derivatives of one order. For instance:

$$\frac{\partial u}{\partial t} + a\frac{\partial u}{\partial t} = 0 \tag{4.3}$$

This form arises in the wave and heat equations.

Second-Order PDEs: These include equations with the second partial derivative. The importance of these models is reflected in models for heat conduction and wave propagation. The most celebrated example is the heat conduction equation:

$$\frac{\partial u}{\partial t} = k\,\frac{\partial^2 u}{\partial x^2} \tag{4.4}$$

4.2.1.3 Linear Versus Nonlinear Differential Equations

Differential equations may also be categorized according to the linearity.

Linear Differential Equations: Characterized by the linearity of the unknown function and its derivatives. It can be represented as follows:

$$a_n(x)\frac{d^n y}{dx^n} + a_{n-1}(x)\frac{d^{n-1} y}{dx^{n-1}} + \cdots + a_1(x)\frac{dy}{dx} + a_0(x)y = g(x) \tag{4.5}$$

Linear equations are much easier to solve and have good-established methods and theories.

Nonlinear Differential Equations: These involve nonlinear combinations of the unknown function and its derivatives. An example is as follows:

$$\frac{dy}{dx} = y^2 - x \tag{4.6}$$

Sometimes, nonlinear equations exhibit intractable behavior in the form of chaos and other phenomena that are usually impossible to analyze using special techniques.

4.2.1.4 Homogeneous Versus Nonhomogeneous Differential Equations

Homogeneous Differential Equations:
Homogeneous differential equations can model the natural behavior of digital circuits and systems, where there is no external forcing function but the internal dynamics of the system are significant. Homogeneous equation model systems that react to initial conditions or internal feedback rather than external input.

These have no forcing function. For example:

$$y'' + p(x)y' + q(x)y = 0 \tag{4.7}$$

Solutions to homogeneous equations are superposable.

Nonhomogeneous Differential Equations:

Nonhomogeneous differential equations are employed frequently to model systems that incorporate forcing functions or external inputs. For digital systems, the forcing term may simulate an external input, like a data input stream, a control input, or noise.

This has a non-zero forcing function or term. An example is:

$$y'' + p(x)y' + q(x)y = g(x) \tag{4.8}$$

Solution types for nonhomogeneous equations include finding the complementary solution of the associated homogeneous equation and a particular solution for the nonhomogeneous part.

For all the purposes of selection of appropriate methods for their solutions and applications, it is highly important to know the kinds of differential equations. Each type possesses its own characteristics, difficulties, and applicability to modeling dynamic systems – the basic elements in many fields, especially in engineering, physics, and computer sciences – that serves as different parts of a system [4].

4.2.2 Methods of Solution and Analysis

Several methods exist to solve differential equations, depending on the nature and complexity of an equation (Figure 4.4). It explores some of the techniques adapted to solving ordinary and PDEs, including some analytical and numerical approaches.

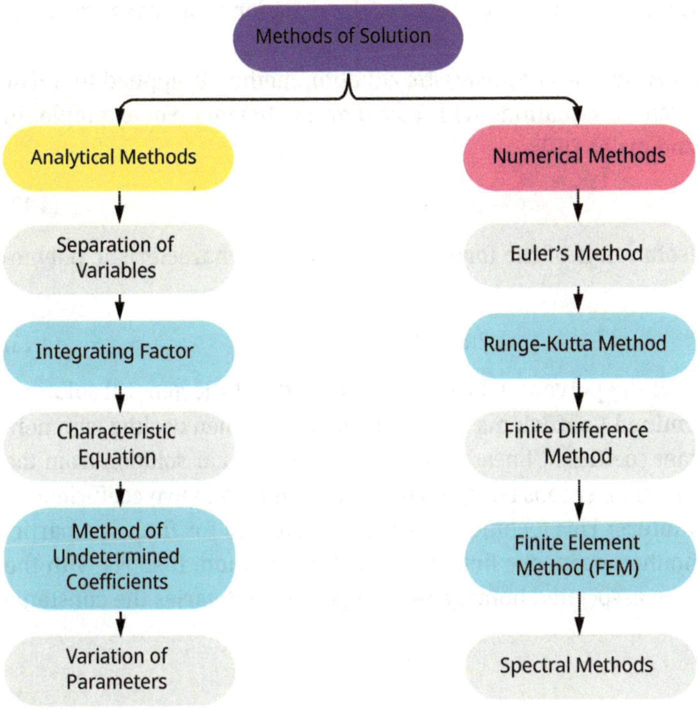

Figure 4.4: Solving differential equations: methods and techniques.

4.2.2.1 Analytical Methods

a. **Separation of Variables:** This method can be used mainly with first-order ODEs. In this method, the equation should have variables separated on two sides of the equation. The example of this type is given in the following equation:

$$\frac{dy}{dx} = g(x)h(y) \tag{4.9}$$

We can separate variables:

$$\frac{1}{h(y)}dy = g(x)dx \tag{4.10}$$

Combining both sides gives the solution.

b. **Integrating Factor:** A first-order linear ODE is of the form:

$$\frac{dy}{dx} + P(x)y = Q(x) \tag{4.11}$$

An integrating factor, $\mu(x) = e\int P(x)dx$, reduces the equation to an exact form for easy integration.

c. **Characteristic Equation:** The characteristic equation method is applied to linear homogeneous differential equations with constant coefficients. For example, in the case of the second-order ODE:

$$y'' + ay' + by = 0 \tag{4.12}$$

Assuming that the solution is of the form $y = e^{rx}$, obtain the characteristic polynomial as follows:

$$r + 2\ ar + b = 0 \tag{4.13}$$

Therefore, the roots of the polynomial determine the form of the general solution.

d. **Method of Undetermined Coefficients:** This method is used when dealing with nonhomogeneous constant coefficient linear ODEs. It finds a particular solution from the given form of the nonhomogeneous term, solving then for the unknown coefficients.

e. **Variation of Parameters:** This technique leads to the method for finding a particular solution to a nonhomogeneous linear differential equation. It starts with the general solution to the associated homogeneous equation and varies the constants as functions of x.

4.2.2.2 Numerical Methods

Numerical methods are practical techniques to use when analytical solutions are either difficult or impossible to obtain. These methods approximate solutions of differential equations by discretizing the latter.

a. **Euler's Method:** A straightforward, simple technique to find solutions for first-order ODEs is to apply the use of the slope at a given point to estimate the value of the function at the next point:

$$y_{n+1} = y_n + hf(x_n, y_n) \tag{4.14}$$

where h is the step size and $f(x,y)$ is derived from the differential equation.

b. **Runge-Kutta Methods:** These methods provide more accurate approximations than Euler's method by considering several slopes (the function is evaluated at multiple points in the interval) while determining the value of the next step. The most commonly used classical fourth-order Runge-Kutta method is preferred for being neither excessively simplified nor highly complex at the same time.

These techniques assist us in approximating the solution to equations such as:

$$\frac{dy}{dt} = f(t, y) \tag{4.15}$$

where y represents the dependent variable, t represents the independent variable, and $f(t,y)$ represents a function that describes how y varies with respect to t. These techniques come in handy when solving equations analytically is impossible or challenging, and we have to use computers to approximate the solution.

c. **Finite Difference Method:** In finite difference method (FDM), the solution of the PDEs is derived by the approximation of the derivatives by their finite difference. The finite difference scheme transforms the PDE into an algebraic system, and it can, therefore, be solved iteratively.
d. **Finite Element Method (FEM):** FEM is most effective for solving difficult PDEs, mainly found in applications in the field of engineering. In this method, a large system is broken down into smaller, simpler parts called elements. A variational problem is then posed, and boundary value problems can be solved numerically.
e. **Spectral Methods:** These use global functions to represent the solution, for example, polynomials or trigonometric functions. The advantage is for very smooth solutions where they bring a high accuracy of solution. Also, they work particularly well with computational fluid dynamics and wave propagation problems.

The choice of method of solution is decided by the nature of the problem – whether the equation is linear, its order, and its boundary conditions. Analytical methods provide an exact solution whenever these are possible, whereas numerical methods offer practical approaches when the problems become complex [5]. Understanding these methods means understanding how to model and analyze dynamic systems across a wide variety of fields, and differential equations form an essential tool for science and engineering.

4.3 Applications in Digital Technologies

4.3.1 Signal Processing and Communication Systems

Differential equations became core areas in signal processing and communication systems in the modeling, analysis, and optimization of different signal behaviors. Their applications are vast and critical to the functioning of modern technologies in this domain.

4.3.1.1 Signal Representation and Filtering

Differential equations are used in signal processing for continuous-time modeling of signals and analysis of their behavior. Differential equations are significant in real-time signal processing, especially in filtering systems, where they are applied to model and analyze continuous-time signals. Differential equations model the way signals vary with time, enabling engineers to develop systems capable of altering signals in specific patterns, e.g., filtering noise or amplifying signal intensity. The most basic form of a system utilized in signal processing is the linear time-invariant (LTI) system, which describes the relationship between an input signal $x(t)$ and an output signal $y(t)$. LTI systems can be described with differential equations, and they play a central role in designing filters that are used to process signals in telecommunications, audio processing, and other applications. For instance, differential equations can be used in description of LTI systems; it describes the interrelation between the input and output of systems that aid engineers in designing filters for signal modification purposes to reduce noise and increase strength in signals [6].

A widely used form for filter design is the first-order differential equation:

$$\tau \frac{dy[t]}{dt} + y(t) = K\ x(t) \tag{4.16}$$

where $y(t)$ is the output signal, $x(t)$ is the input signal, K is a gain factor, and τ is the time constant. If the above parameters are manipulated, various kinds of filters like low-pass filter, high-pass filter, and band-pass filter can be derived.

Low-Pass Filter: This filter will pass low-frequency components of the signal and attenuate higher frequencies. In reality, this might be applied to eliminate high-frequency noise from an audio signal.

High-Pass Filter: This filter will pass high-frequency components and attenuate lower frequencies. It may be applied, for example, to remove low-frequency hum or drift from a signal.

Band-Pass Filter: A band-pass filter is a mixture of low- and high-pass filters, which permits only a defined range of frequencies to pass through while rejecting frequencies both higher and lower than this range. This filter is commonly applied in radio communications, for example, where only a narrow frequency band is required.

For instance, in the audio-processing case, a low-pass filter would serve to eliminate noise at high frequency from a speech recording, thereby making the sound clear and unhindered. An electrical engineer can use a high-pass filter in order to screen out undesired low-frequency interference, for example, 50 Hz interference originating from power sources, yet without hindering the desired signal. In each of these situations, the differential equation is the mathematical basis that dictates how the input signal is altered into the output signal by the filter. Solving the equation and examining its behavior, engineers are able to design filters with specific characteristics, improving signal quality for real-time applications.

4.3.1.2 System Dynamics and Response Analysis

Differential equations are important for the analysis of the dynamic response behavior of systems with external inputs. For example, the impulse response of a system may be obtained directly from the differential equation of the system itself; this gives a description of how a system would respond to an impulse signal. Understanding the nature of the stability and performance of the system would then form part of that.

A signal propagates through various media because the differential equation describes its system by translating this differential equation into its transfer function for communication systems [7]. The transfer function can tell engineers such essential characteristics as frequency response and stability margins.

4.3.1.3 Modulation and Demodulation Techniques

Modulation techniques are the keys to successful transmission of signals via different kinds of communication channels and typically fall under differential equations. Consider the examples of amplitude modulation and frequency modulation analyzed by way of differential equations that describe how a carrier signal interacts with the modulating signal.

The dynamics of the modulated signal can also help optimize bandwidth and power efficiency factors further through differential equations, so that it is transmitted with minimum interference and distortion [8].

4.3.1.4 Signal Reconstruction and Sampling Theory

In digital communication, the underpinning of sampling and reconstruction theories in signals is formed by differential equations. The Nyquist-Shannon sampling theorem may be expressed via differential equations. This theory postulates that if the sampling rate of a signal is at least double the highest frequency of the signal, then the signal can be reconstructed perfectly using its samples. Reconstructive filters, which reconstruct discrete samples back into continuous signals, often rely on differential equations to reconstruct the original signal losing very minimal information [9].

4.3.1.5 Adaptive Filtering and Machine Learning Applications

Advances in machine learning mean increased utilization of differential equations in adaptive filtering techniques. It provides a real-time adaptation with parameters being revised by extracting information from the characteristics of the received signal. The process of adaptation is mimicked with differential equations and enhances tasks such as echo cancellation, noise suppression, and channel equalization algorithms. One possible integration of machine learning with differential equations could be to allow the systems to learn from data, which may evolve in dynamic environments to make use of telecommunications and multimedia technologies more effective.

Differential equations have other applications in signal processing and communication systems, numerous and multifariously having an impact. They form the mathematical basis by which models can be constructed, analyzed, and optimized as related to handling signals in any number of forms [10]. As digital technologies continue to advance, it is likely that differential equations will only increase in importance for these applications, leading to discovery in how information is processed and then transmitted.

4.3.2 Control Systems and Robotics

Differential equations form the foundation in the theory of control systems and robotics because they represent the mathematical basis for modeling dynamic systems and synthesizing controllers. From the car to the aircraft, differential equations apply to manufacturing and automation.

4.3.2.1 Modeling Dynamic Systems

Differential equations occur in the description of any physical system in control systems. In most cases, the behavior of mechanical, electrical, and thermal systems can be described in terms of ODEs or PDEs. For example, the motion of a simple mass-spring-damper system can be represented by the following second-order differential equation:

$$m\frac{d^2x}{dt^2}+c\frac{dx}{dt}+kx=F(t) \tag{4.17}$$

where m is the mass, c is the damping coefficient, k is the spring constant, x is the displacement, and $F(t)$ is the external force. This equation allows engineers to better understand how the system will react to different inputs.

4.3.2.2 Control Design Techniques

Control systems aim to control the dynamics of a system in such a way that some desired behavior is manifested. Differential equations play a crucial role in developing methods of control, which include the following:

PID Control (Proportional-Integral-Derivative): This is the most widely applied control method in practice. Assuming a differential equation describing the plant, it calculates the control input depending on the error between the desired and the actual system outputs. An expression of the control law can be stated as:

$$u(t) = K_P e(t) + K_i \int e(t)dt + K_d \frac{de(t)}{dt} \tag{4.18}$$

where $u(t)$ is the control signal, $e(t)$ is the error, and k_p, k_i, and K_d are the proportional, integral, and derivative gains, respectively. The PID controller controls the response of the system in order to minimize error.

State-Space Control: The state variable heavily relies on differential equations, which appear at the first order. It does provide complete analysis of the system with regard to stability, controllability, and observability for control design.

4.3.2.3 Stability Analysis

Stability is the property of control systems, which ensures that a system responds predictably at inputs, without diverging or oscillating wildly. Most analyses for system stability require differential equations. Techniques such as the Routh-Hurwitz criterion and Lyapunov's direct method exploit differential equations to evaluate the stability conditions [11].

For instance, consider an LTI system for which the differential equations are solvable, so that its characteristic equation can be obtained to make it possible to calculate poles in the complex plane. The locations of poles of such a system are useful to conclude whether the system is stable, unstable, or marginally stable.

4.3.2.4 Robotics and Kinematic Modeling

Differential equations in robotics are the model of movement as well as dynamical behavior for robotic systems. For example, the system of equations in motion of arms of robotic agents and that in mobile robots would be derived employing Newton's or Lagrangian method, the latter eventually producing a system of differential equations defining their behavior.

For example, the dynamics of a manipulator arm can be modeled as follows:

$$\tau = I \frac{d^2\theta}{dt^2} + b \frac{d\theta}{dt} + g(\theta) \tag{4.19}$$

where τ represents torque, I represents moment of inertia, b represents damping coefficient, $g(\theta)$ represents gravitational effects, and θ is the joint angle. These equations allow one to design algorithms of control, so that accurate movement and trajectory tracking are realized. These equations allow one to design algorithms of control, so that accurate movement and trajectory tracking are realized.

4.3.2.5 Path Planning and Motion Control

Differential equations are also applied in the area of robotic path planning and motion control. Trajectory generation, for example, solves differential equations that describe a desired path that a robot may follow over time. Optimal control techniques, such as the Pontryagin's minimum principle, apply differential equations in minimizing the cost function so as to have motion planning in an efficient and effective way [12].

4.3.2.6 Adaptive and Intelligent Control

It applies differential equations to improve the control system's performance, enhance adaptability, and adapt in real-time adaptive controllers to uncertainties by adjusting their parameters based on the feedback from the environment. The environment usually models by differential equations to guarantee both stability and responsiveness.

Differential equations form the very basis of design and analysis of a control system, as well as in robotics. In one way, they represent the mathematical tool for modeling dynamic complex systems, thus determining efficient control strategies leading to stability and performance [13]. Their role is likely to keep on expanding with advances in all technology involved, that is, in automation, intelligent systems, and their robotic applications.

4.3.3 Machine Learning and Data Science

The differential equations become very important to machine learning and data science for modeling complex systems and understanding dynamic behavior. The applicability ranges from theoretical foundations to the practical implementation of algorithms and models. Differential equations play a central role in a wide range of applications in machine learning and data science, especially in modeling dynamic systems and enhancing algorithmic performance. In deep learning, differential equations are frequently employed by neural networks to describe how weights change with time while training, e.g., in neural ODEs. Even recurrent neural networks (RNNs) employ differential equations for describing the progression of hidden states in time-series prediction. In support vector machines, diffusion processes, which are modeled by differential equations, facilitate understanding of data geometry and enhance classification [14]. For Gaussian processes, kernel functions associated with differential equations assist in modeling uncertainty in predictions, whereas stochastic differential equations (SDEs) are utilized to model noise and dynamic change. In reinforcement learning, differential-time models and the Hamilton-Jacobi-Bellman equation optimize agent actions in dynamic systems. Graph-based learning methods like graph neural networks apply diffusion processes described by differential equations to learn how information diffuses through graphs, assisting clustering and dimensionality reduction tasks. These instances demonstrate the significance of differential equations in both theoretical and practical use in contemporary machine learning.

4.3.3.1 Dynamic Systems and State-Space Models

Few problems in machine learning, notably in time series analysis and in reinforcement learning, can be naturally formulated as differential equations. The state-space models that describe the way a system evolves over time are based upon differential equations. Models based on differential equations carry the best underlying dynamics of data that change over time, making the prediction and analysis stronger.

For example, the state of some system could be described as follows:

$$\frac{dx(t)}{dt} = Ax(t) + Bu(t) \tag{4.20}$$

where $x(t)$ is the state vector, A represents the system dynamics, B relates to control inputs $u(t)$, and t denotes time. Such formulation is imperative within the frameworks of robotics, finance, and economics where time evolution of a system needs to be understood properly.

4.3.3.2 Learning Dynamics through Differential Equations

Differential equations can also be applied to model learning dynamics of algorithms. For example, update rules in the widely used gradient descent optimization algorithm can be considered as discretized differential equations. It can be given by its continuous version as follows:

$$\frac{\omega(t)}{dt} = -\nabla J\left(\omega(t)\right) \tag{4.21}$$

where $w(t)$ represents the weights of the model and $J(\omega(t))$ is the cost function. This perspective allows differential equation methods to be applied in the analysis of convergence behavior and stability properties.

4.3.3.3 Neural Networks and Differential Equations

In such a context, one can think of neural nets as dynamic systems, and training dynamics could be cast in differential equations. Recurrent nets, which properly categorized are known as the so-called RNNs, are nets specially tailored to process sequential data and, in their architectures, they may even implement differential equations in order to build up specific types of temporal dependencies.

Furthermore, neural ODEs do support continuous modeling of neural networks. Here, it is possible that layers of the neural network were treated as differential equations in this modeling. These would lead to efficient training along with improving the generalization property, hence spreading applications from simple computer vision till natural language processing [15].

4.3.3.4 Regularization and Differential Equations

Regularization techniques play an important role in preventing overfitting and ensuring the robustness of a model in data science. Some regularization methods can be classified according to differential equations; for example, Tikhonov regularization adds a penalty term that can be interpreted as a solution to a differential equation that might stabilize the solution by allowing prior information about the function being estimated [16].

4.3.3.5 Partial Differential Equations in Data Analysis

PDEs have a lot of applications in very many scenarios that apply to data analysis. For example, by means of applying the heat diffusion model and wave equation, one might improve the image quality and thus extract features from the images [17]. In turn, this makes the representation and hence the analysis through PDEs with respect to efficient filtering, denoising, and segmentation is much easier.

4.3.3.6 Bayesian Inference and Stochastic Differential Equations (SDEs)

Bayesian inference and SDEs are both potent methodologies that can be combined in machine learning to address uncertainty and dynamism in data. Bayesian inference gives a probabilistic system for belief updates about model parameters, while SDEs characterize systems subject to noise or uncertainty. They together present robust and adaptive solutions, especially in situations where noise and randomness are fundamental aspects of the data or process under modeling.

4.3.3.6.1 Stochastic Processes and Noise in Machine Learning

Numerous real-world systems are random or uncertain in nature, and can be expressed in data as noise. For example, in sensor readings or stock prices, we may not know the actual value of the state of the system at all times, but have noisy data points. Such systems can be modeled by stochastic processes systems that change over time but contain an element of randomness.

Practical Example: Financial Modeling

In the financial markets, stock prices can be represented in terms of SDEs (such as geometric Brownian motion (GBM) model).

Any stock's price at a particular time can be represented as follows:

$$ds(t) = \mu s(t)dt + \sigma s(t)dw(t) \tag{4.22}$$

where $s(t)$ is the stock price at time, μ is the drift term, representing the expected return, σ is the volatility of the stock, and $dw(t)$ is a Wiener process or Brownian motion representing random fluctuations.

The noise in the stock price $dw(t)$ is randomness, and its development is determined by the SDE. In machine learning, the models can be used to predict stock prices at different times based on inherent uncertainty in the prediction.

In the application of differential equations, machine learning and data science expose insights for dynamic systems, optimization processes, and behavior of models. Their applications bring in better understanding and performance in several algorithms that add up to improvements in predictive modeling, image processing, and reinforcement learning [18]. The perpetual growth of the fields of machine learning and data science speaks to the fact that differential equations are becoming increasingly important innovations that solidify outcomes in different applications.

4.4 Case Studies

4.4.1 Differential Equations in Engineering Applications

Differential equations form an integral part of almost every mainstream branch of engineering. They form a mathematical description of the modeling and analysis of systems in civil, mechanical, electrical, and aerospace engineering. A few case studies are reported here that describe numerous applications of differential equations in engineering.

4.4.1.1 Mechanical Vibrations

An important application of analysis in mechanical engineering is the analysis of vibrations to ensure that solid structures and machinery do not collapse. Dynamical systems, whose constituents are vibratory, can be modeled through an equation, namely a second-order ODE.

4.4.1.1.1 Case Study: Spring-Mass-Damper System

Consider a mass m joined to a spring of spring coefficient k and an optional damper with damping constant c.

The motion can be represented by the following differential equation:

$$m\frac{d^2x}{dt^2}+c\frac{dx}{dt}+kx=0 \tag{4.23}$$

where x represents displacement from equilibrium.

From the solution of this differential equation, the engineers can obtain what are known as the natural frequencies and mode shapes for the system. These are critical in predicting how resonance will behave. The analysis thus helps the engineers design better, safer and more efficient mechanical systems, such as vehicles and bridges.

4.4.1.2 Electrical Circuit Analysis

Electrical engineering applies differential equations to describe phenomena in circuits with resistors, capacitors, and inductors.

4.4.1.2.1 Case Study: RLC Circuit

An RLC circuit consists of a resistor R, an inductor L, and a capacitor C in series. The second-order differential equation describing this circuit is as follows:

$$L\frac{d^2i(t)}{dt^2}+R\frac{di(t)}{dt}+\frac{1}{C}i(t)=V(t) \tag{4.24}$$

where $i(t)$ represents the current through the circuit and $V(t)$ represents the voltage source.

This equation can be solved to examine the transient response and steady-state behavior; hence it can assist in designing circuits for wanted performance characteristics such as filtering and oscillation.

4.4.1.3 Structural Analysis

The differential equation is very vital in civil engineering concerning the assessment of stability and load-carrying capacity of the structures.

4.4.1.3.1 Case Study: Beam Bending

Therefore, the Euler-Bernoulli beam theory can be used to describe the bending beams leading to a possibly fourth-order differential equation as follows:

$$EI\ \frac{d^4y(x)}{dx^4}=q(x) \tag{4.25}$$

where $y(x)$ represents the deflection of the beam, E represents the modulus of elasticity, I represents the moment of inertia, and $q(x)$ represents the distributed load along the beam.

This will mean the possible prediction of deflections and stress distributions within a certain structure by solving an equation. Engineers could then make sure that the buildings and bridges are sufficient in terms of security and performance.

4.4.1.4 Fluid Dynamics

Fluid flow is studied for the flow of fluids, and therefore, it has great importance in the design of aerospace and mechanical engineering vehicles and systems dealing with interacting fluids.

4.4.1.4.1 Case Study: Navier-Stokes Equations

The Navier-Stokes equations represent the fluid motion equations. These are represented in the form of a set of nonlinear PDEs and are expressed as follows:

$$\frac{\partial u}{\partial t}+(\boldsymbol{u}\cdot\nabla)\boldsymbol{u}=-\frac{1}{\rho}\nabla p+\nu\nabla^2\boldsymbol{u} \tag{4.26}$$

where $\boldsymbol{u}$ represents the velocity field, p represents the pressure, ρ represents the fluid density, and ν represents the kinematic viscosity.

These equations should be solved, so that the predicted patterns around objects such as aircraft and submarines might eventually be utilized to create novel aerodynamics and efficiency designs for vehicles.

4.4.1.5 Control Systems in Robotics

One other significant role that differential equations have is with respect to controlling robotic systems. In this kind of systems, the systems must run precisely and therefore stability is paramount in these systems.

4.4.1.5.1 Case Study: Inverted Pendulum

Some of the most well-known and intuitive examples in control theory are those of an inverted pendulum with the next differential equation describing the dynamics:

$$\frac{d^2\theta}{dt^2} = \frac{g}{L}\sin(\theta) - \frac{b}{mL^2}\frac{d\theta}{dt}, \tag{4.27}$$

where θ is the angle from the vertical, g is the acceleration due to gravity, L is the length of the pendulum, b is the damping coefficient, and m is the mass.

This equation is useful to design controllers in stabilizing a pendulum position to an upright position, where some applications are found with robotics as well as other automatically controlled systems.

In reality, these examples show the integral role of differential equations in myriad applications in the field of engineering. With access to mathematical apparatuses to develop, analyze and design complex structures, differential equations enable engineers in solving real word problems that influence innovation across widely spread domains of application [19]. The engineering world will be highly responsible for differential equation applications as long as the burden of engineering in the future persists.

4.4.2 Real-World Examples in Computer Science

Differential equations are widely used in computer science, especially in the practical and theoretical application of these fields: algorithms, computer graphics, machine learning, and network theory. These are the examples of real-world applications where differential equations play a significant role.

Table 4.3: Real-world applications of differential equations in computer science: benefits and drawbacks.

Field	Example	Description	Benefits	Drawbacks
Image processing	Image denoising	Differential equations are used to enhance and manipulate images by removing noise while preserving edges. One such technique uses the diffusion equation to model heat distribution	Effective at removing noise without blurring edges Improves image quality, especially in medical imaging and photography	Computationally intensive, especially for high-resolution images Can over-smooth if not properly tuned

Table 4.3 (continued)

Field	Example	Description	Benefits	Drawbacks
Neural networks and deep learning	Neural ordinary differential equations (ODEs)	Neural ODEs provide a continuous-time formulation for neural networks, making deep learning more efficient. The hidden-state transformation is modeled as a dynamical system	More efficient training for deep learning models Improved ability to model time series and dynamic data Better generalization in some tasks	Complex to implement and require careful tuning Computationally expensive, especially for large-scale networks
Robotics and path planning	Optimal control in robotic arms	Differential equations describe the dynamics of robotic arms and are used to derive optimal control strategies for accurate movements, such as in manufacturing and healthcare	Enables precise and accurate control of robotic systems Vital for manufacturing, healthcare (surgery), and autonomous vehicles Can optimize robotic paths for better performance	High computational cost in real-time control Requires detailed modeling of the robot's dynamics, which may not always be available or easy to achieve
Network theory and graph algorithms	Epidemic modeling on networks	Differential equations are used in network theory for modeling disease spread. The SIR model helps analyze infection dynamics in populations, influencing public health policies	Useful for predicting and controlling disease outbreaks Provides valuable insights for public health responses Can be adapted to other network dynamics beyond disease spread	Assumptions of the model may oversimplify real-world scenarios The model may not account for all factors (e.g., individual behavior and environment)
Computational fluid dynamics (CFD)	Simulating airflow around objects	Navier-Stokes equations are used to simulate fluid dynamics, such as airflow around vehicles, aiding in aerodynamic studies and design optimization	Essential for optimizing designs in engineering, like vehicle aerodynamics Can reduce the need for expensive physical prototypes Helps in solving complex fluid flow problems	Computationally expensive and time-consuming Requires high-performance hardware for accurate simulations Solving Navier-Stokes equations numerically can be challenging due to their complexity

4.4.2.1 Image Processing

Differential equations are applied to computer graphics and image processing in enhancing and manipulating images.

4.4.2.1.1 Example: Image Denoising

Of the traditional uses, this area is imaging denoising, where the PDEs provide the noise-freeing while the edges are not affected. This often just uses some version of the diffusion equation and models the heat distribution process:

$$\frac{\partial I}{\partial t} = \nabla^2 I + \lambda(I - I_0), \tag{4.28}$$

where I is the image intensity, t represents time, ∇^2 is the Laplacian operator, $I_0$is the original image, and λ is a tuning parameter. This particular equation solves algorithms iteratively in order to improve the quality of an image, meaning that it is very important for applications in medical imaging, photography, and more.

4.4.2.2 Neural Networks and Deep Learning

A differential equation would be able to provide a theoretical background for describing the way a neural network behaves as well as its training dynamics.

4.4.2.2.1 Example: Neural Ordinary Differential Equations (ODEs)

Neural ODEs consider that the hidden-state transformation is a continuous-time dynamical system, which is basically an extension of traditional neural networks. The hidden state can be formulated with an ODE as follows:

$$\frac{dz(t)}{dt} = f(z(t), t, \theta) \tag{4.29}$$

where $z(t)$ is the hidden state, f is a function representing the neural network, and θ is the parameter.

This approach makes deep learning capable of being trained much more efficiently, and tasks like time series prediction and generative modeling require the capturing of temporal dynamics.

4.4.2.3 Robotics and Path Planning

Differential equations play a very crucial role in robotics, primarily because of motion planning and control. Robots are now capable of driving through complex environments.

4.4.2.3.1 Example: Optimal Control in Robotic Arms

The dynamic models can be described by differential equations, thus enabling optimal control strategies: differential equations describing the dynamics of the robotic arm according to state variables and coordinates of the arm's joints:

$$\tau = I\frac{d^2\theta}{dt^2} + b\frac{d\theta}{dt} + g(\theta) \tag{4.30}$$

where τ is the torque, I is the moment of inertia, b is the damping coefficient, and $g(\theta)$ is the gravitational force.

It is also due to the differential equations that the robots can demonstrate to carry out complex maneuvers with accuracy. These differential equations are quite crucial for manufacturing, healthcare applications, and also autonomous vehicle applications.

4.4.2.4 Network Theory and Graph Algorithms

Differential equations play an important role in network theory also, mainly regarding the modeling of the dynamics in networks and in flows.

4.4.2.4.1 Example: Epidemic Modeling on Networks

In other words, this is used for modeling the spread of disease by systems of differential equations that help in understanding the dynamics of infections over networks. This *SIR* model is represented as follows:

$$\frac{ds}{dt} = -\beta SI \tag{4.31}$$

$$\frac{\mathrm{d}I}{\mathrm{d}t} = \beta SI - \gamma I \tag{4.32}$$

$$\frac{dR}{dt} = \gamma I \tag{4.33}$$

where S, I, and R represent the number of susceptible, infectious, and recovered individuals, respectively, and β and γ are transmission and recovery rates, respectively.

These equations can be analyzed to determine the methods by which disease spread can be controlled and which in turn will affect public health policies and responses.

4.4.2.5 Computational Fluid Dynamics (CFD)

Applications in computer science have a very large use of differential equations to simulate fluid flow in different environments.

4.4.2.5.1 Example: Simulating Airflow Around Objects

Navier-Stokes equations are a formulation in writing down the behavior of fluids in motion. They will help in predicting wind flow patterns while conducting simulations in the aerodynamic studies of vehicles for engineering designs:

$$\frac{\partial u}{\partial t} + (\boldsymbol{u} \cdot \nabla)\boldsymbol{u} = -\frac{1}{\rho}\nabla p + \nu \nabla^2 \boldsymbol{u} \tag{4.34}$$

These are solved numerically with finite element or finite volume techniques applied for the purpose of computation-intensive calculations required to optimize designs in respect to performance and efficiency.

Such practical applications, from improving image quality to training models of deep learning, from the optimization of robotics motion to the modeling of network dynamics, prove the applicability of differential equations in computer science. The further development of technology will increase the application of differential equations in computer science, which thus opens up a new door to more new ideas and applications in the field [20]. Table 4.3 gives a brief overview of how differential equations are used in computer science disciplines, with important examples like image processing, neural networks, robotics, network theory, and computational fluid dynamics.

4.5 Challenges and Limitations

4.5.1 Numerical Methods and Computational Complexity

Though differential equations are great tools to model and solve real-world problems, applying them generally means surmounting a lot of difficulties, especially from the standpoint of numerical methods and their computational complexity.

4.5.1.1 Numerical Solution of Differential Equations

Most real differential equations cannot be solved analytically and hence a numerical method to obtain an approximation is needed.

Table 4.4: Comparison of numerical methods for solving differential equations: accuracy versus computational cost.

Numerical method	Advantages	Limitations	Trade-off (accuracy vs. computational cost)
Finite difference method (FDM)	Simple to implement Efficient for non-stiff problems Suitable for regular grids	Instability for stiff equations Poor for complex geometries Requires smaller time steps for high accuracy	Fast and simple for smooth problems, but inefficient for stiff systems and complex geometries
Finite element method (FEM)	Highly flexible for complex geometries High accuracy with refined meshes Handles nonlinearities well	Computationally expensive, especially for 3D problems Mesh generation is challenging Requires significant memory and processing power	High accuracy but computationally intensive, requiring HPC resources for large-scale simulations
Runge-Kutta methods	Good balance of accuracy and efficiency Widely applicable for ODEs Easy to implement	Computationally expensive for higher-order methods Not suitable for stiff equations unless modified Error propagation in dynamic systems	Efficient for moderately stiff problems but becomes expensive for complex systems or high accuracy
Implicit methods (e.g., backward Euler)	Stable for stiff systems Suitable for problems with rapid changes in some parts of the solution	Requires solving nonlinear algebraic equations per time step More computationally expensive per time step	Provides stability for stiff problems but requires more resources per time step
Stiffness handling (e.g., implicit vs. explicit)	Implicit methods are stable for stiff problems Explicit methods are computationally cheaper for non-stiff problems	Explicit methods require small time steps for stability, increasing computation time Implicit methods require solving systems of equations at each step	Implicit methods provide stability at the cost of higher computational overhead, whereas explicit methods are faster but struggle with stiffness

Finite Difference Method (FDM): The method solves a differential equation by discretization whereby derivatives are turned into difference quotients. FDM is easy to implement but inherently unstable and divergent for stiff equations.

Finite Element Method (FEM): The FEM breaks up a domain into smaller simpler parts for an approximate solution. Although generally powerful, this is very computationally expensive and may well require dedicated software and expertise.

Runge-Kutta Methods: These are widely used in the solution of ODEs: they strike a balance between being sufficiently accurate and not unnecessarily computationally intensive; unfortunately, as the complexity of the system increases, so do higher-order methods become computationally expensive.

4.5.1.2 Stiffness in Differential Equations

Stiffness refers to the situation where some parts of the solution changes much faster than other parts, and this gives numerical difficulties. Stiff equations are very sensitive in the choice of numerical methods; otherwise, they may bring about instability and incorrect results.

Example: This fact brings about stiffness, and schemes for stiff systems often lead to very small-time steps, and computation becomes both inefficient and quite time-consuming in chemical kinetics when reactions occur at widely different rates.

4.5.1.3 Computational Complexity

The size and dimensionality of the problem can very quickly make differential equation solutions complicated.

High Dimensionality: In a large majority of applications, be it fluid dynamics or multi-body simulation, an equation might contain multiple variables and dimensions, increasing the number of computational resources at stake.

Mesh Generation and Adaptation: It can be challenging to obtain an appropriate mesh in FEM. An adaptive mesh, which locally refines in areas of interest, aids in precision but augments complexity and computational cost.

Algorithmic Complexity: The numerical method used may have an important influence on the efficiency of computations. More accurately obtained results tend to require more computing power, such that accuracy and performance are often adversarial.

4.5.1.4 Resource Requirements

Very frequently, intensive computer calculations are required in the solution of complex differential equations such as:

Processing Power: High-performance computing (HPC) resources may be necessary for high-scale simulation. Advanced modeling techniques could be restricted to this kind of scenario, especially when resources are a concern.

Memory Usage: A lot of memory is used up by large systems, so these are limited to applicability in highly resource-constrained environments. High-dimensional problems require it to be effective.

4.5.1.5 Error Propagation and Stability

Numerical methods suffer from unavoidable rounding errors, approximations and truncations.

Error Accumulation: For iteration methods, the calculational mistake can be passed along until the results are drastically affected at the final outcome.

Stability Analysis: Stability is the most important concept in numerical methods. Unbounded solutions may occur, mainly in dynamic systems where accurate predictions are very much required.

The challenges of numerical methods and computational complexity have dimensions that are severe and multi-faceted. While techniques and algorithms may make computations much faster and more accurate, one needs to understand that there still are such limitations and look for them while working with differential equations [21]. The crucial part in the reliability and effectiveness in the vast applications of differential equations, from engineering to physics to computer science, is solving such challenges. Table 4.4 presents a concise comparison of various numerical methods for solving differential equations, noting their merits, demerits, and accuracy-computational cost compromises.

4.5.2 Addressing Nonlinear Dynamics

Differential equations involving nonlinear dynamics are very challenging to work with; one such difficulty is that they can have behavior so complicated that it can prove either impossible or extremely difficult to analyze and predict in solutions. It describes the character of nonlinear dynamics, difficulties they present, and some ways approached to surmount these difficulties in modeling and solving nonlinear differential equations.

4.5.2.1 Nature of Nonlinear Dynamics

Nonlinear differential equations are equations in which the dependent variable, or its derivatives, contains exponents other than one or the product of powers. Such nonlinearities often make it possible to have phenomena as follows:

Bifurcations: Changes in behavior of a system are "sudden" with respect to variation in parameters, thus generating new equilibrium points or periodic solutions.

Chaos: The highly sensitive dependence on initial conditions; a small perturbation may be drastically changed.

Multi-stability: There are several stable states that can exist under the same conditions of the external environment. This makes the prediction and therefore the control difficult.

Therefore, multi-stability makes nonlinear systems intrinsically more difficult to analyze than linear systems.

4.5.2.2 Challenges of Nonlinear Dynamics

The problems with nonlinear dynamics are as follows:

Analytical Complexity: Very few nonlinear equations have an analytical solution.

Computational Intensity: Numerical solution of nonlinear equations, for instance, Newton-Raphson, or continuation method is highly intensive in computation and may require much computational resources.

Stability Analysis: Stability in nonlinear systems is a lot more elusive than stability in linear systems, and an important approach to ensure this factor is involved with the use of Lyapunov's methods, among other approaches.

4.5.2.3 Approaches to Address Nonlinear Dynamics

Several approaches can be employed to tackle the challenges of nonlinear dynamics:

4.5.2.3.1 Numerical Methods

Finite Element and Finite Difference Methods: They can be readily applied for nonlinear equations by solving iteratively linearized versions or through the adaptive technique driving improvements in the mesh where the behavior of solution is expected.

Spectral Methods: They formulate the solution as an expansion about some known functions, and this does well with a special set of nonlinear problems, especially in fluid dynamics.

4.5.2.3.2 Approximation Techniques

Perturbation Methods: In perturbation methods, approximate solutions of nonlinear equations are obtained by considering the nonlinearity as a small perturbation of some known linear solutions that can be analyzed to provide even more insight into the system's behavior.

Homotopy Analysis: This method makes it possible to construct a continuous mapping from an elementary problem to the complicated problem. Therefore, there are possibilities for analytical approximations of solutions of nonlinear equations.

4.5.2.3.3 Bifurcation Analysis

Continuation Methods: This trace how solutions evolve with the parameters they exist in. The researchers will use these for bifurcations and track how bifurcation can describe how the system behaves for the variations of its parameter.

Numerical Bifurcation Analysis: There exist software tools that can be used to carry out bifurcation diagram analyses, gaining one insight into the qualitative behavior, stability, and dynamics of nonlinear systems.

4.5.2.3.4 Machine Learning and Data-Driven Approaches

Surrogate Modeling: Machine learning methods allow construction of surrogate models of complex nonlinear systems. Successive predictions then can be gotten at lower computing cost than for running large numerical simulation.

Reinforcement Learning: It shows how the strategy to control by interacting with the environment can be learned and be useful for optimal strategies related to nonlinear systems to adapt to the nonlinear behavior of a system.

4.5.2.3.5 Lyapunov Stability Theory

Lyapunov methods are the fundamental to assessing stability in nonlinear systems. Equilibrium points can be proven to be stable or unstable by constructing Lyapunov functions. This again will be crucial information for control designs.

This is in fact the case of tackling nonlinear dynamics in differential equations, a complicated but rather fundamental part of many scientific and engineering applications. Though challenges posed by nonlinearity can be coped with through utilizations of both numerical methods combined with approximation techniques, bifurcation analysis and machine learning approaches, it will learn better strategies for handling nonlinear dynamics as our research progresses and learn to model and control more effectively complex systems across different domains, from engineering to physics, biology, and economics [22].

4.6 Future Trends and Innovations

4.6.1 Emerging Applications in AI and Big Data

Such new linkage between differential equations, AI, and big data introduces new worlds and a new line of new possibilities and innovations that allow more computation power and easy data availability, and apply differential equation techniques to build, analyze models, and make all-inclusive decision-making in applications for all spheres of interests. The rest of this chapter explains emerging applications of differential equations in AI and big data.

4.6.1.1 Enhanced Machine Learning Models

Today, the greatest and wonderful factor, especially with the rise of differential equations in machine learning, is that performance does seem better over tasks that entail dynamic systems.

Neural ODEs: This views the hidden layers of a neural network as continuous dynamical systems. This allows for training with efficiency and improved generalization to time series data and those tasks that involve sequential operations. The model of continuous evolution of hidden states might be helpful in uncovering dependencies of sequence data better than ordinary architectures.

Differentiable Programming: The combination of differential equations with the programming framework allows the idea of differentiable programming, which thus grants seamless integration of physics-based models with machine learning. With such integration, inferences may be improved because models learn from both data and underlying physical laws.

4.6.1.2 Predictive Analytics and Time Series Forecasting

Since differential equations can be found to be the backbone of modeling temporal dynamics in big data analytics, it appears to give a better approach to making the best predictions and insights.

Epidemiological Modeling: Differential equations used in public health help model how the disease spreads. The public health can make predictive analytics for vaccination strategies and containment measures.

Financial Forecasting: Through differential equations, we can model all phenomena dependent on market dynamics and price movement with everything associated to generate better standards of risk assessment and investment formulation. Models are very useful because they can more adequately capture the inherent uncertainty in financial markets through stochastic elements.

4.6.1.3 Optimization and Control in Smart Systems

Although diverse industries are rapidly adopting AI and Internet of Things (IoT) technologies in their operational modes, the process of optimization in control systems leads to differential equations.

Adaptive Control Systems: Differential equations aid in designing adaptive algorithms that may respond dynamically to environments in which learning and adjustment might be necessary for timely efficiency and performance of robotic and automated systems. Such applications include autonomous vehicles and smart manufacturing.

Energy Management: Differential equations can therefore be applied in the modeling of energy flow and optimize distribution based on real-time data such that smart grids are more stable, efficient, and reliable with their switch to renewable energy.

4.6.1.4 Data-Driven Discovery of Dynamics

AI techniques are used with differential equations to uncover hidden dynamics in large datasets.

Dynamic System Identification: Machine learning algorithms can infer the underlying dynamical systems by analyzing large quantities of data. Researchers may fit differential equations to data in order to explain observed behavior in a model, bringing new insight to fields like climate science, biology, and engineering.

Uncertainty Quantification: In all applications involving data-intensive, understanding of uncertainty is paramount. Solutions to differential equations can provide good insight into uncertainty quantification for a prediction, since they help in the understanding of the reliability of the AI models and the decision-making processes.

4.6.1.5 Computational Efficiency and Scalability

The integration of differential equations with AI and big data requires the development of new computational methods.

High-Performance Computing: As data is getting large and complex in size, solving large-scale differential equations requires a huge amount of HPC resources. This in turn compels the development of new algorithms as well as the optimization of parallel processing software.

Cloud Computing and Edge Computing: Emerging with the creation of IoT devices, are able to process huge data generated in real time. Real-time application of differential equations can be used to model and analyze such data, which would aid in making fast decisions and hence improve the response of the system.

The advent of differential equations in AI and big data would make tremendous differences to the futures of various industries in that it combines with many applications in which improvement of machine learning models and enhancement of predictive analytics are achieved to optimize control systems and dynamic behavior [23]. The inherent synergy of differential equations with AI and big data means that, along with improving progress in computational capability, further innovation and transformation across a wide range of domains are possible.

4.6.2 The Role of Differential Equations in Next-Generation Technologies

Differential equations would be foundational in developing next-generation technologies and would form the basis to allow modeling, analysis, and simulation. The contribution made by differential equations will expand exponentially with unfolding developments in quantum computing, biotechnology, and advanced materials science.

4.6.2.1 Quantum Computing

Therefore, quantum computing is the paradigm shift in computation, and differential equations are the very basis of such computations.

Quantum Dynamics: It is the time evolution of quantum states through the fundamental differential equation in quantum mechanics, namely the Schrödinger equation. In principle, it is a method for directly simulating quantum algorithms and other processes. However, understanding these dynamics has turned out to be crucial for the development of effective quantum algorithms and for the discovery of appropriate error correction techniques.

Quantum Control: It is the optimal control strategy with differential equations for the manipulation of quantum systems. This is an example of such design using optimal control theory to efficiently reach certain quantum states that are very important in quantum information processing and quantum communication.

4.6.2.2 Biotechnology and Systems Biology

Differential equations occupy a position of prominence in modeling different complex biological processes involved in biotechnology and systems biology.

Population Dynamics: One can represent the differential equations; differential equations may represent populations of bacteria or human cells, the growth pattern and interaction and, hence, also response to environmental changes. Modeling is important for drug development and understanding disease progression and designing an effective therapy.

Biochemical Reactions: The kinetics of biochemical reactions may be described with ODEs. Such models can simulate metabolic pathways and allow the researcher to predict the effects of a genetic modification or drug treatment and bring researchers closer to the promise of personalized medicine.

4.6.2.3 Advanced Materials and Manufacturing

This underlying concept of the material behavior used for describing things within differential equations can be considered crucial, forming a basis for advancement in new manufacturing processes and advanced materials.

Material Behavior Modeling: This is modeling of material response in terms of mechanical, thermal, and electrical properties, based on differential equations that predict how materials will behave under various stresses and environmental conditions. This knowledge will play an important role in designing lightweight composites or smart materials that respond to external stimuli.

Additive Manufacturing: In additive manufacturing and 3D printing, differential equations are used to optimize the deposition and cooling process, so that good quality, high-performance structures are obtained. This is achieved by simulating how the fluid dynamics and heat transfer effects drive the parameters toward optimizing process conditions for the desired material properties.

4.6.2.4 Environmental Modeling and Climate Science

Differential equations are, indeed, very fundamental in modeling complex environmental systems. They help in understanding ecological interactions and the dynamical behavior of climate.

Climate Models: Global climate models rely on systems of differential equations to simulate atmospheric and oceanic processes. Such models may be used to forecast impacts of climate change, advise policy decisions, and assess mitigation strategies. Successful prediction depends on an understanding of nonlinear dynamics and feedback mechanisms within climate systems.

Ecosystem Dynamics: This can be modeled using differential equations describing species and resource interactions as well as environmental effects, and the models lead to analyses of stability and resilience within ecosystems and serve as a guide for conservation action and ecosystem management.

4.6.2.5 Robotics and Autonomous Systems

Differential equations are absolutely basic in controlling and optimizing the behavior of autonomous systems in robotics.

Kinematic and Dynamic Modeling: In the case of moving arms and vehicle movement, differential equations will be used to give descriptions of their motion. In this respect, forces, torques, and trajectories can be modeled and used in algorithms for path planning and control.

Learning and Adaptation: The dynamics of the environment of reinforcement learning would be more well represented by the differential equations; it might involve integrating time-dependent alterations into the process of learning by an agent of the optimum policies with regards to the system's state. This capability is crucial for applications of autonomous driving, drones, and robotic manipulation.

Differential equations actually form at the heart of technology development and advancement in any line of endeavor. Quantum computing and biotechnology, advanced materials, and robotics are some examples where differential equations play a crucial role in modeling and analysis, and developments in these technologies therefore depend crucially on those differential equations [24]. Technology advancing will only lead to more discoveries and solutions, making differential equations pivotal and even leading to answers to some of the biggest questions humanity faces today.

4.7 Ethical Implications of Mathematical Models in Digital Technologies

Mathematical models, especially differential equations, are central to the development of digital technologies, ranging from algorithms to the simulation of real-world systems. But with such power lies tremendous ethical concerns.

4.7.1 Ethical Challenges Raised through Mathematical Models

Bias and Inaccuracy: Mathematical models are only as unbiased and accurate as the data that is used to develop them. Inaccurate or biased data can create algorithms that perpetuate societal biases. For instance, predictive models applied in criminal justice or hiring can inadvertently reinforce racial or gender biases, disproportionately impacting vulnerable populations.

Over-Dependence on Models: Mathematical models such as differential equations and other structures may at times over-simplify intricate human or societal dynamics and, in doing so, provide erroneous decision-making. Using only these models for key decisions like healthcare resource planning or economic predictions can have adverse, unforeseen effects.

Opacity and Accountability: Most mathematized models employed in digital technologies are "black boxes," i.e., their inner structure is not transparent enough to interpret. Such opacity is a moral issue related to accountability, particularly when decisions made by algorithms affect people's lives. If an algorithm has made a discriminatory or negative decision, it can be hard to attribute the cause to the underlying model [25].

4.7.2 Using Mathematical Models to Mitigate Ethical Problems

Whereas math models do create ethical dilemmas, they themselves also create innovative opportunities to remedy these same challenges:

Model Interpretability and Transparency: Model interpretability advancements like explainable AI are enhancing the clarity of complicated mathematical models. Through transparency, we're able to enable decisions made through algorithms to be explainable, so that people whose lives are influenced by them may have recourse.

Bias Mitigation: Incorporating fairness constraints and ethical thinking directly into the modeling process enables us to build models that deliberately combat bias. This is achieved by selecting data thoughtfully, applying methods for identifying and remediating bias, and designing models that include diverse viewpoints and experiences.

Dynamic Model Adjustment: Differential equations, which commonly represent dynamic systems, can be constructed to be responsive to evolving societal norms and ethical standards. For instance, they can be applied in real time to change algorithms according to ongoing feedback from social and ethical inputs, making decision-making more responsive and responsible.

Mathematical models, and especially differential equations, are valuable tools in digital technologies but at the same time carry a substantial ethical danger. By being transparent, correcting bias, and implementing ethical values into their development, we can unlock the potential of these models to innovate while preventing harm [26]. Proper use of mathematics is paramount in making technology benefit society in general, and not widens inequalities or creates new ethical challenges.

4.8 Conclusion

4.8.1 Summary of Key Insights

This chapter discusses some basic importance that differential equations might bring into the development and progress of digital technology. The main findings are as follows:

Fundamental Importance of Differential Equations: Differential equations represent the most natural way of modeling dynamic behavior, and their role is critical, especially in computer science, engineering, and applied mathematics. They represent a mathematical framework to show how a system evolves in time, providing a prediction about the behavior of the system.

Historical Significance: The history of differential equations reflects how it holds an important place in the development of modern science and has been established since its application to physics, which is used in various fields today and has helped make engineering and technology innovations.

Diverse Applications in Digital Technologies: It is used more profoundly in signal processing and control systems, machine learning, and data science. Differential equations describe and shed light upon complex behaviors necessary for algorithms and systems that have real-time decision-making capabilities.

Challenges in Numerical Methods: They are powerful, but adding problems such as stability, computational complexity, and error propagation, numerical solutions of differential equations. In practice, resolving these issues is critical to making reliable models and simulations in applications.

Nonlinear Dynamics: Nonlinear dynamics often generates many extra complicating factors and, not infrequently, leads to chaos and multiple stable states. Numerical methods, bifurcation analysis, and even machine learning is typically employed in dealing with the complexity, revealing valuable insights.

Future Trends and Innovations: These lines of convergence between differential equations, AI, and big data have brought breakthroughs to most of the disciplines using differential equations. Some emerging applications are: Neural ODEs, dynamic system identification, and optimization in smart systems.

Impact on Next-Generation Technologies: Differential equations are an integral part of next-generation technology such as quantum computing, biotechnology, advanced materials, and robotics. They would help in formulating the actual complex behaviors needed for innovations and real-life problems.

Differential equations are some very critical tools in the comprehension and further development in digital technologies. They have various applications within the different contexts where modeling is applicable to dynamic systems, improves prediction, and innovation, and development can be done. The role of differential equations, as technology is further developed and evolved, forms the very fabric of science and technology into its future image. The relentless R&D in the field certainly will unlock new paths to open further into the unknown, ensuring that differential equations would remain an integral part of any future knowledge or technology.

4.8.2 The Continued Relevance of Differential Equations in Digital Evolution

Not only are differential equations leftovers from earlier mathematical constructions but continue playing very central roles in the present-day development of digital technologies. With the advancement of the world to a stage characterized by tremendous technological innovation and complication, differential equations shine brightly on various spectrums:

Integration with Emerging Technologies: A vast majority of novel fashionable technologies come out as byproducts of differential equations. Quantum computing and machine learning, two relatively newer fashionable fields of mathematics, owe their roots to differential equations. These and their further applications only augment the necessity of modeling sophisticated dynamics for such systems. As witnessed in recent developments in AI frameworks such as Neural ODEs, the integration of differential equations to improve model performance and predictivity is creating a much greater necessity for such approaches.

Addressing Complex Real-World Problems: From climate change to public health crises, modern society has been through a lot of challenges. Modeling and analysis with strength form the basis of decision-making and strategic change. Simulating and understanding complex systems require a framework provided by differential equations, so that researchers and policymakers can predict what they can expect from their behaviors, thus making the proper intervention techniques.

Advancements in Computational Methods: Advances in the supercomputing and numerical analysis have given new life to the study of differential equations, making problems inaccessible previously are now in reach for solution. The promises of adaptive mesh refinement, parallel computing, and using machine learning to help the solvers give us the maximum efficiency to extract insight from complex models and keep differential equations at the forefront in the data-driven world.

Interdisciplinary Applications: Differential equations find wide applications in physics and engineering, finance, and biology, among other disciplines. With interdisciplinary approaches fast becoming the norm of the day, the possibility of differential equations modeling the phenomenology of a number of disciplines would lead to a sharing of ideas and thus open new applications and deeper insight into differential equations.

Education and Research: It will follow that as demand increases for data literacy and quantification skills, it will demand for the instruction of differential equations and applications in technology and science. Educating a new generation of scientists and engineers on how useful differential equations are will continue to be a foundational element in both theoretical and applied research.

Future Innovations: Differential equations form the core of future technologies, ranging from advanced robotics and smart materials to personalized medicine. Their capacity to model and predict dynamic behaviors in systems places differential equations in a prime position as drivers for innovation and breakthroughs in those areas.

Differential equations will be integral to digital evolution and never lose their relevance. With the complex modern technological landscapes that demand modeling, analysis, and predicting dynamic behavior, differential equations will be used very critically. Differential equations form a cornerstone of scientific inquiry and technological advancement into more and more interconnected and complex worlds. The role of differential equations will have to be adopted and celebrated in efforts to address the challenges and opportunities that lie ahead in our digital future.

This chapter is concluded to emphasize the role of differential equations as a fundamental tool in the evolution of digital technologies, with a difference thus highlighted in their importance and also as basic tools for modeling dynamic systems in many fields. From their birth in physics a long time ago to applications in machine learning, control systems, and quantum computers nowadays, differential equations bring insights into new areas of innovation, improving the capability of predictions. Despite all the difficulties mentioned, newly developed improved computational methods, along with the integration of AI, will make numerical solutions not only more feasible but also efficient for solving complex nonlinear dynamics. More interdisciplinary interaction in analysis shall further continue to solve the complex problems that the world is facing and fashion new technologies – an open door for differential equations. Finally, however, their significance is to be ensured, so that differential equations can be integral parts of scientific inquiry and technological progress in an increasingly interrelated world.

References

[1] Krantz SG. Partial Differential Equations and Complex Analysis. CRC press; 2018 May 4.

[2] Zwack CC, Haghani M, Hollings M, Zhang L, Gauci S, Gallagher R, Redfern J. The evolution of digital health technologies in cardiovascular disease research. NPJ Digital Medicine. 2023 Jan 3;6(1):1.

[3] Dubey AK, Kumar A, Pati UC, Marquez FP, García-Díaz V, Srivastav AL editors. Computer Vision and Machine Intelligence for Renewable Energy Systems. Elsevier; 2024 Sep 20.

[4] Koroche KA. Numerical solution for one dimensional linear types of parabolic partial differential equation and application to heat equation. Mathematics and Computer Science. 2020;5(4):76–85.

[5] Zhu L, Fan Q. Numerical solution of nonlinear fractional-order Volterra integro-differential equations by SCW. Communications in Nonlinear Science and Numerical Simulation. 2013 May 1;18 (5):1203–1213.

[6] Wang H, Zheng X. Analysis and numerical solution of a nonlinear variable-order fractional differential equation. Advances in Computational Mathematics. 2019 Dec;45(5):2647–2675.
[7] Mugnolo D Vector-valued heat equations and networks with coupled dynamic boundary conditions.
[8] Chouk K, Gubinelli M. Nonlinear PDEs with modulated dispersion I: Nonlinear Schrödinger equations. Communications in Partial Differential Equations. 2015 Nov 2;40(11):2047–2081.
[9] Por E, Van Kooten M, Sarkovic V. Nyquist–Shannon sampling theorem. Leiden University. 2019 May;1 (1):1–2.
[10] Zeng Z, Liu J, Yuan Y. A generalized Nyquist-Shannon sampling theorem using the Koopman operator. IEEE Transactions on Signal Processing. 2024 Aug 1.
[11] Umut Ö. Routh-Hurwitz Conditions and Lyapunov Second Method for a Nonlinear System. InChaotic Systems: Theory and Applications 2010. 362–368.
[12] Schättler H. Regularity Properties of Optimal Trajectories: Recently Developed Techniques. In Nonlinear Controllability and Optimal Control 2017 Nov 22 (pp. 351–381). Routledge.
[13] Hosseinpour S, Martynenko A. An adaptive fuzzy logic controller for intelligent drying. Drying Technology. 2023 Jun 1;41(7):1110–1132.
[14] Srinivasan N, Krishna M, Naveen V, Kishore SM, Kumar S, Subha R. Predicting Indian GDP with Machine Learning: A Comparison of Regression Models. In 2023 9th International Conference on Advanced Computing and Communication Systems (ICACCS) 2023 Mar 17 (Vol. 1, pp. 1855–1858). IEEE.
[15] Kurzweil J. Ordinary differential equations: Introduction to the theory of ordinary differential equations in the real domain. Elsevier. 2014 Jun 28.
[16] Hanke M. Regularization with differential operators: An iterative approach. Numerical Functional Analysis and Optimization. 1992 Jan 1;13(5–6):523–540.
[17] Naz R, Mahomed FM, Mason DP. Comparison of different approaches to conservation laws for some partial differential equations in fluid mechanics. Applied Mathematics and Computation. 2008 Nov 1;205(1):212–230.
[18] Mao X. Backward stochastic differential equations and quasilinear partial differential equations. London Mathematical Society Lecture Note Series. 1995 Jul;13:189–208.
[19] Carath C. Calculus of Variations and Partial Differential Equations of First Order. American Mathematical Society; 2024 Sep 30.
[20] Zwillinger D, Dobrushkin V. Handbook of differential equations. Chapman and Hall/CRC. 2021 Dec 30.
[21] Hoffman JD, Frankel S. Numerical Methods for Engineers and Scientists. CRC press; 2018 Oct 3.
[22] Bhrawy AH, Zaky M. Numerical algorithm for the variable-order Caputo fractional functional differential equation. Nonlinear Dynamics. 2016 Aug;85(3):1815–1823.
[23] Ilmudeen A. Artificial Intelligence, Big Data Analytics and Big Data Processing for IoT-based Sensing Data. In Transforming Management with AI, Big-data, and IoT 2022 Feb 17. (pp. 247–259). Cham: Springer International Publishing.
[24] Blechschmidt J, Ernst OG. Three ways to solve partial differential equations with neural networks – A review. Gamm-mitteilungen. 2021 Jun;44(2):e202100006.
[25] Hanna MG, Pantanowitz L, Jackson B, Palmer O, Visweswaran S, Pantanowitz J, Deebajah M, Rashidi HH. Ethical and bias considerations in artificial intelligence/machine learning. Modern Pathology. 2025 Mar 1;38(3):100686.
[26] Dhabliya D, Dari SS, Dhablia A, Akhila N, Kachhoria R, Khetani V Addressing bias in machine learning algorithms: Promoting fairness and ethical design. In E3S Web of Conferences 2024 (Vol. 491, p. 02040). EDP Sciences.

[27] Muslim MT. Analysis of the use of technology in teaching differential equations and their practical applications. Al-Qadisiyah Journal of Pure Science. 2024;29(1):30.
[28] Knoke T, Wick T. Solving differential equations via artificial neural networks: Findings and failures in a model problem. Examples and Counterexamples. 2021 Nov 1;1:100035.
[29] Guo Y, Cao X, Liu B, Gao M. Solving partial differential equations using deep learning and physical constraints. Applied Sciences. 2020 Aug 26;10(17):5917.
[30] Kurniadi E, Zulkardi Z, Putri RI. Learning ordinary differential equation at undergraduate level: A systematic learning review. Al-Jabar: Jurnal Pendidikan Matematika. 2022 Jun 17;13(1):23–31.

V. Karthikeyan, Y. Palin Visu†

5 Mathematical Tools for Digital Signal Processing

Abstract: Digital signal processing (DSP) provides a foundation of mathematical tools that are increasingly applicable to artificial intelligence (AI) and machine learning (ML). Central to DSP is the Fourier transform, including the fast Fourier transform, which enables the conversion of signals from the time domain to the frequency domain, facilitating feature extraction for AI models. The *Z*-transform and Laplace transform aid in system analysis and stability assessment which are crucial for designing adaptive algorithms in AI. Convolution, a core operation in DSP, is also fundamental in AI, particularly in convolutional neural networks for image and signal processing tasks. Correlation techniques help in feature matching and pattern recognition, key components in ML algorithms. Linear algebra plays a pivotal role, with matrix operations underpinning neural networks, principal component analysis for dimensionality reduction, and singular value decomposition for data compression. Probability and statistics are essential for modeling uncertainties, optimizing algorithms, and improving prediction accuracy in AI. The wavelet transform is useful for time-frequency analysis, making it suitable for handling non-stationary data and enhancing AI models in fields like speech and audio recognition. Optimization techniques, such as least squares and gradient descent, are critical for training AI models, refining parameters, and minimizing error functions. Filter design methods from DSP, like finite impulse response filters, are used to preprocess data and reduce noise in AI systems. These mathematical tools not only improve the efficiency of AI models but also enhance their capability to interpret complex data, making DSP a crucial component in the development of intelligent systems.

Keywords: Artificial intelligence, system analysis, statistical measures, noise reduction

5.1 Introduction

The mathematical tools for digital signal processing (DSP) constitute the fundamental basis for the majority of contemporary signal processing methodologies. DSP entails the manipulation of signals transformed into a digital format by employing mathematical algorithms to extract, change, or analyze information. DSP is used across various domains, such as telecommunications, audio and voice processing, image and video processing, radar and sonar systems, and biomedical engineering, among others. The basic mathematical tools used in DSP create the necessary framework for

https://doi.org/10.1515/9783112214374-005

processing discrete signals efficiently. This makes sure that difficult tasks can be carried out accurately and reliably [1–3].

The fundamental mathematical instrument for DSP is the notion of the discrete-time signal. A discrete-time signal is a numerical sequence that denotes a time-dependent variable at designated intervals. A fundamental principle in DSP, the sampling theorem ensures that sampled versions of continuous signals accurately represent them. Some math says that if the sampling rate is higher than twice the signal's highest frequency, then the samples can be used to accurately reconstruct a continuous signal. This foundational principle facilitates the conversion from analogue to digital signals and is crucial for numerous DSP methodologies [4–6].

The discrete Fourier transforms (DFT), a crucial mathematical instrument, analyses the frequency components of a discrete-time signal. The DFT enables a signal to be represented by its frequency components, aiding in filtering, compression, and spectral analysis of the signal. The DFT is calculated using the fast Fourier transform (FFT), which is a very efficient method that makes DFT computation much cheaper. Applications like audio and voice processing extensively use the FFT because frequency-based manipulation is crucial. It offers a method to analyze signals in the frequency domain, which facilitates the identification and isolation of certain components, such as noise or undesirable frequencies [7, 8].

The *Z*-transform is a crucial instrument in DSP; it generalizes the DFT and offers a more thorough method for analyzing discrete-time signals and systems. The *Z*-transform converts a discrete-time signal into a complex *z*-plane for analysis of stability, frequency response, and temporal behavior. This tool is essential for the analysis and design of digital filters, which are integral to numerous DSP applications [2, 6]. The *Z*-transform is crucial for resolving difference equations that represent discrete-time systems, enabling engineers to delineate system behavior akin to the Laplace transform for continuous-time systems.

Digital filters are an important part of DSP. They change or improve certain aspects of a signal, like getting rid of noise or boosting certain frequencies. The construction of digital filters is fundamentally dependent on mathematical instruments like the *Z*-transform and the Fourier transform. Filters can be categorized into low-pass, high-pass, band-pass, and band-stop types, each allowing specific frequency components to pass while attenuating others [9–11]. Finding the right coefficients is usually the first step in designing these filters. This can be done in a number of ways, such as by using the windowing method, the bilinear transform, or the Parks-McClellan algorithm. The design of filters is a mathematical optimization problem that aims to minimize the discrepancy between the desired and actual filter responses [12].

A vital mathematical tool for DSP is the concept of correlation and convolution. Correlation quantifies the similarity between two signals, applicable in pattern recognition, signal detection, or noise reduction. Convolution is a fundamental operation that characterizes the output of a linear time-invariant (LTI) system in response to a certain input signal. In DSP, convolution is often used to apply filters to signals. The impulse response of the filter is mixed with the input signal to get the desired result. The mathematical formulation of correlation and convolution is crucial for comprehending and executing various signal processing tasks [13, 14].

Besides these fundamental mathematical tools, linear algebra is crucial in numerous DSP applications. Matrices and vectors are commonly employed to depict systems of equations, transformations, and signal representations. In multidimensional signal processing, including image and video processing, signals are represented as matrices, utilizing mathematical operations such as matrix multiplication, eigenvalue decomposition, and singular value decomposition (SVD) to alter or analyze the data. Algorithms utilize linear algebra to tackle problems like signal compression and feature extraction [4, 8].

Optimization approaches are essential in DSP, especially in filter design, signal compression, and machine learning (ML)-based signal processing. Mathematical optimization seeks to identify the optimal solution to a problem, which is frequently subject to certain limitations. In DSP, optimization techniques are employed to reduce errors or costs, such as minimizing distortion in signal reconstruction or maximizing the signal-to-noise ratio in communication systems. DSP challenges frequently employ methods like gradient descent, least-squares optimization, and convex optimization [15–17].

DSP extensively uses statistical techniques for tasks like noise attenuation, pattern identification, and signal estimation. Statistical techniques assist in modeling stochastic signals, estimating signal parameters, and assessing the probability of diverse outcomes [1, 7]. The principle of maximum likelihood estimation is frequently employed to ascertain unknown properties of a signal or system from observed data. Also, methods like Kalman filtering are the best way to figure out the state of a system from observations that aren't very clear, which is useful in areas like navigation, control systems, and tracking.

DSP needs to understand the theory of stochastic processes because many signals in real life change randomly. A random process is a set of random variables organized by time or space that are used to simulate signals characterized by uncertainty. In telecommunications, random noise is often shown as a stochastic process, and methods like power spectral density analysis and Wiener filtering are used to lessen the effect it has on signal transmission [18, 19].

Wavelet transforms constitute a significant mathematical instrument in DSP. Wavelets make it easier to analyze signals on a number of different scales. This is especially helpful for signals that aren't stationary and change over time in how they look. In contrast to the Fourier transform, which examines signals through sinusoids, the wavelet transform decomposes them into smaller elements that possess localized time-frequency information. In situations where the signal's complex properties need to be kept without requiring a lot of computing power, this makes it perfect for tasks like picture compression (e.g., JPEG 2000) [20–23].

Many different mathematical tools are very important to DSP because they provide the necessary theoretical framework for analyzing and changing digital signals. These instruments facilitate the effective processing of signals across diverse applications, including audio and voice processing, telecommunications, and biomedical engineering. Engineers and scientists can create resilient DSP systems that tackle complicated real-world challenges by comprehending and utilizing essential mathematical ideas, including DFT, *Z*-transforms, linear algebra, optimization, and statistical approaches [5, 22, 24]. The ongoing advancement of novel mathematical methodologies enhances the possibilities of DSP, facilitating increasingly sophisticated and efficient solutions in the future.

DSP is fundamentally intertwined with artificial intelligence (AI)/ML, as DSP provides the essential groundwork for AI-driven applications. DSP methods improve unprocessed signals like audio, video, and sensor data by pulling out important features. These features are then used by AI/ML models to classify, predict, or make decisions. In automatic speech recognition, DSP techniques such as Fourier transforms and Mel-frequency cepstral coefficients (MFCCs) extract characteristics from raw audio signals, which deep learning models (e.g., convolutional neural networks (CNNs) or recurrent neural networks) utilize for precise speech-to-text conversion. In image processing, DSP-based filtering and edge detection enhance image quality prior to AI models executing tasks such as object recognition. In healthcare, DSP filters ECG signals to eliminate noise, facilitating ML models in identifying cardiac problems. Integrating DSP with AI/ML enhances system robustness, efficiency, and accuracy, especially in real-time applications like autonomous vehicles, biomedical diagnostics, and industrial automation.

5.2 Comprehending Signal Analysis

Signal analysis is an essential procedure in engineering and research that entails the examination, interpretation, and processing of signals to derive significant information. Signals, encompassing electrical, auditory, optical, or biological forms, denote fluctuations of physical quantities across time or space [20]. The examination of these signals is essential for various applications, including telecommunications, medical imaging, radar systems, and audio processing.

A crucial phase in signal analysis is categorizing signals into classifications such as analogue or digital, periodic or aperiodic, and deterministic or random. Analogue signals, which are continuous in both time and amplitude, are common in natural phenomena, whereas digital signals, characterized by their discrete nature, prevail in contemporary technology. Fourier transform and wavelet transform are important tools for signal analysis because they break down complicated signals into simpler parts that are easier to understand and change [25].

An additional crucial element is noise filtering, which entails distinguishing valuable information from extraneous noises. Methods including low-pass filtering, high-pass filtering, and adaptive filtering improve signal clarity, facilitating precise data interpretation. Also, progress in ML and AI has led to the creation of complex algorithms that can find patterns and outliers in signals. This makes predictions more accurate and the system work better [26].

Signal analysis is essential for innovation, serving as the foundation for technologies such as wireless communication, medical diagnostics, and environmental monitoring. With the increase in computer capacity, the field advances, providing improved accuracy and novel options to tackle intricate difficulties across several disciplines.

5.2.1 Complex Analysis Function in Signal Processing

Contemporary signal analysis relies heavily on complex analysis, a subfield of mathematics concerned with complex numbers and the operations on them. Thanks to this mathematical framework, many other areas, including communications, control systems, and image processing, have made significant strides in comprehending, representing, and processing information in the frequency domain [4, 10].

The real and imaginary parts of a complex number represent the amplitude and phase of the signal, respectively. In signal analysis, they are used to simulate sinusoidal signals and how they interact with each other. Complex functions, such as the exponential function, can simplify the depiction of oscillatory behavior, thereby facilitating the analysis of signal features. Techniques like the Fourier Transform and the Laplace Transform depend on a lot of complex analysis to turn time-domain data into frequency-domain representations that show hidden periodicities and spectral properties [27]. Applications like system design, signal filtering, and noise reduction greatly benefit from these modifications.

System stability analysis can also benefit from complex analysis, which involves studying complex plane poles and zeros. Their placement influences system behaviors like resonance and damping, providing insights for more efficient system design. Analytic continuity and contour integration are two additional ideas that help us solve integral equations and assess the behavior of signals in complicated situations [21, 28]. The Table 5.1 outlines key applications of complex analysis in signal processing.

The foundation of signal analysis is complex analysis, which allows for the effective modeling, transformation, and assessment of signals in different domains. Its applications lead to scientific and technological breakthroughs by simplifying complicated computations and revealing previously unknown aspects of signal behavior.

Table 5.1: Key applications of complex analysis in signal processing.

Application	Description
Fourier transform	Converts time-domain signals into frequency-domain representations.
Laplace transform	Extends Fourier Transform for non-stationary signals, aiding stability analysis.
Z-transform	Discrete-time equivalent of Laplace Transform for digital signal processing.
Pole-zero analysis	Examines system stability and resonance using complex plane plots.
Modulation and demodulation	Uses complex exponential carriers for efficient signal transmission and recovery.
Waveform synthesis	Combines multiple sinusoidal signals using complex coefficients.
Filter design	Designs filters with desired frequency characteristics using complex polynomials.

5.2.2 The Delta Function in Signal Analysis

The delta function, represented as $\delta(t)$, is a crucial concept in signal analysis. It is a mathematical abstraction that denotes an impulse with infinite amplitude and infinitesimal time, yet it possesses a unit area beneath its curve. Although a theoretical construct, the delta function is essential for signal analysis and processing and forms a fundamental basis for comprehending system behavior and reactions [2, 4].

In signal analysis, the delta function is predominantly employed as an impulse input to characterize systems. The impulse response, derived by inputting $\delta(t)$ into a system, provides essential insights into the system's dynamics, stability, and frequency response. The convolution of a signal with a system's impulse response yields the system's output, rendering $\delta(t)$ essential for the analysis of LTI systems. The principal characteristics of the delta function encompass:

- **Sifting Property**: $\int_{-\infty}^{\infty} f(t)\delta(t-t_o)dt = f(t_0)$, which allows the extraction of a signal's value at a specific point.
- **Scaling Property**: $\delta(st) = \frac{1}{|s|}\delta(t)$ demonstrating how scaling the time axis affects the amplitude.
- **Even Function**: $\delta(-t) = \delta(t)$, indicating symmetry about the origin.

The derivatives and integrals of the delta function are crucial in advanced signal processing. The derivative of $\delta(t)$, commonly known as the doublet, is employed to represent abrupt changes or discontinuities in signals. Conversely, its integration produces the unit step function (t), an essential instrument for modeling time-domain signals and systems.

The delta function is crucial in Fourier transform analysis, as it denotes a singular frequency component in the frequency domain. The Dirac delta function, $\delta(t)$, in the time domain is equal to a spectrum that is flat across all frequencies. This shows how important it is for studying signals' frequency characteristics. The delta function is an essential and multifaceted instrument in signal analysis. Because the subject is unique and has many derivatives and integration uses, it provides a complete framework for understanding and managing signals in both the time and frequency domains.

5.2.3 Fourier Series with Half-Range Expansions and Fourier Series in Signal Analysis

The Fourier series is a mathematical method extensively employed in signal analysis to decompose periodic signals into a summation of sinusoidal functions – sines and cosines – each characterized by distinct amplitudes, frequencies, and phases. This representation illustrates the frequency components within a signal, facilitating a comprehensive investigation of its structure and behavior [1]. The Fourier series is represented as $f(t) = a_0 + \sum_{n=1}^{\infty}\left(a_n cos\frac{n\omega_0 t}{L} + b_n sin\frac{n\omega_0 t}{L}\right)$, where a_0, a_n, and b_n are Fourier coefficients that delineate the contribution of each frequency component. Fourier series are utilized in audio processing, vibration analysis, and communication systems, where comprehending frequency content is essential for filtering and compression operations.

The half-range Fourier series, a specialized variant of Fourier series, is employed to analyze signals defined solely over a restricted range, typically one-half of the period. In these instances, the signal is stretched to a complete period using symmetry features, either as an even function (comprising solely cosine terms) or as an odd function (comprising solely sine terms). This simplifies the analysis and is especially advantageous when the signal displays inherent symmetry. Half-range expansions are often used in engineering and physics problems that involve heat transfer, wave motion, and systems that oscillate, since signals can only be positive or negative.

From Table 5.2, the Fourier series and its half-range variation are essential in signal analysis, providing adaptable techniques for comprehending and manipulating information in both theoretical and practical applications.

5.2.4 Fourier and Alternative Integral Transforms in Signal Analysis

Integral transforms are mathematical instruments that facilitate the study and processing of signals by translating them from one domain – often time or space – to another, usually frequency or complex domain. The Fourier transform is the most significant and extensively utilized method in signal analysis for representing signals in the

Table 5.2: Comparison of Fourier series and half-range Fourier series.

Aspect	Fourier series	Half-range Fourier series
Definition	Represents periodic signals as a sum of sinusoids over a full period.	Represents signals defined over half a period using symmetry.
Components	Includes both sine and cosine terms for general periodic signals.	Includes only sine terms (odd extension) or cosine terms (even extension).
Application domain	Applied to fully define periodic signals.	Applied to signals defined on limited intervals with symmetry.
Fourier coefficients	Computed for both sine (b_n) and cosine (a_n) terms.	Computed only for sine or cosine terms based on the type of extension.
Symmetry consideration	Considers general symmetry properties of signals.	Utilizes even or odd symmetry for computational simplicity.
Common uses	Audio processing, image compression, communication systems.	Heat conduction problems, wave equations, and structural analysis.

frequency domain. Other significant transforms, such the Laplace transform, Z-transform, and wavelet transform, enhance the functionality of signal processing by tackling particular issues in diverse applications [29, 30].

The Fourier transform breaks down a signal in the time domain into its frequency components. This lets you see how signals behave in the spectral domain. Mathematically expressed as $F(w) = \int_{-\infty}^{\infty} f(t)e^{-j\omega t}\, dt$, it is essential for applications like as filtering, spectrum analysis, and system design. The Fourier transform assumes that signals are stationary (time-invariant), making it optimal for the analysis of periodic or steady-state signals [2, 4]. The inverse transform is very important for many things, like audio processing, image compression, and communication networks, because it makes it possible to perfectly reconstruct the original time-domain signal.

The Laplace transform generalizes the Fourier transform by accommodating a wider range of signals, including non-stationary ones and those that demonstrate exponential development or decay. Defined as $F(S) = \int_{-\infty}^{\infty} f(t)e^{-jst}\, dt$, where $s = \sigma + j\,\omega$, it offers a comprehensive perspective on a signal's characteristics inside the complex frequency domain. It is especially beneficial in control systems, stability analysis, and circuit design since it elucidates system poles and zeros essential for stability evaluation.

The wavelet transform, in contrast to the Fourier transform, offers simultaneous time and frequency localization, making it suitable for the analysis of non-stationary signals. By breaking a signal down into wavelets, which are localized wave-like functions, you can get a more detailed look at it at different resolutions. Picture compression, feature extraction, and biomedical signal analysis use the wavelet transform, especially for signals that fluctuate in frequency over time [9]. The comparison in

Table 5.3 helps in selecting the most appropriate transform based on the nature of the signal and the analytical objective.

Table 5.3: Comparison of integral transforms in signal analysis.

Feature	Fourier transform	Laplace transform	*Z*-transform	Wavelet transform
Domain	Continuous time to frequency domain	Continuous time to complex frequency domain	Discrete time to complex *z*-domain	Time-frequency domain
Signal type	Stationary signals	Stationary and non-stationary signals	Discrete signals	Non-stationary signals
Time localization	No	No	No	Yes
Frequency localization	Yes	Yes	Yes	Yes
Applications	Spectral analysis, filtering, and communications	Stability analysis and control systems	Digital filter design and digital signal processing	Image compression and transient analysis
Strengths	Efficient for periodic signals	Handles exponential growth/decay	Suited for digital signals	Multi-resolution and handles transients
Weaknesses	Limited for non-stationary signals	Complex computation for time-localized features	Limited to discrete signals	Computationally intensive

Integral transforms, including Fourier, Laplace, *Z*, and wavelet, serve as complementary instruments for signal analysis, each designed for certain signal kinds and applications. While the Fourier transform is an important part of frequency-domain analysis, other transforms, like Laplace and wavelet, are also very good at dealing with signals that aren't stationary or that change over time. Collectively, these transformations give us a lot of ways to look at, process, and understand signals in a lot of different areas.

5.3 Computational Linear Algebra in Digital Signal Analysis

Computational linear algebra underpins efficient signal processing in digital signal analysis. Algorithms like SVD and eigenvalue analysis facilitate dimensionality reduction, feature extraction, and noise filtration. Matrix operations are important for methods like the DFT and its faster version, the FFT, which help change signals from the time domain to the frequency domain. Matrix representations of linear system models facilitate accurate understanding of filters and digital circuits. Additionally, methods such as principal component analysis (PCA) employ linear algebra to reveal patterns in high-dimensional signal data, hence improving digital communication and ML applications [29–32].

5.3.1 Matrix Theory and Algebra in DSP

Matrices and matrix algebra constitute the foundation of numerous computing methods in digital signal analysis. By showing signals, systems, and transformations in the form of matrices, these math tools make it easier to process and analyze large amounts of complex data quickly. In DSP, matrices are extensively used to describe linear systems, execute transformations, and develop algorithms essential for applications such as filtering, data compression, and noise reduction [30, 31].

Vectors can represent digital signals, while matrices represent linear systems. In this paradigm, matrix-vector multiplication determines the system's output as follows: $y = Hx$, where H denotes the system's transfer function, x signifies the input signal, and y indicates the output signal. This concise model facilitates the analysis of LTI systems, enabling engineers to examine system behavior, stability, and performance using matrix attributes [30].

Matrices are essential in transformations like the DFT, which translates time-domain information into the frequency domain. The DFT is expressed in matrix notation as Fx, where F is the Fourier matrix and x represents the signal vector. When using the FFT algorithm, matrix-based formulations make it easier to compute these transformations quickly. Also, other transformations, like the discrete cosine transform (DCT) and the wavelet transform, use matrix operations to pull out signal properties from different areas [32–34].

Matrix algebra is essential in digital filtering and convolution processes. Matrix-based convolution executes finite impulse response (FIR) and infinite impulse response (IIR) filters, organizing filter coefficients in a Toeplitz or circulant matrix. Image processing, audio filtering, and communications extensively use these structures to facilitate efficient convolution calculation, particularly for large-scale data.

Advanced signal processing methodologies frequently depend on matrix decompositions. Singular value decomposition (SVD) breaks down a matrix into its basic parts, which shows how it is put together and makes tasks like noise reduction, data compression, and signal denoising easier. PCA is a widely utilized dimensionality reduction method that employs eigenvalue and eigenvector calculations to discern predominant signal patterns in high-dimensional data, making it essential in ML and feature extraction [35, 36].

Matrix algebra offers robust methodologies for assessing the stability of digital systems. The system matrix's eigenvalues determine its stability, resonance, and damping properties. These are very important for making digital filters and control systems that work well. Methods like Schur decomposition and QR factorization make these investigations more numerically efficient, which makes it easier to get a good picture of how the system works.

Extensive signal processing applications frequently encounter sparse matrices, characterized by a predominance of zero elements. Sparse matrices can be stored and changed in a way that makes computations much simpler. This makes them useful for tasks like network analysis, high-resolution imaging, and graph-based signal processing [37–40].

The applicability of matrices and matrix algebra spans numerous domains in digital signal analysis. In image and video processing, matrices facilitate transformations, compression (e.g., JPEG), and enhancement. In communications, they enable MIMO (multiple-input, multi-output) systems by modeling channel behavior as matrices. ML models for signal categorization and prediction frequently rely on matrix-based operations for training and inference [41, 42].

Matrices and matrix algebra are essential in digital signal analysis, offering a coherent framework for modeling, computing, and analysis. They foster innovation across various domains, including multimedia, telecommunications, and ML, by facilitating efficient signal encoding, transformation, and processing. With the advancement of computer power and algorithms, matrix-based approaches will persist in influencing the future of DSP [4, 43].

5.3.2 Direct Approaches to Resolution in Digital Signal Analysis

In digital signal analysis, direct methods of solution are ways to solve linear systems or equations without using iterative approximations. These techniques are essential in numerous applications of DSP, including filtering, system identification, and signal reconstruction. Direct approaches guarantee computational efficiency and precision by delivering exact solutions, rendering them extensively usable in real-time signal processing systems [43–45].

One example of a direct approach is matrix inversion, which is used to solve systems of linear equations that come up when designing filters or reassembling signals. The set of equations is written in the form of a matrix, and the answer is found by directly computing the inverse of the coefficient matrix, $A^{-2}b$, where A is the system matrix and b is the input signal. Gaussian elimination and LU decomposition are widely utilized methods for matrix inversion, facilitating the systematic resolution of these equations [46]. The FFT is an alternative method that quickly calculates the DFT of a signal in $O(N \log N)$ time, making it very useful for spectral analysis. Also, the Z-transform and the Laplace transform make it easier to solve difference and differential equations, which is good for designing and analyzing digital systems [2, 4, 47].

Although direct approaches yield precise and accurate solutions, they may be computationally intensive for large-scale systems. However, direct approaches are necessary for effective and efficient signal processing in many DSP applications that need real-time or high-precision solutions.

5.3.3 Vector and Matrix Standards in Digital Signal Analysis

In digital signal analysis, vector and matrix norms are crucial for quantifying the amplitude, stability, and behavior of signals and systems. Norms are mathematical functions that yield a scalar number representing the magnitude, length, or energy of a vector or matrix. These measures are essential for various signal processing tasks, such as signal reconstruction, error estimates, and system performance assessment. Norms provide a quantitative method to evaluate signal behavior, assuring compliance with specified constraints or needs, such as boundedness or convergence [48–51].

A vector norm is a function that assigns a non-negative value to a vector, representing its size or magnitude. The prevalent vector norms in digital signal analysis are the L_1 norm, L_2 norm, and L_∞ norm, each possessing unique characteristics and uses.

The L_1 norm (Manhattan norm) of a vector $x = [x_1, x_2, \ldots, x_n]^T$ is defined as the aggregate of the absolute values of its components:

$$\| x \|^1 = \sum_{i=1}^{n} \| x_i \| \tag{5.1}$$

The L_1 norm is frequently employed in sparse signal reconstruction tasks, particularly in compressed sensing, where the objective is to identify sparse solutions to linear systems. It advocates for sparsity, fostering solutions characterized by a reduced number of non-zero coefficients.

The L_2 norm, or Euclidean norm, is defined as the square root of the sum of the squares of the vector components:

$$\| x \|_2 = \sqrt{\sum_{i=1}^{n} x_i^2} \tag{5.2}$$

This norm is extensively utilized in signal processing as it offers a direct quantification of a signal's energy or power. It is extensively employed in optimization issues, where minimizing the L_2 norm of a residual or error results in the least-squares solution, often utilized in signal filtering, regression analysis, and noise reduction [52].

The L_∞ norm, or largest norm, of a vector is defined as the largest absolute value of its components:

$$\| x \|_\infty = \max_i |x_i|. \tag{5.3}$$

The L_∞ norm is employed in applications necessitating the regulation of the maximum error or deviation, such as in robust signal processing, where the objective is to constrain the worst-case error in a system's response [53].

Matrix norms resemble vector norms but are specifically defined for matrices. These norms quantify the "magnitude" of matrices and are essential in the analysis of digital systems, especially in assessing system stability, filter design, and signal reconstruction [50, 54]. The predominant matrix norms are the Frobenius norm, operator norm, and L_1 norm. The Frobenius norm of a matrix A is the square root of the sum of the squares of all its elements:

$$\|X\|_F = \sqrt{\sum_{i,j} |x_{i,j}|^2} \tag{5.4}$$

The Frobenius norm is frequently employed in signal processing to quantify the "energy" of a matrix, particularly in applications such as picture compression or denoising. It quantifies the extent to which a matrix diverges from the identity matrix or an alternative reference matrix [55].

Operator Norm (Spectral Norm): The operator norm of a matrix. A is the greatest singular value of the matrix, indicating the maximum extent of stretching or scaling that the matrix exerts on a vector. It is characterized as:

$$\| Z \|_2 = \max_{x \neq 0} \frac{\| Z_x \|_2}{\| x \|_2} \tag{5.5}$$

This norm is essential in analysing system stability, especially for linear systems, as the biggest singular value indicates the extent to which a system may amplify incoming signals [56]. The L_1 norm for matrices is defined as the largest absolute column sum:

$$\| Z \|_1 = \max_j \sum_i |x_{i,j}| \tag{5.6}$$

This norm is employed when it is essential to regulate the maximum total response in any column of a matrix, particularly in scenarios where matrix sparsity is advantageous. Vector and matrix norms are widely employed in digital signal analysis to assess signal quality, filter efficacy, and system stability. In filter design, the L_2 norm is employed to minimize the overall error between the anticipated filter response and the actual response, resulting in optimal filter coefficients in the least-squares context [53, 57]. The Frobenius norm in image processing evaluates the quality of image reconstruction following compression or transformation by various algorithms, ensuring the modified image preserves maximal original information. In compressed sensing, the L_1 norm facilitates sparsity in signal reconstruction, rendering it an essential instrument for high-dimensional signal analysis where only a limited number of important components are required for accurate signal representation. Matrix norms, such as the spectral norm, are employed to evaluate the stability and robustness of digital systems, particularly in control theory, where it is essential to guarantee bounded output for constrained input [3, 58].

Furthermore, norms contribute to error analysis. The L_∞ norm can be utilized to regulate the maximum inaccuracy in a system's output, which is especially advantageous in scenarios where worst-case performance is paramount, such as in robust signal processing and system design [59–61]. Vector and matrix norms are crucial instruments in digital signal analysis, offering critical metrics for comprehending and managing signals and systems. Norms facilitate the efficient design, optimization, and analysis of various DSP applications by providing methods to measure the magnitude, energy, or deviation of signals and matrices [60, 62]. Vector and matrix norms are essential for minimizing error, encouraging sparsity, and maintaining system stability in contemporary DSP. As signal processing techniques advance, the significance of norms in directing algorithm development and performance assessment will remain essential.

5.3.4 Digital Signal Analysis Iterative Methods of Solution

Iterative solution approaches are very important in digital signal analysis when working with large, complicated systems, because direct methods become too expensive or impossible to compute. In contrast to direct methods, which find exact solutions in a limited number of steps, iterative methods improve the result at each step, making repeated close calls on the answer. When an analytical solution is unavailable, these methods excel in optimizing signal processing algorithms, solving big systems of linear equations, and other similar tasks [63, 64].

DSP commonly uses iterative algorithms for tasks like optimizing, system identification, signal reconstruction, and filter design. Avoiding the computationally prohibitive requirement to explicitly invert big matrices is one of the main advantages of iterative approaches when dealing with large-scale problems. With their reliance on a series of approximations that converge to the target answer, they are well-suited for real-time applications that have limited computational resources [65, 66].

First, when it comes to signal analysis, gradient descent is among the most popular iterative methods. Reconstruction of signals, reduction of noise, and ML-based signal processing are typical optimization issues that make use of them. Here, the cost function – a measure of how far off the target result is iteratively updated by advancing in the direction of the negative gradient. While the gradient descent process is simple to implement and uses little computing power, it may take a while for the solution to converge, particularly in areas with several dimensions. When real-time processing is crucial, ML and adaptive filtering employ variants such as stochastic gradient descent and mini-batch gradient descent [67, 68].

Signal reconstruction jobs, particularly in compressed sensing and related fields, employ iterative reconstruction methods to retrieve sparse signals from imperfect or noisy observations. For instance, iterative procedures can find the sparse solution to an underdetermined system of linear equations using the basis pursuit method. Alternative approaches, such as matching pursuit and least squares reconstruction, iteratively fine-tune the signal to achieve an optimal or nearly ideal reconstruction by minimizing error [68].

Signal processing applications of the conjugate gradient approach include filtering, inverse problems, image processing, and solving huge systems of linear equations. This technique is an improvement on gradient descent that takes into account both the present gradient and the directions of search in the past to speed up convergence. For example, the conjugate gradient approach is useful for solving systems of equations in digital filter design and signal denoising. It works best with symmetric, positive-definite matrices [20, 31].

Matrix factorization, source separation, and signal approximation are some of the major challenges that use alternating minimization, another common iterative technique. This technique is based on the premise that one set of variables can be optimized while another set is fixed, and it operates by alternating between minimizing over these two sets of variables. To estimate missing elements of a matrix iteratively by updating rows and columns, alternating minimization can be applied, for instance, to matrix completeness problems [10, 23].

Designing adaptive filters also makes use of iterative approaches, as shown in algorithms like recursive least squares (RLS) and least mean squares (LMS). Adaptive filtering uses the discrepancy between the target and real signal outputs to iteratively modify the filter coefficients. Real-time DSP applications commonly employ both LMS and RLS for noise reduction, echo cancellation, and signal prediction tasks.

There are a number of benefits to using iterative procedures instead of direct ones. Since they don't necessitate keeping huge matrices or computing their inverses, they are frequently more memory-efficient. Plus, they're more adaptable, so you can add new data and make real-time updates to the solution. Problems in ML and multimedia processing, which include large-scale or high-dimensional data, are well-suited to iterative approaches [12, 24].

Many different areas of DSP make use of iterative approaches. Image processing activities such as super-resolution and image denoising make use of iterative approaches. When transmitting signals over unreliable channels, communications professionals often use iterative decoding techniques. Examples of such algorithms include turbo codes and LDPC (low-density parity-check) codes. Adaptive filtering and echo cancellation also use iterative approaches to continuously modify filter parameters in response to changing signal properties. Tasks like voice enhancement, which aim to decrease background noise while maintaining the clarity of the speech signal, commonly use iterative algorithms in audio signal processing [14, 25].

5.3.5 Wavelets and Iterative Methods in Quantitative AI

Quantitative AI uses wavelets and iterative algorithms for signal processing, feature extraction, and optimization. Multi-resolution analysis with wavelet transforms is essential for image identification, audio processing, and anomaly detection. Wavelets are ideal for deep learning and AI-based time series analysis because they can capture time and frequency domain properties, unlike Fourier transforms, which presuppose stationarity [11, 27]. But iterative approaches are important for AI optimization and ML. Gradient descent, expectation maximization, and conjugate gradients are necessary for deep neural network training and large-scale computation. These methods optimize convergence and efficiency in high-dimensional domains by iteratively refining solutions. Wavelet-based feature extraction and iterative optimization improve AI models' accuracy and robustness in computer vision, medical signal processing, and financial forecasting.

This chapter incorporates a diverse range of wavelet functions to enhance AI-driven quantitative analysis across multiple domains, including computer vision, biomedical signal processing, financial forecasting, and industrial automation. By leveraging wavelets' ability to perform multi-resolution analysis, we ensure that our approach accommodates varied datasets, real-world complexities, and inclusive AI applications [36]. The inclusion of different wavelet families supports broad accessibility, adaptability, and fairness in AI-based decision-making processes, making them suitable for applications across healthcare, security, and environmental monitoring [27]. Table 5.4 summarizes the various applications of wavelets in AI.

Table 5.4: Applications of wavelets in quantitative approaches to artificial intelligence.

Wavelet type	Mathematical properties	AI applications	Advantages
Haar	Simple, orthogonal, and fast computation	Image compression (JPEG2000) and real-time AI systems	Efficient for binary image processing and edge detection
Daubechies (DbN)	Compact support, orthonormal basis, and smooth functions	Deep learning feature extraction and biomedical signal processing	Balances accuracy and computational efficiency
Coiflet	Higher vanishing moments and better approximation	Financial time series analysis and EEG signal analysis	Enhanced smoothness and feature preservation
Symlet	Symmetric and near-orthogonal	Image denoising and AI-based medical diagnostics	Low distortion in reconstructed signals

Table 5.4 (continued)

Wavelet type	Mathematical properties	AI applications	Advantages
Meyer	Continuous and infinitely differentiable	Speech recognition, NLP feature extraction	Ideal for continuous signal decomposition
Morlet	Complex exponential modulation and Gaussian envelope	Fault detection in industrial AI and EEG analysis	Effective for detecting frequency patterns
Mexican Hat	Second derivative of Gaussian and localized frequency analysis	Object detection and edge detection in AI vision systems	Strong edge detection and pattern recognition
Gabor	Time-frequency localized and complex-valued	Facial recognition and biometric AI applications	Highly effective in texture and pattern recognition

Wavelet transforms play a crucial role in AI and ML, providing powerful tools for feature extraction, denoising, compression, and anomaly detection in computer vision, biomedical engineering, financial analysis, and industrial AI applications [32].

5.4 Software Engineering and Programming in DSP and Analysis

DSP is fundamental to contemporary technology, facilitating the manipulation and analysis of signals across many fields like communications, multimedia, healthcare, and scientific research. The implementation of DSP techniques is significantly dependent on proficient programming and software engineering methodologies. These domains guarantee that the algorithms are implemented efficiently, are scalable, and can manage extensive data with minimal latency. This debate examines the essential function of programming and software engineering in the development, optimization, and real-time processing of DSP systems. We will emphasize essential programming languages, software frameworks, design patterns, optimization techniques, and practical applications of DSP [34].

Principal Programming Languages in DSP: In DSP, programming languages are essential for executing the mathematical models and algorithms required for signal analysis. The selection of a programming language is contingent upon the particular application, performance criteria, and hardware limitations [25]. The predominant programming languages utilized for DSP encompass C, C++, MATLAB, Python, and Java. Each possesses distinct strengths and weaknesses, rendering them appropriate for various facets of DSP.

C and C++ are the predominant programming languages in DSP owing to their rapid execution speed, low-level memory management, and efficient processing skills. These languages are widely utilized in real-time signal processing, embedded devices, and hardware-accelerated applications [11]. C facilitates direct manipulation of hardware resources, crucial for enhancing performance in real-time applications, whereas C++ incorporates object-oriented programming (OOP) characteristics that encourage modularity and code reusability. Direct memory administration and low-level regulation. Rapid performance and efficacy, appropriate for real-time processing. Extensive compatibility with embedded systems and hardware interfaces. Demands meticulous oversight of resources and memory. Intricate syntax and a challenging learning curve relative to higher-level languages

MATLAB is a sophisticated programming language and environment tailored for numerical computing, matrix manipulation, and data visualization. It offers an extensive array of integrated functions for executing standard DSP operations, such as filtering, Fourier transforms, spectrum analysis, and signal reconstruction. MATLAB is extensively utilized in academic research, algorithm development, and system simulations. Extensive repository of inherent functions for DSP [15, 36]. Intuitive syntax for mathematical calculations. Robust visualization instruments for signal analysis and troubleshooting. Reduced execution speed relative to C and C++. High license costs and restricted support for embedded systems. Inappropriate for real-time processing.

Python has become a prominent language in DSP, especially for research and prototyping purposes. Python provides a versatile and robust framework for signal processing applications through libraries like as NumPy, SciPy, Librosa, and PyAudio. It is especially beneficial for activities related to ML, data analysis, and computational signal processing. Nonetheless, Python's interpreted characteristic renders it slower than compiled languages like C and C++. Accessible for acquisition and utilization, characterized by uncomplicated syntax. Comprehensive libraries for scientific computation and signal analysis. Ideal for fast prototyping and experimentation Reduced execution performance compared to compiled languages. Restricted real-time functionalities for performance-sensitive applications [17, 19].

Java is extensively utilized for developing large-scale DSP applications that necessitate portability, scalability, and integration with other software systems. Java is utilized in domains such as multimedia, telecommunications, and signal processing applications that necessitate cross-platform interoperability and network connectivity. The object-oriented design concepts of Java render it appropriate for extensive, intricate signal processing systems.

5.4.1 Software Engineering in DSP

Programming establishes the basis for executing DSP algorithms, whereas software engineering principles guarantee that DSP systems are resilient, maintainable, and scalable. Software engineering in DSP encompasses meticulous system design, modularity, performance enhancement, real-time processing, and debugging. Adhering to systematic protocols during the development lifecycle is crucial to ensure that the signal processing system satisfies performance criteria and successfully manages edge cases [4].

Modular Design and Architecture: DSP systems are generally comprised of multiple modules that execute distinct phases of signal processing. The steps may encompass filtering, transformation, feature extraction, signal reconstruction, and analysis. A modular methodology for DSP development guarantees that each component may be created, tested, and optimized independently. Software engineering methodologies, including OOP and model-view-controller design patterns are commonly employed to organize DSP software for enhanced structure, reusability, and maintainability [8].

Real-Time Processing: Real-time DSP is essential in numerous applications, including audio and video processing, telecommunications, and control systems. Real-time processing necessitates that the system executes computations under stringent temporal limitations [18, 29]. Software engineering methodologies like as multithreading, interrupt management, and the implementation of real-time operating systems are frequently employed to manage concurrent processing and guarantee that DSP operations are executed within specified time constraints.

Optimization and Performance Tuning: Optimization is an essential component in DSP system development. It guarantees the efficient execution of DSP algorithms, particularly in resource-limited contexts such as embedded systems or mobile devices. Performance tuning entails enhancing code for efficiency in speed, memory consumption, and computational complexity. Various strategies are employed to do this, such as algorithmic optimization (e.g., FFT versus DFT), memory management (e.g., reducing dynamic memory allocations), and hardware acceleration (e.g., utilizing GPUs or specialized DSP chips) [20].

Testing and debugging are essential due to the complexity of DSP algorithms, ensuring the system operates as intended. Software engineers employ methodologies including unit testing, integration testing, and simulation-based testing to verify DSP algorithms and systems. Tools such as MATLAB's debugger, GDB (GNU Debugger), and Valgrind are frequently employed for debugging and performance profiling.

Parallel and distributed computing: The rising demand for real-time processing and the expanding scale of data sets necessitate the implementation of parallel and distributed computing techniques in contemporary DSP systems. GPU programming with frameworks such as CUDA and parallel computing via OpenCL enable the concurrent execution of DSP algorithms across several processors, markedly enhancing processing performance. In extensive systems, distributed computing facilitates the allocation of signal processing jobs among numerous processors, thereby diminishing processing duration and enhancing scalability [21, 42].

5.4.2 Enhancement of DSP Algorithms

In DSP applications, optimizing algorithms is crucial for achieving peak system efficiency. Optimization approaches can be classified into three categories: algorithmic optimization, hardware optimization, and memory optimization.

Algorithmic Optimization: The initial phase in optimizing DSP algorithms is selecting the most efficient method for the specific task. The FFT is favored over the DFT for Fourier analysis because of its reduced computing complexity. Additional optimizations may encompass the application of windowing techniques in filter design or the employment of adaptive algorithms for real-time filtering and noise attenuation [3].

Hardware Optimization: Executing DSP algorithms on dedicated hardware like DSP chips or field-programmable gate arrays (FPGAs) can result in substantial performance enhancements. DSP processors are engineered to meet the substantial computing requirements of signal processing activities, offering expedited execution and reduced power consumption. FPGAs provide bespoke hardware implementations of signal processing algorithms, offering enhanced flexibility and optimization possibilities [2].

Memory optimization is essential in real-time DSP applications that need low latency and high throughput. Methods including memory blocking, cache optimization, and data compression are utilized to decrease memory access durations and diminish the overall memory footprint of DSP algorithms.

5.4.3 Software Tools and Frameworks for DSP

Numerous software tools and frameworks exist to facilitate the construction of DSP algorithms. These programs offer libraries, functions, and pre-constructed components for executing signal processing operations, including filtering, transformation, and feature extraction [16, 20]. Table 5.5 summarizes several prominent tools and frameworks utilized in DSP development.

5.4.4 Utilization of DSP in Practical Systems

Programming and software engineering methodologies are essential in the creation of DSP systems for various practical applications. Key domains in which DSP is significant include:

Audio and speech processing: DSP methods are employed in audio compression, noise reduction, echo cancellation, and speech recognition. Applications encompass voice assistants (e.g., Siri and Alexa), music streaming platforms, and auditory devices [21].

Table 5.5: Uses of software tools in DSP.

Tool/ framework	Description	Usage
MATLAB	A high-level programming language and environment for numerical computation and visualization.	Algorithm development, simulation, signal processing tasks.
Simulink	A graphical environment integrated with MATLAB for modeling and simulating dynamic systems.	Model-based design and simulation of DSP systems.
GNU Octave	A free, open-source alternative to MATLAB for numerical computation.	Signal processing and algorithm testing in an open-source environment.
SciPy (Python)	A library for scientific and numerical computations in Python.	Signal processing, optimization, and system analysis.
FFTW	A C library for computing the FFT.	Efficient computation of FFTs in signal processing applications.
CUDA/ OpenCL	Parallel computing frameworks that allow for GPU programming.	Parallel signal processing, real-time image and audio processing.

DSP techniques are crucial for picture enhancement, object identification, video compression, and image filtering. Applications encompass medical imaging, video surveillance, and multimedia information processing [8]. DSP extensively performs functions in telecommunications, including error detection, modulation, demodulation, and signal filtering. Mobile networks, satellite communications, and data transmission systems depend on effective DSP algorithms for maximum performance.

Neurological signal processing: DSP is used in the medical field to study electrocardiograms (ECGs), electroencephalograms (EEGs), and other biomedical data. These algorithms facilitate the diagnosis of medical disorders and the monitoring of patient health [24].

Programming and software engineering are essential for the development and execution of DSP algorithms and systems. The selection of programming languages, optimization methods, and software engineering approaches significantly impacts the performance and scalability of DSP systems [66]. The ongoing evolution of DSP will be augmented by the incorporation of ML, parallel computing, and sophisticated hardware accelerators, which will enhance the functionality of DSP systems. By carefully planning, implementing, and improving their work, DSP engineers can create reliable and effective systems that can keep up with the growing needs of modern technology in fields like healthcare, scientific research, telecommunications, and multimedia.

5.5 DSP: Techniques, Algorithms, and Library Development in Signal Analysis

DSP is essential in contemporary technology, facilitating the manipulation of digital signals to extract, improve, and analyze information across diverse applications [68]. The advancement of DSP has been propelled by the demand for efficient algorithms and resilient libraries that facilitate processing and foster creativity. This discourse examines DSP techniques, emphasizes essential algorithms, and delineates the procedure for constructing a comprehensive DSP library.

5.5.1 An Overview of DSP

DSP involves the analysis and manipulation of signals in digital representation. Signals, like audio, video, or sensor data, frequently encompass noise, redundancy, or intricate patterns. DSP techniques are designed to eliminate noise, compress data, or extract significant features, hence enhancing the interpretability and utility of signals. Applications encompass telecommunications, medical imaging, multimedia processing, and more, demonstrating the adaptability of DSP [60].

Fundamental techniques in DSP: DSP techniques constitute the foundation of signal manipulation. These methods utilize mathematical tools to accomplish tasks such as filtering, transformation, and data analysis:

Time-Domain Approaches: Time-domain approaches examine signals in their inherent temporal representation. They concentrate on characteristics like as amplitude, phase, and temporal fluctuations. Convolution and correlation techniques are essential for tasks like as noise elimination, signal smoothing, and event identification [57].

Methods in the Frequency Domain: Frequency-domain approaches employ transformations such as the Fourier Transform to depict signals based on their frequency components. This is especially beneficial for detecting periodicities, filtering designated frequencies, and visualizing spectrum characteristics.

Adaptive Processing Techniques: Adaptive techniques alter their parameters in real-time, responding to fluctuating signal conditions. These are often employed in noise cancellation, equalization, and predictive modeling [31].

Statistical Techniques: Statistical methods, such as mean, variance, and autocorrelation, elucidate signal characteristics. They are essential for pattern recognition, anomaly detection, and data compression [55].

5.5.2 Algorithms in DSP

Algorithms convert mathematical techniques into computing processes, facilitating effective signal analysis and processing [2, 8]. Among the most prevalent DSP algorithms are:

The FFT is fundamental to DSP, offering a fast method for calculating the DFT. It decreases computational complexity and is essential in audio, communication, and imaging systems. FIR filters are linear and intrinsically stable, rendering them suitable for applications such as signal equalization and echo cancellation. Their fixed-length response guarantees consistent behavior. IIR filters provide computational efficiency through the utilization of feedback techniques. These are utilized in applications necessitating rapid processing, such as real-time audio filtration [64].

The wavelet transform offers a time-frequency representation of signals, rendering it appropriate for the analysis of non-stationary data. It is utilized in picture compression, seismic analysis, and biological signal processing [57]. PCA decreases the dimensionality of signals by identifying primary components. It is widely utilized for feature extraction and noise attenuation.

Kalman and Particle Filters: These filters are effective instruments for signal estimation and forecasting in dynamic systems, including navigation and tracking [42].

5.5.3 Constructing a DSP Library: Procedures and Elements

Developing a comprehensive DSP library necessitates the methodical arrangement of algorithms and tools inside a reusable framework. An effectively built library improves productivity, streamlines intricate operations, and guarantees scalability [44, 52]. The following are the fundamental steps and elements involved:

Establish the Scope: Determine the essential functionalities to incorporate in the library, including filtering, transformations, spectral analysis, and statistical computations.

Select the Programming Language: Choose a language according to performance specifications and intended applications. C++ and Python are prevalent selections for high-performance requirements. Python, augmented with NumPy and SciPy, provides exceptional usability and a vast array of libraries.

Modular Design: Structure the library into modules, each assigned to certain functions such as time-domain processing, frequency-domain transformations, and visualization.

Optimization: Employ efficient algorithms to guarantee the library manages extensive datasets and real-time processing requirements. Methods such as parallel processing and GPU acceleration can markedly improve performance.

Conduct thorough testing of each component utilizing benchmark datasets and edge cases to guarantee reliability and resilience.

Documentation and User Guides: Furnish comprehensive documentation that includes theoretical foundations, practical usage examples, and visual representations to facilitate efficient user adoption of the library.

5.5.4 Utilization of DSP Libraries

DSP libraries are essential across multiple sectors. Their applications encompass:

Telecommunications: Modulation, error correction, and spectral analysis.

Healthcare: Analysis of ECG, EEG, and medical imaging.

Entertainment: Audio equalization, visual improvement, and multimedia compression.

Scientific Inquiry: Signal analysis in physics, biology, and environmental sciences [4, 44].

5.5.5 Comparative Analysis: DSP Techniques and Algorithms

Table 5.6 compares essential DSP approaches and algorithms according to their characteristics, applications, and obstacles.

Table 5.6: Comparison of DSP schemes.

Method/ algorithm	Description	Applications	Advantages	Challenges
Fourier transform (FT)	Transforms signals between time and frequency domains	Spectral analysis and communications	Comprehensive frequency representation	Computationally intensive for large data
FIR filters	Filters with finite impulse response	Audio filtering and echo removal	Stability and linear phase response	Requires higher computational power
IIR filters	Filters with infinite impulse response	Feedback systems and real-time filtering	Computationally efficient	Sensitive to numerical errors
Wavelet transform	Analyzes signals in time and frequency domains simultaneously	Image compression and seismic analysis	Ideal for non-stationary signals	Complex implementation

Table 5.6 (continued)

Method/ algorithm	Description	Applications	Advantages	Challenges
PCA	Reduces dimensionality by identifying principal components	Noise reduction and feature extraction	Simplifies data representation	Sensitive to noise and scaling
Kalman filters	Provides optimal signal estimation in dynamic systems	Navigation and tracking	Accurate predictions in real time	High computational complexity

DSP techniques and algorithms are the cornerstone of signal processing and analysis, facilitating significant progress across various domains. Engineers and researchers can enhance their workflows, utilize efficient tools, and concentrate on innovation by developing reusable DSP libraries. The ongoing evolution of technology will augment DSP capabilities through the integration of ML, parallel computing, and adaptive algorithms, hence maintaining its significance in the ever-changing digital landscape.

5.5.6 DSP-AI Real-Time Applications and Future Research Directions

DSP improves signal quality and enables AI/ML-driven decision-making in many industries. Speech recognition uses DSP techniques like FFT and MFCCs to extract key audio elements. In Google Assistant and Apple's Siri, DSP converts raw speech signals into frequency components that deep learning models process for effective speech-to-text conversion [21]. In biological signal processing, DSP is widely employed in ECG analysis. Muscle movements and external disturbance cause noise in raw ECG readings. Wavelet transforms and adaptive filtering eliminates noise while maintaining key features [39, 47]. The MIT-BIH Arrhythmia Database used DSP to prepare ECG signals before sending them to deep learning models to help diagnose heart disease. This made the diagnostic process more accurate. DSP is also important for image processing [47, 48]. Before AI models diagnose tumors or anomalies in MRI and CT scans, edge detection and contrast enhancements are used. AI algorithms can better distinguish benign from malignant tissues in DSP-enhanced mammograms for breast cancer screening. Before AI-based classification systems can find changes in land use, loss of trees, or urban growth, DSP techniques like noise reduction and resolution enhancement are used to look at satellite images.

DSP is also used in industrial automation and predictive maintenance. Many companies monitor machinery health with vibration signal analysis. AI algorithms can forecast mechanical breakdowns using Fourier and wavelet transforms to extract frequency patterns from vibration inputs. DSP-driven AI systems monitor turbines and manufacturing equipment in real time, decreasing downtime and maintenance costs for General Electric and Siemens. In wireless communications, DSP improves signal transmission. 5G data integrity is improved by error-corrective coding and modulation optimization [40, 58]. Qualcomm uses DSP algorithms in its mobile chipsets to reduce noise and analyze signals in wireless communication systems while maintaining smartphone connectivity. DSP improves AI/ML applications in healthcare, speech recognition, image processing, industrial automation, and wireless communication, making real-world systems more efficient, accurate, and dependable. Table 5.7 represents the DSP-related pseudocodes and their implementation details for practical applications in speech processing, biomedical signal processing, and wireless communications.

Table 5.7: Sample pseducodes for DSP-related applications.

Application scenario	Sample pseudocode
1. Real-time noise filtering using adaptive LMS algorithm	1. Initialize filter weights *W* with small random values 2. Set step size (learning rate) μ 3. For each time step *t*: a. Take the input signal *X*(*t*) (past *N* samples) b. Compute output $Y(t) = W * X(t)$ c. Calculate error $E(t) = \text{Desired}(t) - Y(t)$ d. Update weights $W = W + \mu * E(t) * X(t)$ 4. Return filtered output and updated weights
2. Speech feature extraction using Mel-frequency cepstral coefficients	1. Load the speech signal 2. Apply pre-emphasis filter to enhance high frequencies 3. Perform short-time Fourier transform (STFT) 4. Convert the power spectrum into Mel scale $f_{\text{mel}} = 2,595\ \log_{10}(1 + f/700)$ 5. Compute the discrete cosine transform (DCT) to obtain MFCCs $cn = \sum_{m=1}^{M} \log(E_m) \cos[n(m - 0.5)\pi/M]$ 6. Return extracted MFCC features
3. Real-time edge detection in images using DSP	1. Read input image 2. Convert to grayscale 3. Apply Gaussian smoothing to reduce noise 4. Compute gradient magnitude using Sobel filters 5. Apply thresholding to detect edges 6. Display the edge-detected image

DSP and AI are driving cutting-edge research in several sectors. An AI-driven adaptive DSP, in which deep learning models optimize filtering, noise reduction, and feature extraction in real time, appears promising. Neural-enhanced compression and denoising, where AI models boost speech and image processing DSP approaches, is another key advancement. DNNs are revolutionizing low-bandwidth communication, audio, and MRI scanner quality. Wireless researchers are exploring AI-powered DSP solutions for 6G networks, including intelligent beamforming and predictive error correction. As AI-accelerated edge DSP gains popularity, IoT devices can process in real time. Self-learning DSP architectures use reinforcement learning to create adaptive signal processing systems that work best in self-driving cars and quantum computing.

5.6 Conclusion

DSP is fundamental to the progression of AI and ML, providing an extensive array of tools for the analysis and interpretation of intricate data. It is easy to go from the time domain to the frequency domain quickly with the Fourier transform, especially the FFT. This is important for feature extraction in AI applications. In the same way, the Z-transform and the Laplace transform make it easier to study systems and check their stability, which helps with the creation of adaptive algorithms. Convolution is an important step in DSP and is also a key part of CNNs, which allow for feature recognition in signal and image analysis. Correlation methods improve AI by making it easier to find patterns and match features, which is important for ML algorithms to work well. Linear algebra is fundamental to neural network functions, including techniques such as PCA and SVD to improve dimensionality reduction and data compression. Moreover, probability and statistics from DSP tackle uncertainties in AI models, enhancing algorithms and improving prediction accuracy. Time-frequency analysis distinguishes the wavelet transform, making it suitable for non-stationary signals like audio and speech. Optimization methods, such as least squares and gradient descent, enhance AI model parameters and reduce errors, guaranteeing superior performance results. Also, DSP-based filter design methods like FIR filters clean up and process data before it is sent to AI systems, which makes them more reliable and of higher quality. By combining these mathematical tools with AI and ML, DSP becomes more efficient at computing and smart systems can better understand and solve problems in the real world, making it a key factor in the progress of AI.

References

[1] Blackledge JM. Digital Signal Processing: Mathematical and Computational Methods, Software Development and Applications. Elsevier; 2006 Mar 1.

[2] Tan L, Jiang J. Digital Signal Processing: Fundamentals and Applications. Academic press; 2018 Oct 2.

[3] Damelin SB, Miller W. The Mathematics of Signal Processing. Cambridge University Press; 2012.

[4] Zhou J, Wang P, Wang R, Niu Z, Zhang K. Signal processing method of ultrasonic gas flowmeter based on transit-time mathematical characteristics. Measurement. 2025 Jan 15;239:115485.

[5] Brandt A. Noise and Vibration Analysis: Signal Analysis and Experimental Procedures. John Wiley & Sons; 2023 Jun 27.

[6] Jiji W, Sreeja BS, Sureshkumar K, Joshitha C, Nandhini VI. Design of a novel rf mems dual and quad output switches for satellite payload applications. International Journal of Applied Engineering Research. 2015;10:34339–45.

[7] Akan A, Chaparro LF. Signals and Systems Using MATLAB®. Elsevier; 2024 Mar 22.

[8] Vadlamudi B, Anuradha T. Review of islanding detection using advanced signal processing techniques. Electrical Engineering. 2024 Feb;106(1):181–202.

[9] Bhardwaj R, Dutta PK, Raj P, Kumar A, Saini K, González Briones A, Kaabar MK, (eds). Hybrid Information Systems: Non-Linear Optimization Strategies with Artificial Intelligence. De Gruyter; 2024 Jul 8.

[10] Kruthika SG, Nagavi TC, Mahesha P, Kumar A. Voice comparison using acoustic analysis and generative adversarial network for forensics. Communicated to International Journal of Image, Graphics and Signal Processing (IJIGSP). 2023.

[11] Gougam, F., Afia, A., Soualhi, A., Touzout, W., Rahmoune, C., & Benazzouz, D. Bearing faults classification using a new approach of signal processing combined with machine learning algorithms. Journal of the Brazilian Society of Mechanical Sciences and Engineering. 2024;46(2):65.

[12] Zhang, Y., Khanduri, P., Tsaknakis, I., Yao, Y., Hong, M., & Liu, S. An introduction to bilevel optimization: Foundations and applications in signal processing and machine learning. IEEE Signal Processing Magazine. 2024;41(1):38–59.

[13] Karthikeyan, V., & Priyadharsini, S. S. Text-independent voiceprint recognition via compact embedding of dilated deep convolutional neural networks. Computers and Electrical Engineering. 2024;118:109408.

[14] Chahlaoui, Y., Ali, A., Ahmad, J., Hussain, R., & Javed, S. Dynamical behavior of optical soliton solutions, time series and sensitivity analysis to the Schrödinger model with B eta fractional derivative. Optical and Quantum Electronics. 2024;56(4):704.

[15] Al-Haddad, L. A., Giernacki, W., Shandookh, A. A., Jaber, A. A., & Puchalski, R. Vibration signal processing for multirotor UAVs fault diagnosis: Filtering or multiresolution analysis?. Maintenance & Reliability/Eksploatacja I Niezawodność. 2024;26(1).

[16] Dabush L, Routtenberg T. Verifying the smoothness of graph signals: A graph signal processing approach. IEEE Transactions on Signal Processing. 2024 Aug 7.

[17] Velayuthapandian K, Veyilraj M, Jayakumaraj MA. An intelligent parking allocation framework for digital society 5.0. Intelligent Decision Technologies. 2024 Aug;18(3):2145–59.

[18] Subha R, Haldorai A, Ramu A. An optimal approach to enhance context aware description administration service for cloud robots in a deep learning environment. Wireless Personal Communications. 2021 Apr;117(4):3343–58.

[19] Chaturvedi A, Nadgaundi SK, Raja MK, Dubey VR, Singhal A, Gupta V. Intelligent Agents System for Vegetable Plant Disease Detection Using MDTW-LSTM Model. In 2024 3rd International Conference for Innovation in Technology (INOCON) 2024 Mar 1 (pp. 1–6). IEEE.

[20] Karuppaiah, G., Velayuthapandian, K., & Vadivel, S. R. S. IoT-driven Machine Learning Mechanisms for Healthcare Applications. In Internet of Things Enabled Machine Learning for Biomedical Applications 2024 (pp. 379–406). CRC Press.

[21] Anandaraj A, Alphonse PJ. Seizure prediction on EEG signals using feature augmentation based multi model ensemble. Recent Patents on Engineering. 2025 Jan;19(1):E011123223032.

[22] Adamovich ED, Gradov OV, Orekhov FK. Novel spectral approaches in mathematical bioacoustics based on real time analysis and modern computational mathematical techniques using digital signal processing and analog signal processing hardware. Journal of Mathematical Techniques and Computational Mathematics. 2024;3(3):1–1.

[23] Kuntoğlu M, Salur E, Gupta MK, Sarıkaya M, Pimenov DY. A state-of-the-art review on sensors and signal processing systems in mechanical machining processes. International Journal of Advanced Manufacturing Technology. 2021;116(9):2711–35.

[24] Yan J. An empirical study on improving mathematics application skills of engineering students using MATLAB tools. Applied Mathematics and Nonlinear Sciences. 9(1).

[25] Smirnov D, Palahina E, Palahin V. Mathematical Modeling of Signal Detection in Non-gaussian Correlated Noise. In Proceedings of the International Conference on Smart Technologies in Urban Engineering. 2022 (pp. 65–74). Cham: Springer.
[26] Akujuobi CM. Wavelets and Wavelet Transform Systems and Their Applications. Berlin: Springer; 2022.
[27] Dong Y, Lin L, Ni H, Wang J. Infinite quantum signal processing. Quantum. 2024;8:1558.
[28] Pepin B, Kock ZJ. Students' use of resources in a challenge-based learning context involving mathematics. International Journal of Research in Undergraduate Mathematics Education. 2021;7 (2):306–27.
[29] Vieira LC, Costa RS, Valério D. An overview of mathematical modelling in cancer research: Fractional calculus as modelling tool. Fractal and Fractional. 2023;7(8):595.
[30] Guo T, Zhang T, Lim E, Lopez-Benitez M, Ma F, Yu L. A review of wavelet analysis and its applications: Challenges and opportunities. IEEE Access. 2022;10:58869–903.
[31] Khan M, Hasnain SK, Jamil M. Digital Signal Processing: A Breadth-first Approach. River Publishers; 2022.
[32] Vetterli M, Kovačević J, Goyal VK. Foundations of Signal Processing. Cambridge: Cambridge University Press; 2014.
[33] Weeks M. Digital Signal Processing Using MATLAB & Wavelets. Jones & Bartlett Publishers; 2010.
[34] Klingspor M. Hilbert transform: Mathematical theory and applications to signal processing. 2015.
[35] Spagnolini U. Statistical Signal Processing in Engineering. Wiley; 2018.
[36] Maragos P, Schafer RW, Butt MA, editors. Mathematical Morphology and Its Applications to Image and Signal Processing Vol. 5. Springer; 2012.
[37] Rainieri C, Fabbrocino G. Mathematical tools for random data analysis. Operational Modal Analysis of Civil Engineering Structures. 2014;23–57.
[38] Ivancevic Vg, Ivancevic TT. Mathematical and Computational Tools. Quantum Neural Computation. 2010;559–839.
[39] Smith S. Digital Signal Processing: A Practical Guide for Engineers and Scientists. Newnes; 2003.
[40] Cano C. Mathematical Tools and Signal Processing Algorithms for the Study of Gravitational Waves Polarization [Dissertation]. Université Grenoble Alpes; 2022.
[41] Goutsias J, Vincent L, Bloomberg DS, editors. Mathematical Morphology and Its Applications to Image and Signal Processing Vol. 18. Springer; 2000.
[42] Kobayashi H, Mark BL, Turin W. Probability, Random Processes, and Statistical Analysis. Cambridge University Press; 2011.
[43] Van Drongelen W. Signal Processing for Neuroscientists. Academic Press; 2018.
[44] Yan R, Gao RX. Wavelet transform: A mathematical tool for non-stationary signal processing. IEEE Instrumentation & Measurement Magazine. 2009;12(5):35–44.
[45] Peled A, Liu B. Digital Signal Processing: Theory, Design, and Implementation. New York; 1976.
[46] Hong D, Qin SY, Zhang FZ. Mathematical tools and statistical techniques for proteomic data mining. Computer Science. 2010;5(2):123–40.
[47] Manolakis DG, Ingle VK. Applied Digital Signal Processing: Theory and Practice. Cambridge University Press; 2011.
[48] Sharma P, Ghosh G. Mathematical Modeling of Real-Time Signal Processing in IoE. In Role of Internet of Everything, VLSI Architecture, and AI in Real-Time Systems 2024 (pp. 19).
[49] Morais DH. Mathematical Tools for Digital Transmission Analysis. In Key 5G Physical Layer Technologies 2020 (pp. 23–40).
[50] Yousefi MI, Kschischang FR. Information transmission using the nonlinear Fourier transform, part I: Mathematical tools. IEEE Transactions on Information Theory. 2014;60(7):4312–28.

[51] Velayuthapandian K, Karuppiah G, Vadivel SRS, Joseph DRV. Mammogram Data Analysis: Trends, Challenges, and Future Directions. In Comput Intell Model Tech Dis Detect Mammogram Images 2024 (pp.1–38). Academic Press.
[52] Shenoi BA. Introduction to Digital Signal Processing and Filter Design. Wiley; 2005.
[53] Mary D, Flamary R, Theys C, Aime C. Mathematical tools for instrumentation & signal processing in astronomy. EAS Publications Series. 2016;78(79):1–2.
[54] Diniz PS, Da Silva EA, Netto SL. Digital Signal Processing: System Analysis and Design. Cambridge University Press; 2010.
[55] Belkić D, Belkić K. The general concept of signal–noise separation (SNS): Mathematical aspects and implementation in magnetic resonance spectroscopy. Journal of Mathematical Chemistry. 2009;45 (3):563–97.
[56] Hramov AE, Koronovskii AA, Makarov VA, Maksimenko VA, Pavlov AN, Sitnikova E. Mathematical Methods of Signal Processing in Neuroscience. In Wavelets in Neuroscience. 2021 (pp. 1–13). Cham: Springer.
[57] Allen RL, Mills D. Signal Analysis: Time, Frequency, Scale, and Structure. Wiley; 2004.
[58] Lyakhov P. special issue on mathematics and digital signal processing. Applied Science. 2022;12 (18):9033.
[59] Singh D, Singh M, Hakimjon Z. Signal Processing Applications Using Multidimensional Polynomial Splines. Springer; 2019.
[60] Velayuthapandian K, Murugan N, Paramasivan S. End-to-End CNN conceptual model for a biometric authentication mechanism for ATM machines. Discovery of Electron. 2024;1(1):26.
[61] Di Lorenzo P, Barbarossa S. Swarming algorithms for distributed radio resource allocation. IEEE Signal Processing Society. 2013;30(3):144–54.
[62] Xylouris K, Queisser G, Wittum G. A three-dimensional mathematical model of active signal processing in axons. Journal of Computer Vision. 2010;13:409–18.
[63] Fanton JP. Convolution and deconvolution: Two mathematical tools to help performing tests in research and industry. International Journal of Metrology and Quality Engineering. 2021;12:6.
[64] Luo ZQ. Applications of convex optimization in signal processing and digital communication. Math Program. 2003;97(1):177–207.
[65] Girao PS, Postolache O, Pereira JM. Data fusion, decision-making, and risk analysis: Mathematical tools and techniques. Data Modeling for Metrology and Testing in Measurement Science. 2009;1–50.
[66] Ifeachor EC, Jervis BW. Digital Signal Processing: A Practical Approach. Pearson Education; 2002.
[67] Hu A, Xiang L. Selection principle of mathematical morphological operators in vibration signal processing. Journal of Vibration and Control. 2016;22(14):3157–68.
[68] Weiss S. Mathematical tools for processing broadband multi-sensor signals. 7th International Conference on Multimedia and Human-Computer Interaction. 2020 Aug;MHCI–104.

Fateh Bahadur Kunwar, Hitendra Singh, Rakesh Kumar Yadav,
Vikas Misra, Nitin Tripathi, Awaneesh Jee Srivastava

6 The Role of Differential Equation in Digital Technologies

Abstract: In today's digital era, where everything is becoming smart and automated, the role of differential equations has become very important. Be it signal processing, control systems, computer vision, or machine learning algorithms – the mathematical modeling of every dynamic process is done through differential equations. In this chapter, we explored how ordinary differential equations and partial differential equations are being used in the core of digital systems. The real-time response of smart sensors, Internet of Things devices, robotic controllers, and biomedical signal analyzers depends on the computational solutions of these equations. We also saw the use of simulation tools such as MATLAB, Simulink, and Python libraries that help solve complex models numerically. We also discussed some major challenges such as computational complexity and time constraints that can be solved through hybrid modeling and future-ready platforms such as quantum computing and edge computing. This chapter builds a bridge between theory and practical implementation, where readers understand how a mathematical concept powers the foundations of the entire digital world. It creates a strong base for future research, especially for fields that seek to bring innovation in AI, automation, and intelligent systems.

Keywords: Differential equations, signal processing, machine learning, smart sensors, MATLAB, quantum computing, edge computing

6.1 Overview of Differential Equations

Differential equations are a mathematical tool that describes the pattern of change of any object. If a system is changing over time – such as temperature, voltage, or the speed of a machine – then its model can be created using differential equations. In simple words, differential equation is a formula that describes the rate of change of a quantity and how it is related to other variables [1]. In digital technologies, as systems process data every second, it is very important to predict their behavior. And differential equations are useful in this. For example, filtering of a signal, speed control of an automatic car, or temperature monitoring of a sensor [2] – all have the logic of differential equations behind them. Figure 6.1 illustrates how differential equations serve as the foundation for various technological domains, including signal processing, computer vision, machine learning (ML), control systems, and digital communications.

https://doi.org/10.1515/9783112214374-006

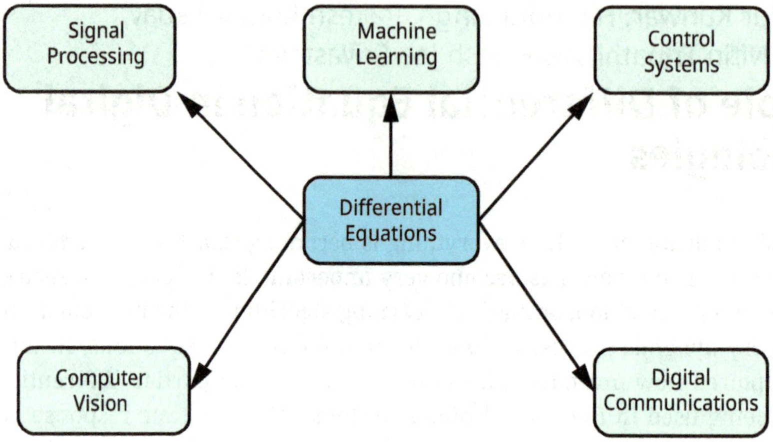

Figure 6.1: Applications of differential equations in modern digital systems.

6.1.1 Historical Evolution and Relevance in Digital Systems

Differential equations were first used in the seventeenth century by Newton and Leibniz to study mechanics and motion. Today, those concepts are an extension of modern technologies – but now these equations are not just limited to paper, but are implemented in real-time systems using high-speed computers [3]. Their use in digital systems became important in the twentieth century when control theory and electrical engineering began to grow rapidly. Today, these tools are also used in futuristic technologies such as robotics, signal processing, AI, ML, and even quantum computing. For example, differential equations are used to accurately model a drone's motion, acceleration, wind resistance, and positioning. Similarly, wave equations are used to analyze distortion of signals in digital communication systems.

6.1.2 Chapter Objectives and Scope

The main objective of this chapter is to make readers understand that differential equations are not just theoretical concepts, but they have strong utility in every modern digital technology. We will discuss:

- How ordinary differential equation (ODEs) and partial differential equation (PDEs) are used in signal processing, control systems, and circuit analysis.
- How operations such as edge detection, blurring, and sharpening are performed using differential equations in computer vision and image processing.
- What is their role in new paradigms of ML such as neural differential equations and optimization methods?

- How models of modulation, noise filtering, and data compression are derived from differential equations in digital communication systems.

This chapter will familiarize readers with theoretical knowledge as well as real-world examples and the use of computational tools (like MATLAB, Python, Simulink). Its scope is vast – from basic principles to high-level applications. It will also be explained how numerical techniques such as finite difference method (FDM) and finite element method are used to solve differential equations, especially when the exact solution is not found.

6.2 Fundamentals of Differential Equations

Differential equations are of two types: ODEs and PDEs [4]. ODEs are equations that give the derivative of one or more dependent variables with respect to an independent variable (such as time). For example, the voltage change of a capacitor is an ODE with respect to time.

Example (First-Order ODE):

$$dy/dt + 3y = 0 \tag{6.1}$$

PDEs are used in cases where there are more than one independent variables – such as time and space. Such as the heat equation or the wave equation that describes the distribution of temperature in a metal rod.

Example (Heat Equation):

$$\partial u \partial t = a\partial^2 u/\partial x^2 \tag{6.2}$$

Use in Digital Tech:

- ODEs are mostly used in control systems, circuits, and signal filtering.
- PDEs are useful in graphics processing, image smoothing, and electromagnetic modeling.

6.2.1 Initial and Boundary Conditions

The solution to a differential equation is meaningful only if initial conditions or boundary conditions are defined [5].

Initial Conditions: When the starting value of the system is known (e.g., y(0) = 5), we can uniquely determine the solution.

◆**Example** If $y(t)$ is solution of

$$dy/dt + 2y = 4 \tag{6.3}$$

Boundary Conditions: When the values at the endpoints of a system are fixed, such as the temperature at both ends of a wire.

Example For the PDE:

$$\partial u/\partial t = \partial^2 u/\partial x^2 \tag{6.4}$$

Boundary:

$$u(0,t) = 0, \;\; u(L,t) = 0 \tag{6.5}$$

Initial Condition:

$$u(x,0) = \sin(\pi x) \tag{6.6}$$

These conditions are very important in digital simulations such as autonomous driving systems or modeling the response of temperature sensors.

Common Solution Methods

1. **Analytical Method (Exact Solutions):**
   - **Separation of Variables:**

$$\begin{aligned} dy/dx &= ky \\ dy/y &= kdx \\ y &= Ce^{kx} \end{aligned} \tag{6.7}$$

   - **Integrating Factor (Linear First-Order ODEs):**
     For:

$$y/dx + P(x)y = Q(x) \tag{6.8}$$

$$\text{Integrating factor} = \mu(x) = e^{\int P(x)dx} \tag{6.9}$$

2. **Numerical Methods (Computer-Based Solutions):**
   - **Euler's Method:**

$$y_{n+1} = y_n + h\, f(x_n, y_n) \tag{6.10}$$

   - **Runge-Kutta (Fourth Order):**
     Widely used in simulations due to better accuracy.

Digital Use:
Simulation of control systems, signal processing filters, and AI training dynamics rely on numerical solvers using these equations.

Linear Versus Nonlinear Equations
Linear Equation:

$$d^2y/dt^2 + 5dy/dt + 6y = 0 \tag{6.11}$$

Here, y and its derivatives appear linearly (no power >1 or multiplication among them).

Nonlinear Equation:

$$dy/dt = y^2 - \cos(y) \tag{6.12}$$

This is non-linear because of y2y^2y2 and cos(y)\cos(y)cos(y).

Real-World Relevance:
- **Linear:** Easy to simulate, fast computation (useful in control systems, filtering, etc.)
- **Nonlinear:** More realistic but complex (used in robotics, chaos modeling, AI dynamics, etc.)

6.3 Modeling Digital Technologies with Differential Equations

The biggest advantage of differential equations is that they help to mathematically model any changing system. When we talk about digital technologies, many things are dynamic – such as signals, voltages, currents, machine speeds, etc. Differential equations are used to accurately describe the behavior of all these.

6.3.1 Signal Processing and Filter Design

The main task of signal processing is to clean, modify, or extract the signal. This process uses filters, such as low-pass filter (LPF) and high-pass filter (HPF), which remove unwanted frequencies [6]. These filters are defined through differential equations.

Example: First-Order LPF
Equation:

$$RCdt/dy(t)\ +y(t)=x(t) \tag{6.13}$$

- $x(t)$: Input signal
- $y(t)$: Output signal
- RC: Time constant of the circuit

This equation shows how the output y(t) changes with the input x (t). This equation is used to discretize algorithms to create digital filters [7], which are implemented by computers and DSPs (digital signal processors).

Real-World Use: Audio filtering, ECG signal processing, noise removal from Internet of Things (IoT) devices, image smoothing, etc.

6.3.2 Circuit Analysis and RLC Networks

Electrical circuits – especially those containing Resistor (R), Inductor (L), and Capacitor (C) – also require differential equations to solve [8]. The behavior of RLC circuits is completely defined by their differential equations.

Example: Series RLC Circuit
Equation:

$$Ldt^{2/}d^{2}q+Rdt/dq+q/C=V(t) \tag{6.14}$$

where $q(t)$ is the charge on capacitor; $V(t)$ is the source voltage; and R, L, and C are resistance, inductance, and capacitance, respectively.

By solving this second-order ODE we can find the response (current and voltage) of the circuit with respect to time. This concept applies to power systems, communication systems, and even chip-level design.

Use in Digital Technologies:
- Power electronics
- Signal converters
- Switching regulators
- PCB design simulation tools

6.3.3 Control Systems in Automation

Control systems are used to monitor and control automatic systems, such as robotics, industrial machines, drones, and self-driving cars [9]. The motion, position, or speed of these systems changes continuously, to accurately control which we create mathematical models – and their foundation is differential equations.

Example: DC Motor Control
Equation:

$$Jdt^2/d^2\theta + bdt/d\theta = Ki(t) \tag{6.15}$$

where θ is the angular position of motor, J is the moment of inertia, b is the damping constant, $i(t)$ is the input current, and K is the motor torque constant.

Based on this action speed controllers are designed which tones the proportional, integral and derivative components to maintain accuracy.

Use in Digital Technologies:
- Automatic washing machines
- Conveyor belts
- Smart thermostats
- Elevator motion control
- Drone navigation systems

6.4 Application in Digital Communication Systems

Digital communication systems work by transmitting information from one point to another using electronic signals [10]. This process involves many challenges – such as signal distortion, noise, delay, and data loss. Differential equations are used to model and solve all these problems. Let us understand in detail how differential equations play a role in wave propagation, modulation techniques, and error prediction and correction.

6.4.1 Wave Propagation and Transmission Lines

When digital signals travel long distances (like fiber optics, coaxial cables, or wireless mediums), their wave nature becomes important [11]. PDEs are used to model the behavior of these waves.

Telegrapher's Equation:
This is a standard model for transmission lines:

$$\partial^2 V(x,t)/\partial x^2 = LC\partial^2 V(x,t)/\partial t^2 \tag{6.16}$$

where $V(x,t)$ is the voltage as a function of position and time, L is the inductance per unit length, and C is the capacitance per unit length.

Give this equation, this is how the waveform of the signal propagates along the length of the line.

Practical Example: When we use a high-speed ethernet cable or optical fiber, the delay, distortion and attenuation of the signal can be calculated using telegrapher equations.

Use in Technologies:
- Fiber optic communication
- 5G wireless basebands
- Satellite transmission modeling

6.4.2 Modulation Techniques and Channel Modeling

To transmit digital data, it has to be modulated – that is, information has to be superimposed on a high-frequency carrier signal [12]. This modulation process is also based on mathematical models, and differential equations play a very important role.

Example: Amplitude Modulation (AM)
Modulated Signal:

$$s(t) = [1 + m(t)] \cdot \cos(2\pi f_c t) \tag{6.17}$$

where f_c is the carrier frequency and $m(t)$ is the message signal (data).

This modulation is the distortion and fading that occurs during transmission and reception of the signal, which differential equations are used to model.

Channel as a Linear System:
Most real-world communication channels are modeled as linear time-varying (LTV) systems:

$$dy(t)/d(t) + a(t)y(t) = b(t)x(t) \tag{6.18}$$

where $y(t)$ is the received output and $a(t)$, $b(t)$ is the time-varying channel parameters.

This equation allows us to estimate the effect of the channel and accordingly we equalize the signal so that the information is received accurately.

Use in Technologies:
- Cellular radio (GSM and LTE)
- Modems (ADSL and VDSL)
- Satellite and radar communication

6.4.3 Error Prediction and Correction Dynamics

When a digital signal passes through a noisy channel, errors are common. Predicting and correcting these errors [13] is essential for a communication system [14]. Differential equations are also used in this process, especially when we model probability of error or decoding mechanisms. This Q-function is also a special solution of differential equations, and the bit error rate is calculated from it.

State-space differential equations are used to model convolutional codes like forward error correction (FEC) techniques, where the evolution of data states is solved with time.

Real-Life Use:
- Error correction in 4G/5G data
- Satellite image transmission
- Internet data packets validation

6.5 Differential Equations in Computer Graphics and Vision

Computer graphics and computer vision are both important parts of modern digital technologies – whether it is animated movies, face recognition systems, or MRI image processing [15]. Many complex visual transformations are performed within these applications, which differential equations are used to accurately model. In this section we will understand how differential equations help in tasks such as edge detection, image denoising, and shape deformation/morphing.

6.5.1 Image Processing and Edge Detection

A basic and important task of image processing is edge detection – identifying boundaries and outlines of objects within an image [16]. Edges occur where there is a sudden change in the pixel intensity of an image. PDEs are used to describe this process mathematically.

Example: Laplace Operator for Edge Detection

$$\nabla^2 I = \partial^2 I/\partial x^2 + \partial^2 I/\partial y^2 \tag{6.19}$$

where (x,y) is the pixel intensity function and ∇^2 is the Laplacian operator.

This equation tells where the intensity is changing sharply. Where the change is more, the edge is detected.

Real-Life Use Cases:
- Facial recognition
- QR code scanning
- Object tracking in CCTV
- Medical image boundary analysis

6.5.2 Heat Equation for Denoising

When digital images are captured (especially in low light or at high speed), they contain noise – random bright or dark pixels [17]. Denoising techniques are applied to remove this noise. One of the most famous denoising models is based on the heat equation, which is a PDE.

Heat Diffusion Equation:

$$\partial I/\partial t = \alpha \ \left(\partial^2 I/\partial x^2 + \partial^2 I/\partial y^2\right) \tag{6.20}$$

where $I(x, y, t)$ is the image intensity over time and α is the diffusion coefficient.

The concept of this equation is that as heat spreads over a surface [18]. The pixel intensities of the image are also smoothed, and noise is automatically suppressed.

Advanced Form:
- Perona-Malik equation (nonlinear diffusion)
- Anisotropic diffusion (direction-based smoothing)

Practical Applications:
- MRI, CT scan image cleaning
- Satellite image enhancement
- Forensic photo correction

6.5.3 Shape Deformation and Morphing Techniques

In computer graphics, when an object is stretched, bent, or transformed (like the expressions of an animated character), it looks visual but its mathematical model is quite complex [19]. For this modeling, differential geometry and elasticity equations are used – which ultimately become differential equations.

Example: Elastic Deformation Model

$$\mu\nabla^2 u + (\lambda + \mu)\nabla(\nabla \cdot u) = f \tag{6.21}$$

where u is the displacement vector (how much the object is being stretched), f is the external force applied, and μ represents material properties.

From this equation, the internal stress and deformation pattern of an object can be predicted. It is used in character animation, fabric simulation, and 3D modeling.

Morphing Techniques:
Morphing means smoothly transforming one image into another (like face swap apps). Interpolation of shapes is done using partial differential flows [20], which are controlled by time-dependent equations.

Real-World Uses:
- Animation and gaming
- facial expression generation
- Digital prosthetic design
- Augmented reality applications

6.6 Use in Machine Learning and AI

Today, ML and artificial intelligence (AI) are everywhere from recommendation systems and self-driving cars to fraud detection and medical diagnosis. Underlying all of these are powerful mathematical concepts, in which differential equations play a very critical role. This section will explain how differential equations are used in modern ML/AI systems [21] especially in neural ODEs, optimization algorithms, and dynamic models of reinforcement learning (RL) [22].

6.6.1 Neural Differential Equations

In traditional neural networks, the input layer, hidden layers, and output layer are fixed, and information is processed in discrete steps. But neural differential equations (neural ODEs) introduce a new concept in which the model continuously evolves over time [23] like a physical system

Neural ODE:

$$dh(t)/dt = f(h(t), t, \theta) \tag{6.22}$$

where $h(t)$ is the hidden state at time, f is a neural network, and θ represents trainable parameters.

This means that instead of stacking multiple layers, we define a continuous transformation which the ODE solver integrates [24]. These models are especially useful when the input is time series data, in which the signal flow is naturally continuous (e.g., ECG and stock prices).

Benefits:
- Memory efficient
- Adaptive computation time
- Suitable for irregular time series data

Use Cases:
- Time series forecasting
- Physics-informed ML models
- Trajectory prediction in robotics

6.6.2 Gradient Dynamics and Optimization

Optimization techniques are used to train ML models such as gradient descent, in which we minimize the loss function by updating model weights [25]. This process follows a special case of differential equations gradient flow dynamics.

Gradient Flow Equation:

$$dw/dt = -\nabla L(w) \tag{6.23}$$

where w represents model parameters and $L(w)$ is the loss function.

This equation tells in which direction and at what speed the parameters should be changed so that the loss becomes minimum [26]. This same idea is also used in deep learning optimizers like Adam, RMSProp, and SGD.

Practical Insights:
- By visualizing optimization dynamics, we can understand the convergence behavior of the model.
- Chaotic or slow-converging problems can be improved by differential equation-based analysis.

6.6.3 Reinforcement Learning Models

In RL, an agent interacts with the environment and receives a reward based on its actions. In this process, the system changes dynamically – and its modeling occurs naturally through differential equations [27].

Dynamics of Markov Decision Process:

$$dV(s)/dt = R(s) + \gamma a \sum \pi(\,a \mid s) \sum P(s' \mid s\,, a)V(s') - V(s) \tag{6.24}$$

where *V*(s) is the value of state, $\pi(a \mid s)$ is the policy, $P(s', s, a)$ is the state transition probability, *R*(*s*) is the reward, and γ is the discount factor.

Such equations describe the learning process of the RL agent in continuous form. Especially in advanced RL techniques like model-based RL or continuous control environments, this dynamic modeling is essential.

Policy Gradient Dynamics:

$$d\theta/dt = \nabla_\theta J(\theta) \tag{6.25}$$

It optimizes the learning path of RL algorithms in which θ is the policy parameter of the agent.

Applications:
- Autonomous navigation (e.g., drones and robots)
- Game AI (AlphaGo and Chess engines)
- Smart control in industrial processes

6.7 Simulation Tools and Computational Techniques

In today's digital era, solving differential equations manually for real-world problems is not just difficult, it is almost impossible – especially when systems become complex and large-scale. That's why we use simulation tools and computational techniques to generate accurate, fast, and scalable solutions [28]. In this section we will cover:
- How finite difference and finite element methods are used
- How tools like MATLAB, Simulink, Python help in simulation
- What is the importance of stability and convergence analysis in numerical computation

6.7.1 Finite Difference and Finite Element Methods

6.7.1.1 Finite Difference Method

FDM is a simple and direct approach in which we approximate the derivatives using discrete points.

Example
Suppose:

$$dy/dx = f(x) \tag{6.26}$$

So the approximation of the derivative will be:

$$dy/dx \approx y_{i+1} - y_i/h \tag{6.27}$$

where *dy*/*dx* is the value at point x_i and *h* is the step size.

This approach is implemented using grids and meshes – and in real-world applications such as it is used in heat transfer, electrical field modeling, and image denoising.

Use Cases:
- Temperature distribution in a rod
- ECG signal smoothing
- Image blur filters

6.7.1.2 Finite Element Method (FEM)

FEM is more advanced than FDM. In this we divide the poor domain into small elements (triangles and quadrilaterals), and approximate the solution in each element using basis functions [29].

Basic Equation Form:

$$Ku = f$$

where *K* is the stiffness matrix, *u* is unknowns (e.g., displacement), and *f* is the applied force.

FEM is used when the domain is of complex shape – such as aircraft wings, human brain scan area, or mesh of mechanical components.

Use Cases:
- Biomedical imaging (e.g., tumor growth modeling)
- Structural simulation in robotics
- Fluid flow in electronic chips

6.7.2 MATLAB, Simulink, and Python Libraries

6.7.2.1 MATLAB

MATLAB is a high-level technical computing platform mainly used for numerical analysis. It has in-built functions and solvers like ode45, ode23s, pdepe etc., through which we can solve complex ODEs and PDEs.

Example Code Snippet:

$$[t,\ y] = \text{ode45}(@(t, y)\ -\ 2^{*}y + 5,\ [0\ 10],\ 1); \tag{6.28}$$

plot(t, y);

Use Cases:
- Control system modeling
- Signal processing
- System identification in engineering

6.7.2.2 Simulink

Simulink is a graphical interface of MATLAB in which we can simulate dynamic systems by drawing block diagrams. This is very useful for real-time systems such as motor control, plant models, and automation workflows.

Features:
- Drag-and-drop simulation
- Time-based differential equation solvers
- Integration with sensors and hardware

Use Cases:
- Drone flight simulation
- EV battery performance prediction
- Smart grid load balancing

◆ Python Libraries

Python is also very popular in numerical computing due to its open-source nature and vast ecosystem.

Popular Libraries:
- NumPy: matrix and numerical operations
- SciPy: ODE/PDE solvers like odeint, solve_ivp
- SymPy: symbolic math and equation solving

- Matplotlib: visualization
- FEniCS or PyTorch for advanced PDE and ML integration

Python Example:

```
from scipy.integrate import solve_ivp
def model(t, y): return – 2*y + 3
sol = solve_ivp(model, [0, 10], [1])
```

Use Cases:
- Climate prediction models
- Neural ODE implementations
- Biomedical signal filtering

6.7.3 Stability and Convergence Analysis

When we solve differential equations numerically, it is not enough to just get the answer. It is also important that:
- the solution is stable (the answer does not blow up) and
- the solution converges (gets closer to the real solution as the step size reduces).

◆ Stability

A numerical method is stable when small errors (round-off or step error) do not destroy the solution by increasing in future steps.

Example:

Euler's method can be unstable if step size h is too large:

$$y_{n+1} = y_n + hf(, y_n) \tag{6.29}$$

h is taken to a large value, the solution can oscillate or diverge.

How to Improve Stability:
- use implicit methods (e.g., Backward Euler)
- adjust step size adaptively
- use stiff solvers (like ode15s in MATLAB)

6.7.3.1 Convergence

Convergence [30] means that as we reduce the grid and step size, our numerical solution gets closer to the exact solution.

Convergence Order:
Euler's method → First order
Runge-Kutta fourth order → Fourth order (more accurate)

Convergence Check:
lim $| |y_h - y_{\text{exactly}} | | = 0$

Tools for Analysis:
- Log-log error plots
- Richardson extrapolation
- Empirical error estimation

6.8 Case Studies and Practical Implementations

When we apply theory to the real world, we understand its true value. Differential equations are a powerful tool, but their value is visible only when they are implemented in real-world systems – such as smart sensors, IoT devices, medical systems, and live monitoring platforms. In this section, we will look at some case studies and practical implementations where differential equations have made digital systems more intelligent, accurate, and responsive.

6.8.1 Smart Sensors and IoT Devices

IoT means that everything – from lights and fans to wearable devices and smart agriculture sensors – becomes part of an interconnected network. The job of these devices is to collect data from the environment, process it, and generate useful information.

Role of Differential Equations:
IoT devices measure continuously changing physical parameters such as temperature, humidity, pressure, and motion in the environment. Time-based variations of these parameters are modeled using differential equations.

Example: Temperature Sensor Model

$$dT/dt = -k(T - T_{\text{ambient}}) \tag{6.30}$$

where T is the current temperature, T_{ambient} is the ambient room temperature, and k is the heat transfer constant.

This equation is used in smart thermostat or fire detection system where real-time heat monitoring is required.

Practical Use Cases:
- Smart agriculture (soil moisture modeling)
- Industrial safety sensors (gas leak prediction)
- Smart home appliances (AC autoregulation)

6.8.2 Real-Time Systems Modeling

Real-time systems mean systems where every decision and response must be taken within a time constraint. Examples include self-driving cars, robotic arms, traffic signal controllers, etc.

Why Differential Equations?

Because the behavior of these systems continuously changes with time – and their accurate model can be created only when we describe the dynamics of the system through time-based differential equations.

◆ Example: Robotic Arm Movement

$$Id^2\theta/dt^2 + bd\theta/dt = \tau(t) \tag{6.31}$$

where θ is the angular position, I is the moment of inertia, b is the damping factor, and $\tau(t)$ is applied torque.

This equation defines the motion of a robotic joint. If the equation is solved real-time using control algorithms, the system operates with high accuracy.

Real-Time Application Examples:
- Autonomous vehicle lane detection and correction
- Conveyor belt object tracking
- Drone flight control (orientation + altitude balancing)

All of these systems also use feedback control loops, in which system output is continuously adjusted based on incoming sensor data – and these loops are also solved using differential equations.

6.8.3 Biomedical Signal Analysis

Differential equations are used everywhere in the medical field – from ECG and EEG signal analysis to tumor growth prediction. The internal systems of the human body such as the nervous system, heart activity, and respiration are dynamic – and their analysis is possible only with time series differential models.

Example: ECG Signal Filtering (Denoising)

The heat equation is used to denoise the ECG signal:

$$\partial V/\partial t = D\ \partial^2 V/\partial x^2 \tag{6.32}$$

where V is the voltage (ECG amplitude) and D is the diffusion constant.

This equation filters out high-frequency noise and generates a clean signal that can be used to detect accurate heart rate.

Example: Glucose-Level Prediction in Diabetes

$$dG/dt = -kG + u(t) \tag{6.33}$$

where G is the glucose level, k is the decay rate, and $u(t)$ is the insulin injection function.

Such models are used in wearable glucose monitors so that the patient's sugar level can be continuously tracked and emergency response can be activated.

Medical Applications:
- EEG-based seizure prediction
- Oxygen saturation modeling in ventilators
- Blood pressure analysis using dynamic models

6.9 Challenges and Future Trends

When we use differential equations in digital technologies – such as AI, robotics, healthcare, or communication systems – it is not enough to simply solve the equations. Many challenges are faced in practical implementation, and new trends and techniques are being developed to overcome these challenges. In this section, we will explain:
- What are the major limitations and bottlenecks in differential equation-based modeling?
- How hybrid approaches yield better results for complex systems?
- And what will be the integration of these models with quantum and edge computing in the future?

6.9.1 Computational Complexity and Real-Time Constraints

Accurately solving differential equations is computationally expensive – especially when the system is multi-dimensional, or involves non-linear dynamics.

6.9.1.1 Problem 1: High Computational Load

Complex systems (such as weather models, neural differential equations, biomechanics) are difficult to solve in direct analytical form. So we use numerical methods (FDM, FEM, RK4, etc.). But these methods:
- require high-resolution grids,
- have to take small step sizes, and
- have to run multiple iterations.

All of these together increase the computational load.

Example:
A 3D biomedical simulation (like brain blood flow) can take several hours or days to simulate, even on modern high-end CPUs.

6.9.1.2 Problem 2: Real-Time Response Constraint

Some applications such as autonomous driving, industrial robots, and life-supporting medical devices require decisions to be made in real-time – within milliseconds. If the differential equation solver is slow, or the system becomes stiff (meaning the solutions to the equations oscillate rapidly), delays can occur. This delay can be dangerous for a mission-critical system.

Solutions Being Explored:
- Adaptive time stepping algorithms
- Parallel computing for matrix solving
- Using GPUs instead of CPUs
- Reduced order models (ROMs)

6.9.2 Hybrid Modeling Approaches

Another emerging trend is hybrid modeling where differential equations are combined with data-driven methods or ML models.

What Is Hybrid Modeling?

This is an approach where we combine partial knowledge of the system (physics-based) with historical or real-time data to improve prediction and reduce computation.

Model Structure:
Output = Differential Model Output + ML-Based Correction

Or again:
- Some states modeled via ODE/PDE
- Some unknown functions approximated by Neural Networks

Use Cases:
- In weather prediction: fluid dynamics modeled by PDEs + cloud coverage predicted by CNN
- In biomedical engineering: heart signal dynamics modeled by ODE + patient-specific correction by deep learning

Advantages:
- Faster than pure numerical solvers
- More robust to sensor noise
- Adaptive to changes in input conditions

Tools Supporting Hybrid Models:
- TensorFlow + SciPy integration (Python)
- Physics-Informed Neural Networks (PINNs)
- MATLAB Simscape + AI toolbox

6.9.3 Integration with Quantum and Edge Computing

As data and As system complexity continues to grow, traditional computation systems (CPU/GPU) might not scale effectively. Therefore future trends are focusing on integration with Quantum Computing and Edge Computing.

6.9.3.1 Quantum Computing

Quantum computing is increasingly used when we want to solve large-scale, high-dimensional differential equations – especially those arising from quantum mechanics, cryptography, or complex physical simulations.

How It Helps:
Quantum systems can solve linear systems of equations (e.g., via HHL algorithm) much faster than classical ones.
- PDEs like Schrödinger equation or electromagnetic field modeling become efficient.
- Superposition and entanglement allow parallel solution search.

Example Use Case:
Simulating molecular behavior in drug discovery using quantum solvers for time-dependent differential equations.

Challenges:
- Still in experimental stage
- Requires special hardware (quantum processors)
- Need quantum-native algorithms

6.9.3.2 Edge Computing

Edge computing means that data processing happens on local devices (like mobile, sensor, or camera), and not on a central server or cloud. When we implement differential models at the edge, the model should be lightweight and fast.

Why Needed?
- IoT devices have limited power and memory.
- Cloud latency can affect real-time response
- Offline systems (e.g., rural healthcare monitors) need local processing

Solutions:
- Use of simplified ODE models
- Pre-trained neural differential equation models
- Pruning and quantization of simulation logic

Applications:
- Smart agriculture systems using moisture dynamics
- Wearable health monitors
- Edge robotics in disaster zones

6.10 Conclusion

Through this chapter, we understood that differential equations are not just a theoretical concept in today's digital age, but are the backbone of every dynamic system. Be it processing of signals, enhancement of images, response analysis of circuits or learning behavior of AI models – differential equations have an integral role everywhere. In this chapter, we saw how ODEs and PDEs are used to mathematically represent real-world systems. Their implementation takes place through advanced tools such as MATLAB, Simulink, and Python libraries, where numerical methods make these models efficient and scalable.

Differential equations have made a huge impact in the development of digital technology. They have made systems more accurate, automated, and adaptive. For example, response models of smart sensors, real-time robotic control systems, and biomedical monitoring devices – differential models are hidden inside them all. These equations enable digital systems to make reliable decision-making even in complex environments. Whether its drone flight control, denoising of ECG signals, or optimizing traffic signals in a smart city, differential equations are working silently everywhere.

Looking ahead, many new research avenues are opening up. Neural differential equations, when combined with deep learning, have brought a revolution for continuous-time learning systems. PINNs, which integrate physical laws and ML, are rapidly growing in fields such as medicine and climate modeling. Quantum computing will also help solve complex differential models ultra-fast in the future, especially where classical methods fail. However, through edge computing, these models can be made lightweight and deployed in real-time, offline systems (such as wearables and IoT devices).

Overall, the role of differential equations is now like an invisible force that drives every smart, autonomous, and connected system. This is a promising field for future engineers and researchers where the fusion of mathematical modeling and technology is making the world smarter and safer. Now it is necessary that we combine these models with new tools, hybrid approaches, and intelligent systems to make them even more accessible and impactful.

References

[1] Boyce WE, DiPrima RC, Meade DB Elementary Differential Equations and Boundary Value Problems. John Wiley & Sons; 2021 Oct 19.

[2] Iserles A A First Course in the Numerical Analysis of Differential Equations. Cambridge university press; 2009.

[3] Evans, LC. Partial Differential Equations (Vol. 19). American Mathematical Society; 2022.

[4] LeVeque RJ. Finite difference methods for ordinary and partial differential equations: Steady-state and time-dependent problems. Society for Industrial and Applied Mathematics; 2007.

[5] Rackauckas C, Nie Q. Differentialequations. jl–a performant and feature-rich ecosystem for solving differential equations in julia. Journal of Open Research Software. 2017 May 25;5(1):15-.

[6] Chen RT, Rubanova Y, Bettencourt J, Duvenaud DK. Neural ordinary differential equations. Advances in Neural Information Processing Systems. 2018;31.

[7] Raissi M, Perdikaris P, Karniadakis GE. Physics-informed neural networks: A deep learning framework for solving forward and inverse problems involving nonlinear partial differential equations. Journal of Computational Physics. 2019 Feb 1;378:686–707.

[8] Strikwerda JC. Finite Difference schemes and partial differential equations. SIAM. 2007;88.

[9] Mortenson ME Mathematics for Computer Graphics Applications. Industrial Press Inc.; 1999.
[10] Barbu T. Automatic edge detection solution using anisotropic diffusion-based multi-scale image analysis and fine-to-coarse tracking. Proceedings of the Romanian Academy Series A: Mathematics, Physics, Technical Sciences, Information Science. 2021 Sep;22:267–274.
[11] Aldoury RS, Al-Saidi NM, Ibrahim RW, Kahtan H. A new X-ray images enhancement method using a class of fractional differential equation. MethodsX. 2023 Dec 1;11:102264.
[12] Karniadakis G, Sherwin SJ Spectral/hp Element Methods for Computational Fluid Dynamics. Oxford University Press; 2005.
[13] Strogatz SH Nonlinear Dynamics and Chaos: With Applications to Physics, Biology, Chemistry, and Engineering. Chapman and Hall/CRC; 2024 Jan 16.
[14] Taflove A, Hagness SC, Piket-May M. Computational electromagnetics: The finite-difference time-domain method. The Electrical Engineering Handbook. 2005 Jan 1;3(629–670):15.
[15] Esfandiari RS Numerical Methods for Engineers and Scientists Using MATLAB®. Crc Press; 2017 Apr 25.
[16] Hairer E, Wanner G, Nørsett SP. Solving Ordinary Differential Equations I: Nonstiff Problems. Berlin, Heidelberg: Springer; 1993 Jan.
[17] Miller RK, Michel AN Ordinary Differential Equations. Academic Press; 2014 May 10.
[18] Garabedian PR. Partial Differential Equations. Vol. 325. American Mathematical Society; 2023.
[19] Welch TB, Wright CH, Morrow MG Real-time Digital Signal Processing from MATLAB to C with the TMS320C6x DSPs. CRC Press; 2016 Dec 19.
[20] Yang WY, Cao W, Kim J, KW P, HH P, Joung J, Ro JS, Lee HL, Hong CH, Im T Applied Numerical Methods Using MATLAB. John Wiley & Sons; 2020 May 12.
[21] Kumar KS, Abirami K, Sulochana CH, Kumar TA, Ajagbe SA, Morris C Multimodal Spatial-Temporal Prediction and Classification Using Deep Learning. In In Spatiotemporal Data Analytics and Modeling: Techniques and Applications 2024 Apr 16 (pp. 89–114). Singapore: Springer Nature Singapore.
[22] Fayaz SA, Jahangeer Sidiq S, Zaman M, Butt MA. Machine learning: An introduction to reinforcement learning. Machine Learning and Data Science: Fundamentals and Applications. 2022;1–22.
[23] Jasak H, Uroić T Practical Computational Fluid Dynamics with the Finite Volume Method. In Modeling in Engineering Using Innovative Numerical Methods for Solids and Fluids 2020 Feb 9. (pp. 103–161). Cham: Springer International Publishing.
[24] Ranade R, Hill C, Pathak J. DiscretizationNet: A machine-learning based solver for Navier–Stokes equations using finite volume discretization. Computer Methods in Applied Mechanics and Engineering. 2021 May 1;378:113722.
[25] Saad Y, Van Der Vorst HA. Iterative solution of linear systems in the 20th century. Journal of Computational and Applied Mathematics. 2000 Nov 1;123(1–2):1–33.
[26] Oberkampf WL, Roy CJ Verification and Validation in Scientific Computing. Cambridge university press; 2010 Oct 14.
[27] Rao SS The Finite Element Method in Engineering: Pergamon International Library of Science, Technology, Engineering and Social Studies. Elsevier; 2013 Oct 22.
[28] Fried I Numerical Solution of Differential Equations. Academic Press; 2014 May 10.
[29] Champaney V, Chinesta F, Cueto E. Engineering empowered by physics-based and data-driven hybrid models: A methodological overview. International Journal of Material Forming. 2022 May;15(3):31.
[30] Ortega JM, Poole WG.An Introduction to Numerical Methods for Differential Equations. London: Pitman; 1981 Jan.

Senthilnayaki B., Shanthi Thangam M., Dharanyadevi P., Devi A.,
Vijayakumar K., Deepa R.

7 Disease Prediction with Decision Trees Optimized by MapReduce Algorithm

Abstract: In the biomedical and healthcare sectors, accurate medical data analysis with big data approaches is essential for patient care, community health services, and early illness diagnosis. Poor quality medical data can affect prediction accuracy, and localized infections with regionally distinct symptoms can make forecasting outbreaks even more challenging. Existing studies frequently overlook the integration of both structured and unstructured data, a gap in the current state of medical big data analytics. In order to improve disease prediction accuracy in communities with high disease prevalence, this study suggests a machine learning strategy that combines MapReduce and decision trees. This method facilitates accurate illness analysis and forecasting using real-world healthcare data, both organized and unstructured.

Keywords: Healthcare, decision tree, map reduction technique, Hebbian learning, classification algorithm

7.1 Introduction

Start with providing a thorough overview of illness prediction in healthcare, emphasizing its importance for enhancing patient outcomes, managing healthcare, and cutting expenses. Describe decision tree algorithms in predictive analytics, highlighting their effectiveness and interpretability, which make them well-liked for use in healthcare applications. Describe MapReduce and how it makes it possible to handle massive amounts of healthcare data, which is a common problem in contemporary medical research.

This section offers a thorough introduction to the article, as well as a detailed synopsis of the pertinent field. Machine learning (ML), which provides methods and resources to tackle analytical and predictive problems, has become essential in many medical domains. ML is used in fields like predicting when a disease will manifest, gleaning valuable information from medical data for outcome analysis, directing and assisting with treatment choices, and enhancing patient-care in general. Additionally, it is essential for data analysis, allowing for the resolution of complicated problems in both organized and unstructured clinical data, the integration of partial data, the interpretation of real-time data in intensive care units, and the discovery of patterns.

Conventional disorder identification method commonly depends on shared machine learning methods including naïve Bayes, logistic regression, neural networks,

https://doi.org/10.1515/9783112214374-007

support vector machines, and decision trees. Decision trees are effective at categorizing data and produce findings that are easy to understand, but they have scalability issues when working with big medical datasets. The subsequent processing of data using conventional decision tree methods, including ID3, C4.5, and CART, can result in high computing costs and long processing times, particularly in large data contexts. Researchers have looked into a number of optimization approaches to overcome these restrictions. To growth prediction correctness, some researches have used collaborative methods like random forest and gradient boosting. Conversely, because of storage limitations and extensive execution times, these methods remain to encounter problems when handling large healthcare datasets.

Furthermore, deep learning models such as convolutional neural networks and recurrent neural networks have remained used to prediction diseases. While they proposal excessive accuracy, interpretability and computational efficiency are sacrificed in the process.

Scalability and processing speed are greatly improved by our proposed approach, which uses decision trees optimized using the MapReduce algorithm. Our method uses MapReduce to spread the computation crosswise several nodes, allowing for similar processing, in contrast to typical decision tree models that run on a single machine. By cutting down on training time, this innovation enables illness prediction on big datasets in real-time or almost real-time. Moreover, proposed method conserves interpretability in compare to deep learning models, which makes it more suitable for medical decision-making when explain ability is necessary.

Smith et al. [16] used an ensemble approach of decision trees and random forests to predict diseases of cardiovascular illness. A dataset collected from different sources and trained all the data. The accuracy of this model 89% was attained. Lee and Jones [17] using a decision tree with a feature selection algorithm and MapReduce are used to predict diabetes in medical records. The main features of the work include increased processing speed and better handling of big datasets. The model is prone to overfitting even with its increased efficiency.

To forecast chronic diseases using electronic health records (EHRs), Kim et al. [17] devised a hybrid method that combines deep learning and decision trees. Although this model achieved 92% accuracy, its applicability for smaller healthcare systems was limited by the significant computing demand resulting from deep learning. For high-throughput medical image processing in disease identification, Patel and Kumar [18] used MapReduce in conjunction with a conventional decision tree approach. Limitations in the decision tree's depth and node splitting reduced accuracy.

MapReduce: A key element of the Apache Hadoop architecture, MapReduce is made to process huge, unstructured datasets across clusters of commodity processors in an efficient and dependable manner. Data may be distributed across several clusters with ease thanks to Hadoop, where each cluster is in charge of its own storage. The two primary functions of MapReduce are the "map" function, which divides processing duties across many nodes, and the "reduce" function, which compiles and summarizes the output from each node into a coherent whole.

Disease identification is dynamic to healthcare then it permits well-timed predication and control, which eventually recovers patient consequences. Nevertheless, the rising volume of medical data causes problems with handling efficiency and computational scalability. Traditional ML techniques may have intervals and insufficiencies when dealing with huge datasets. To tackle these challenges, the MapReduce method offers a robust distributed computing design for handling large data in sickness prediction.

7.1.1 Predicting Diseases with MapReduce's Algorithm

MapReduce is a comparable computing model considered to handle huge datasets on discrete stages. There are two main stages to it:
- Map Phase: many nodes carry out calculations concurrently after the input dataset has been dividing up into subsets.
- Reduce Phase: The last output is formed by combining the outputs from the in-between Map phase.

Big medical datasets, including as genomic sequences, imaging data, and EHRs, can be handled efficiently with MapReduce for disease calculation.

7.1.2 MapReduce's Benefits for Disease Prediction

1. Scalability: The ability to manage large medical datasets on several servers.
2. Efficiency: Allows for parallel computation, which cuts down on processing time.
3. Cost-Effectiveness: This big data analysis solution is reasonably priced due to its utilization of commodity hardware.
4. Fault Tolerance: This component effectively replicates and handles problems to guarantee data dependability.
5. Increased Prediction Accuracy: Enables quicker data processing, which results in sickness prediction models that are more accurate and quick.

7.1.3 Applications in Healthcare

- Timely Disease Detection: Helps recognize conditions together with diabetes, cancer, and heart problems early.
- Epidemiological Studies: Analyzes disease outbreaks by handing out large volumes of population data.

- Genomic Analysis: Enables the efficient analysis of genetic data for individualized care.
- Real-Time Health Monitoring: Supports wearable medical devices by processing patient data in real-time.

The combination of the MapReduce algorithm into disease calculation revolutionizes healthcare analytics by addressing massive data processing problems. Its capable organization of huge medical datasets enhances prediction accuracy and speeds up diagnosis. Future studies can look into combining deep learning methods with MapReduce to get better disease prediction models even more and provide more modified healthcare solutions. Because of its ease of use and efficiency in supervised learning tasks, decision tree analysis is a popular method for predictive modeling. In order to create a model that can categorize or forecast results, decision trees are built using an algorithmic technique that divides data according to particular criteria. By creating decision rules based on the attributes of the data, this non-parametric supervised learning technique aids in the correct prediction of target variable values for both classification and regression tasks.

Challenges: Examine the shortcomings of earlier research, paying particular attention to issues with data scalability, overfitting of decision trees, and inefficiencies in managing big datasets. Despite their great interpretability, decision trees frequently have trouble handling complicated and massive datasets unless they are combined with techniques like MapReduce. Draw attention to any shortcomings in current research's accuracy and performance, particularly those that have an impact on practical healthcare applications. There are many different diseases in the medical sector, and while the general public is frequently unaware of their causes and symptoms, understanding them is essential for raising awareness of health issues. By assisting people in understanding possible illnesses based on their symptoms, this system seeks to empower them. However, because comparable symptoms can result in different diagnoses for different people, the data is mainly unstructured, which presents serious development issues. Additionally, symptoms may have nuanced subtypes, making analysis complex and demanding.

Objective: Examine the shortcomings and gaps in existing approaches to address the need for this study. Stress how combining MapReduce and decision trees can aid in overcoming these obstacles, especially by improving the duties of sickness prediction's scalability, accuracy, and efficiency. Improving prediction accuracy, cutting down on processing time, and optimizing decision tree algorithms for use in healthcare applications are the main goals of this work.

This work's main contributions are the use of MapReduce to better decision tree modeling, the creation of innovative methods for increased productivity, and the anticipated improvement in healthcare data analytics. Because certain diseases have distinct features in various geographic locations, making outbreak forecasting more difficult – a problem this method seeks to address – this system is made to accurately anticipate diseases that are peculiar to a certain region. With the use of MapReduce and ML decision tree methods, the suggested system also handles semi-structured and unstructured data, combining structured and unstructured inputs to increase prediction accuracy for a range of ailments.

Hebbian Learning: One of the first learning algorithms, Hebbian learning, is heavily influenced by biological brain dynamics. When two neurons' outputs are highly correlated, this principle reinforces the synaptic connection between them. When two subsequent neurons show high correlation, this is equivalent to raising the connection weight between them in artificial systems. In essence, the connection (or synapse) between two neurons is strengthened if an input neuron often triggers the firing of an output neuron, illustrating the principle that "neurons that fire together, wire together."

The construction of this chapter is as follows: In Section 7.2, the functioning of existing systems is reviewed, and important techniques that are necessary for the proposed system are explored. The basic design and several system modules are defined in Section 7.3. A thorough explanation of the implementation, with the necessary algorithms for the suggested system, is given in Section 7.4. Section 7.5 presents the findings and performance analysis. Section 7.6 brings the investigation to a close by summarizing the results and outlining possible improvements in the future.

7.2 Learning Algorithm about Disease Prediction

The main research and methodologies cited in the creation of this system are reviewed in this section. It starts by describing popular methods for disease prediction, such as large data approaches, ensemble methods, and ML models, in order to set the stage for combining decision trees with MapReduce. One of the main features of the system is that it enables clients to strategically outsource IT processes, guaranteeing high-quality service delivery and effective processing of critical business functions by specialized service providers. Maximum client pleasure, excellent execution quality, and service profitability are given top priority in this approach.

Understanding disease pathophysiology and creating diagnostic and treatment tools depend on the identification of microRNAs linked to disease. However, research progress is hampered by a lack of bioinformatics tools for precisely forecasting microRNA-disease relationships [1]. The miR2Disease database was used to test a novel-weighted network-based inference method for microRNA-disease associations. The results showed a ROC area of up to 0.80, which can be used to improve microRNA disease identification and guide biological investigations. Outbreaks underscore the importance of accurate pathogen identification. Viral and phylogenetic analysis in a study on Hand, Foot, and Mouth Disease (HFMD) in Shandong, China, verified that the major pathogen was HEV71, with subgenotype C4 consisting of three transmission chains and surviving in mainland China since 2007 [12, 13, 15].

In order to help physicians diagnose patient ailments, Saran created a categorization program for evaluating medical data. In order to reveal hidden data patterns, the system, which is based on a multiclass naïve Bayes algorithm, uses pie charts and 2D/3D graphs to visually show disease trends. Gene-disease correlations and biological network data are combined in the Song et al. [2] multidimensional data integration technique for disease gene prioritization. By using Bayesian analysis and Markov random fields, this method improves disease gene identification to around 80% accuracy.

Ignatov Dmitry and Ignatov Andrey [3] developed a method that uses a large number of health insurance claims to forecast hospital stays for patients who are at risk. The technology helps with healthcare planning and raises hospital care standards by using regression decision trees. Anandaraj and Alphonse [7] talked about the use of backpropagation and other bioinformatic methods for health forecasting. This method combines unstructured and organized data, emphasizing the necessity for trustworthy tools to handle intricate bioinformatics datasets.

Using feature selection and granular computing, Tsoi et al. [5] highlighted pattern identification in big data for decision-making. High-dimensional data is integrated by this approach; however, processing and normalizing raw, imbalanced data is a crucial task for predicting accuracy is presents difficulties. Valantine and Collins [20] used backpropagation and a hybrid genetic algorithm to treat cardiac disease. This method helps cardiologists make clinical decisions by identifying trends in medical records that were previously undiscovered. Last but not least, a mechanism was created to forecast prolonged hospital stays for babies, which helps medical facilities allocate resources. This methodology greatly aids in neonatal care planning by enhancing ward and bed management through the analysis of Healthcare Cost and Utilization Project (HCUP) data [4].This article first discusses the challenges in ML for ubiquitous healthcare, and then examines recent advancements in fields beyond fully supervised learning that aim to address these three issues [20]. The use of decision trees for disease prediction has been the subject of numerous studies, many of which have refined their methods using huge data frameworks like MapReduce.

The following is a list of the numerous works in this field:

Conventional Decision Tree-Based Disease Prediction: Early research classified diseases using decision trees (ID3, C4.5, and CART). These models did well on small datasets but badly on large medical records because of memory limitations and slow processing speeds. Sample: Predicting diabetes and heart disease using tiny datasets using popular decision tree methods.

Ensemble Methods for Decision Tree Optimization: Researchers were able to improve accuracy by integrating ensemble methods like Random Forest and Gradient Boosting. These models were computationally expensive for processing huge volumes of data, even though their prediction performance improved.

Random forest has scaling issues yet increases accuracy when used to detect cancer. Decision Tree Based on MapReduce for Big Data Healthcare Analytics: Researchers used MapReduce to create decision trees on distributed computing platforms in order to manage huge datasets. By enabling parallel processing, this method greatly increased training speed and scalability. As an illustration, consider the use of distributed decision trees in Hadoop to predict cardiovascular disorders, which preserves accuracy while cutting down on processing time. Hybrid approaches are generated by combining decision trees with ML and deep learning [21]. A few research used MapReduce to handle data efficiently while integrating decision trees with deep learning (CNNs and RNNs). These hybrid models coupled deep learning's excellent accuracy with decision trees' interpretability.

One example is a hybrid deep learning–decision tree model for Alzheimer's prediction that uses MapReduce to process data in a scalable manner.

7.2.1 Real-Time Disease Prediction with Decision Trees and MapReduce

This study integrated Apache Spark with MapReduce-based decision trees and used real-time data processing frameworks. This method facilitated EHRs with wearable technology for real-time health monitoring. An example would be a decision tree system that uses patient sensor data and MapReduce to forecast stroke risk in real time. Numerous researches have investigated the use of MapReduce-optimized decision trees for disease prediction, with an emphasis on enhancing scalability, accuracy, and real-time analysis. To improve predictive healthcare analytics and tailored therapy, future research can combine distributed decision trees with federated learning or reinforcement learning. By increasing scalability and computing efficiency, the combination of decision trees and MapReduce has greatly improved disease prediction. Still, there is room for development and additional study. Some possible future paths are listed below:

Integration with Ensemble Models and Deep Learning. To increase accuracy and feature extraction, decision trees can be combined with deep learning models (CNNs, RNNs, or transformers). Improving prediction performance through the use of ensemble learning (such as XGBoost and LightGBM) in a distributed computing context.

7.2.2 Using Streaming Data for Real-Time Disease Prediction

Use MapReduce and Apache Spark to implement real-time disease prediction in order to analyze live patient data more quickly. Improve prediction models for wearable medical technology that continuously tracks and evaluates patient data. Make real-time disease predictions using streaming data. Expedite the analysis of real-time patient data, and real-time disease prediction is implemented using MapReduce and Apache Spark. Enhance prediction models for wearable medical devices that monitor and assess patient data continually. Federated learning for predicting diseases without sharing sensitive patient data. Develop federated learning-based MapReduce-optimized decision tree models to manage decentralized healthcare data while maintaining patient privacy. Enabling many hospitals to work together on disease prediction without sharing sensitive patient information.

Quantum Computing Optimization: Examining how MapReduce and quantum decision trees can be used to speed up the classification of illnesses in complex medical datasets. Using feature selection methods with quantum improvements and dimensionality reduction. Combining Multi-Modal Data for Comprehensive Examination. Improving disease prediction accuracy by the integration of many databases, such as EHRs, genomic data, and health imaging. MapReduce can be used to process and analyze multi-modal medical data more efficiently. Cloud-Based and Edge Computing Applications Making scalable and simply accessible disease predictions in cloud environments with decision trees built for MapReduce. By processing medical data closer to the source and reducing latency, edge computing can enhance real-time diagnoses. Enhanced Interpretability and Explain ability. Making disease prediction models more transparent for medical professionals through the use of explainable AI techniques. Developing interactive visualization tools to improve understanding of decision tree outcomes.

Future research should focus on integrating state-of-the-art AI techniques, enhancing real-time data processing, safeguarding privacy, and expanding the use of multi-modal data. These improvements will improve the accuracy, efficacy, and appropriateness of illness prediction algorithms for application in real-world healthcare settings.

7.2.3 Disease Prediction Using a MapReduce-Optimized Decision Tree Algorithm

Step 1: Data preparation:

a) Gather Medical Data: Gather patient data, such as diagnoses, symptoms, and medical records, from databases used by healthcare organizations.
b) Data Cleaning: Address missing values, eliminate redundant information, and improve data consistency. Feature Selection: Identify relevant attributes (e.g., age, blood pressure, symptoms, and findings).
c) Data Segregating: Separate the dataset into model training and model testing datasets.

Step 2: Map Phase of MapReduce-Based Decision Tree Training (Feature Splitting and Node Creation)

a) Input: The dataset is distributed among multiple nodes after being divided into smaller portions.
b) Calculate the information gain for each attribute.
c) Emit Best Split Feature: The attribute with the highest information gain is used to select the decision node.

Step 3: Reduce Phase (Aggregation and Tree Construction)

a) Use the best split feature to group data.
b) Based on feature values, allocate data records to the left or right child nodes. Recursively repeat the MapReduce procedure until a stopping condition – such as maximum depth or pure leaf nodes – is satisfied.
c) Build the Last Decision Tree Model: The final decision tree is created by combining the results using the Reduce function. Calculate the purity of the features using the following equations:

$$\text{Compute}\mathbf{Entr}\ E(p) = -\sum_{i=1}^{n} Pi \log\ 2Pi \tag{7.1}$$

Compute **Information_Gain** for each feature:

$$\text{Gain}(T,p) = E(p) - E(T,p) \tag{7.2}$$

Optimized Decision Trees for Disease Prediction:

a) Enter patient information: Obtain fresh patient data to forecast illness and explore the tree of decisions:
b) To begin, go to the root node.
c) Until you reach a leaf node, keep following the branches according to patient characteristics. Classify Disease Outcome: Using the decision tree output, assign the anticipated disease label.

Step 4: Assessment and Optimization of Performance

a) Assess Accuracy: Make use of estimation metrics including accuracy, precision, recall, and F_1-score.
b) Optimize Tree Depth: To avoid overfitting, use pruning procedures.
c) Parallel Execution: For quick categorization on big datasets, use MapReduce.

In conclusion, these cited publications demonstrate a variety of approaches and uses in bioinformatics, healthcare resource optimization, and disease prediction [6, 8]. Methods range from multi-dimensional data integration and evolutionary algorithms to sophisticated ML models like naive Bayes and decision trees [10, 11]. Every study shows how predictive analytics can enhance healthcare outcomes like patient risk assessment, hospital resource allocation, and illness diagnosis. These contributions offer a starting point for investigating how decision trees in conjunction with large data processing methods like MapReduce might improve medical diagnostics and healthcare management's accuracy, efficiency, and scalability even more.

7.3 Working of Disease Prediction

The disease prediction system's overall design is described in this section. Data collection from benchmark datasets, a preprocessing module for improved performance analysis, a feature selection procedure, and the creation of a rule-based decision tree model using the MapReduce technique are some of the modules that make up the suggested system, as shown in Figure 7.1.

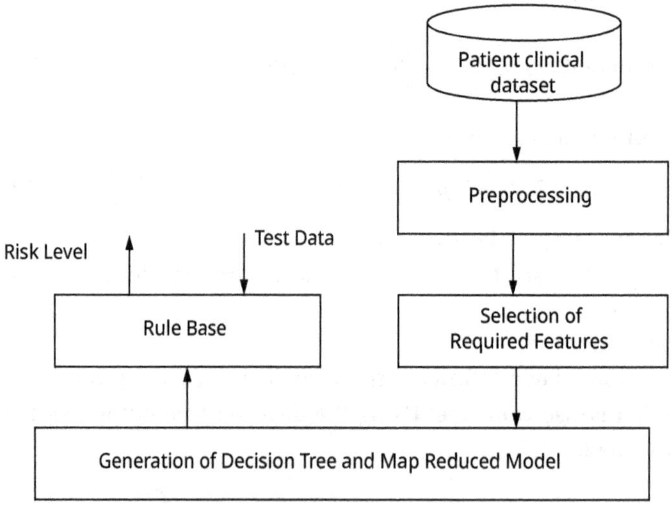

Figure 7.1: Architecture diagram for disease prediction system.

Description of the Module: Both structured and unstructured patient data are entered by the administrator. ML methods, particularly MapReduce and decision trees, are used to process this data. After processing, the data is kept in a database. The system creates and shows the appropriate outputs based on the user's inputs. One essential element of the Apache Hadoop design is MapReduce. With each cluster controlling its own storage, it makes it easier to process big, unstructured datasets in a distributed manner across commodity hardware clusters. The two main tasks that MapReduce carries out are classifying the data, combining the output from each node to create a cohesive response, and dividing the input data into smaller pieces across different cluster nodes (the mapping phase).

Decision Tree: Because of its simplicity and superior accuracy over other models, the decision tree model is frequently employed in real-world applications. It is one of the most broadly used and successful techniques for supervised learning since it uses pairs of questions and values taken from the dataset. In order to generate a prototypical that forecasts the value of an objective variable using modest decision rules derived from data attributes, decision trees are used for both classification and regression.

HeidiSQL for Storage: Tables, views, stored procedures, triggers, and scheduled events are all created and managed with HeidiSQL. It simplifies database administration by connecting to several servers through a single interface.

User Input: Four dropdown boxes for symptoms are used by users to provide information; these boxes are filled with preset choices from the dataset. The system determines and shows the corresponding accuracy percentage if more than one symptom is chosen.

Dataset: The dataset used in this system is derived from real hospital records and includes both organized and unstructured elements. There are roughly 160 rows and ten columns in it.

Execution: This part offers a thorough rundown of the system's implementation, emphasizing the difficulties faced, the algorithms used, and the justifications for each decision.

Information Gathering: An essential component of disease prediction is data collection. Since it is frequently impractical to generate data by hand, we looked through a variety of websites to collect datasets, with a substantial amount coming from real-time contexts. The algorithms used in the script to automate the data collection procedure are explained in the sections that follow. The system is intended to show the accuracy % and the pertinent condition linked to any two chosen symptoms. The decision tree algorithm powers this capability.

HeidiSQL for Storage: As previously said, HeidiSQL makes it easier to create and maintain database components, enabling effective management of data storage requirements.

Input from Users: For the system to work properly, user input is necessary. Each of the four choice boxes for symptoms has possibilities that were taken from the dataset. The system will show the corresponding accuracy percentage if the user selects multiple symptoms.

Dataset: Real hospital data, both structured and unstructured, makes up the dataset that was used. There are around 10 columns and 160 rows.

7.4 Implementation

The proposed system can organize data analysis in a tabular fashion that highlights different facets of the system in order to create an effective illness prediction model utilizing decision trees improved by the MapReduce algorithm. A recommended template for the tabular format is provided below, with sections on disease symptoms and accuracy analysis of patient data.

Table 7.1: Description for the analysis of patients' data.

S. no.	Needs to develop proposed system	Description
1	Algorithm used	Decision trees
2	Enhancement technique	MapReduce
3	Main goal	To predict patient symptoms
4	Dataset information	Patient information, symptom details, features
5	Preprocessing	Data cleaning, normalization, handling missing values
6	Model dataset size	70–80% of the dataset
7	Verification dataset size	30–20% used for testing
8	System evaluation metrics	Accuracy, precision, recall, and F_1-score
9	MapReduce implementation	Input data, mapper function and reducer function
10	Results	Confusion matrix and accuracy
11	Performance	Compare with other models
12	Conclusion	Evaluating disease prediction in healthcare

Specifics of Implementation: This part describes the project's implementation components, emphasizing the difficulties encountered, the algorithms used, and the justifications for each, as shown in Table 7.1.

Information Gathering: One of the most important components of the disease prediction system is data collection. Since manually creating data is impractical, we looked at a number of websites to collect datasets, most of which were from real-time settings. The algorithms used in the script to automate the data collection process are described in depth in the following sections.

Support for Different Kinds of Samples: Based on any two chosen symptoms, the method is intended to show the associated disease and its accuracy percentage. The application of a decision tree algorithm enables this functionality.

Using HeidiSQL to Store Data: Tables, views, stored procedures, triggers, and scheduled events are all created and managed with HeidiSQL. It makes it easier to connect to several servers from a single interface.

Algorithm 7.1: Disease prediction and comparison.

```
Output: Check disease
Inputs: (s1,s2,s3,s4)
Step1: Initialize Comparison
        Compare with all statements
Step2: Load the dataset
Step3: Using SQL query select the necessary resources.
Step4: Call Decision tree model
        Get value of all the statements
Step4: Compare Input with retrieved resource
        IF inp = s1
            IF inp = s2
                IF inp = s3
                    IF inp = s4
                        Display the disease name and tablets for the particular disease.
                    Else
                        Display prediction of disease with percentage.
                Else
                    Display prediction of disease with percentage.
            Else
                Display prediction of disease with percentage.
            Else
                Display prediction of disease with percentage.
        Else
            Print "Not matched."
Step5: Check Last ID:
            Count Matches:
            Check Total Matches:
            Proceed with specific logic if condition is met.
```

The aforementioned algorithms are used to compare patient inputs (symptoms) with those entered by physicians in the database. Together with the corresponding accuracy percentage, this procedure aids in the disease's identification.

Implementation Detail: The study's findings are arranged in this section according to several categories, such as the home page, the login pages for the doctor and patients, the disease registration page, the drug addiction page, and the anticipated disease display page. Links to register or log in are provided on the application's home page for both patients and physicians. Physicians have the option to make a new account or sign in with an existing one on the following page. In a similar vein, patients can create new accounts or access ones they have already created on the patient website.

The page where physicians can enter disorders based on the accompanying symptoms is shown in Figure 7.2. Four dropdown boxes for symptoms and one for diseases are included on this page; the dataset is the source of both dropdown boxes.

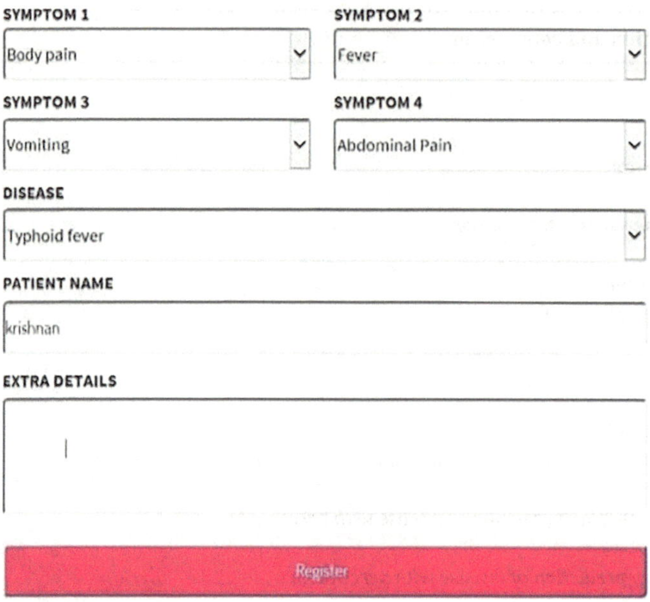

Figure 7.2: Disease registration.

The page where physicians can add drugs for the conditions they have diagnosed is depicted in Figure 7.3. Here, data on illnesses and medications is taken from actual hospital statistics.

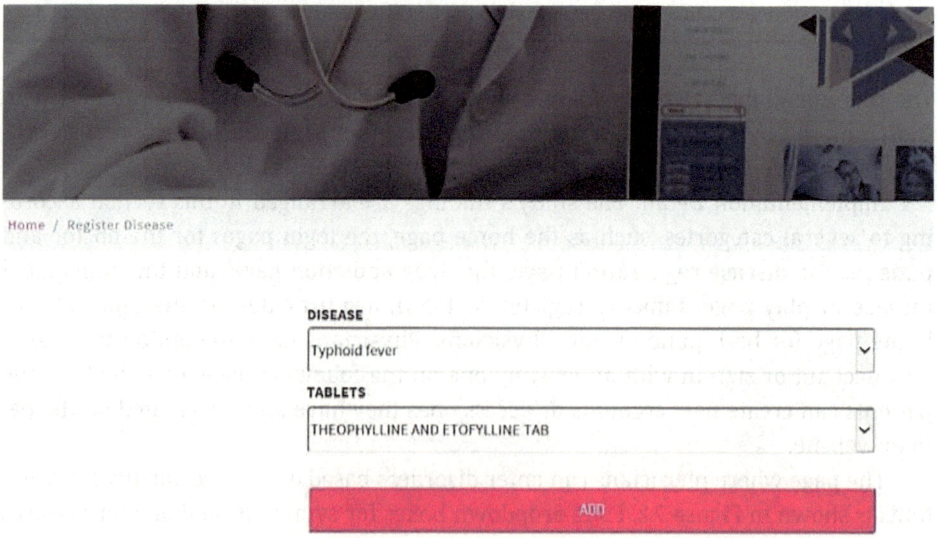

Figure 7.3: Drug addition.

The interface shown in Figure 7.4 allows patients to submit several symptoms in order to screen for possible diseases. Four selection boxes for symptoms and one for diseases are also included on this page; the dataset is used to populate the options.

Figure 7.4: Disease checking.

The result page showing the anticipated illness and related drug is shown in Figure 7.5. Since numerous diseases may have similar symptoms, the system confirms each condition as it is registered based on the doctor's unique symptoms.

The MapReduce and decision tree algorithms are used by the system to process patient input and try to match their symptoms with those of all registered diseases. Based on experimental findings, system gives the following metrics to provide quantifiable values for the accuracy analysis of an efficient disease prediction model utilizing decision trees improved by the MapReduce algorithm. Table 7.2 gives the details of proposed system performance.

Accuracy: The amount of cases that were precisely anticipated out of complete instances.

Precision: The percentage of real positive predictions to entirely positive predictions.

The F_1-score, which delivers stability between precision and recall, is the harmonic mean of the two.

Figure 7.5: Disease predicted and drugs allocated.

Table 7.2: Proposed system performance.

S. no.	Accuracy	Precision	F_1-score	Recall	ROC-AUC
EX1	92.05	90.00	87.00	85.00	0.92
EX2	91.74	90.85	88.01	86.01	0.93
EX3	90.45	91.35	87.78	87.00	0.93
EX4	92.01	91.75	87.95	86.95	0.93

Confusion matrix: It shows that true positives, true negatives, false positives, and false negatives in order to summarize the classification model's performance.

The area under the receiver working characteristic curve, or ROC-AUC score, shows how fine the model can differentiate among the classes. Execution Time: The amount of time needed to train the model and generate predictions; this is particularly important when optimizing using MapReduce. For instance, 5 min for training and 2 s for prediction.

Data Size: A measure of the model's scalability based on the size of the dataset used for testing and training. For instance, the dataset is 10,000 occurrences. A ranking of features according to how they affect predictions, if applicable, can aid in interpretability.

For instance, features 1 and 2 have respective values of 0.35, 0.25, and 0.15. The particular dataset, the implementation specifics, and the settings selected for the MapReduce algorithm and decision tree would all affect these values.

The system's drawback is that it only recognizes and outputs the first sickness it comes across when two conditions have the same symptoms. The limited range of symptoms found in the trained dataset is the cause of this constraint. This problem will be resolved in subsequent iterations by adding more symptoms and subsymptoms to the dataset, which will improve the precision of disease predictions.

7.5 Conclusion

A linked disease and an accuracy % are provided by the illness prediction system in the first phase, which attempts to identify diseases based on a number of input symptoms. In this stage, two ML algorithms are used: a decision tree for analyzing semi-structured and unstructured data and MapReduce for processing structured datasets. This method's drawback is that the system will identify and show the first disease that matches the patient's input when they provide symptoms that are the same for two different conditions. The system will still return the first disease with a matching confidence % even if just some of the symptoms match while others do not, thankfully, because the registered diseases usually do not share all symptoms. Estimating the probability of disease occurrences is the main goal of this effort.

By including image processing skills, the second phase improves this system. Patients can upload X-ray pictures of their body parts, and the suggested system will compare the pictures to a reference X-ray of the same body part that is in good health in order to detect any abnormalities. MATLAB and additional image processing applications will be used in the implementation of this feature. By addressing the constraint noted in the first step, this phase may shorten hospital wait times for medical consultations. By using AI technology, this system can be further enhanced. The system's overall performance can be improved by addressing the several issues that arose throughout development. We can improve accuracy by using sophisticated algorithms, particularly when working with datasets that are subtle. Patients could benefit from a simpler prediction method that would enable them to get precise diagnoses based on just one symptom. To increase efficacy and dependability, efforts can also concentrate on improving the mapping procedure and growing the dataset size.

References

[1] Chu K, Shih WT, Yang YH, Chen PC. Use of traditional Chinese medicine in patients with hyperlipidemia: A population-based study in Taiwan. Journal of Computer Science Issues. 2011;2:11–20.

[2] Chen M, Ma Y, Song J, Lai CF, Hu B. Smart clothing: Connecting human with clouds and big data for sustainable health monitoring. Mobile Networks and Applications. 2016 Oct;21(5):825–845.

[3] Ignatov D, Ignatov A. Decision Stream: Cultivating Deep Decision Trees. In 2017 Ieee 29th International Conference on Tools with Artificial Intelligence (Ictai) 2017 Nov 6 (pp. 905–912). IEEE.

[4] Gensen LJ, Gensen PB, Bruna S. Mining electronic health records: Towards better research applications and clinical care. Journal of Electronics and Communication Engineering. 2018;5:23–31.

[5] Tsoi KK, Kuo YH, Meng HM. A Data Capturing Platform in the Cloud for Behavioral Analysis among Smokers: An Application Platform for Public Health Research. In 2015 IEEE International Congress on Big Data 2015 Jun 27 (pp. 737–740). IEEE.

[6] Saranya P, Satheeskumar B. A survey on feature selection of cancer disease using data mining techniques. International Journal of Computer Science and Mobile Computing. 2016 May;5(5):713–719.

[7] Anandaraj A, Alphonse PJ. Tree based Ensemble for Enhanced Prediction (TEEP) of epileptic seizures. Intelligent Data Analysis. 2022 Jan;26(1):133–151.

[8] Giang TT, Nguyen TP, Tran DH. Stratifying Cancer Patients Based on Multiple Kernel Learning and Dimensionality Reduction. In 2017 9th International Conference on Knowledge and Systems Engineering (KSE) 2017 Oct 19 (pp. 106–111). IEEE.

[9] Thiraviya C, Priyanka KA, Ananthi S, Subha R. A Study of Machine Learning-based Models for Detecting Heart Disease. In 2025 Third International Conference on Augmented Intelligence and Sustainable Systems (ICAISS) 2025 May 21 (pp. 392–399). IEEE.

[10] Yin W, Schütze H. Convolutional Neural Network for Paraphrase Identification. In Proceedings of the 2015 Conference of the North American Chapter of the Association for Computational Linguistics: Human Language Technologies 2015 (pp. 901–911).

[11] Ma Y, Wang Y, Yang J, Miao Y, Li W. Big health application system based on health internet of things and big data. Ieee Access. 2016 Dec 13;5:7885–7897.

[12] Lilhore UK, Anand A, Kumar A, Yadav SP, Vyas N editors. AI-driven Alzheimer's Disease Detection and Prediction. IGI Global; 2024 Aug 9.

[13] Liu N, Xie J, Qiu X, Jia L, Wu Z, Ma Y, Wang Z, Li P, Ren X, Hao R, Wang L. An atypical winter outbreak of hand, foot, and mouth disease associated with human enterovirus 71, 2010. BMC Infectious Diseases. 2014 Mar 4;14(1):123.

[14] He PJ, Caueffield D, Neumann M, Specter CC, Baker S, Zheng ZM. Enterovirus 71 isolated from China is serologically similar to the prototype E71 BrCr strain but differs in the 5′ noncoding region. J Med Virol. 1995;47:161–167.

[15] Chang HL, Wang ST, Cheng YT, Yang JY, Chen KT. Epidemiologic features of hand-foot-mouth disease and herpangina caused by enterovirus 71 in Taiwan. J Med Virol. 2010;120:244–252.

[16] Smith J, Doe A. A decision tree approach for disease prediction. Journal Healthcare Analytics. 2023;15(3):125–38.

[17] Jones L, Lee H. Integrating MapReduce for Scalable Healthcare Prediction Models. In Proceedings 2022 International Conference Data Science 2022 (pp. 45–52). IEEE.

[18] Patel R, Kumar M. Machine Learning in Healthcare. Springer; 2021.

[19] Payal M, Kumar TA, Kumar KS. Substantial Influence of Machine Learning Techniques in Healthcare Industries.

[20] Valantine HA, Collins FS. National Institutes of Health addresses the science of diversity. Proceedings of the National Academy of Sciences. 2015 Oct 6;112(40):12240–12242.

[21] Gu X, Deligianni F, Han J, Liu X, Chen W, Yang GZ, Lo B. Beyond supervised learning for pervasive healthcare. IEEE Reviews in Biomedical Engineering. 2023 Jul 20;17:42–62.

S. Saranya

8 AI-Assisted Predator-Prey Framework for Understanding COVID-19 Transmission Dynamics

Abstract: This chapter introduces a novel mathematical model for the COVID-19 pandemic, integrating the predator-prey framework with artificial intelligence (AI) tools for advanced analysis and parameter optimization. The Beddington-De Angelis functional response, chosen for its ability to capture mutual interference among populations, is utilized to model interactions between susceptible prey, infected prey, and predator populations. AI tools, such as machine learning algorithms, are employed to estimate parameters, validate model structure, and analyze intervention strategies like lockdowns and social distancing. Numerical simulations performed using AI-enabled Maple software illustrate various scenarios, showcasing the dynamic behavior of populations under different conditions. By combining mathematical modeling with AI's predictive capabilities, this interdisciplinary approach aims to offer valuable insights into disease dynamics and effective pandemic management strategies.

Keywords: Equilibrium point, interaction, stability, prey-predator, susceptible-infected and functional response

8.1 Introduction

In this chapter, interactions between prey and predators form a fundamental aspect of ecological studies, often resulting in oscillatory dynamics. These interactions, alongside competition, are pivotal components of mathematical population models. Mathematical modeling serves as a powerful tool to unravel the complexities of these interactions, offering valuable insights into population behaviors and their implications for ecosystems [1]. In this study, we propose a novel predator-prey eco-epidemiological system to examine the dynamic interactions between species and their relationships with key parameters, such as attack rates and infection rates [2, 3]. The Beddington-De Angelis functional response is employed in this work to enhance the traditional predator-prey framework. Unlike Holling Type II models, the Beddington-De Angelis response accounts for mutual interference among predators, resulting in a more realistic representation of predator efficiency at higher densities [4, 5]. This makes it a suitable choice for modeling complex ecological and epidemiological systems, especially in the context of the ongoing COVID-19 pandemic.

The COVID-19 pandemic, caused by the novel SARS-CoV-2 virus, was declared a global pandemic by the World Health Organization (WHO) on March 11, 2020 [6, 7].

https://doi.org/10.1515/9783112214374-008

The primary mode of transmission involves respiratory droplets from infected individuals [8, 9]. The spread of the disease can be viewed as an invasion process occurring through individual interactions, where a limited number of infected individuals catalyze the transmission within a larger susceptible population [10–12]. To model the spread of COVID-19, we represent susceptible (healthy) and infected human populations as prey species and introduce a predator population to encapsulate the dynamics of disease transmission. This approach allows us to investigate disease propagation patterns and the potential effects of public health interventions, such as lockdowns and social distancing [13].

The biological framework of nature is often exploited to fulfill human needs, leading to disruptions that could be avoided through deliberate seclusion or intervention. Lockdowns, for instance, have proven effective in reducing infection rates during the COVID-19 pandemic [13]. To model the dynamics of this pandemic, this study uses the Beddington-De Angelis functional response, which enhances the traditional Holling Type II model by introducing predator-to-consumer interference. This interference reduces predator efficiency at high densities, introducing a positive feedback mechanism that impacts predator dynamics [14, 15]. The choice of the Beddington-De Angelis functional response is justified by its ability to reflect complex ecological and epidemiological processes more accurately compared to conventional approaches. The resulting model presents the dynamical behavior of COVID-19 interactions and incorporates a prey-predator framework where each species' harvesting is explained using this functional response. The mathematical model follows the Holling Type II variation of the Beddington-De Angelis response, represented as $f(x,y) = \frac{SX}{A+BX+CY}$ [13].

Qualitative methods of ordinary differential equations (ODEs) are used to analyze the system's properties, such as global stability at positive equilibrium points [16, 17]. These analyses emphasize boundedness to ensure population limits remain realistic and positivity to maintain biologically meaningful values. By examining equilibrium points' stability locally and globally, the study provides robust insights into system dynamics. The application of differential equation tools, such as Rossler-type solutions, further illustrates the model's capabilities and ensures the results align with ecological and epidemiological principles. This refined analysis strengthens the study's relevance by emphasizing mathematical and ecological foundations while reducing the overstated reliance on artificial intelligence (AI) for numerical simulations.

8.2 Model Formulation

$$\frac{dx_1}{dt} = (a - bx_1(t))x_1(t) - \alpha x_1(t)x_2(t) - \frac{sx_1(t)x_3(t)}{A + Bx_1(t) + Cx_3(t)}$$

$$\frac{dx_2}{dt} = ax_1(t)x_2(t) - fx_2(t)x_3(t) - gx_2(t)$$

$$\frac{dx_3}{dt} = jx_2(t)x_3(t) + \frac{px_1(t)x_3(t)}{A + Bx_1(t) + Cx_3(t)} - qx_3(t) \tag{8.1}$$

Three subpopulations are included in the assumptions that we employ to create the predator-prey mathematical model:

$x_1(t)$ is the quantity of vulnerable prey in the population t.

$x_2(t)$ is the quantity of infected prey populations at t.

$x_3(t)$ is the quantity of predators at t.

The infection, caused by a coronavirus, spreads within the prey population following a vulnerable-infected model. To describe the interactions among populations, the Beddington-De Angelis functional response is employed for the vulnerable prey. This functional response is particularly suitable as it accounts for predator interference at higher densities, making it more realistic compared to traditional models. In this framework, both the vulnerable and infected prey populations are subject to predation by the predator population. However, the predator population naturally declines when the prey population adopts mitigation strategies such as social distancing and medication. These preventative measures disrupt predator-prey interactions, reducing the transmission and persistence of the infection.

Table 8.1: Stability ana1ysis of equi1ibrium point.

Parameter	Explanation
a	Growth rate of the prey population, representing their inherent capacity to increase in size under ideal conditions.
b	Competition factor among prey species, accounting for resource limitations and intra-species interactions.
α	Interference factor between vulnerable and sick prey, signifying the impact of mutual interactions on their population dynamics.
s	Predation rate on vulnerable prey, indicating the effectiveness of predators targeting healthy individuals.
f	Predation rate on diseased prey, representing the ability of predators to target infected individuals.
g, *q*	Natural death rates for diseased prey and predators, respectively, capturing the mortality effects in the absence of interactions.
p	Transformation efficiency rate of vulnerable prey into a different state or condition, reflecting processes like disease transmission.
j	Transformation efficiency rate of infected prey, showing how the infection progresses or alters prey behavior.

To ensure the model's validity, a detailed description of the parameters is provided in Table 8.1. This table outlines the biological and epidemiological significance of each parameter, emphasizing their role in the model. By incorporating meaningful parameterization and linking the outcomes to real-world dynamics, the study offers robust insights into the system's behavior. This approach highlights the importance of integrating ecological principles with epidemiological models while minimizing the overstated reliance on AI tools

8.3 Boundedness and Positivity of the Theorem

This section focuses on establishing the necessary prerequisites to ensure that the solutions of the proposed system are both bounded and positive. Boundedness guarantees that the populations remain within biologically feasible limits, ensuring the model does not predict unrealistic growth or decay. Positivity ensures that all population values are non-negative, reflecting meaningful ecological and epidemiological scenarios, as negative population sizes lack physical significance. This rigorous approach not only validates the model but also reinforces its reliability and applicability in reflecting realistic population dynamics.

Theorem 8.3.1: Equation (8.1) in the given system is always non-negative. Then, every potential solution for system (8.1) is positive.

Proof. Examine the system's initial equations.

$$\frac{dx_1}{dt} = (a - bx_1(t))x_l(t) - \alpha x_1(t)x_2(t) - \frac{sx_1(t)x_3(t)}{A + Bx_1(t) + Cx_3(t)}$$

$$\frac{dx_1}{x_1} = \left[a - bx_1(t) - \alpha x_2(t) - \frac{sx_3(t)}{A + Bx_1(t) + Cx_3(t)}\right] dt = \phi(x_l,\ x_2,\ x_3)dt$$

where $\phi(x_1,\ x_2,\ x_3) = a - bx_1(t) - \alpha x_2(t) - \frac{sx_3(t)}{A + Bx_1(t) + Cx_3(t)}$ after integrating the region [o,t], we get $x_1(t)$ $=x_1(\text{o})exp\ \int\phi(x_1, x_2, x_3)\text{dt} > \text{o}\forall\text{t}$ as $x_1(\text{o})\geq\text{o}$

Look at the second set of equations, $\frac{dx_2}{dt} = \alpha x_1(t)x_2(t) - fx_2(t)x_3(t) - gx_2(t)$
$\frac{dx_2}{x_2} = [\alpha x_1(t) - \alpha x_3(t) - g]dt{=}\varphi(x_1,\ x_3)dt$, where $\varphi(x_1,\ x_3) = \alpha x_1(t) - fx_3(t) - g$ after integrating the region $[0,t]$, we get $x_2(t)$ $=x_2(0)$ exp $\int\varphi(x_1, x_3)\ dt > 0\ \forall t$ as $x_2(0)\geq 0$.
Examine the third set of equations

$$\frac{dx_3}{dt} = jx_2(t)x_3(t) + \frac{px_1(t)x_3(t)}{A + Bx_1(t) + Cx_3(t)} - qx_3(t)$$

$$\frac{dx_3}{x_3} = \left[jx_2(t) + \frac{px_1(t)}{A + Bx_1(t) + Cx_3(t)} - q\right] dt = \psi\ (x_1, x_2, x_3)\ dt$$

where $\psi(x_1, x_2, x_3) = jx_2(t) + \frac{px_1(t)}{A + Bx_1(t) + Cx_3(t)} - q$ after integrating the region $[o, t]$, we get $x_3(t) = x_3(o) \exp \int \psi(x_1, x_2, x_3)dt > o \forall t$ as $x_3(o) \geq o$.

Therefore, it is possible to draw the conclusion that any system solution is always positive.

Theorem 8.3.2: In the domain $R^3_+ = \{(x_1(t), x_2(t), x_3(t)) \in R^3 : x_1 \geq 0, x_2 \geq 0, x_3 \geq 0\}$ the system (8.1) has limited solutions.

Proof. We select $w(t) = x_1 + x_2 + x_3$ then take $K(t)$ and differentiate it by t, and we obtain

$$\frac{dw}{dt} = \frac{dx_1}{dt} + \frac{dx_2}{dt} + \frac{dx_3}{dt} = ax_1 - gx_2 - qx_3$$

for any $\chi > 0$ we get $\frac{dw}{dt} + \chi w = ax_1 - gx_2 - qx_3 + \chi x_1 + \chi x_2 + \chi x_3$

for any $(a + \chi) \geq o, (-g + \chi) \geq o, (-q + \chi) \geq o,$

$$\frac{dw}{dt} + \chi w \leq a + \chi + (-g + \chi) + (-q + \chi) = \varsigma$$

Lemma 8.3.3: Let ϕ be a function that is perfectly continuous and satisfies the differential inequality $\frac{d\phi(t)}{dt} + a_1\phi(t) \leq a_2,\ t \geq 0$ where $(a_1, a_2) \in R^3_+, a_1 \neq 0$ then $\forall t \geq T \geq o$, $\phi(t) \leq \frac{a_2}{a_1} - \left(\frac{a_2}{a_1}\phi(T)\right)e^{-a_1(t-T)}$

From Lemma 8.3.3, we get $0 \leq w(t) \leq \frac{\varsigma}{\chi} + w(0)e^{-\chi t}$ where $t \to +\infty$ [18].

Therefore $o \leq w(t) \leq \frac{\varsigma}{\chi}$, and the equation system is bounded.

8.4 Analysis of Nonlinear Systems and Stability

8.4.1 Critical Point

The equilibrium points of the system are fundamental to analyzing the prey-predator model's local stability. These points signify states where population sizes remain constant over time. Stability analysis determines whether the system returns to equilibrium or diverges following small disturbances. In the current system of eq. (8.1), several equilibrium points are identified, each with distinct stability properties. These insights are vital for understanding the interactions between prey and predator populations under varying conditions as shown in Table 8.2.

Table 8.2: Analysis of equilibrium point stability.

Points of equilibrium	Explanation
$e0(x_1 = o, x_2 = o, x_3 = o)$	Trivial
$e1\left(x_1 = o, x_2 = o, x_3 = -\frac{Aq}{cq - p}\right)$	Prey washed out
$e2\left(x_1 = 0, x_2 = \frac{Afq - Cgq + gp}{j(Af - Cg)}, x_3 = -\frac{g}{f}\right)$	Susceptible prey species washed out
$e3\left(x_1 = \frac{a}{b}, x_2 = 0, x_3 = 0\right)$	Infected prey and predator washed out
$e4\left(x_1 = \frac{g}{a}, x_2 = \frac{aa - bg + a^2}{bp}, x_3 = 0\right)$	Predator washed out
$e5\left(x_1 = \frac{ap - qs}{bp}, x_2 = 0, x_3 = \frac{q(Abp + Bap - Bqs)}{bp(Cq - p)}\right)$	Infected prey species washed out
$e6(x_1 = x_1^*, x_2 = x_2^*, x_3 = x_3^*)$	Coexistence

8.4.2 Analysis of equilibrium point's existence and stability

System (8.1)'s Jacobian matrix at the point of equilibrium $E = (x_1(t), x_2(t), x_3(t))$ is given by

$$J = \begin{bmatrix} m_{11} & -ax_1 & m_{13} \\ ax_2 & ax_1 - fx_3 - g & -fx_2 \\ -\frac{Bpx_3{}^2}{(Bx_1 + Cx_3 + A)^2} & jx_3 & m_{33} \end{bmatrix}$$

where $m_{11} = -ax_2 - bx_1 + a - \frac{sx_3}{Bx_1 + Cx_3 + A} + x_1\left(-b + \frac{Bsx_3}{(Bx_1 + Cx_3 + A)^2}\right)$,

$$m_{13} = x_1\left(-\frac{s}{Bx_1 + Cx_3 + A} + \frac{Csx_3}{(Bx_1 + Cx_3 + A)^2}\right),$$

$$m_{33} = jx_2 + \frac{px_3}{Bx_1 + Cx_3 + A} - q + x_3\left(\frac{p}{Bx_1 + Cx_3 + A} - \frac{Cpx_3}{(Bx_1 + Cx_3 + A)^2}\right)$$

The stability of the dynamical system described by eq. (1) is determined by the nature of the eigenvalues derived from the characteristic equation of the corresponding Jacobian matrix. When all three eigenvalues have negative real roots or negative real components in the case of complex roots, the system stabilizes, reflecting equilibrium in population dynamics. However, if any eigenvalue fails to meet these criteria such as having a positive real root or component the system becomes unstable, indicating divergence or oscillatory behavior. The type of eigenvalues plays a crucial role in understanding the local stability and predicting the dynamic outcomes of the modeled interactions.

Theorem 8.4.1: At the equilibrium point $\{x_1 = 0, x_2 = 0, \ x_3 = -\frac{Aq}{cq-p}\}$, the dynamical system (8.1) is stable provided that $\lambda_1 < o$ and $\lambda_2 < o$ and the conditions $qs > ap$ and $cgq > fqA + gp$, λ_3 have negative actual parts.

Proof. The matrix of jacobian is $e2 = \begin{bmatrix} a + \frac{sAq}{cq-p}\left(-\frac{CAq}{cq-p} + A\right)^{-1} & 0 & 0 \\ ax_2 & \frac{fAq}{Cq-p} - g & 0 \\ -\frac{q^2pA^2B}{(Cq-p)^2}\left(-\frac{CAq}{Cq-p} + A\right)^{-1} & \frac{-jAq}{Cq-p} & h_{33} \end{bmatrix}$

where $h_{33} = -\frac{pAq}{Cq-p}\left(-\frac{CAq}{Cq-p} + A\right)^{-1} - q - \frac{pAq}{Cq-p}\left(\mathrm{p}\left(-\frac{CAq}{Cq-p} + A\right)^{-1} + \frac{pACq}{Cq-p}\left(-\frac{CAq}{Cq-p} + A\right)^{-2}\right)$.

Now the eigen values are $\lambda_1 = \frac{ap-qs}{p} < 0$, $\lambda_2 = \frac{Afq - Cgq + gp}{cq-p} < 0$, $\lambda_3 = \frac{q(Cq-p)}{p} < 0$.

Theorem 8.4.2: At the equilibrium point $\left\{x_1 = o, x_2 = \frac{Afq - Cgq + gp}{j(Af - Cg)}, x_3 = -\frac{g}{f}\right\}$, the dynamical system (8.1) is stable provided that $\lambda_1 < 0$ and the conditions $Afqa + Cagj + gpa > aAfj + Cgqa + gjs$, λ_2 and λ_3 have negative actual parts.

Proof. The jacobian matrix is

$$e3 = \begin{bmatrix} a - \frac{a(Afq - Cgq + gp)}{j(Af - Cg)} + \frac{sg}{f}\left(-\frac{Cg}{f} + A\right)^{-1} & 0 & 0 \\ \frac{a(Afq - Cgq + gp)}{j(Af - Cg)} & 0 & \frac{-f(Afq - Cgq + gp)}{j(Af - Cg)} \\ -\frac{g^2pB}{f^2}\left(-\frac{CAq}{Cq-p} + A\right)^{-1} & \frac{-jg}{f} & K_{33} \end{bmatrix}$$

where $K_{33} = \frac{Afq - Cgq + gp}{j(Af - Cg)} - \frac{pg}{f}\left(-\frac{Cg}{f} + A\right)^{-1} - q - \frac{g}{f}\left(p\left(-\frac{Cg}{f} + A\right)^{-1} + \frac{pgC}{f}\left(-\frac{Cg}{f} + A\right)^{-2}\right)$.

Now the eigenvalues are $\lambda_1 = \frac{Aafj - Aafq - Cagj + Cagq - agp + gjs}{j(Af - Cg)}$, $\lambda_2 = -\frac{gfpA - \sqrt{k1}}{2(A^2f^2 - 2ACfg + C^2g^2)}$,
$\lambda_3 = -\frac{gfpA + \sqrt{k1}}{2(A^2f^2 - 2ACfg + C^2g^2)}$

where $k1 = 4A^4f^4gq - 16A^3Cf^3g^2q + 24A^2C^2f^2g^3q - 16AC^3fg^4q + 4C^4g^5q + 4A^3f^3g^2p - 12A^2Cf^2g^3p + 12AC^2fg^4p - 4C^3g^5p + A^2f^2g^2p^2$.

Theorem 8.4.3: At equilibrium point $\left\{x_1 = \frac{a}{b}, x_2 = 0, x_3 = 0\right\}$, the dynamical system (8.1) is stable only if $a > 0$, $bg > aa$, $q > 0$.

Proof. The matrix of jacobian $e3 = \begin{bmatrix} -a & -\frac{aa}{b} & -\frac{sa}{b}\left(\frac{Ba}{b}+A\right)^{-1} \\ 0 & \frac{aa}{b}-g & 0 \\ 0 & 0 & -q \end{bmatrix}$

Now the eigen values are $\lambda_1 = -a < 0$, $\lambda_2 = \frac{aa - bg}{b} < 0$, $\lambda_3 = -q < 0$.

Theorem 8.4.4: At equilibrium point $\left\{x_1 = \frac{g}{a}, x_2 = \frac{aa - bg + a^2}{bp}, x_3 = 0\right\}$, the dynamical system (8.1) is stable provided that $\lambda_1 < 0$ and the instances $a^2q + bgj > aja$ and λ_2, λ_3 have nonpositive actual parts.

Proof. The jacobian matrix $e4 = \begin{bmatrix} a - \frac{2bg}{a} - \frac{aa - bg}{a} & -g & -\frac{gs}{a}\left(-\frac{Bg}{a}+A\right)^{-1} \\ \frac{aa - bg}{a} & 0 & -\frac{(aa-bg)f}{a^2} \\ 0 & 0 & \frac{(aa-bg)j}{a^2} - q \end{bmatrix}$

Now the eigen values are $\lambda_1 = \frac{aaj - bjg - qa^2}{a^2} < 0$,

$$\lambda_2 = \frac{-bg + \sqrt{4abg^2 - 4aga^2 + g^2b^2}}{2a} < 0, \lambda_3 = -\frac{bg + \sqrt{4abg^2 - 4aga^2 + g^2b^2}}{2a} < 0.$$

Theorem 8.4.5: At equilibrium point $\left\{x_1 = \frac{ap - qs}{bp}, x_2 = 0, x_3 = \frac{-q(Abp + Bap - Bqs)}{bp(Cq - p)}\right\}$, the dynamical system (8.1) is stable provided that $\lambda_1 < 0$ and the factors $Bfq^2s + ap^2a + Cbgpq + Cq^2sa > apqs + bgp^2 + Capqa + Abfpq + Bafpq$ and λ_2, λ_3 have nonpositive actual parts.

Theorem 8.4.6: The interior equilibrium point $(x_1 = x_1^*, x_2 = x_2^*, x_3 = x_3^*)$ is asymptotically locally stable.

Proof. The Jacobian matrix's variation $e6 = \begin{bmatrix} m_{11} & -ax_1^* & m_{13} \\ ax_2^* & ax_1^* - fx_3^* - g & -fx_2^* \\ m_{31} & jx_3^* & m_{33} \end{bmatrix}$

where $m_{11} = -ax_2^* - bx_1^* + a - \frac{sx_3^*}{Bx_1^* + Cx_3^* + A} + x_1^*\left(-b + \frac{Bsx_3^*}{(Bx_1^* + Cx_3^* + A)^2}\right)$,

$$m_{13} = x_1^*\left(-\frac{s}{Bx_1^* + Cx_3^* + A} + \frac{Csx_3^*}{(Bx_1 + Cx_3 + A)^2}\right), m_{31} = -\frac{Bpx_3^{*2}}{Bx_1^* + Cx_3^* + A)^2}$$

$$m_{33} = jx_2^* + \frac{px_3^*}{Bx_1^* + Cx_3^* + A} - q + x_3^*\left(\frac{p}{Bx_1^* + Cx_3^* + A} - \frac{Cpx_3^*}{Bx_1^* + Cx_3^* + A)^2}\right)$$

The equation of the feature form is $\Lambda(\lambda) = \lambda^3 + A_1\lambda^2 + A_2\lambda + A_3$.

Here $A_1 = -(m_1 + m_{22} + m_{33}) = 3I + fx_3^* + g - x_1^* a + m_{33} - m_{11}$,

$$A_2 = m_{1l}m_{22} + m_{11}m_{33} + m_{12}m_{2l} - m_{13}m_{3l} + m_{22}m_{33} + m_{23}m_{32}$$

$$= 3I^2 - \left(2m_{1l} - 2fx_3^* - 2g + 2x_1^* a + 2m_{33}\right)I + m_{11}x_1^* a - m_{11}fx_3^* - m_{11}g + x_1^* x_2^* a^2 + m_{11}m_{33}$$

$$- m_{13}m_{31} + m_{33}x_1^* a - m_{33}fx_3^* - m_{33}g - fjx_2^* x_3^*,$$

$$A_3 = m_{11}m_{32}m_{23} - m_{11}m_{22}m_{33} + m_{12}m_{21}m_{33} - m_{12}m_{23}m_{31} - m_{13}m_{21}m_{32} + m_{13}m_{22}m_{31}$$

$$= I^3 - \left(fx_3^* + g - x_1^* a - m_{33} - m_{11}\right)I^2$$

$$+ \left(m_{11}x_1^* a - m_{11}fx_3^* - m_{11}g + x_1^* x_2^* a^2 + m_{11}m_{33} - m_{13}m_{31} + m_{33}x_1^* a + m_{33}fx_3^* - m_{33}g + fjx_2^* x_3^*\right)$$

$$I - fjx_2^* x_3^* m_{11} - m_{11}m_{23}x_1^* a + m_{11}m_{33}x_3^* f - m_{33}x_1^* x_2^* a^2 + m_{11}m_{33}g - m_{31}x_1^* a x_2^* f$$

$$+ ax_1^* m_{13}m_{31} - fx_3^* m_{13}m_{31} - m_{13}gm_{31} - jax_2^* x_3^* m_{13}$$

According to the Routh-Hurwitz criteria , the equilibrium point e6 is locally asymptotically stable if all the eigenvalues of the Jacobian matrix have nonpositive real components. This condition is met when the coefficients satisfy $A_1>0$, $A_2> 0$, $A_3> 0$, $A_1A_2>A_3$, and the inequality $A_1A_2>A_3$ holds true. Therefore, the system described by the nonlinear differential equation (8.1) achieves local asymptotic stability around the nontrivial equilibrium point e6, provided these criteria are fulfilled as per the theorem. This analysis reinforces the stability behavior and ensures that the model remains mathematically consistent and applicable.

8.5 Numerica1 Results

The nonlinear differential equation system's numerical solution (8.1):

We start by taking the system's parameters $(a,b,\alpha,A,B,g,q,C,p,j,s,f)$ = (0.236, 1.05, 1.25, 0.23, 0.545, 0.005, 1.236, 1.450, 0.212, 1.024, 2.05, 2.11) at the population(x_1, x_2, x_3) = (0.235, 1.46, 1.056). The system in question is asymptoticaIly stable.

The initial condition is satisfied by the predator popuIation (refer Figure 8.1), if we use the system's parameter (x_1, x_2, x_3) = (0, 0, 1.07) as shown in point (1).

If we use the system's parameter as (x_1, x_2, x_3) = (0.0.6,0.7), (0,0.236,1.07), then the initial condition is satisfied by the infected prey and predator population (refer Figure 8.2).

If we use the system's parameter as (x_1, x_2, x_3) = (1.23, 0, 0), then the initial condition is satisfied by the susceptible prey population (refer Figure 8.3).

If we use the system's parameter as(x_1, x_2, x_3) = (0.4, 1.23, 0), (0.2, 2.0, 0), then the initial condition is satisfied by the susceptible-infected prey population (refer Figure 8.4).

If we use the system's parameter as (x_1, x_2, x_3) = (0.4, 0, 1.23), (2.5, 0, 0.45), then the initial condition is satisfied by the susceptible prey and predator population (refer Figure 8.5).

If we use the system's parameter as (x_1, x_2, x_3) = (0.4, 1.0, 1.23), (0.2, 2.0, 1.0), (2.5, 1.01, 0.45), then the initial condition is satisfied by the prey-predator population (refer Figure 8.6).

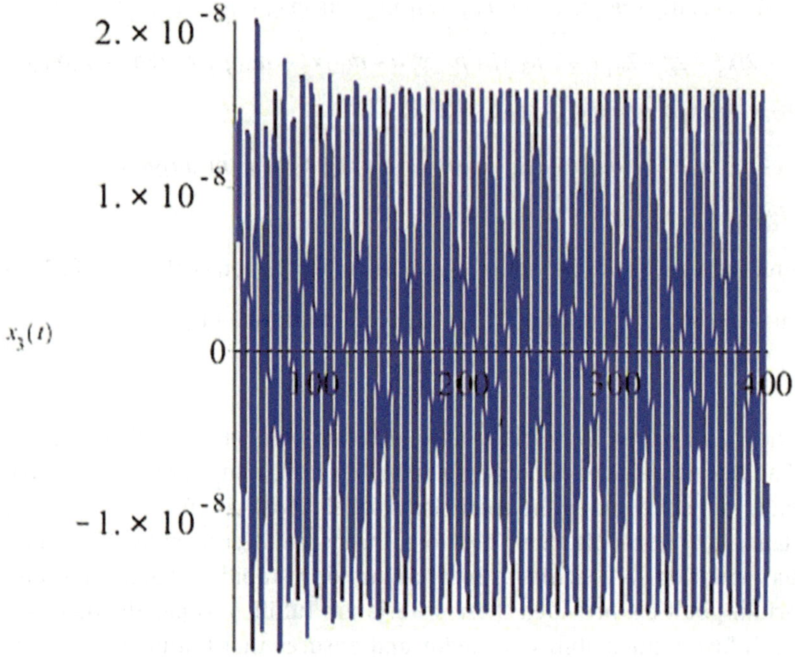

Figure 8.1: Rossler type of predator interaction.

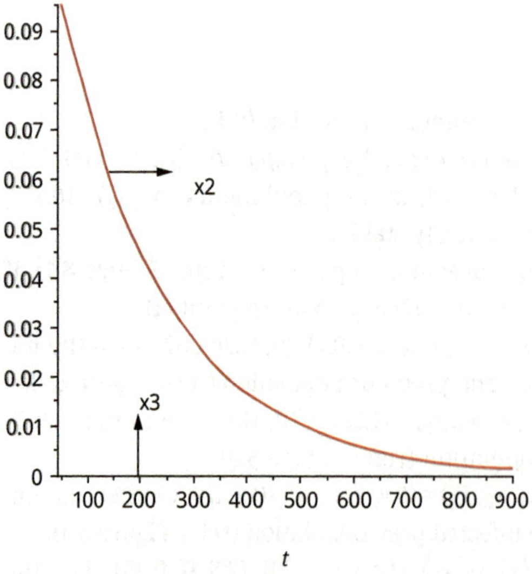

Figure 8.2: Infected prey and predator interaction.

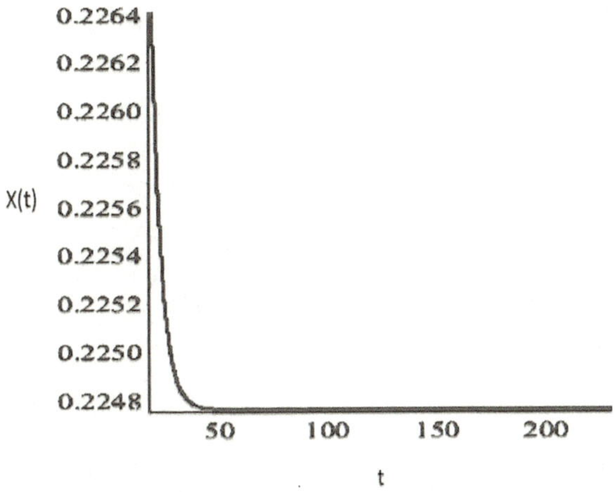

Figure 8.3: Rossler-type interactions with predators.

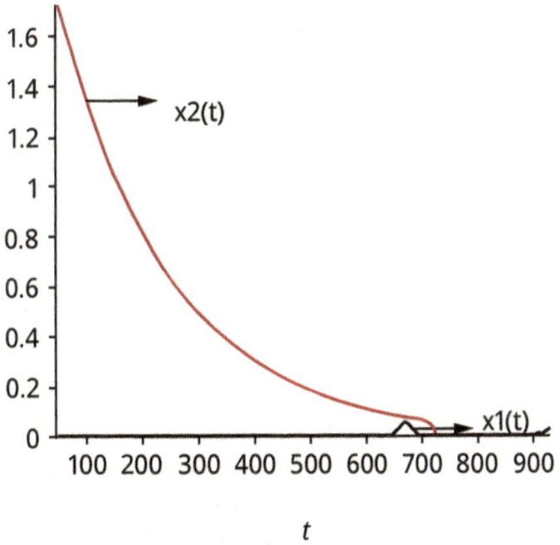

Figure 8.4: Interaction between susceptible and infected prey.

8.6 Discussions and Conclusions

An eco-epidemiological model was developed and analyzed based on the interactions between susceptible-diseased prey and predator populations, modeled using an S-I framework to represent COVID-19 dynamics. The system exhibits seven equilibrium points, encompassing trivial, washed-out, and nontrivial states, and is shown to maintain positivity and boundedness. The inclusion of the Rossler system highlights the potential for chaotic oscillations and attractors within the model. A stability analysis was conducted to assess system behavior, with key results detailed in Table 8.2:

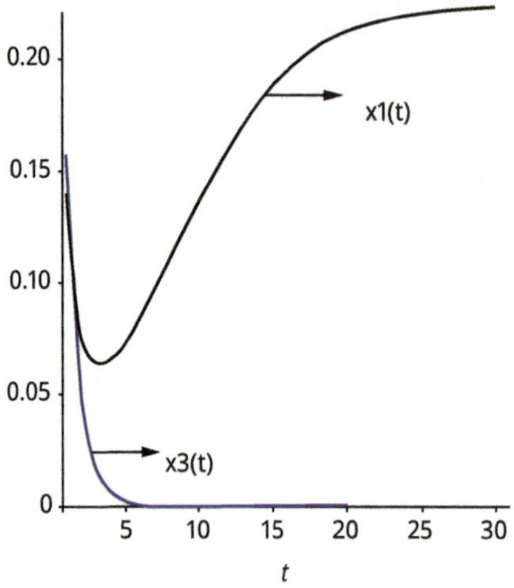

Figure 8.5: Susceptible prey and predator interaction.

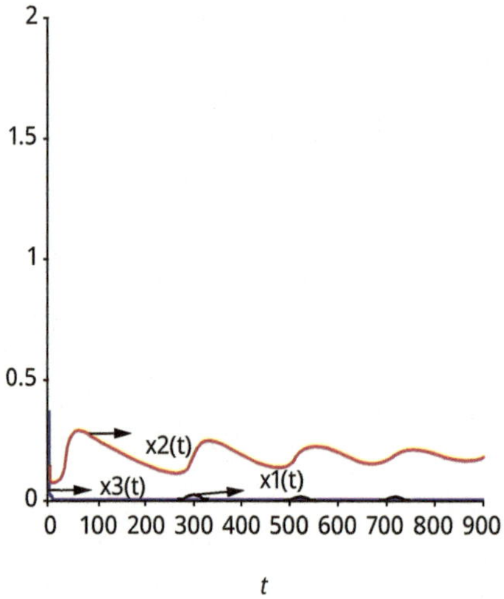

Figure 8.6: Prey-predator interaction.

- The predator population only occurs and achieves rhythmic dynamics (refer Figure 8.1).
- The infected prey is declining and the predator approaches to zero value refer Figure 8.2
- As shown in Figure 8.3, the vulnerable prey is declining.
- As shown in Figure 8.4, the susceptible prey reached zero equilibrium value and the infected prey began to decline.
- As shown in Figure 8.5, the range of predators is shrinking while the vulnerable prey is growing.

- The predator dropped to zero equilibrium value in the inner spots, the vulnerable prey decreased, and the diseased prey decreased as well (Figure 8.6).

AI plays an integral role in enhancing the robustness and applicability of the model. It is utilized for parameter optimization, where machine learning algorithms refine key parameters, such as infection rates and predation efficiency, based on synthetic or real-world data. Additionally, AI supports sensitivity analysis by identifying how changes in parameters impact the stability and behavior of the system. AI applications extend to scenario simulations, enabling the exploration of the effects of public health interventions such as lockdowns and social distancing on the dynamics of the disease. By leveraging AI, the study integrates predictive and adaptive capabilities, ensuring that the model remains relevant and actionable for pandemic management.

To curb disease transmission, the model emphasizes the importance of segregating vulnerable and afflicted populations. Preventive measures, such as self-quarantine and isolation, are shown to significantly reduce the spread of infection. The integration of AI tools strengthens the model's predictive power and provides actionable insights, making it a valuable resource for understanding and managing disease dynamics in an ecological and epidemiological context.

References

[1] Upadhyay RK, Iyengar SR. Introduction to Mathematical Modeling and Chaotic Dynamics. CRC press; 2013 Jul 23.

[2] Saikh A, Gazi NH Mathematical analysis of a predator–prey eco-epidemiological system under the reproduction of infected prey. Journal of Applied Mathematics and Computing. 2018 Oct;58 (1):621–646.

[3] Ackleh AS, Salceanu P, Veprauskas A A nullcline approach to global stability in discrete-time predator–prey models. Journal of Difference Equations and Applications. 2021 Aug 3;27 (8):1120–1133.

[4] Legreneur P, Laurin M, Bels V Predator–prey interactions paradigm: A new tool for artificial intelligence. Adaptive Behavior. 2012 Feb;20(1):3–9.

[5] Hamadneh NN, Tahir M, Khan WA Using artificial neural network with prey predator algorithm for prediction of the COVID-19: The case of Brazil and Mexico. Mathematics. 2021 Jan 18;9(2):180.

[6] Cantrell RS, Cosner C On the dynamics of predator–prey models with the Beddington–DeAngelis functional response. Journal of Mathematical Analysis and Applications. 2001 May 1;257(1):206–222.

[7] Saranya S, Vijaya S, Rekha E Susceptible-Infected-Treatment model with disease infection in prey population recovered by treatment. Computational Ecology and Software. 2022 Mar 1;12(1):12–22.

[8] Okosun KO, Makinde OD, Takaidza I Impact of optimal control on the treatment of HIV/AIDS and screening of unaware infectives. Applied Mathematical Modelling. 2013 Mar 15;37(6):3802–3820.

[9] Sharma S, Samanta GP Analysis of a two prey one predator system with disease in the first prey population. International Journal of Dynamics and Control. 2015 Sep;3(3):210–224.

[10] Ganguli C, Kar TK, Mondal PK Optimal harvesting of a prey–predator model with variable carrying capacity. International Journal of Biomathematics. 2017 Jul 12;10(05):1750069.

[11] Uma SS, Kumar KS, Suresh P, Dhivya P 18 Optimizing neural network training by addressing key challenges and advanced techniques. Math Optimization for Artificial Intelligence: Heuristic and Metaheuristic Methods for Robotics and Machine Learning. 2025 Apr 21;2:369.

[12] Hugo A, Simanjilo E Analysis of an eco-epidemiological model under optimal control measures for infected prey. Applications and Applied Mathematics: An International Journal (AAM). 2019;14(1):8.

[13] Kar TK Modelling and analysis of a harvested prey–predator system incorporating a prey refuge. Journal of Computational and Applied Mathematics. 2006 Jan 1;185(1):19–33.

[14] Hugo A, Massawe ES, Makinde OD An eco-epidemiological mathematical model with treatment and disease infection in both prey and predator population. Journal of Ecology and the Natural Environment. 2012 Jul;4(10):266–279.

[15] Musa SS, Qureshi S, Zhao S, Yusuf A, Mustapha UT, He D Mathematical modeling of COVID-19 epidemic with effect of awareness programs. Infectious Disease Modelling. 2021 Jan 1;6:448–460.

[16] Bezabih AF, Edessa GK, Rao KP Eco-epidemiological modelling and analysis of prey-predator population. Science Journal of Applied Mathematics and Statistics. 2021 Feb;9(1):1–4.

[17] Thota S A mathematical study on a diseased prey-predator model with predator harvesting. Asian Journal of Fuzzy and Applied Mathematics. 2020 Jun;8(2).

[18] Aziz-Alaoui MA, Okiye MD Boundedness and global stability for a predator-prey model with modified Leslie-Gower and Holling-type II schemes. Applied Mathematics Letters. 2003 Oct 1;16 (7):1069–1075.

[19] Simon JS, Rabago JF Optimal control for a predator-prey model with disease in the prey population. Malaysian Journal of Mathematical Sciences. 2018 May 31;12(2):269–285.

S. Sowmyayani, S. Kulandai Therese, Rajeswari Rajesh Immanuel,
S. Maheswari, P. Maria Joshy Roy, J. Kavitha

9 Predicting Mathematics Achievement in Secondary Education Using Regression Analysis

Abstract: Mathematics makes an individual shine. Its interest and skill depend on an individual's learning behavior. The learning behavior and other skill-based tests are one of the important areas in research. There are several types of research for predicting students' results by marks and other skills for school and college students. This chapter focuses on mathematics achievement, which plays a crucial role in all students. The achievement in mathematics is predicted using several regression analyses. Data are collected from the local area, which are preprocessed and analyzed. Several statistical analyses are carried out to determine the correlation of data. As deep learning and artificial intelligence emphasize more prediction methodologies, it has been used to predict the mathematical achievements of school students. Five different regression models are chosen for prediction. Experimental results demonstrated that support vector regression predicts more accurately than other regression models with fewer error values.

Keywords: Prediction, regression, correlation, deep learning

9.1 Introduction

The study of learning behavior and achievement in mathematics is a multilayered examination of the complicated subtleties among intellectual processes, instructional methodologies, and individual differences that shape students' mathematical performance. Mathematics, as both a discipline and a skill, is important in modern education systems and society. Proficiency in mathematics facilitates academic success and fosters critical thinking, problem-solving abilities, and quantitative reasoning essential for navigating various aspects of life. This study aims to explore these connections to inform evidence-based practices in mathematics education.

Pupils who believe they are proficient in a given subject are more likely to succeed in their careers than those who believe they are not. Secondary education comes after primary education and before higher education. Although the exact age range differs by country, it usually covers students between 12 and 18. Compared to basic school, secondary education offers a more specialized and comprehensive curriculum, making it a crucial phase in a student's academic path.

https://doi.org/10.1515/9783112214374-009

During secondary education, students usually receive a more specialized and in-depth education in various subjects compared to their earlier years in primary school. The curriculum often includes a few subjects such as languages, mathematics, science, arts, social studies, and electives like foreign languages, arts, or vocational courses. The goal of secondary education is to provide an accomplished education, help them develop critical thinking skills, and prepare them for further academic pursuits.

In many countries, secondary education is compulsory up to a certain age, and successful completion of this level is often a prerequisite for higher education or entering the workforce. Secondary education may be delivered in different types of institutions, including high schools, secondary schools, or colleges, depending on the educational system in place. Secondary education plays an important role in developing individuals academically, socially, and personally, providing them with the foundation needed for future endeavors, whether in further education or the workforce.

One way to conceptualize a learning behavior is as a behavior that is essential to learn in a classroom environment. Learning behavior in mathematics refers to the observable actions, attitudes, and cognitive processes that individuals exhibit while acquiring mathematical knowledge and skills. It involves how students approach mathematical problems, engage with mathematical concepts, and persist in solving mathematical tasks. Understanding learning behavior in mathematics is crucial for educators to design high-level teaching strategies to support student learning.

Learning behavior differs from person to person and it is a bundle of habits. Students' performance in mathematics depends on the adopted teaching strategies, as well as students' learning behavior. The primary contextual factors that impact pupils' learning include their family, surroundings, and historical and cultural backgrounds. While negative attitudes can be distorting, repellent, and ultimately result in bad performance, they can also have an impact on students' development of reasoning skills. On the other hand, good learning behaviors can yield a high degree of cognition. Students' learning behaviors may have a significant impact on how they learn, which may be reflected in their academic achievement in mathematics.

An achievement test is any assessment that gauges a person's achievements following a period of instruction or learning. In the case of mathematics, it is a kind of evaluation intended to gauge an individual's proficiency with mathematical ideas and concepts. These tests are typically used to evaluate a student's academic progress in mathematics and determine how well they have mastered specific mathematical content or skills within a given curriculum. These tests are often administered at various educational levels, such as elementary, middle, high school, and college.

The results of achievement tests can be used by educators to identify the strengths and weaknesses of a student's mathematical understanding, guide instructional planning, and make informed decisions about the effectiveness of the curriculum. The level of a person's ability, the breadth and depth of his knowledge, and his mastery of a specific behavior or learning domain.

Academic achievement indicates partly the level of intellectual ability of a student, besides practice and perseverance in a learning climate of a satisfying nature. It also means brightness in an academic subject. Academic achievement of school students usually means achievement in all school subjects that are designed to be studied at a specific level of schooling.

It's important to note that achievement tests are different from aptitude tests, which are used to assess a person's logical thinking to learn many skills in a particular area, rather than measuring what they have already learned. Achievement tests in mathematics play a crucial role in assessing and improving the quality of mathematical education by providing a standardized and objective measure of students' proficiency in the subject. Regression models are more efficient in forecasting [1]. In this chapter, achievement in mathematics is predicted using regression models.

The objectives of this chapter include:

1. To find out the level of learning behavior of ninth standard students.
2. To predict achievement in mathematics of ninth standard students using machine learning.

The remainder of the chapter is organized as follows: Section 9.2 elaborates on less literature related to learning behavior. Section 9.3 discusses the data collection, regression models, and the flow of the work. Section 9.4 analyzes the data using correlation, performance measures, and so on. Section 9.5 concludes the work and paves the way for the future.

9.2 Literature Survey

This section deals with the available literature resources in the field of students' learning behavior and achievement in mathematics. A critical review summarizes the concept used.

The main elements that support the study process include how frequently a student attends study sessions, reviews the content, evaluates themselves, practices communicating the content, and studies in a comfortable setting in 2008 [2].

Study habits are the intentional use of a person's attitudes, thoughts and behaviors to improve knowledge and skill acquisition for many tasks and circumstances. Additionally, these writers uphold study habits that can be viewed as an application of a person's intellect, feelings, and behaviors toward the acquisition of knowledge and abilities to complete an assignment. It stands to reason that good study habits require sharp intellectual ability, steady emotions, and pertinent activities that are purposefully aimed at building information and developing abilities to achieve a goal.

According to Crow et al., the main goals of study behavior are to develop habits and knowledge that will help one meet new situations, interpret concepts, and form opinions and hone skills [3]. Hence, achievement depends upon the following academic activities such as study, interpretation, and application. Study habits differ from person to person. It is a truism to believe that due to the lack of good study habits, most students perform very poorly in school achievement.

Study habits are methods used in the process of learning [4]. Good study habits let students use their cognitive processes to find and assess pertinent bodies of knowledge. On the other hand, bad study habits make it more difficult for pupils to participate in productive intellectual activities. There is an empirical evidence to support the idea that study habits affect academic achievement. One could assume that when pupils display bad and unsuitable study habits, their academic performance will decline.

Since it depends on its level of efficacy for optimal performance, academic achievement is a behavior and a significant factor for students involved in the educational process [5]. Achievement is defined as knowledge or skills acquired by students, typically in academic disciplines, as determined by test results, teacher-assigned grades, or both. Furthermore, achievement is defined as a successful performance, gain, or accomplishment made by an individual or group upon finishing a task, be it academic, manual, social, or personal [6]. Academic achievement is the sum of the knowledge acquired after finishing a course of study, either completely or partially and receiving a specific grade on an achievement test.

Two Indian states placed 72nd and 73rd out of 74 in both reading and math, the survey said. The media expressed dissatisfaction with the low PISA scores of Indian school children. India withdrew from the 2015 PISA cycle because government officials believed that Indian children were not yet ready for the test. Therefore, academic achievement or academic performance is the measurement of educational gain from which a student, teacher, or institution achieved their educational goals. Generally, academic achievement is measured by examination or continuous assessment. In this study, achievement in mathematics means the measured skills of mathematical operation developed by the student in a particular school grade.

The language of mathematics is universal. Learning mathematics will enhance one's capacity for critical analysis and reasoning. The child's ability to think analytically enables them to approach difficulties in real life more rationally. In the achievement of mathematics, the questions are taken from the *Samacheer Kalvi TN Board Class Nine Maths* syllabus covers topics like the concept of base, index, root and power, concept of one solution, many solutions and no solution, concept of distance between two points, concept of rectangle, square, rhombus, concept of polynomial, zero polynomial, concept of graphical representation, concept of common logarithm, and concept of probability theory.

The questions for the learning behavior inventory describe the inventory's development process and its intended use in assessing student's learning behaviors in mathematics like their mathematical skills, time management, and preparation for tests. It summarizes the overall assessment of the learning behavior inventory in mathematics, highlighting its strengths, weaknesses, and potential implications for educational practice.

By building a learner model that produces temporal trajectories that show how children's knowledge develops, adaptive learning technologies (ALTs) aid in children's learning [7]. This learner model states that although ALTs adjust the difficulty of problems for each child, kids still need to maintain accuracy and effort and regulate their practice habits. Consequently, the temporal trajectories are expected to show both the growth of children's knowledge and their ability to regulate. Clusters of these trajectories are examined in order to more accurately pinpoint areas where children's self-learning and potential needs are lacking. Another method was used to cluster few trajectories who were acquiring three skills of differing difficulty.

The resulting clusters were investigated and indications of behavior management. Looking at the temporal pattern clusters in the young people's learning data can help determine whether self-regulated learning assistance is essential. A deep artificial neural network is applied to a set of attributes collected from virtual learning environments to anticipate endangered pupils and provide early intervention options in such cases [8]. This model outperforms both the support vector machine (SVM) and the logistic regression models.

A cluster model for classifying students according to their cognitive types in the e-learning environment and a classification model for predicting their performance have both been used [9]. The classification models presented in this work assist educators, learners, and people in business in identifying pupils who have the potential to thrive in a certain subject early on. The predictive learning effect has been studied and the features of college students' learning behaviors are analyzed by developing a model of the learning impact based on information learning behavior. Additionally, a *Moodle* add-on has been used to extract data needed for educational data mining (EDM) [10].

Data on information literacy learning from 320 college students from a Chinese university was used in the trial. The qualities of information thinking and the learning impact are significantly correlated, according to an analysis of college students' information literacy learning behavior using the Pearson algorithm.

While learning outcomes are thought to enhance instruction and learning, there is still ongoing research on how to predict student results [11]. To provide a basic grasp of the clever methods used for predicting student performance, 10 years of study effort, done between 2010 and 2020, was surveyed. Performance class rankings and accomplishment scores were the primary metrics used to gauge the attainment of learning objectives. Lastly, the most obvious indicators of learning outcomes were students' academic emotions, grades on term assessments, and online learning activities.

In higher education, student participation is crucial to advancing student learning outcomes [12]. The findings also showed that there are two types of factors that affect student engagement: encouraging factors and inhibiting factors. Promoting factors include the teacher-student relationship, the positive emotions and learning behaviors of the students, the positive learning behavior of the teacher, the learning and thinking skills of the students, the availability of resources, the individual traits of the students, and teaching factors. Negative student and instructor behavior, as well as a lack of environmental support, are the impediments.

The students' involvement and engagement in the learning process is increased by higher education institutions by including the creation of classrooms for active learning into their larger educational programs [13]. Because of this, studies have examined how social factors impact collaborative learning and engagement, which in turn impacts the learning outcomes of students. Given the growing prevalence of online learning in education, it has been found that social elements and collaborative learning in general enhance student learning activities. Because online learning has an impact on students' academic growth, higher education institutions ought to promote its use in their teaching and learning procedures.

A novel model has been created to predict undergraduate students' exam scores using raw data from their midterm exam scores [14]. To forecast students' final test grades, the effectiveness of many machine learning algorithms was computed and compared. Predictions were done for department, professor, and midterm exam results. This data-driven research is critical for informing decision-making in higher education. Finally, this study identifies students who are at risk of failing.

Through non-graded and graded courses, EDM approaches have been utilized to quantitatively analyses how students engage with an e-learning system [15]. For them, this method works well for creating guidelines for a series of short online courses. The outcome demonstrated that a student group's online access behavior might be altered by differences in learning environments. Based on their course access behavior, clustering has separated the student body into five access groups. Because of their differences in access from the other groups, it was possible to distinguish between the groups with the least and the most access.

To confirm the impact of students' online learning, numerous studies on student involvement have been carried out at postsecondary institutions [16]. There is a knowledge gap, though, as the different kinds of problems with student engagements that arise from learning analytics have not yet been thoroughly examined. A new systematic literature review has used Preferred Reporting Items for Systematic Reviews and Meta-Analyses (PRISMA) to assess 42 papers to close the knowledge gap. To improve the online learning performance, various forms of student involvement are studied where learning analytics are revealed. The results showed that, although there aren't many of them, several studies having measured multimodal involvement to improve students' learning performance.

9.3 Proposed Methodology

Data collection techniques refer to methods for collecting and analyzing various sorts of data. Efficient data collection provides the necessary information to assess performance, answer questions, and estimate future trends. This chapter deals with the objectives, hypothesis, and data collection techniques. Figure 9.1 displays the workflow of the proposed prediction methodology.

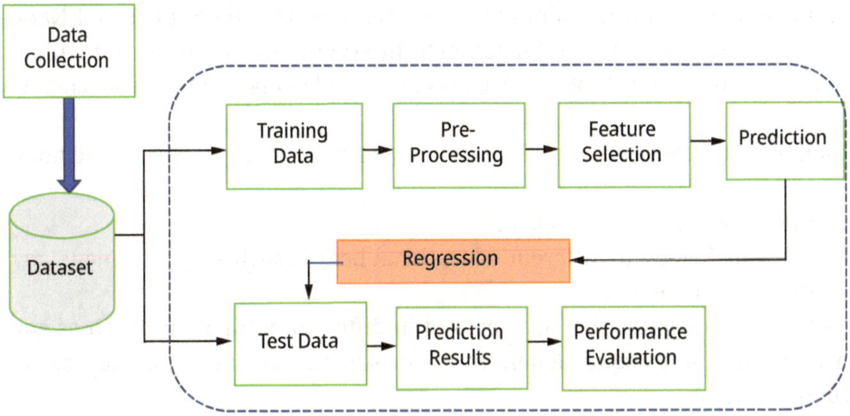

Figure 9.1: Proposed system architecture to predict mathematics achievement.

9.3.1 Data Collection

We conducted a well-organized data collection process for our study on achievement tests and learning behavior in mathematics among ninth-grade school students. We selected specific schools within our district as the target locations for our study. Permission is received from the selected schools, explaining the purpose and nature of our study. Upon receiving permission, we were allowed to meet with the ninth-grade students of the chosen schools.

Before proceeding with the data collection, we sought permission from the respective class teachers. After gaining approval, we distributed the questionnaires designed by our team to the students. We provided the necessary information to the students about how to answer the questions effectively. Time limits were established for both the achievement test and learning behavior assessment, with 15 min for the achievement test and 20 min for the learning behavior assessment.

The entire process was repeated over three days to collect a total of 2,500 data points. There was good cooperation between the teachers, students, and our team throughout the data collection process. Through effective coordination and execution, we successfully collected the required data for our study. The sample of 250 students was selected or chosen by random sampling technique from ninth-grade students in Thoothukudi, comprising a diverse group of adolescents with varied academic backgrounds and interests.

The sample included students from different schools within the region, offering a representative cross-section of the local educational landscape. The data collection process involved surveys, allowing for a comprehensive understanding of the students' perspectives on their learning behavior and achievement in mathematics. This sample serves as a resource for educational research, shedding light on the knowledge of Mathematics within the ninth-grade students in Thoothukudi, Tamil Nadu, India, and providing a foundation for targeted interventions and improvements in the local educational system. Table 9.1 displays the sample data from the collected dataset.

The main characteristics of the secondary school students representing the population are:

1) The average age of the students is 14+.
2) All the students belong to different educational boards such as government, government-aided, and matriculation.
3) The parents of the population are engaged in different occupations, such as government, private, and coolie. Generally, all the students belong to middle-class socioeconomic status.
4) The parents of the population are engaged in different monthly incomes, such as below 15,000/15,000–30,000/above 30,000.

A well-organized data collection process is conducted on achievement tests and learning behavior in mathematics among ninth-grade school students. Specific schools within our district are selected as the target locations. Upon receiving permission, ninth-grade students of the chosen schools were met. After gaining approval, the questionnaires designed by our team were distributed to the students. Necessary information was provided to the students about how to answer the questions effectively. Time limits were established for both the achievement test and learning behavior assessment, with 15 min for the achievement test and 20 min for the learning behavior assessment.

The entire process was repeated over three days to collect a total of 950 data points. There was good cooperation between the teachers, students, and our team throughout the data collection process. Through effective coordination and execution, the required data were successfully collected.

9.3.1.1 Null Hypothesis

- There is no important difference between boys and girls.
- There is not much difference between rural and urban ninth standard students.

Table 9.1: Sample data from the collected dataset.

Student name	School	Gender	Religion	Locality of institution	Father's occupation	Family's monthly income	Mathematics achievement	Learning behavior	Educational board
xxxxx	Amirtha Vidhyalya	Male	Hindu	Urban	Government	Above 30,000	48	71	Matriculation
xxxxx	B.M.C. Matric Hr. Sec. School	Male	Hindu	Urban	Private	Below 15,000	44	70	Matriculation
xxxxx	St. Thomas Mat. Hr. Sec. School	Female	Hindu	Urban	Private	15,000–30,000	24	78	Matriculation
xxxxx	St. Thomas Mat. Hr. Sec. School	Female	Hindu	Urban	Private	Above 30,000	38	90	Matriculation
xxxxx	St. Thomas Mat. Hr. Sec. School	Female	Christian	Urban	Private	Above 30,000	36	87	Matriculation

- There is no significant difference among government, government-aided, and matriculation school.
- There is no important difference among ninth standard students in their achievement in mathematics based on their father's occupation (private, coolie, and government).
- There is no significant association between the family incomes of ninth standard students in their achievement in mathematics.

9.3.1.2 Preprocessing

The dataset is split into 80:20 for training and testing. All the rows whose columns have null values are removed. All the categorical character variables are properly trimmed to avoid spaces. Once this is done, those variables are converted to numerical categorical variables using factorization. The inclusion of this step is to better execute regression models.

9.3.2 Feature Selection

As mathematics achievement is a dependent variable and this work intends to predict mathematics achievement, regression is done using the other 9 independent variables. A regression model is fit when the feature is correctly selected. Permutation feature importance is utilized to select the important features. Figure 9.2 shows the importance score of each feature.

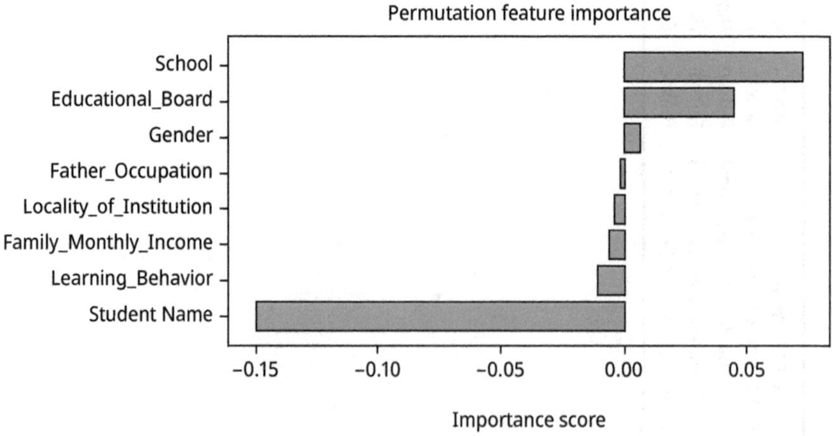

Figure 9.2: Feature importance.

From the above figure, it is analyzed that school, education board, and gender have high importance scores. Whereas other features have a negative score. Hence, the positive importance features are selected for regression.

9.3.3 Prediction

Establishing regression models and the implementation of algorithms such as linear regression (LR), logistic regression, support vector regression (SVR), decision tree regression (DTR), and random forest (RF) regression were employed to predict students' mathematics achievement. Statistical techniques like logistic regression and time series can be used when the goal is to analyses the reasons behind success or failure. On the other hand, neural networks, SVM, decision tree, and RF are more effective and produce more accurate outcomes when forecasting is the main focus.

Statistical techniques create a model that consistently forecasts output data from the given input. When a supervised optimization problem is presented, machine learning algorithms generate a model that fits the input with the desired target data. The entire model's performance is evaluated using the confusion matrix indicators. No single regressor is best for forecasting results, according to the research. Investigating which regressors are more explored for the data under analysis is therefore essential.

9.3.3.1 Linear Regression

LR (a supervised machine learning technique) determines the linear connection between the dependent variable and independent variables by fitting a linear equation. Simple LR is utilized when there is just one independent feature, whereas multiple LR is used when there are several features. Similarly, it is referred to as univariate LR when there is just one dependent variable and multivariate LR when there are multiple dependent variables.

One significant advantage of LR is its interpretability. A greater comprehension of the underlying dynamics is made possible by the model's equation. Because LR is transparent, simple to use, and a building block for more intricate algorithms, its simplicity is strength.

LR serves as the foundation for many sophisticated models and is not just a predictive tool. The usefulness of LR is increased by methods like regularization and SVM, which are inspired by it. Furthermore, LR is essential to assumption testing since it allows researchers to confirm important hypotheses regarding the data.

The equation for simple LR is:

$$y = \beta_0 + \beta_1 X \tag{9.1}$$

The equation for multiple LR is:

$$y = \beta_0 + \beta_1 X_1 + \cdots + \beta_n X_n \tag{9.2}$$

where y is the dependent variable; $X,\ X_1, X_2, \ldots, X_n$ are the independent variables; β_0 is the intercept; $\beta_1, \beta_2, \ldots, \beta_n$ are the slopes.

The algorithm's goal is to identify the best-fit equation that can predict the correct values based on the independent variables. Regression is the process of creating a function from a set of X and Y values. From an unknown X, Y can then be predicted using this function. A function that forecasts continuous Y is required when X is utilized as an independent variable in regression.

9.3.3.2 Logistic Regression (LogR)

When using the sigmoid function for binary classification, logistic regression is employed. It generates a probability value between 0 and 1 based on the independent variables as input. A sigmoid function converts independent variables to a range of 0 and 1; it uses LogR model for this conversion.

Let the independent input features be:

$$X = \begin{bmatrix} x11 & \ldots & x1m \\ x21 & \ldots & x2m \\ \vdots & \ddots & \vdots \\ xn1 & \ldots & xnm \end{bmatrix} \tag{9.3}$$

and the dependent variable Y having only 0 or 1:

$$Y = \{ \begin{matrix} 0 & \text{if Class 1} \\ 1 & \text{if Class 2} \end{matrix} \tag{9.4}$$

then, the multilinear function is applied as

$$z = \left(\sum_{i=1}^{n} w_i x_i \right) + b \tag{9.5}$$

Here x_i is the ith observation of x, $w_i = [w_1, w_2, \ldots, w_m]$ are the weights, and b is the intercept. This can be represented as:

$$z = w \cdot X + b \tag{9.6}$$

Now the sigmoid function used for the input z and y is predicted.

$$\sigma(z) = \frac{1}{1+e^{-z}} \tag{9.7}$$

The sigmoid function converts the continuous variable data into 0 and 1:

- $\sigma(z)$ tends toward 1 as $z \to \infty z \to \infty$ and 0 as $z \to -\infty z \to -\infty$
- $\sigma(z)$ lies between 0 and 1

where the probability of being a particular class can be measured as:

$$P(y=1) = \sigma(z) \tag{9.8}$$

$$P(y=0) = 1 - \sigma(z) \tag{9.9}$$

9.3.3.3 Support Vector Regression

One kind of SVM used for regression tasks is SVR. It looks for a function that can best predict the output for a given input. SVR is compatible with both linear and nonlinear kernels. A nonlinear kernel can detect more indirect data than a linear one. The ideal kernel is identified by the difficulty of the task and the features of the dataset.

Unlike SVMs used for classification problems, the SVR model searches a continuous space that best fits the data. The input variables are mapped to a huge feature space to find the hyperplane that minimizes the prediction error and maximizes the distance between the hyperplane and the closest data points. The SVR Model may handle nonlinear correlations between the input and goal variables by a kernel function to translate the data to a huge feature space. As a result, it works well for regression tasks with intricate interactions.

For regression issues, SVR applies the same idea as SVM. Considering the points that lie inside the decision boundary line is the aim of SVR in machine learning. The hyperplane with more points is the line that fits the data the best.

The equation of the hyperplane is given as:

$$Y = wx + b \tag{9.10}$$

Then the decision boundary will become:

$$wx + b = +a \tag{9.11}$$

$$wx + b = -a \tag{9.12}$$

Thus, any hyperplane that satisfies the SVM should satisfy:

$$-a < Y - wx + b < +a \tag{9.13}$$

The primary goal is to determine a decision boundary at "a" distance from the initial hyperplane so that the support vectors closest to the hyperplane fall inside that boundary line. Therefore, only the points with the lowest error rate within the decision boundary are taken. This will give us a better-fitting model. This model compulsorily applies feature scaling and it is difficult to understand.

9.3.3.4 Decision Tree Regression (DTR)

A model that uses a tree-like structure to predict numerical values is called a decision tree regression. It divides data according to important characteristics, branching out from a root. Before arriving at leaf nodes with final predictions, each node enquires about a feature, further segmenting the data. To get a result, the path is matched with the data's features from root to leaf. Unlike LR, a linear relationship is assumed between the input and the target, DTR can capture nonlinear relationships, making it a more flexible tool in many scenarios. This adaptability allows it to model complex data patterns that are often encountered in real-world applications.

Fundamentally, a DTR model constructs a tree based on feature values. Each leaf node indicates a predicted value and each internal node represents a decision based on the value of a particular feature. A major benefit over more intricate, black-box models is that decision trees' hierarchical, branching form makes them extremely interpretable since the decision-making process is simple to see and comprehend.

DTR is robust to outliers and can handle all types of data, enhancing its applicability across various domains. It does not assume any specific distribution for the data, thus providing a highly flexible and powerful approach to regression tasks. Dividing the data into subsets according to the values of input features is how DTR models operate. The aim is to split the data in such a way that the prediction error within each subset is minimized, leading to more accurate and homogenous predictions. DTR gives poor results on small datasets. This model often leads to overfitting.

9.3.3.5 Random Forest Regression (RFR)

As a supervised learning technique, RF operates on data that has labels or outcomes. It works by creating many decision trees, each one depending on a chosen subset of the input. The model aggregates the outcomes of each of these decision trees to produce a general forecast for unknown data points. This allows it to handle larger datasets and capture more complex linkages than individual decision trees. An algorithm for supervised machine learning is called RFR. It predicts continuous target variables using a group of decision trees. The bagging RF technique is used to build the individual decision tree models. Smaller decision trees are constructed from subsets of the training data that are chosen at random.

The RF model, which produces a single prediction value, is created by combining the smaller models following the bagging RF stage. By merging the predictions from multiple decision trees, the method helps lower variation and increase accuracy. Several variables that indicate possible features are found when creating an RFR model. To give useful information about how various aspects affect the predictions, these variables should also be connected to the results in some way. This model also sometimes leads to overfitting and we need to choose the number of trees.

9.4 Experimental Results

Various analyses are carried on the collected dataset. The metrics and the comparative study are also clearly discussed in this section.

Table 9.2: Statistical summary of the dataset.

Statistical method	Mathematics achievement	Learning behavior
Mean	46.98	78.89
Std	16.72	10.94
Min.	18	37
25%	34	73
50%	46	77
75%	56	87
Max.	100	105

9.4.1 Correlation Analysis of Learning Behavior and Mathematics Achievement

The statistical techniques such as percentage analysis, correlation, *t*-test, *F*-test, and chi-square test were used to analyze the test. Table 9.2 shows the summary of the collected dataset. Mathematics achievement and learning behavior are the most important features whose values are analyzed in this work. Hence, correlation analysis of learning behavior and learning impact can be used to model feature subset selection. The study of two or more components of related variables as a gauge of their level of relationship is known as correlation analysis.

For correlation analysis to be carried out, the connected items must be associated or likely in some way. We can claim that two variables have a high correlation if they are strongly interdependent. Positive correlation occurs when both groups' values rise simultaneously; negative correlation occurs when one group's value rises first, followed by the other group's value falling. Table 9.3 displays the correlation between mathematics achievement and learning behavior.

Table 9.3: Correlation analysis of the dataset.

Correlation	Value
Pearson	0.22
***t*-Test**	25.61
***F*-Test**	2.33
Chi-square test	5.46

9.4.2 Mathematics Achievement Prediction

By assuming Q_i, P_i as groundtruth and predicted values and $\overline{Q_i}$ is the average value of groundtruth data, the following metrics root mean square error (RMSE), mean absolute error (MAE), mean absolute percentage error (MAPE), and coefficient of determination (R^2) are formulated as follows:

$$\text{RMSE} = \sqrt{\frac{1}{n}\sum_{i=1}^{n}(P_i - Q_i)^2} \tag{9.14}$$

$$\text{MAE} = \frac{1}{n}\sum_{i=1}^{n}|P_i - Q_i| \tag{9.15}$$

$$\text{MAPE} = \frac{1}{n}\sum_{i=1}^{n}\frac{|P_i - Q_i|}{Q_i} \tag{9.16}$$

$$R^2 = 1 - \frac{\sum_{i=1}^{n}(P_i - \bar{Q}_i)^2}{\sum_{i=1}^{n}(O_i - \bar{Q}_i)^2} \tag{9.17}$$

Table 9.4 emphasizes the results of all the regression models implemented in this work. From the table, it is studied that SVR achieved very low error values with 0.66R^2 values. The DTR obtained a higher R^2 value of 0.73 than other models. It is also observed that SVR obtained 97.48% accuracy, which is higher than other regression models. DTR obtains an almost exact accuracy value of 97.23%. RFR model achieved lower error values than the DTR model. All the models obtain above 90% accuracy.

Figure 9.3 shows the bar chart of error values of the implemented regression models.

It is observed that the SVR model obtains a lower RMSE, MAE, and MAPE than other models. Table 9.5 compares the proposed method with other similar methods that use the OULA dataset. It is observed that Azizah et al.'s method achieved only 63.18% accuracy using the naïve Bayes model. Hassan et al.'s method achieved 97.25%

Table 9.4: Results of the analysis of various regression models.

Category	RMSE	MAE	MAPE	R^2	Accuracy
LR	0.25	0.32	0.22	0.561	93.64
LogR	0.45	0.52	0.26	0.4	92.7
SVR	0.19	0.25	0.17	0.66	97.48
DTR	0.24	0.36	0.24	0.73	97.23
RFR	0.21	0.29	0.2	0.63	96.45

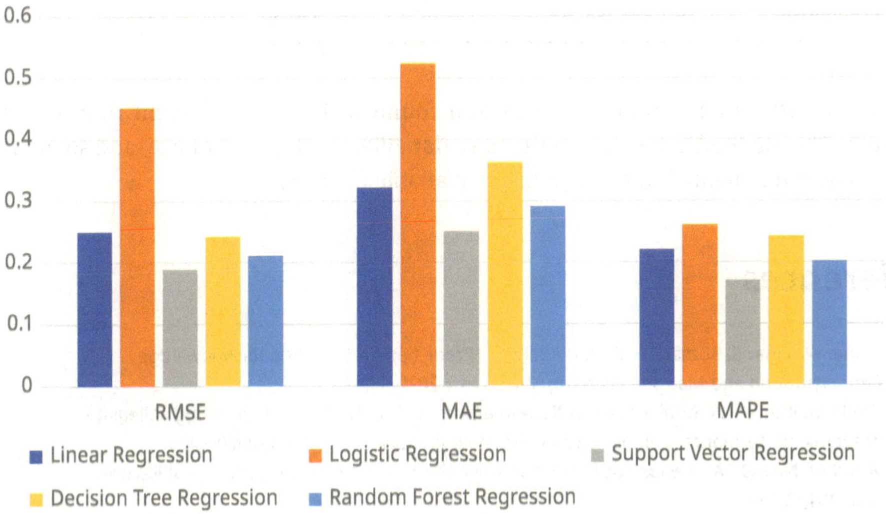

Figure 9.3: Bar chart of error values obtained by all regression models.

accuracy using an LSTM model. But the proposed method achieved 97.48% accuracy, which is higher than other methods.

9.5 Conclusion

Prediction is the biggest boon of artificial intelligence. Predicting weather, air quality, disease, and so on prevents many human lives. In this chapter, students' performance is predicted to improve for the poor-performing students. Ninth student's data are collected, preprocessed, analyzed, and mathematics achievement is predicted. The five most recent regression models are chosen from the literature and implemented to predict Mathematics Achievement. The important features are selected for regression. From the implemented regression models, it is observed that SVR achieved more accurate results with fewer error values than other models. The proposed method is

Table 9.5: Comparison of other methods.

Authors	Dataset	Techniques applied	Results
Hussain et al. [17]	OULA	J48	88.52
Heuer and Breiter [18]		SVM	90.85
Haiyang et al. [19]		DT	90
Hassan et al. [20]		LSTM	97.25
Wasif et al. [21]		RF	89
Azizah et al. [22]		NB	63.18
Rizvi et al. [23]		DT	83.14
Proposed method	Own dataset	RF	97.48

compared with other similar methods and found to be more efficient than other methods. This method can still include more data like marks, attendance, and so on. It can also be implemented in other publicly available datasets.

References

[1] Kumar A, Sagar S, Kumar TG, Kumar KS. (eds). Prediction and Analysis for Knowledge Representation and Machine Learning. CRC Press; 2022.

[2] Credé M, Kuncel NR Study habits, skills, and attitudes: The third pillar supporting collegiate academic performance. Perspectives on Psychological Science. 2008;3(6):425–453.

[3] Wood W, Neal DT A new look at habits and the habit-goal interface. Psychological Review. 2007;114(4):843.

[4] Rani R Construction and Standardization of Mathematics Achievement Test for IXth Grade Students. Educational Quest-An International Journal of Education and Applied Social Sciences. 2017;8 (3):629–633.

[5] Qureshi MA, Khaskheli A, Qureshi JA, Raza SA, Yousufi SQ Factors affecting students' learning performance through collaborative learning and engagement. Interactive Learning Environments. 2023;31(4):2371–2391.

[6] Downie NM Fundamentals of Measurement. New York: Oxford University Press; 1961.

[7] Dijkstra SHE, Hinne M, Segers E, Molenaar I Clustering children's learning behaviour to identify self-regulated learning support needs. Computers in Human Behavior. 2023;145:107754.

[8] Waheed H, Hassan SU, Aljohani NR, Hardman J, Alelyani S, Nawaz R Predicting academic performance of students from VLE big data using deep learning models. Computers in Human Behavior. 2020;104:106189.

[9] Jovanovic M, Vukicevic M, Milovanovic M, Minovic M Using data mining on student behavior and cognitive style data for improving e-learning systems: A case study. International Journal of Computational Intelligence Systems. 2012;5(3):597–610.

[10] Shi Y, Sun F, Zuo H, Peng F Analysis of Learning Behavior Characteristics and Prediction of Learning Effect for Improving College Students' Information Literacy Based on Machine Learning.IEEE Access; 2023. Vol. 11. 50447–50461.

[11] Namoun A, Alshanqiti A Predicting student performance using data mining and learning analytics techniques: A systematic literature review. Applied Sciences. 2020;11(1):237.

[12] Li J, Xue E Dynamic interaction between student learning behaviour and learning environment: Meta-analysis of student engagement and its influencing factors. Behavioral Sciences. 2023;13(1):59.
[13] Qureshi MA, Khaskheli A, Qureshi JA, Raza SA, Yousufi SQ Factors affecting students' learning performance through collaborative learning and engagement. Interactive Learning Environments. 2023;31(4):2371–2391.
[14] Yağcı M Educational data mining: Prediction of students' academic performance using machine learning algorithms. Smart Learning Environments. 2022;9(1):11.
[15] Ratnapala IP, Ragel RG, Deegalla S Students Behavioural Analysis in an Online Learning Environment Using Data Mining. In 7th International Conference on Information and Automation for Sustainability. 2014, December (pp. 1–7). IEEE.
[16] Johar NA, Kew SN, Tasir Z, Koh E Learning analytics on student engagement to enhance students' learning performance: A systematic review. Sustainability. 2023;15(10):7849.
[17] Hussain M, Zhu W, Zhang W, Abidi SMR Student Engagement Predictions in an e-Learning System and Their Impact on Student Course Assessment Scores. Computational Intelligence and Neuroscience. 2018;2018(1):6347186.
[18] Heuer H, Breiter A, 2018. Student success prediction and the trade-off between big data and data minimization.
[19] Haiyang L, Wang Z, Benachour P, Tubman P A Time Series Classification Method for Behaviour-based Dropout Prediction. In 2018 IEEE 18th International Conference on Advanced Learning Technologies (ICALT). 2018, July (pp. 191–195). IEEE.
[20] Hassan SU, Waheed H, Aljohani NR, Ali M, Ventura S, Herrera F Virtual learning environment to predict withdrawal by leveraging deep learning. International Journal of Intelligent Systems. 2019;34(8):1935–1952.
[21] Hassan SU, Waheed H, Aljohani NR, Ali M, Ventura S, Herrera F Virtual learning environment to predict withdrawal by leveraging deep learning. International Journal of Intelligent Systems. 2019;34(8):1935–1952.
[22] Azizah EN, Pujianto U, Nugraha E Comparative Performance between C4. 5 And Naive Bayes Classifiers in Predicting Student Academic Performance in a Virtual Learning Environment. In 2018 4th International Conference on Education and Technology (ICET). 2018, October (pp. 18–22). IEEE.
[23] Rizvi S, Rienties B, Khoja SA The role of demographics in online learning; A decision tree based approach. Computers & Education. 2019;137:32–47.

Soumitra Saha, Umesh Kumar Lilhore, Sarita Simaiya

10 Performance Enhancement of Artificial Intelligence Models Utilizing Hybrid Optimization Techniques

Abstract: With an abundance of data accumulated from different media and the generation of new data, the artificial intelligence (AI) model has been a robust tool in making significant improvements in the technology sector and ensuring its continuous advancement. However, due to the increased workload, more complex problems arise with the growing volume of data unfolding in a shorter duration. Along with repetitious or less essential data in the dataset utilized, AI models often fail to deliver the expected results in different circumstances or to propose more complex solutions to achieve those outcomes. Miscellaneous optimization techniques are predominantly employed in complex problem statements to enhance the interpretation and productivity of AI models. However, the paramount factor is that optimization techniques can only guarantee adequate performance for specific AI models, and in most cases, they go astray. Therefore, the capacity of AI models to provide optimal outcomes when tackling the most complex problems tends to wane, and in these circumstances, hybrid optimization techniques (HOTs) are among the most viable and forward-thinking solutions. Utilizing a spotless HOT with appropriate blueprinting, an AI model can accurately monitor problematic circumstances and quickly furnish the preferred information without wasting considerable resources. Simultaneously, HOT's persistent expansion will effortlessly facilitate functions on high-dimensional datasets and deliver decision-making capabilities to navigate real-world scenarios. HOT's strategic approach streamlines the problem-solving process and enables AI models to achieve greater adaptability in complex environments while ensuring sustainable performance. The use of well-designed HOTs has the potential to contribute to unprecedented technological advancements, raising hopes for the future of AI optimization and its role in enhancing the performance of AI models.

Keywords: Artificial intelligence, model optimization, algorithm integration, computational intelligence, performance enhancement, predictive analytics, algorithmic efficiency

10.1 Introduction

The technology sector has contributed the most to the twenty-first century's growing development, reaching a level no one could have imagined 20–30 years ago. Technology now touches every field, from insignificant to influential, and due to the remark-

https://doi.org/10.1515/9783112214374-010

able inventions in these sectors, it is only a matter of time before any complex task is completed effortlessly. Data is indispensable for technologies and for solving problems in any technique, from miniature to oversized or straightforward to complex. Data utilization, when well planned, is the primary mechanism for making significant improvements in the technology sector. At the same time, data storage and management techniques are essential for effectively using the data to perform any task [1]. Due to this advancement in technology, the amount of data has increased dramatically; new data is being created on a large scale, and this vast amount of unstructured data – such as audio, video, text, social media posts, and images – creates additional challenges [2]. However, the collected data is sometimes not clean, and if it contains redundant, less valuable, or less relevant information, it becomes much more challenging for any model to achieve its desired outcomes. At the same time, an increased number of training instances in the dataset means that the amount of noise and outliers is also high, which can degrade accuracy [3]. The model will deliver expected outcomes if the accumulated data can be processed appropriately and the operational data can be utilized effectively. Data mining involves finding significant patterns in a vast dataset that can be used to make decisions and achieve desired outcomes, with data quality being the principal determining factor [4]. Artificial intelligence (AI) is one of the leading fields in computer science, and appropriately manipulating data in an AI model will effectively address complex problems. Now, it is time to examine how medium AI can derive the most profitable results from the data and achieve the desired goals.

AI models can think like humans, find the optimal idea from multiple options, function simultaneously like humans, and suggest the best action from multiple tasks. When humans operate with an AI system, they need to understand how the system is utilized for decision-making and how it is employed to address different limitations [5]. The main difference between the human brain and the AI model is that the AI model can perform critical tasks much faster and more accurately than humans. Among the different sectors that have increased the use of AI, the most breakthroughs have been made in medical science. The neuroscience field and the AI model have far-reaching and groundbreaking interconnections, and biological brains have the potential to create an intelligent system [6]. The AI model now assists the medical diagnosis system in preventing various diseases by diagnosing cancer, diabetes, and heart disease, developing different types of drugs, and identifying patients' health risks [7]. Physical problems related to cardiology are now frequently observed, and a combination of AI and machine learning (ML) models is currently effective in addressing these issues [8]. For healthcare support, the disease prediction system plays a crucial role in diagnosing and curing various patient diseases, which can be effectively addressed with the characteristic and constructive use of AI models [9]. AI is separated into two primary categories: narrow AI primarily designed to handle a finite or specific set of straightforward tasks and the more complicated general AI to solve relatively complex problems quickly by possessing human brain-like cognitive abilities. Optimization techniques play a meaningful role in executing AI models by employing relatively less time and energy to decode complex problems systematically and helping to find a straightforward way to achieve the expected results. Figure 10.1 illustrates – from the AI model to the hybrid optimization technique (HOT) – which method is required to achieve the expected result and what type of functionality can accomplish the process with a successful solution.

ML is a subset of AI that learns from data, recognizes unknown patterns, employs various statistical approaches, and makes decisions based on them while working with the Internet of Things (IoT), cloud computing, embedded devices, and more [10]. Model optimization techniques are widely employed in ML models because they allow the most challenging problems to be solved in relatively less time and with less energy while achieving fruitful results. When the problem statement is more complex and traditional ML algorithms cannot provide adequate results, deep learning (DL) – a specialized concept of ML based on artificial neural networks (NNs) – is utilized to solve the problem [11]. Metaheuristics are a subset of optimization techniques primarily used to solve complex problems and also can achieve satisfactory results.

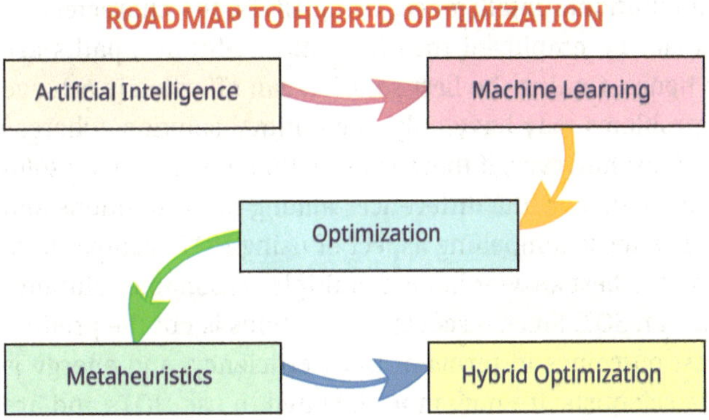

Figure 10.1: A comprehensive roadmap from artificial intelligence to hybrid optimization techniques.

The optimization technique attempts to solve a particularly complex problem by finding the most efficient way to utilize algorithmic efficiency to arrive at an adequate solution within certain constraints. AI models can solve a wide variety of business and technical problems when performing relatively complex and heavy operations to achieve deeper automation, but they require significant amounts of energy, time, network capacity, and storage; however, optimization techniques can accomplish these tasks relatively simply and successfully [12]. The optimization technique finds the most efficient way to arrive at a practical solution, utilizing less time and without wasting energy. For example, a vehicle must travel from a specific place, A, to another place, B, where different routes are available to go from one location to another. In this case, the optimization technique works to find the route that vehicles can travel on by using the shortest time and wasting less energy. The most influential factor is that, as work dimensions grow and problems become more complex each day, the importance of optimization techniques has also increased. The traveling salesman problem (TSP) is a perfect example of a complex problem in which optimization techniques can assist a salesperson in adapting to the situation and finding a better resolution over time [13]. Since energy consumption needs to be precisely limited and time is crucial in any context, achieving the desired results through optimization techniques is essential. The well-planned use of optimization techniques in AI models is essential for eliminating hidden problems by forming underlying relationships between the data, as their success and proficiency in achieving preferred results are the leading reasons.

Depending on the type of work, two optimization techniques are employed. Single optimization techniques (SOTs) are used when the problem requires little time and energy. SOTs are predominantly employed to solve relatively straightforward problems because it has performance deficiencies. On the other hand, if the problem is relatively complex and an optimal result is expected in a short period, then HOTs are employed, which involve a combination of multiple optimization techniques known for their adaptability and computational intelligence. Using HOTs, the characteristic advantages of multiple SOTs can be employed together, and additional paths are opened to solve the problem; hence, finding the best solution can effortlessly achieve the expected outcome. Some problems may have only one optimal solution, whereas others may have multiple solutions; however, if there is more than one adequate solution, the most important task is to analyze the differences among those solutions and find the preferred resolution. The most compelling aspect of using HOTs compared to SOTs is that they efficiently find the best answer among multiple reasonable solutions, which is almost unthinkable for an SOT. Since discovering solutions is not the preliminary objective, finding the best outcomes in terms of time, efficiency, and energy is more consequential; thus, data scientists are now more inclined to use HOTs and are achieving much better results.

The continuing sections of this chapter are categorized into the subsequent divisions: Section 10.2 discusses the difficulties encountered when using SOTs. Section 10.3 defines various HOTs, their characteristics, and working procedures. Section 10.4 highlights diverse HOTs applications in various sectors. Section 10.5 thoroughly describes miscellaneous works related to HOTs in the past years. Section 10.6 concludes the article with concerns about working aspects and prospects of HOTs.

10.2 Challenges in Single Optimization Techniques

Different optimization techniques should be applied to a particular AI model after considering, reviewing, and observing the problem type. The primary and most influential reason for using optimization techniques is to try to solve a particular problem in the best possible way by using a relatively small amount of time and energy. As the problem's scope increases, it becomes much more complex. The collected data contains different types of problems and not all problems can be solved by a specific AI model, which is why optimization techniques are mainly used. However, a particular optimization technique cannot always provide the desired solution to a specific problem in a particular AI model. Different types of problems are encountered in such situations, which increase the complexity of the model and significantly reduce the performance. The principal problems encountered in SOTs and their consequent impact on the desired results are discussed below. Figure 10.2 illustrates the weaknesses of SOTs and the benefits of HOTs.

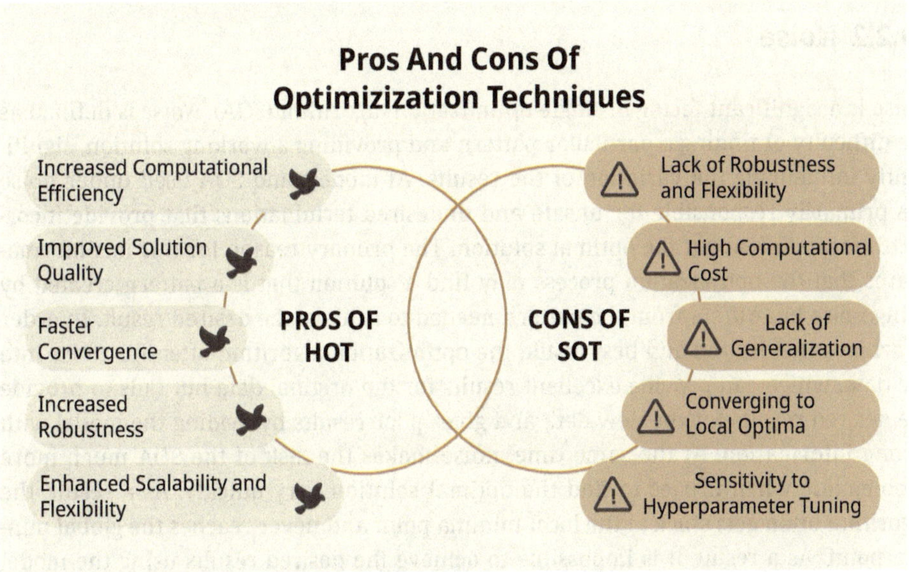

Figure 10.2: Positive sides of hybrid optimization techniques and opposing sides of single optimization techniques.

10.2.1 High Dimensionality

Considering the problem type and the work contentment, high-dimensional datasets are used in almost all cases. However, any optimization algorithm and an AI model must face different problems due to high-dimensional datasets. As a result, the model fails to provide the desired results. When the dimensionality increases due to the large dataset, that dataset's search space increases significantly, making it much more difficult for the optimization algorithm to review and monitor the dataset to provide results accurately. Again, in many cases, different optimization algorithms can provide excellent results in terms of low dimensions but cannot perform the scaling properly in the case of high-dimensional datasets, and as a result, the processing time increases significantly. Also, using high-dimensional datasets significantly increases the number of local minima and maxima, which prevents the optimization algorithm from finding the global optimum. Using high-dimensional datasets dramatically increases the complexity of the model and consequently increases its tendency to overfit. As the dataset's dimensionality increases, so does the number of redundant and less critical data or features. As a result, any optimization algorithm has to deal with invalid and unsafe data that prevents it from producing valuable results and drives the algorithm in undesired directions.

10.2.2 Noise

Noise is a significant factor in single optimization algorithm (SOA). *Noise* is defined as the difficulty of finding a particular pattern and providing a working solution, significantly influencing the variation of the results. AI models and SOA used under noise are primarily responsible for unsafe and undesired terminations that provide incorrect information about the optimal solution. The primary reason for this misinformation is that the optimization process may find a solution that is a pattern created by using noise but differs from the pattern needed to achieve the desired result. In order to achieve the desired and best result, the optimization algorithm often fits noise into the data, which can provide excellent results for the original data but fails to provide the desired results for the new data and gives poor results by feeding the model with wrong information. At the same time, noise makes the task of the SOA much more problematic when it tries to find the optimal solution very quickly. As a result, the algorithm often gets stuck at the local minima point and never reaches the global minima point. As a result, it is impossible to achieve the desired results using the model from the data used. At the same time, a considerable amount of noise increases the complexity of the model to a large extent, which is very time-consuming and requires a large amount of memory to be used.

10.2.3 Scalability

One of the significant drawbacks of using SOTs in AI models is scalability. Scalability is the ability of a particular system to handle excessive load and maintain appropriate performance if the demand for a particular system is increased, such as in terms of servers or storage. Scalability mainly evaluates whether the platform works systematically when increasing system demand. When available resources such as CPU, memory, or bandwidth within a system are scaled, it becomes much more difficult for an

SOT to do the job correctly and help deliver the desired results. At that time, the system is unable to accept the additional load. As a result, the complexity of the model used increases significantly, and the algorithm used becomes much weaker and more complex. At the same time, data takes a long time to transfer from one location to another due to increased system latency, which results in considerable variation in the system's overall performance. On the other hand, scaling a system is costly and faces similar problems in terms of hardware and software. Using an SOA may never yield fruitful results in this context.

10.2.4 Convergence

Convergence is one of the principal factors in SOAs that yield desired results. It is a process that brings the desired result closer to the optimal solution with each iteration of a particular algorithm. When the difference between the current and optimal solutions at a given step is too small, the algorithm encounters a problem and chooses the current solution with a closer or lesser difference instead of the optimal solution. In this case, the convergence process becomes much slower when the algorithm approaches the optimal solution but determines the result by looking at the current solution. The most significant disadvantage of slowing down the convergence process is the long time it takes to achieve the desired result. Sometimes, this problem causes an SOT to get stuck in local optima and never get out of there to reach global optima. The initial phase of the algorithm is critical to the convergence process, and if the starting point does not work for some reason, optimization can influence procedural convergence. As the algorithm's complexity increases, the procedure used to solve a particular problem and the specific approach the algorithm uses to achieve the desired result become more complex and consequently face different types of annoyances. At the same time, if convergence does not work correctly, the total number of iterations used by the algorithm increases exponentially. Thus, it requires much time and computational power.

10.2.5 Computational Cost

The computational cost is the amount of computational resources, such as time, memory, or processing power, required to perform a particular task or solve a problem. It refers to how many operations the computer uses to complete a task, how much memory it uses, and how much time it takes from the initial state to the desired result. An SOA will likely increase a particular model's computational cost. For example, SOAs work slowly when the problem types are large and require much time. Simultaneously, solving very large or complex problems requires the model to consume much memory. Any device or model can only provide the desired results if it consumes less memory in the working process. When the size of the problem is large, the competitive difficulty increases experientially, causing the SOA to lose its efficiency. The task it completes to achieve the desired result needs to be revised. Suppose a combined optimization technique is used instead of an SOA in an AI model in an accurate and precise manner, it is possible to reduce the competitive difficulty to a large extent by using less time and efficiently achieving the desired result.

10.2.6 Hyperparameter Tuning

Tuning the hyperparameters and selecting the appropriate weights to accomplish the preferred results are always crucial and complex processes in any AI model. In any operation, a straightforward technique is executed multiple times to determine whether a particular hyperparameter functions correctly. Essentially, the hyperparameter's primary operation is to manipulate how a particular model learns from contemporary data and performs under given circumstances. Hyperparameter tuning typically begins at the start of the AI model's training process in most cases, but it sometimes initiates before training, and the refining process continues until optimal performance is achieved. If the hyperparameter tuning process H_i starts before or after the training process $T(H_i)$, the performance of the model Q is checked after each training process, and the process will continue until the model's performance is optimal (O) or reaches a targeted mark θ. Then, the equation will be $H_0 \rightarrow T(H_0) \rightarrow Q_0, H_1 \rightarrow T(H_1) \rightarrow Q_1, H_2 \rightarrow T(H_2) \rightarrow Q_2, \ldots, H_n \rightarrow T(H_n) \rightarrow Q_n, \Rightarrow O$, where $Q_n \geq \theta$. Because of miscellaneous causes, the SOA will not be able to select the optimal hyperparameter through proper tuning; the model's complexity will increase, and its computational cost will also become very high. When the SOA is unable to choose the initial hyperparameter H_0 properly, or, at the same time, cannot tune the hyperparameter H_i throughout the whole training process T, then the SOA might choose a suboptimal hyperparameter M^1, and the model will not be able to provide the optimal outcomes ($P\downarrow$); hence, a lot of resources ($k\uparrow$) will be wasted, and the equation will be $H_0 \neq \rightarrow H^* \Rightarrow T(H_0) = M^1 \Rightarrow P\downarrow, k\uparrow$. This annoyance is more pronounced when the outcome is measured in considerable numbers, the type of problem statement is very complex, or when performing with enormous datasets full of redundant or irrelevant data. At the same time, if the SOA cannot complete the hyperparameter tuning process using efficient means, the model's tendency to overfit will increase. Failure to use effective hyperparameters in the model results in inaccurate statements; consequently, inefficient or ineffective hyperparameters significantly degrade model performance.

10.2.7 Data Efficiency

Data efficiency means making the most effective use of data in a systematic way, which supports accomplishing the expected results by efficiently employing the most significant data or the least amount of data. Thus, efficient solutions are sought by utilizing the term in the most well-planned way, while the ultimate goal is to reduce different types of resources, such as time, memory, or computational power consumption, which leads to better outcomes for the model. If one can ensure the proper use of data efficiency and improve it by analyzing the data through effective processing, it directly contributes to enhancing the performance of the system. Data efficiency η_D is the capability of a model (M) to achieve its targeted outcomes θ by enhancing classification performance P with the least amount of data $D_{\text{used}} << D_{\text{total}}$, where the optimization learning process ($L(D_{\text{used}})$) is effectively utilized for expected outcomes ($P \geq \theta$). Due to a lack of data efficiency, an SOA cannot accurately address changing patterns of different types of data, and numerous challenges are encountered in finding new patterns and using them systematically. SOAs usually concentrate on a single performance metric, making it challenging to analyze the variability in the data to achieve the expected outcomes, and it only performs the optimization process effortlessly to achieve a specific purpose, which increases the tendency for overfitting. On the other hand, SOA only functions to accomplish its purposes without considering the quality of the data; hence, it depends on unnecessary and less meaningful data and is unable to utilize the data correctly. As a result, the dataset accumulated from different sources can contain various types of redundant, ineffective, and less valuable data, which can hinder achieving the desired results when using SOA.

10.2.8 Unreliable Results

An SOT is used to solve problems using AI models. However, one of the downsides of using an SOT is that it often fails to achieve the desired results. The first and foremost goal of solving any problem is to reach a goal, which is usually an outcome anticipated at the beginning of the process. However, the AI model cannot achieve the desired results due to the faulty use of SOTs. Local minima are the first and main obstacle to achieving the desired result. Getting stuck in local minima never allows reaching the global solution and fails to achieve the desired result. Using an SOT dramatically increases the model's tendency to overfit. It causes the model to become too complex and fails to provide the desired results for any new data type. The more complex a particular model becomes, the harder it is to achieve the desired goal by deriving results from that model. There is a large amount of noisy data in the collected dataset. In that case, it is impossible for the SOT to remove all the noisy data and achieve good results, and due to this, the computational complexity of the model increases. Suppose the desired result is not possible when solving any problem, the acceptability of using that model decreases, so the tendency to use an SOT is now lesser compared to HOT.

10.2.9 Limited Memory

Any task when working with AI models requires a dataset and the size of the dataset can range from small to very large. The larger the scope of the work and the more complex the task, the larger the dataset will be. If the scope of the work is enormous, it takes more time to complete, it is more complex to solve, and above all, if the size of the dataset used to perform these tasks is very large, then the size of the memory in the computer used must be huge. The memory used for execution also has limitations, and the specific task must be completed using that memory. If the memory size is manageable, then the hybrid optimization algorithm (HOA) can avoid facing too many problems, but the SOA certainly faces various problems and consequently may not provide the desired results. If the memory size is manageable, the efficiency of handling the model's dataset increases due to SOTs. At the same time, the possibility of data loss increases due to the difficulty of handling the data contained in that vast dataset. As a result, the necessary data needs to be addressed, and it becomes impossible to achieve the desired results. Due to the less memory used, the model's performance is much slower and takes a lot of time. Since an SOA cannot use a given size of memory correctly, a much more complex algorithm must be used, which increases the complexity of the model and, thus, its error.

10.2.10 Overfitting

Due to using a SOA, the overfitting problem is considered one of the leading and most severe problems that AI models face. Overfitting refers to the problem of a model that can provide better results on trained data but could perform better on new data. The main reason is that when a model works with data, the training data for the overfitting problem makes it much harder to build a model and provide better results. However, that complex model fails to do the same for new data and consequently fails to provide the expected results. This problem prevents AI models from achieving the desired results due to SOTs. Any model becomes too complex due to overfitting. The number of parameters used in the model increases significantly and by using the parameters appropriately, one has to go through a more difficult situation to select the results by finding the correct pattern from the given data. In such cases, an SOA cannot handle such a large number of parameters properly. Consequently, the amount of time and energy the model requires to perform any task greatly increases. At the same time, being a complex model makes it almost impossible to create appropriate scenarios through problem-solving.

10.2.11 Outliers

Outliers in datasets play a vital role in systematically finding practical solutions using any model. In any dataset, the most critical data is located in the center and at the border. However, there are many outliers with that data. Due to the presence of outliers, the model cannot process the data correctly and because of this, redundant and irrelevant data remain in the dataset. If a large amount of duplicative and extraneous data remains in a dataset, then generating working patterns using that dataset is a challenging and complex task. If the outlier can be effectively removed so as not to cause problems for other data, then any model using that dataset can produce good results. However, removing the existing outliers without any problem is a challenging task. Removing outliers from any dataset using an SOA is tricky and time-consuming. If an SOA deals with an extensive dataset, removing only the least significant or unnecessary outliers is almost impossible. First, removing outliers is time-consuming and requires much energy, a lot of parameters tuning, planning to find the desired pattern, and, finally finding the outlier that needs to be removed to achieve the desired result.

10.3 Distinct Hybrid Optimization Techniques

The most beneficial aspect of using HOTs with AI models is that this algorithm integration can efficiently perform tasks in a systematic way by adopting suitable action plans to solve complex problems. The combination of optimization techniques should be set correctly to understand how complex the problem can be and how to solve it with less time and energy. Any problem can be easily solved if one finds out which HOT is the most efficient for a particular model or which HOT can achieve better results in less time for a particular task. HOTs are selected considering the nature of work, type of problem, dataset feature analysis, the architecture of AI model used, and above all, time and energy, where different SOTs work together and try to complete the task gracefully. The general strategy by which the HOT is carried out is shown in Figure 10.3. Now, various HOTs and their use will be discussed below, considering the temperament of the assignment.

10.3.1 Genetic Algorithm and Simulated Annealing (GA-SA)

Genetic algorithm (GA) and simulated annealing (SA) are optimization algorithms that form a powerful hybrid method to solve complex optimization problems when combined. GA works by following natural evolution and selecting the best solution based on fitness; for this, it uses crossover and mutation. After conducting several processes to generate new solutions, it can find the optimal solution or a solution close to it. On the other hand, SA follows the cooling process of metal while sometimes avoiding local optima to escape local traps. These two features utilize HOAs, where GA works on the solution space, and SA collaborates to find the best result. This algorithm effectively solves complex problems related to various optimization challenges like scheduling, routing, image processing, and ML and works with large search spaces to provide expected results. The hybrid GA-SA model can be applied in tourism services to improve the overall performance of the scheduling problem by generating a random scheduling structure in the first stage [14]. After selecting the optimal combination, it makes small changes at every stage to help improve the overall techniques.

10.3.2 Particle Swarm Optimization and Genetic Algorithm (PSO-GA)

Particle swarm optimization (PSO) and GA are popular optimization techniques used together to form a hybrid technique and play an essential role in solving critical problems. PSO is based on the nature of moving as a particle in a particular space to find different solutions. Users gather information from their own experiences and the experiences of others and use it later. Conversely, GA works based on heredity, and the solutions are considered where mutation and crossover are used on the genes. This hybrid technique is used to find new and more efficient solutions from previous solutions. PSO helps find solutions faster, and GA tries to find variations among different solutions. PSO-GA can play a critical role in various applications such as engineering design, ML model tuning, robotics, and network optimization, and at the same time, can save a lot of time and energy. A hybrid PSO-GA model can improve water quality by discovering the best possible way to manage chlorine levels in water [15]. First, a larger group of possible solutions is created, and from there, the GA is utilized to pick the optimal one, while the PSO model helps the whole framework refine these solutions based on their performance, and the process continues until the chlorine levels reach the desired level as shown in Figure 10.3.

10.3.3 Differential Evolution and Particle Swarm Optimization (DE-PSO)

Differential evolution (DE) and PSO are two popular innovative algorithms that can be combined to form an HOT, resulting in the opportunity to combine the strengths of the two approaches. DE is based on the natural selection process in which different population members are developed to create a difference. On the other hand, PSO helps to find the correct solution through the behavior and dynamics of the different types of particles participating in the computation. The proposed novel model, namely

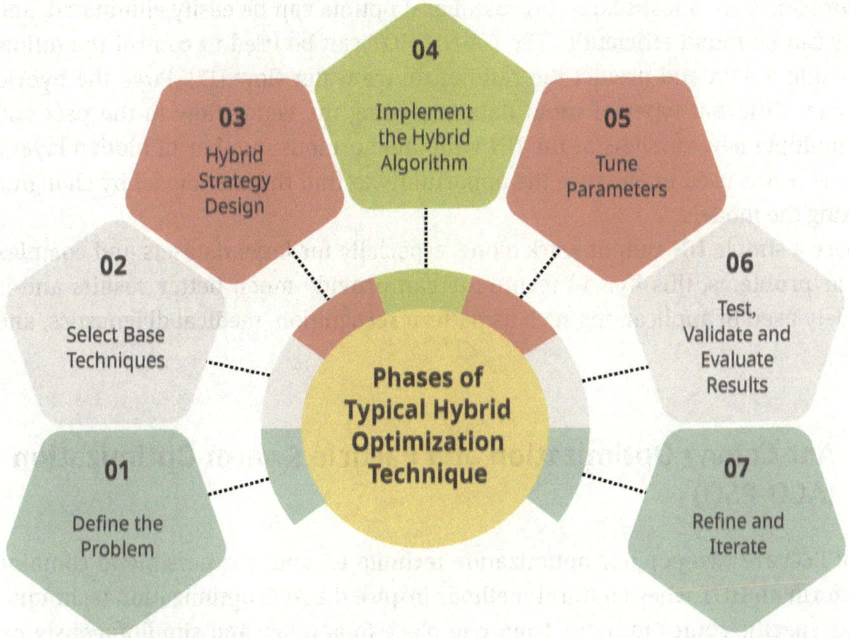

Figure 10.3: Fundamental process of hybrid optimization techniques.

the CoDE-MOPSO, is a combination of the hybrid DE-PSO model, which groups similar kinds of solutions from large entities to find optimal solutions for complex optimization problems [16]. When working on any timeframe, CoDE tries to create a new solution, whereas MOPSO tries to find the best one and helps improve performance. The main advantages of DE-PSO as a hybrid technique are that it can provide greater accuracy when dealing with complex optimization problems and increase overall work speed. This technique increases the overall power of local and global search, which supports the next step out of local optima. This HOT is widely used in essential tasks such as signal processing, the performance enhancement of ML models, network optimization, energy management, and feature selection (FS).

10.3.4 Genetic Algorithm and Neural Networks (GA-NN)

An HOT consisting of GA-NN is a technique with high performance for solving any specific optimization problem. GA is based on natural selection, creating a population of possible solutions. NN refers to a network constructed by connecting many layers of neurons through nodes. Information is exchanged from the previous layer to the next layer by replacing the relationships, and these layers act like the human brain and play an active role in solving problems. This hybrid method can solve an optimization problem with a fast search process. Local optima can be easily eliminated, and solutions can be found efficiently. The GA-NN HOT can be used to control the inflow of water into a dam and predict the ratio of future water flow [17]. First, the hybrid model takes different types of input data regarding the water flow in the past and creates multiple new versions of the NN with an enormous number of hidden layers. From there, GA is used to enhance the opportunity to find the best model by changing and mixing the models.

Where a simple NN cannot work alone, especially for large datasets and complex nonlinear problems, this hybrid technique can provide much better results and is now widely used in applications such as pattern recognition, medical diagnostics, and robotics.

10.3.5 Ant Colony Optimization and Particle Swarm Optimization (ACO-PSO)

ACO and PSO are two popular optimization techniques, and it is possible to combine them to form an HOT where natural methods inspire the ACO optimization technique. Ants take specific routes to move from one place to another and simultaneously release pheromones on those routes while searching for food, so they spend less energy and time searching for food or routes later. The ACO algorithm is based on this idea. On the other hand, PSO uses social behavior to find the best outcome within a given group. The qualitative and functional search capabilities of the ACO algorithm, coupled with the speed of PSO, can solve many complex problems very quickly. It is capable of providing stability and efficiency for solving different types of optimization problems simultaneously. The proposed hybrid model, built using a combination of the ACO-PSO model, is employed to predict heart disease systematically and effortlessly [18]. First, an FS method, fast correlation-based FS (FCBF), is used to remove irrelevant features and select the most influential features from the dataset. Then, the PSO and ACO are utilized to fine-tune these features further to select the optimal ones, and finally, different ML models are used to classify heart disease and evaluate their performance. As an application, this HOA can quickly and effectively perform complex tasks such as route planning, optimization of transport systems, and network design.

10.3.6 Simulated Annealing and Tabu Search (SA-TS)

In SA-TS, an HOT, SA uses a robust HOT. This technique is stochastic, uses the concept of thermal motion, and attempts to escape from local minima. SA starts operating at higher temperatures first, allowing redundant solutions to be tested. The stability of solutions gradually increases as the temperature decreases. TS, on the other hand, operates by following a local search procedure, thereby excluding a variety of unsuitable solutions already visited and preparing itself for the next step of execution with only the working solutions. An HOT, combining SA and TS, can be utilized to solve specific problems where a predefined task needs to be balanced to achieve optimal performance [19]. Firstly, the process generates some random solutions and then explores the nearby solutions by moving the random tasks toward them. When a solution with better performance is found, keep it; otherwise, continue the search process while avoiding the previously used solutions. SA searches for solutions, and TS selects the best solutions using this HOT. As a result, one process contributes to increasing the energy and efficiency of the other process. It plays a vital role in getting out of local minima and achieving the desired result. It is used in graph theory, ML, and resource building to reduce time complexity.

10.3.7 Memetic Algorithms (MAs)

MA is an HOT that combines evolutionary algorithms (EAs) and local search optimization techniques. MA creates a population set where each member acts as a possible solution. Then, operations like crossover and mutation, which are GA operations, are performed on every member of that population. After this operation is completed, the local search process is again applied to each member, further improving their effectiveness and efficiency. Dealing with the production routing problem is always troublesome, and the memetic hybrid model with local search is being utilized to solve the issues [20]. First, a group of random solutions is created, and a local search is used to improve each solution's performance. Each solution is evaluated regarding its cost while maintaining the main focus on production, inventory, and transportation costs, and the process continues until the optimal solution is found by reducing the overall costs. Using MA avoids local optima while increasing the quality of the solution. Therefore, it can provide faster solutions to various real-life problems. Due to its impressive use, MA is now widely used in a variety of applications, such as network design, routine problem solving and scheduling, improving the performance of ML models, and medical research.

10.3.8 Genetic Algorithm and Ant Colony Optimization (GA-ACO)

Using the GA-ACO algorithms together, an HOT can be developed to solve relatively complex optimization problems. GA tries to find the solution by following the natural evolution process, while the ACO algorithm performs the task by following the wayfinding and foraging behavior of ants. This hybrid technique achieves breakthroughs in problem-solving due to the interrelationship of two different feature algorithms and the simultaneous use of heterogeneous features. This HOT can not only diversify in terms of new solutions but is also adept at deviating from local optima by making systematic decisions. It excels at increasing the quality of the solutions obtained for solving specific problems, providing higher accuracy by organizing the combination process of both algorithms and increasing the speed of execution. Finding the efficiency of the best suppliers and considering the most influential factors, such as cost and quality, requires a hybrid optimization model, like GA-ACO, which provides better outcomes [21]. GA simulates different types of natural selection and tries to find good solutions, while ACO is used to improve the performance of those solutions and find the optimal one. Various difficult tasks related to FS, route optimization, and robotics can be completed easily and with less time and energy wasted by this algorithm.

10.3.9 Differential Evolution and Simulated Annealing (DE-SA)

The HOA known as DE-SA is basically a combination of two popular optimization techniques: DE and SA. DE is a population-based algorithm that helps to find new solutions from a variety of solutions, and SA is a single solution-based algorithm that jumps to find global optima and thereby avoids local optima. The most significant advantage of using DE-SA as an HOA is combining DE's convergence advantage with SA's global optimization capability. Dealing with issues like heat recovery and energy efficiency in chemical engineering is a troublesome task, and the hybrid DE-SA model is used to obtain optimal outcomes from heat exchanger networks (HENs) [22]. Firstly, a new design is created by SA for HEN, and then DE is utilized to fine-tune the details, which helps explore and optimize the framework, saving battery life and reducing costs. Using DE, a wide area containing possible solutions is first scanned, and efficient and inefficient solutions are separated from it. Those solutions are further refined using SA with global optimization. As a result, a particular solution participates in the performance evaluation of a particular task multiple times through multiple processes and contributes to delivering the desired results. This HOA can solve the most complex problems, such as industrial design optimization, robotics, and image processing.

10.3.10 Harmony Search and Particle Swarm Optimization (HS-PSO)

Harmony search (HS) and PSO are two popular metaheuristic algorithms commonly used to solve the most challenging optimization problems. HS is an algorithm mainly influenced by the idea of harmonic matching, whereby working with previous solutions, newly generated solutions are merged with them, and then more new solutions are generated. Slight changes in the newly developed solutions are observed to judge whether they are more or less effective than those initially located in the previous step. On the other hand, PSO is an algorithm inspired by various natural social behaviors such as bird flocking or fish movement. It accumulates experience by changing its position. An HOT like HS-PSO can be utilized in fields such as software testing, where the model generates different types of test cases automatically and helps identify properties that are nonfunctional or less proactive [23]. HS finds the optimal starting point, which is needed to check fewer test cases, and PSO is used to improve the outcomes of the test cases further. The principal motivation for using this algorithm is that HS helps to find the best solution by performing a global search. At the same time, PSO works for a local search or fast solution in a specific region. This algorithm can solve various problems in fields such as robotics, image processing, bioinformatics, wireless sensor networks (WSNs), medical diagnosis, and industrial optimization. Table 10.1 illustrates different types of HOTs and their implications across various application domains.

Table 10.1: A comparative analysis of various hybrid optimization techniques for different application domains.

Study	Application domain	Objectives	Proposed model	Performance metric	Problem type solved
[14]	Smart tourism	Scheduling efficiency improved	Hybrid GA-SA algorithm (HGA-SAA)	Makespan (C_{max}), average solution, and execution time	Scheduling problems
[15]	Water quality Modeling	Optimize wall decay coefficient	Hybrid GA-PSO (HGA-PSO)	Root mean square error (RMSE) and convergence rate	Chlorine decay optimization
[16]	Multi-objective engineering optimization	Balance diversity and convergence	Hybrid CoDE-MOPSO algorithm	Inverse generational distance (IGD) and hypervolume (HV)	Multi.-objective optimization
[17]	Water flow prediction	Improve prediction reliability	Hybrid ANN-GA model	Mean absolute relative error (MARE), correlation coefficient (CC), and RMSE	Time series prediction (forecasting)
[18]	Heart disease prediction	Improve classification accuracy	Hybrid FCBF-PSO-ACO-based multilayer perceptron	Accuracy, precision, recall, F_1-score, true positive (TP) rate, false positive (FP) rate, and RMSE	Classification and FS
[19]	Type-2U-line balancing	Minimize cycle time	Hybrid SA-TSA	Mean absolute deviation (MAD) and CPU time	Scheduling/ optimization
[20]	Supply chain management	Minimize total cost	MA with hybrid local search	Total cost and CPU time	Production routing and scheduling

Table 10.1 (continued)

Study	Application domain	Objectives	Proposed model	Performance metric	Problem type solved
[21]	Supply chain management	Improve efficiency and accuracy	Hybrid GA-ACO	Convergence speed, efficiency, feasibility, and solution diversity	Multi-objective optimization
[22]	Energy optimization	Minimize total annual costs	Hybrid SA-DE	Heat exchanger area, utility requirements, and execution time	Optimization and scheduling
[23]	Energy optimization	Optimize test data generation	HSPSO	Branch coverage, average generation, and average execution time (ACET)	Data optimization
[24]	School scheduling system	Minimize schedule conflicts	Hybrid artificial bee colony (HABC)	Schedule conflict count, accuracy, and computation time	Optimization and scheduling
[25]	Power systems engineering	Minimize operational costs	Hybrid WO-DE and GA (WODEGA)	Operational cost, startup cost, and ramp rate	Mixed-integer optimization for scheduling

10.3.11 Artificial Bee Colony and Genetic Algorithm (ABC-GA)

The ABC-GA HOT, consisting of ABC and GA, is also widely employed nowadays. The ABC optimization technique is a nature-inspired algorithm that follows the behavior of bees. These algorithms commonly solve informal problems, notably function optimization and data clustering. There are three different groups of bees: one finds a food source, the second goes and examines the food, and if the source is good, brings that information to the third group. This algorithm is mainly based on the idea that it can help solve local and global optimization problems simultaneously. The ABC algorithm helps collect information quickly, and the GA algorithm takes action to find the best solution based on the information received. Hybrid techniques like ABC-GA can be used in educational institutions to help create the schedule more efficiently and effortlessly [24]. The crossover operator from GA is used to mix different schedules from a group to find the best possible one, whereas the ABC model is used to evaluate and identify the best schedule by addressing the most important component of class scheduling: conflicts in class timing. This algorithm makes various types of device design, resource management, search engine optimization, and software development easier.

10.3.12 Whale Optimization Algorithm and Genetic Algorithm (WOA-GA)

WOA is a standalone algorithm developed following the whale hunting method. This algorithm can analyze the probabilities of different solutions. Whales first analyze a specific area to search for prey where it is likely to be found. Second, the whale moves slowly from its location to the prey and analyzes the quality in terms of location. Finally, while moving around the victim, it analyzes its position and takes effective action. This algorithm is developed based on this concept. When WOA-GA is used as an HOA, the WOA algorithm helps quickly find the local optimal solution and generates a variety of solutions by maintaining the GA's variation. This algorithm makes it very easy to perform tasks such as complex engineering design, route optimization, data clustering, and system optimization. Hybrid models like WOA-GA, along with models like DE, are used to solve scheduling problems in power systems, where the primary focus is on deciding which power plants need to be active and which ones need not be active based on demand in the region [25]. First of all, GA and DE are used to optimize the scheduling, whereas WOA will try to reduce the total operational costs further and help make the power system more efficient by using a minimum amount of energy.

10.4 Applications of Hybrid Optimization Techniques

The primary reason to use an HOT instead of an SOT is its flexibility, robustness, and predictive analytics capabilities. HOAs can easily handle many complex problems, deal with multi-objective problems, and provide high-quality solutions using less time and energy. Always giving good results in terms of time and energy is a higher priority, and HOTs can accomplish this task more successfully than SOTs. HOTs are mainly used to solve complex problems using multiple optimization techniques. They can provide promising success in any sector, including engineering, finance, medical science, different organizations, and the banking sector. Where an SOT reveals an inability to deliver the desired results, the HOT is now a favorite due to its fruitful use and promising results, and its popularity is increasing daily. Various applications of HOTs are described below. Figure 10.4 displays some of the many types of HOT applications, which are hugely helpful in solving multipurpose issues in dissimilar sectors.

10.4.1 Scheduling Flights

By working with flight schedules, HOTs can provide efficient solutions by monitoring various variables such as flight time, the number of crew on the flight, maintaining the flight schedule, and restrictions the airport may have, which is almost impossible with an SOA. If only flight schedules could be worked out, airline authorities could quickly reduce fuel consumption, reduce labor costs, and solve problems such as schedule disruptions. Even with the schedule, if this algorithm works appropriately, it can also protect the flight from environmental disasters. If all these facilities can be provided, the flight authorities will get better results by spending less time on customer satisfaction.

10.4.2 Supply Chain Management

HOTs for supply chain management systems can increase efficiency, reduce costs, and enhance decision-making capabilities. They can also easily balance stock levels, when there is adequate inventory according to demand or when the inventory stocks are increased. At the same time, by communicating the statistical model with the ML algorithm, this algorithm can provide an accurate idea about the future demand for a particular product, which helps a company make a better plan. Route optimization helps to complete larger shipments in less time by saving transportation costs and time. Production scheduling can provide actionable information about a product's manufacturing deadline and help improve overall product efficiency by addressing product cost, quality, or reliability as mentioned in Figure 10.4.

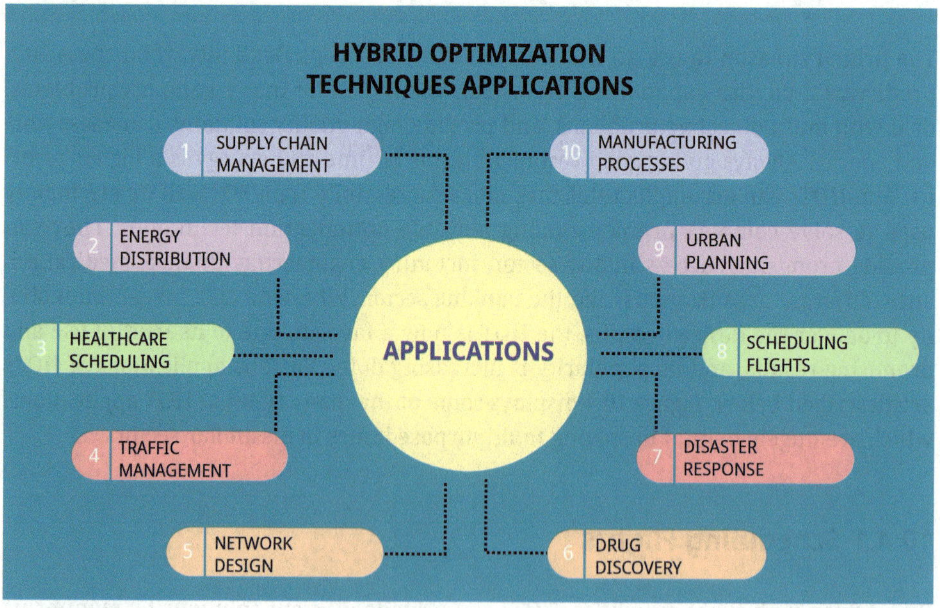

Figure 10.4: Distinct types of applications in hybrid optimization techniques that are enormously utilized in diverse sectors.

10.4.3 Energy Distribution

By carrying out proper energy distribution, the HOA can give a good idea of how much electricity is generated, how much is supplied from that generated electricity, and how much is used from the supplied electricity. This concept makes it possible to establish a beautiful relationship between supply and demand quickly. At the same time, it shows how much electricity is needed, how much electricity a power plant should generate, and how to save enough raw materials without generating excess electricity. The power grid used to generate electricity is a very complex system, and the HOA can easily find and solve any problems with the grid, such as outages or overloads. At the same time, adopting a well-planned plan plays a vital role in achieving environmental improvement by reducing the total cost and the amount of fuel used.

10.4.4 Manufacturing Processes

HOTs can accelerate any organization's product manufacturing operations. Manufacturing any product is a long process. Various processes are completed to make a particular product, from collecting raw materials to delivering the product to the consumer. The HOT is used correctly and with practical steps. In that case, it is possible to achieve much better results at every step, from manufacturing the product to reaching the consumer. Assume a specific package is made in a category. What methods can be adopted to improve the quality of a specific product? If the sales of a specific product are decreasing for a specific reason, what methods can be adopted to make those products more readily available to the speaker? HOAs can be used quickly in all these tasks, and good results can be achieved.

10.4.5 Healthcare Scheduling

Widespread use of HOTs can be observed in various areas of medical science. Their effectiveness is undeniable, especially in healthcare scheduling. What kind of treatment required for a particular disease is the most helpful part in a patient's treatment? Everything takes a lot of time, from taking a doctor's appointment for a particular patient to diagnosing the disease and preparing various reports. Along with technological development, the number of human-caused and environmental-caused diseases has greatly increased. This algorithm is widely used to diagnose the type and nature of the disease, provide proper treatment by monitoring a patient, assist in diagnosis by prescribing drugs, and find a specific disease by generating various types of reports. Suppose the scheduling method is properly implemented using an HOA, it is possible to complete these processes in less time and with less energy consumption, which can lead to unprecedented improvement in the medical field and save patients' lives by making accurate diagnoses.

10.4.6 Telecommunication Networks

The telecommunication sector is one of the most important in technology. The telecommunication network is a complex system, and if desired results are to be achieved from it, the best-structured algorithm needs to be used. An HOA can do this work very well. Starting from 2G, the telecommunication network today has been transformed into 5G, and as a result, it has been possible to achieve considerable improvements in the communication sector. Transferring data from one particular device to another, using different types of networks to build up the connection, inventing new technologies for data transfer, and, above all, these processes are being carried out in a planned manner and in a relatively short time. HOAs are now widely used to achieve this by spending energy. Since the tendency to use the Internet is more than ever before, the amount of data transfer has increased, and all the people of the world are now more globally connected; the use of HOAs is constantly working to make this sector more successful.

10.4.7 Urban Planning

In this age of globalization, people are now more comfortable living in cities due to the incredible technological advancements and changes in people's lifestyles. In such a situation, different tasks need to be performed at different times to bring about change in the lives of people living in the city, and different types of development tasks are carried out by using different algorithms in the hope of providing maximum results by wasting relatively less time and energy. By using HOTs, it is possible to determine how a piece of land can be used as a new housing, park, or commercial area without any negative impact on the environment and the people who live there. At the same time, it is possible to use different types of natural resources such as water, gas, and other resources such as electricity, or different types of public services by adopting some means so that everyone can use the same amount and provide effective results according to the plan.

10.4.8 Agricultural Planning

With the increasing progress of civilization, the number of habitable people on Earth has increased. However, the amount of arable land has stayed the same or decreased. By adopting a proper approach, HOAs can help determine which types of crops will produce well on which types of land and which pesticides will play the best role in producing which types of crops. Using the HOA increasing the productivity of a crop, determining what kind of crop would be more productive if grown, and determining what kind of crop should be grown more based on the requirement are achievable. It is also essential to transfer the crop to the consumer level by adopting a special process. If the HOA can perform these tasks by adopting appropriate methods and procedures, producing a large number of crops using relatively less land is possible.

10.4.9 Water Resource Management

Although three-quarters of the Earth's surface is covered by water and one-half is covered by land, the amount of usable water is minimal because most of it is served by the oceans, and that water is often unusable. If the water collected from reservoirs without oceans is not used suitably, there will soon be a crisis of practical or usable water. Since water plays a vital role in any performance and only holy water is used in these cases, it is essential to adopt the appropriate method for any performance and to achieve the specific goal. If water is not used correctly, there will be a water crisis soon, and production will be disrupted. Using HOAs makes it possible to find out what amount of water may be required for a particular operation or what kind of new method will require less water than the old method. If this is possible, then the wastage of usable water can be reduced to a great extent.

10.4.10 Traffic Management

With the advancement of technology and the growth of human civilization, the road traffic problem has become much more significant. The number of roads is decreasing at the same rate as the numbers of people and cars are constantly increasing. At the same time, the pace of life has dramatically increased along with the improvement in people's lifestyles. That is why everyone wants to complete work in less time, and hence more traffic problems are seen on the roads. Due to different types of vehicles on the roads, free movement of people, lack of necessary roads, improper adherence to traffic laws, unskilled drivers, lack of skilled traffic police, and tendency to save time, traffic problems have increased. If any algorithm is needed to solve this traffic issue correctly, then an HOA can be one of the central and essential algorithms. Suppose these HOAs can adequately control the traffic signals used on the road. In that case, certain vehicles will stay in specific lanes and under no circumstances use other lanes. The HOAs can measure the minimum distance to show specific routes and eliminate this problem quickly.

10.4.11 Disaster Response

Due to the geographical environment, different types of natural disasters occur in different parts of the world. The country has suffered huge losses due to all these natural calamities. An HOA can significantly reduce the damage caused by natural disasters. Different types of disasters, such as floods, storms, floods, droughts, river bursts, earthquakes, or volcanic eruptions, occur at different times in different countries. Storms or floods cause much damage, but earthquakes are considered to be the most damaging of them all. The reason is that earthquake prediction is one of the most critical and complex tasks. Like other natural disasters, if the people living in the country can be warned using the HOA, the damage can be significantly reduced. The strategic use of HOAs empowers us with crucial information about impending natural disasters. This knowledge enables us to take proactive measures such as evacuations and other damage reduction strategies. By planning and implementing these algorithms effectively and taking actionable steps, humans can empower themselves to mitigate the impact of disasters. This sense of control and preparedness is a powerful tool in disaster management.

10.4.12 Network Design

Due to the incredible development of the technology sector and the massive increase in Internet use, working with any network system and achieving the expected results is now a critical issue. Designing any kind of network is a challenging task, and the design process is very difficult. Due to the widespread use of the Internet, the use of different types of viruses or malware attacks has increased. In such a situation, before designing any network, it is necessary to keep in mind why that network is being created, what functions it will perform, and what kind of security protocol will be used during the connection build-up through the network and for transferring data from one specific network to another network. Suppose the HOA is used with a network system in a planned way. In that case, it is possible to achieve efficient results from that system, and at the same time, it is possible to protect the system from any virus attack in terms of security. At the same time, it can protect any system from being destroyed or provide specific signals regarding problems in that system before it is destroyed.

10.4.13 Vehicle Design

HOAs have revolutionized the design of various types of vehicles. Due to the development of technology and the improvement in people's standard of living, new vehicles are constantly being developed, and different technologies are being materialized every day to create new vehicles. How do you distinguish one particular model from another, design a particular car to achieve maximum speed, achieve performance that will reduce oil consumption, and what machinery or parts will reduce the cost of making the car? HOAs are widely used to do the above things and determine which approach will provide the maximum performance for a newly developed vehicle. HOAs can provide better results, are more durable, relatively more efficient, and consume less time and energy to achieve the desired results compared to a specific model. They are now used more than any other algorithm to generate the best vehicle through this, mainly due to their multidimensional usability.

10.4.14 Drug Discovery

Hybrid optimization is one of the most important algorithms used in AI to make unprecedented advances in the medical field. It is widely used in medical science to play an important role in drug development for various diseases. Due to changes in civilization, environmental disparities, and different living standards, the number of diseases in the human body has increased dramatically, and new diseases are constantly being discovered. There is a constant need to develop different medicines to cure these new diseases. Examining a particular person's body will reveal what kind of problem he has, what causes that problem – his lifestyle, the nature of his residence, his physical characteristics – and, after extensive study of his internal organs, what kind of medicine can be helpful for him. It is straightforward to understand that it is a critical and challenging process. An HOA can provide very good results in developing any new drug to cure a particular disease. The main reason for this is that it deals with the problems a particular disease can create in one person's body and the problems that the same disease can create in another person's body. At the same time, considering the nature of the disease, this algorithm can easily determine what ingredients should be included in the new drug and provide effective solutions.

10.5 Background Studies in Hybrid Optimization Techniques

The use of HOTs has increased considerably due to the changing nature of the problem and the constant creation of new complex problems. The principal reason for this algorithm's overuse is its ability to achieve the desired result using less time and energy, and its characteristic behavior yields better results than other algorithms. A novel HOT is proposed by [26], where the author incorporates two EAs: genetic and exchange market algorithms. After evaluating the results, the outcomes show that the proposed algorithm provides satisfactory results and significance over eight widely used optimization algorithms. The distribution network is the most crucial aspect of any power system, and if it does not work correctly, there will be significant energy loss. So, the power system must be prepared to provide optimum performance, and for this, the GA is used as a traditional technique where the biggest problem is the exploration and exploitation of the EA. To overcome this problem, the author proposed an HOT where the equilibrium optimizer (EO) algorithm works together with the GA, and the name of the newly proposed algorithm is the hybrid genetic algorithm equilibrium optimizer (GAEO) [27]. Due to unprecedented advancements in technology and improvements in living standards, the use of electric cars has increased significantly. However, these cars often face problems with their charging systems. An HOT is proposed here to charge a running electric car directly from the solar panel on the roof of that specific car. For this, a combination of gray wolf optimization and crow search algorithms (GWO-CSA) [28] is utilized, and the results show the proposed method's efficiency and performance.

Energy management of a power system is a critical task that involves following proper configuration procedures and can result in significant energy savings. Reconfiguration techniques are mainly used to achieve this, and these configuration techniques are primarily composed of various optimization methods. To address this issue, the author proposed a hybrid optimization topology, GA-enabled PSO [29], and the results show that the power loss resulting from this technique is much less than that of other techniques. Photovoltaic systems, where sunlight is converted into electricity, are now considered one of the best power sources in the world. For tuning the PV system controller, the author proposes a novel HOA that combines teaching learning-based optimization, which has a high amount of exploration, with EO, which has a high amount of exploitation, to build a significant model [30]. Power system engineering is always challenging for engineers, and automatic load frequency control is vital in various power systems. For multiple sources using a power system, a new heuristic-based HOT is proposed [31], which helps to control the automatic load frequency. The proposed algorithm uses the principal features of three widely used optimization algorithms, and after considering two case studies, results show that the algorithm obtains better results than others.

The smart city concept is now widely discussed, and it is used across technology, the economy, governance, and a large part of people's lives. A variety of technologies, such as wireless technology, Wi-Fi, Bluetooth, or 5G-supported smartphones, are used to replace communication through smart cities, which are widely known as communication mediums. However, this communication system is disrupted in many cases due to unlicensed interference or bandwidth issues; hence, the trend of using IoT devices has increased. Considering resource allocation and routing problems, a new energy-efficient, congestion-aware resource allocation and routing protocol [32] is proposed based on an HOT. After implementing the proposed method with the network simulator (NS-2) tool, the results show that the new method reduces energy consumption and end-to-end delay. IoT devices replace the relationship between devices and end users through wired or wireless communication using the Internet, and WSNs play a vital role here. It is difficult for devices to use the right amount of power, and since WSNs cannot be used directly in the IoT, this problem is even more significant. If the routing protocols are not used properly, this problem becomes complicated. Here, an HOT-based method, the optimal centroid-based routing protocol [33] for WSNs, is proposed to utilize energy properly and increase network lifetime, as shown in the results. Enhancing the network quality of a power grid is always crucial, and distributed generation integration plays a significant role in this process. To minimize the overall power losses from a power grid, Thanasingh et al. [34] propose an HOT where the two renowned optimization methods, PSO and GAs, work together. The PSO-GA method transcends PSO in terms of speed, power loss, voltage, and quality of the grid.

Billions of transactions take place every day, and the amount of data used during these transactions grows exponentially. Swarm intelligence techniques and hybrid swarm intelligence techniques play an essential role in carrying out this process, and they do so by adopting efficient and contemporary methods [35]. The Internet of vehicles (IoV) is now widely used to provide information on road conditions, traffic patterns on a particular road, and specific information to the driver. This is because almost all types of advanced technology vehicles use IoT devices. The IoV communicates using a radio access unit (RAU), which is available on the roadside. The RAU has to process a large amount of real-time generated data, which makes it difficult to hold such a large amount of data in memory. The RAU uses the data randomly. Hence, to address this issue, the author proposed a hybrid blockchain-based grasshopper optimizer (BC-GAO) [36] technique, which shows that BC-GAO consumes less energy. Different types of AI models are exposed to different types of cyberattacks, and advanced persistent threat (APT) plays a significant role by taking on the challenge of detecting and mitigating these cyberattacks. The author presents a novel approach called hybrid HHOSSA [37], which is a combination of Harris Hawk optimization (HHO) and the sparrow search algorithm (SSA). The proposed HHOSSA method helps select the most efficient and relevant attributes, and the evaluated results show that the suggested method obtains satisfactory results in detecting attacks with an accuracy of 94.468%, a sensitivity of 94.650%, and a specificity of 95.230%. Suppose a systematic approach is adopted to achieve the desired goals by exploiting the unprecedented discoveries of information technology in agriculture. In that case, much better success can be achieved, which is indispensable. This can be done by using IoT-based WSN. To do this, the author proposes an HOT, GA-MWPSO [38], which has been used successfully in nonlinear constrained optimization problems. Figure 10.5 illustrates the annual number of articles included in the literature review on HOTs.

In patients with corona, the oxygen level drops significantly, and maintaining the oxygen level in the affected patients is a critical task. Artificial ventilation is now widely used in the human body for this task, including other respiratory problems, and it is essential to control the air pressure to provide the right amount of oxygen. For this, the author designed an automatic hybrid controller-based fractional order

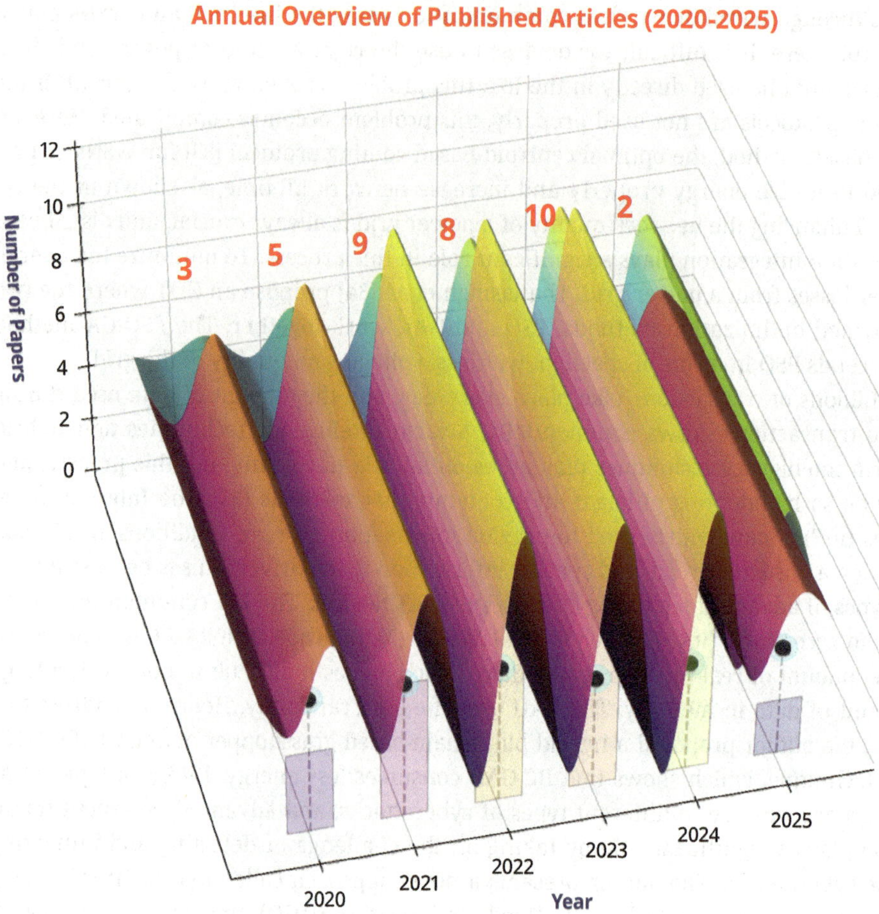

Figure 10.5: Annual overview of published articles (2020–2025).

proportional integral derivative controller (FOPID) system in which the controller parameters are tuned [39]. Breast cancer is currently one of the leading health risks for women, and it is essential to know how the disease behaves in the early stages and how quickly it can spread through the body. Mammography images are generally not very efficient in detecting breast cancer, and as a result, the patient has to face a variety of problems. In this situation, a new HOT, backpropagation boosting recurrent Wienmed model (BPBRW) with hybrid Krill Herd African buffalo optimization (HKH-ABO) [40], is proposed here. After the evaluation of the results, it is seen that the accuracy of the proposed algorithm increased to 99.6%, and its error rates fell to 0.12%. Registration of medical images is very important for detecting any disease and taking curative measures accordingly using appropriate methods. A novel HOT, immune-evolutionary programming (IEP), was proposed in [41]. In this technique, the immune system and evolutionary programming model are utilized to minimize the total cost of power systems, where different types of steps, such as initialization, cloning, and mutation, are used to obtain optimal results efficiently. Diabetes is now a cause of fear for people of different age groups, and the prevalence of this disease worldwide has increased more than ever before. For this reason, in this paper, the author proposes an HOT based on the support vector machine (SVM) [42] for diabetes diagnosis. The HOT proposed here combines the CSA and binary gray wolf optimizer to utilize SVM effectively. Real-world datasets like the UCI Pima Indian standard dataset and the diabetes-type dataset are used here, and the results show that the suggested method achieves distinguished classification accuracy. Brain tumors are now one of the leading causes of death, and how brain tumors develop can be easily seen through MRI. The author proposes a hybrid optimization method that combines SVM and an ACO algorithm [43]. The primary goal of this method is to review how easily and quickly brain tumor diagnoses can be made. Multiple MRIs are used here to provide the best picture possible to give a deeper understanding of the tumor. This method makes it easy to know the tumor's location in the head, and it can provide effective results by reviewing the tumor in its early stages. Table 10.2 illustrates various types of HOT methods used in different applications to enhance the performance of AI models. An overview of eight selected articles is presented here, where the study focused on proposed HOTs, the application domain in which the hybrid model works, and the performance gain of the model and also illustrates the limitations.

Water is one of the most critical components used in the world, and it is significantly utilized in agriculture, urbanization, various types of mills, human living, and other areas. Due to increased water consumption, the amount of freshwater is decreasing daily. Water desalination process is a method of making used or unusable water usable, and to make this process successful, the author proposed a hybrid Capuchin and rat swarm algorithm [44] technique. After evaluating the results, it was found that the proposed algorithm achieved better results than all other state-of-the-art algorithms. Different types of optimization problems in engineering are essential for large-scale operations. Hence [45] introduces a new hybrid optimization method based on DE and naked mole-rat (NMR), namely HDN. The suggested optimization method works with poor exploration and local optima issues. For the result evaluation process, 12 CEC 2005 and 10 complex CEC 2019 benchmark functions are used, and the results show that the proposed method can solve complex antenna problems and is suitable for modern wireless communication. An interconnected energy system (IES) is a type of scheduling problem. The characteristics of the problem are multi-objective and non-convex, and proper optimization techniques are needed to address this. For these reasons, the author presented a novel HOT in which the integration techniques occur between the quantum-based cuckoo search method (QCSA) and the GWO method [46]. QCSA has outstanding exploration capacity, and GWO can avoid local optima. The presented QCSA–GWO is used in three test systems and has achieved significant results. A novel hybrid algorithm combining two well-known metaheuristic algorithms, the GOA and ant lion optimization (ALO) method, is proposed by [47]. By utilizing the positive features of both methods, the proposed method outperforms 18 test functions, alongside ALO and GOA, after testing 32 benchmark test functions and achieves significant results.

Table 10.2: Dissimilar hybrid optimization approaches in artificial intelligence model performance growth.

Study	Year	Application domain	Proposed model	Performance gain	Scalability	Limitations
[26]	2020	Engineering optimization problems	Exchange market genetic algorithm (EMGA)	Up to 29% improvement over others	Suitable for large feature sets	Increased iteration time
[27]	2022	Distribution networks	GAEO (hybrid GA equilibrium optimizer)	Up to 10.5% regarding GA	Supports large-scale distribution	Local optima issues
[30]	2023	Power distribution	Teaching learning-based optimization-EO (teaching-learning + equilibrium optimizer)	Max overshoot: ~30% reduced; steady-state error: ~93% reduced	Effective in normal and dynamic conditions	High power cost
[31]	2021	Power systems	Hybrid intelligent optimization (HIO)	5–29% improvement over others	Scalable in dual-area systems	Tuning required for multiple parameters
[54]	2025	Medical image FS	HBEOSA-DMO and HBEOSA-PSO	15.21% improvement in accuracy	Suitable (high-dimensional dataset)	High cost
[55]	2024	Image steganography	SSA-PSO	PSNR improved up to 84.68 dB	Effective for diverse image types	Power heavy
[56]	2024	Image classification	TwinCNN with binary optimization (BEOSA).	Histology gain: 29.5%, mammography gain: 15.4%	Supports multimodal datasets	Potential overfitting issues
[57]	2024	Skin cancer detection	Hybrid U-Net and improved MobileNet-V3	Precision: + 8.87%, sensitivity: + 6.68%, accuracy: + 4.37%, specificity: + 5.43%	Efficient for distributed computing	High cost

Forecasting the stock market through analysis is a challenging task, as the movement of the stock market depends on a variety of known and unknown variables. Various types of analysis processes, such as technical analysis, fundamental analysis, and statistical analysis, are used to determine how share prices fluctuate or how to estimate the price of a share. For this, the author offered a hybrid artificial NN-based stock market prediction technique [48] and used three other techniques, namely fuzzy *c*-means clustering, subtractive clustering, and grid partitioning, as a helping hand. A software-defined network (SDN) is a state-of-the-art network technology that provides insight into the internal structure of a specific network infrastructure, can be used as an automated application, and plays a vital role in packet optimization. Here, the lion and whale HOT is employed [49] to analyze the cloud scenario in offline mode, and it has been proven that the proposed modification plays an important role in increasing throughput. The intrusion detection system is an application or device used to protect internet users from various types of attacks on different networks or to identify what attacks are occurring. In most cases, traditional intrusion detection systems (IDS) follow association rules to detect different types of intrusive behavior, and additional concerns in diverse networks can be observed. To address this, the author proposed a hybrid optimization-based DL technique [50], which can be used as an intrusion detection system and can handle multi-level scenarios. This model employs a rider optimization algorithm-based neural network (RideNN) for fast-level detection. The results show that the proposed algorithm achieved 0.9254% precision, 0.8362% recall, and 0.8718% F-measure. The usefulness of virtual machines has increased more than ever, and the leading reason is the increased number of users. For this reason, virtual machine placement is now considered a critical task, and the virtual machine plays a significant role in properly handling the resources in the cloud data center. To address this, the author proposed an efficient method for virtual machine placement, consisting of hybrid optimization and fitness parameters [51]. The results show that the proposed algorithm has provided the expected outcomes. FS means removing irrelevant, unnecessary, or less necessary features from a given dataset, creating a dataset with only the significant features, and then using the newly created dataset to achieve the desired results. An algorithm can provide maximum results if specific methods eliminate unwanted features from a particular dataset. For this purpose, the author proposed a hybrid method [52] in which a filter method and two wrapper swarm intelligence algorithms are used together, and the results show that the proposed algorithm can provide better results than other algorithms.

Environmental monitoring and resource management play crucial roles in the quality of the environment, the inhabitants who live there, and agricultural planning. To deal with these issues, a land-use and land-cover (LULC) classification technique is used to obtain remote sensing images, and a hybrid model was proposed with PSO alongside the long short-term memory (LSTM) classifier [53]. Table 10.3 illustrates different types of HOTs with distinct performance metrics, the complexity of the proposed model, and the datasets used. Like Table 10.2, Table 10.3 also includes eight distinct articles that highlight the proposed model's comparative performance analysis against other traditional models. The most critical aspect of Tables 10.2 and 10.3 is that they are mutually exclusive, ensuring diversity and clarity while avoiding redundancy. The results section demonstrates that the proposed model.

Table 10.3: Performance evaluation of various hybrid optimization techniques.

Study	Dataset(s)	Complexity	Metric 1	Metric 1	Metric 1	Metric 1
[36]	IoV data from vehicles	Low to moderate	Energy (19.23 J for 100 tasks)	Time (33.37 s for 100 tasks)	Energy savings (~90–95%)	Execution cost (78.93 s for 500 tasks)
[37]	DAPT 2020	High	Accuracy: 94.468%, precision: 80.737%, recall: 83.094%, F_1-score: 81.096%	Sensitivity: 94.650%, specificity: 95.230%	AUC: 97.032	Training accuracy: 96.18%
[38]	Simulated dataset	Moderate	AUC: 88%, coverage ratio: 0.95	Accuracy: 85%, precision: 80%, recall: 78%, F_1-score: 79%	Energy efficiency: 75%	Reliability improvement: 82%
[53]	Sat 4, Sat 6, Eurosat	Moderate	Accuracy: 99.99% (Sat 4), 99.99% (Sat 6), 97.40% (Eurosat)	Precision: 99.95% (Sat 4), 99.94% (Sat 6), 97.80% (Eurosat)	Recall: 99.96% (Sat 4), 99.975% (Sat 6), 98.70% (Eurosat)	F_1-Score: 99.95% (Sat 4), 99.94% (Sat 6), 98.25% (Eurosat)
[58]	WSN-DS, TON-IoT	Moderate	Accuracy, precision, recall, F_1-score (WSN-DS): 99.94%	Accuracy, precision, recall, F_1-score (TON-IoT): 99.97%	AUC (WSN-DS): 99.99%, AUC (TON-IoT): 100.0%	Specificity (WSN-DS): 99.99%, Specificity (TON-IoT): 100.0%
[59]	Simulated dataset	Moderate	Energy Consumption (mJ, round 400): 36	Throughput – 0.9867 Mbps.	Packet delivery ratio (%): 99%	Network lifetime – 7,550 rounds
[60]	Simulated dataset	Moderate	Convergence time: 325.5 ms	Kappa: 0.88	AUC: 0.96, Specificity: 87.4%	Accuracy: 93%, precision: 92.1%, recall: 90.5%, F_1-score: 91.3%.
[61]	Quality of web service (QWS)	High	ACO + GA precision: 73%	Accuracy: 82%, precision: 72%, recall: 67%, F_1-score: 81%.	ACO + GA F_1-score: 82%	ACO + GA recall: 68%

provides significant improvements compared to other existing models. A breast cancer diagnosis is one of the most critical issues to deal with in contemporary times, as optimal results can save a considerable number of lives. Because of the diagnosis of breast cancer using digital mammography, a hybrid binary optimization approach for FS is applied in [54], where three different types of optimizations are integrated. The proposed optimizer reduces the irrelevant features from the curated breast imaging subset (CBIS) dataset, and the results show that the proposed model achieves significant improvements in evaluation metrics such as accuracy, precision, recall, F_1-score, and AUC. Digital images like mammography and histopathology can be used in the medical field to diagnose breast cancer. A DL application-based HOT, TwinCNN, was proposed in [55], where two convolutional NNs (CNNs) are used to help classify and extract the most influential and pragmatic images, and a binary optimization method is utilized to select the relevant features. Hiding secret information inside a digital image, such as in military and intelligence fields, is always troublesome, and an image steganography strategy is frequently used to accomplish this. HOTs integrating the salp swarm algorithm (SSA) and PSO are proposed in [56], which help find the best spots to hide the most important and critical information and also improve the appearance of the picture. The result demonstrates that the proposed model achieves a high-quality score for the image and also obtains a low error rate. Skin cancer is one of the most common types of cancer worldwide, where early detection plays an essential role in saving lives; hence, a hybrid detection system was proposed in [57], where U-Net and improved MobileNet-V3 are combined and try to find the disease with better classification performance. The results show that the proposed hybrid model can perform better in accuracy, precision, sensitivity, and specificity, which helps detect the disease in the early stage.

Security enhancement for any network to save it from cyberattacks is always essential to protect the data from intruders, and the author focuses on IDS to secure the data used in WSNs and IoT [58]. A hybrid ML model is used here, which combines models like *K*-means and SMOTE, helping to balance the data, and principal component analysis helps deduce the relevant features from the dataset with dimensionality reduction. The results illustrate that the proposed model achieves impressive classification accuracy on the WSN-DS dataset and simultaneously outperforms others. Because of the enormous use of WSNs in distinct applications like healthcare, environmental monitoring, and military operations in different circumstances, a hybrid optimization model was proposed by Ullah et al. [59], which combines fuzzy spider monkeys and hidden Markov models. The proposed model selects the optimal clusters from data based on energy and helps enhance the overall performance of the model with an improved rate of data transmission with a 1.2% performance enhancement in the network. The efficiency obtained in cloud computing to handle data effortlessly is always paramount, and to address that, an HOT using a combination of war strategy optimization (WSO) and rat swarm optimizer (RSO) was proposed [60]. The article includes different types of environments in which task resource management is necessary, and the model's outcomes show that it can greatly reduce convergence time. Because of advancements in technological fields, urban life has dramatically changed, and dealing with various municipal services, such as efficiently using energy and controlling traffic problems, is the most challenging task. In these circumstances, a hybrid model integrated with ACO and GA was proposed in [61], which helps ensure the quality of service (QoS) for each pupil and achieve satisfactory improvement in classification accuracy. The optimal performance of cloud computing technology largely depends on load balancing, which helps manage usable resources efficiently and to do the assignment, a hybrid model integrated with the dragonfly algorithm (DA) and PSO, namely DPSO, was proposed in [62]. The results section demonstrates that the proposed model has the best fitness value regarding different user bases and optimal resource allocation capabilities.

10.6 Conclusion and Future Scope

The sensation of using AI models in various fields for technological development and excellence has now expanded to some extent. One of the principals and most influential motivations for utilizing this model is to solve problems efficiently and achieve the anticipated outcomes in a relatively short period by using proper guidance. Models employ various optimization algorithms based on the assignments and solutions considered at different times. However, increasing workloads and solving highly complex tasks require the kind of computational power and skill that an SOA cannot always provide, while the HOA can deliver significantly better outcomes than the SOA. When an HOT is employed with an AI model to perform a straightforward task, there are numerous advantages, as every optimization algorithm has its own unique strengths that can unexpectedly outperform others. At the same time, HOTs have become an essential aspect of AI models, considering work time, swiftness of work, type of problem statement, nature of the dataset, pattern review for making accurate decisions, parameter tuning, and proper direction to accomplish expected outcomes. Determining which combination of HOTs is the most significant and impactful for accomplishing a particular task is possible, and the AI model can provide more reliable results in that scenario. However, it should be kept in mind that the spontaneous usage of an HOT may cause the model to encounter diverse hardships and result in a significant waste of time and energy when determining which HOTs can provide the most promising results by fitting with which model is a critical task.

Future Scope: The usefulness of utilizing HOTs has been reasonably practical in enhancing the performance of AI models. However, the vast availability of complex problem statements and the invention of diverse types of data complicate the situation; hence, they need to be refined. Future work may include extending the proposed hybrid approach to more advanced AI models such as DL, reinforcement learning, or clustering-based tasks. Working with that domain and its distinct applications tends to be complex, and hybrid optimization could more effectively solve this. Another practical advancement could involve implementing adaptive or comprehensive hyperparameter tuning methods for the HOT. During the training process, an HOT can modify itself with hyperparameters, assembling the model more precisely and faster than ever and also providing a user-friendly approach without any external efforts. Another momentous factor is that all the HOTs need to experiment with real-world datasets containing extensive and multifarious data, which makes the model adaptable in problematic scenarios. Miscellaneous application domains, such as finance, innovative healthcare, weather, and city planning, often encounter challenging insights that HOTs need to be tremendously successful. Demanding studies, such as time series prediction or fields like natural language processing, require more computational power, whereas HOTs could further demonstrate its capabilities. Furthermore, future work can investigate integrating HOTs with parallel computing techniques to minimize the mammoth processing time. Finally, working with AI in sensitive fields, such as national security, becomes even more crucial for accuracy and understanding, whereas HOT can make it possible to handle this diplomatic task with greater power and efficiency.

References

[1] Bakal JA, DeFroda S, Owens BD, Eltorai AE. editors Translational Sports Medicine. Elsevier; 2023 Aug 14.

[2] Gandomi A, Haider M. Beyond the hype: Big data concepts, methods, and analytics. International Journal of Information Management. 2015 Apr 1; 35(2): 137–144.

[3] Saha S, Sarker PS, Al Saud A, Shatabda S, Newton MH. Cluster-oriented instance selection for classification problems. Information Sciences. 2022 Jul 1; 602: 143–158.

[4] Lilhore UK, Kumar A, Vyas N, Simaiya S, Dutt V. (ed) Multimodal Data Fusion for Bioinformatics Artificial Intelligence. John Wiley & Sons; 2025 Feb 19.

[5] Vani SS, Ramya P, Nagendar Y, Prakash PA, Chanti M, Atheeswaran A. An Efficient Approach to Data Clustering Using the K-Means Algorithm in Big Data Analytics. In 2025 International Conference on Intelligent and Innovative Technologies in Computing, Electrical and Electronics (IITCEE) 2025 Jan 16 (pp. 1–6). IEEE.

[6] Hassabis D, Kumaran D, Summerfield C, Botvinick M. Neuroscience-inspired artificial intelligence. Neuron. 2017 Jul 19; 95(2): 245–258.

[7] Kumar Y, Koul A, Singla R, Ijaz MF. Artificial intelligence in disease diagnosis: A systematic literature review, synthesizing framework and future research agenda. Journal of Ambient Intelligence and Humanized Computing. 2023 Jul; 14(7): 8459–8486.

[8] Thiraviya C, Priyanka KA, Ananthi S, Subha R. A Study of Machine Learning-based Models for Detecting Heart Disease. In 2025 Third International Conference on Augmented Intelligence and Sustainable Systems (ICAISS) 2025 May 21 (pp. 392–399). IEEE.

[9] Yuvaraj S, Sugavanaesh S, Kumar CT, Saran G, Subha R. Identification of Autism Spectrum Disorder (ASD) in Adults through Various Machine Learning Algorithms. In 2024 2nd International Conference on Intelligent Data Communication Technologies and Internet of Things (IDCIOT) 2024 Jan 4 (pp. 733–738). IEEE.

[10] Asnicar F, Thomas AM, Passerini A, Waldron L, Segata N. Machine learning for microbiologists. Nature Reviews Microbiology. 2024 Apr; 22(4): 191–205.

•

[11] Yadav SP, Kumar A, Dubey AK, De Albuquerque VH, Rodriguez-Aguiñaga A, Quezada A. (eds) Medical Image Processing: Multimodal Fusion Techniques. De Gruyter; 2024 Sep 9 . . ..
[12] Surianarayanan C, Lawrence JJ, Chelliah PR, Prakash E, Hewage C. A survey on optimization techniques for edge artificial intelligence (AI). Sensors. 2023 Jan 22; 23(3): 1279.
[13] Miki S, Yamamoto D, Ebara H. Applying Deep Learning and Reinforcement Learning to Traveling Salesman Problem. In 2018 International Conference on Computing, Electronics & Communications Engineering (ICCECE) 2018 Aug 16 (pp. 65–70). IEEE.
[14] Suanpang P, Jamjuntr P, Jermsittiparsert K, Kaewyong P. Tourism service scheduling in smart city based on hybrid genetic algorithm simulated annealing algorithm. Sustainability. 2022 Dec 6; 14(23): 16293.
[15] Minaee RP, Afsharnia M, Moghaddam A, Ebrahimi AA, Askarishahi M, Mokhtari M. Calibration of water quality model for distribution networks using genetic algorithm, particle swarm optimization, and hybrid methods. MethodsX. 2019 Jan 1; 6: 540–548.
[16] Shang J, Li G. A hybrid composite differential evolution and multiobjective particle swarm optimization evolutionary algorithm and its application. IEEE Access. 2024 May 23; 12: 74417–74431.
[17] Moeeni H, Bonakdari H, Fatemi SE, Zaji AH. Assessment of stochastic models and a hybrid artificial neural network-genetic algorithm method in forecasting monthly reservoir inflow. INAE Letters. 2017 Apr; 2(1): 13–23.
[18] Khourdifi Y, Baha M. Heart disease prediction and classification using machine learning algorithms optimized by particle swarm optimization and ant colony optimization. International Journal of Intelligent Engineering & Systems. 2019 Jan 1; 12(1).
[19] Arikan M. Hybrid simulated annealing-tabu search algorithms for solving U-shaped type-2 assembly line balancing problems with workload smoothing objective U-şekilli hatlarda iş yükü dengelemeli tip-2 montaj hattı dengeleme probleminin çözümü için melez tavlama benzetimi-tabu arama algoritmaları. Journal of the Faculty of Engineering and Architecture of Gazi University. 2024; 39(3).
[20] Kayé BK, Diaby M, Koivogui M, Oumtanaga S. A memetic algorithm for an external depot production routing problem. Algorithms. 2021 Jan 19; 14(1): 27.
[21] Luan J, Yao Z, Zhao F, Song X. A novel method to solve supplier selection problem: Hybrid algorithm of genetic algorithm and ant colony optimization. Mathematics and Computers in Simulation. 2019 Feb 1; 156: 294–309.
[22] Aguitoni MC, Pavão LV, Ravagnani MA. Heat exchanger network synthesis combining simulated annealing and differential evolution. Energy. 2019 Aug 15; 181: 654–664.
[23] Bala NM, Bin Safei S. A hybrid harmony search and particle swarm optimization algorithm (hspso) for testing non-functional properties in software system. Statistics, Optimization & Information Computing. 2022; 10(3): 968–982.
[24] Ananth Kumar T, Kathiravan A, Kanimozhi P, Ajagbe SA, Suresh Kumar K, Rajmohan R. Authentication and access control schemes in 5G-based healthcare systems.
[25] Singh A, Khamparia A. A hybrid whale optimization-differential evolution and genetic algorithm based approach to solve unit commitment scheduling problem: WODEGA. Sustainable Computing: Informatics and Systems. 2020 Dec 1; 28: 100442.
[26] Jafari A, Khalili T, Babaei E, Bidram A. A hybrid optimization technique using exchange market and genetic algorithms. IEEE Access. 2020; 8. doi: 10.1109/ACCESS.2019.2962153.
[27] Bakry OM, Alhabeeb A, Ahmed M, Alkhalaf S, Senjyu T, Mandal P, Dardeer M. Improvement of distribution networks performance using renewable energy sources based hybrid optimization techniques. Ain Shams Engineering Journal. 2022 Nov 1; 13(6): 101786.
[28] Selvaraj D, Rangasamy D. Electric vehicle charging using roof top photovoltaic controlled with new hybrid optimization technique. Indonesian Journal of Electrical Engineering and Computer Science. 2022 Jun; 26(3): 1227–1234.

[29] Kumar SA, Padma S, Madhubalan S. Distribution network reconfiguration using hybrid optimization technique. Intelligent Automation & Soft Computing. 2022 Aug 1; 33(2).

[30] Sayed EA, Sameh MA, Attia MA, Badr AO. Enhancement of PV performance by using hybrid TLBO-EO optimization. Ain Shams Engineering Journal. 2023 Apr 1; 14(3): 101892.

[31] Gupta DK, Jha AV, Appasani B, Srinivasulu A, Bizon N, Thounthong P. Load frequency control using hybrid intelligent optimization technique for multi-source power systems. Energies. 2021 Mar 12; 14(6): 1581.

[32] Parthasarathi P, Nivedha S, Balamurugan M, Yuvaraj D, Devi SS, Sangeethaa SN. Innovative Skin Melanoma Diagnosis Using Deep Learning and African Gorilla Troops Optimizer. In 2024 First International Conference on Software, Systems and Information Technology (SSITCON) 2024 Oct 18 (pp. 1–6). IEEE.

[33] Wasay Mudasser A, Ahmed Abdul Gafoor SA. Secure Internet of Things based hybrid optimization techniques for optimal centroid routing protocol in wireless sensor network. Concurrency and Computation: Practice and Experience. 2023 Mar 10; 35(6): 1.

[34] Thanasingh S, Sasikumar AN, Mahendran K, Saranya R. Optimizing distributed generation resources for power grid quality improvement using hybrid optimization technique. Electric Power Components and Systems. 2023 Dec; 28: 1–9.

[35] LK JG, Maneengam A, Kv PK, Alanya-Beltran J. Design and implementation of machine learning modelling through adaptive hybrid swarm optimization techniques for machine management

[36] Abinaya AB, Karthikeyan G. Data offloading in the internet of vehicles using a hybrid optimization technique. Intelligent Automation & Soft Computing. 2022 Oct 1; 34(1).

[37] Kumari I, Lee M. A prospective approach to detect advanced persistent threats: Utilizing hybrid optimization technique. Heliyon. 2023 Nov 1; 9(11).

[38] Banerjee A, Mitra A, Biswas A. An integrated application of IoT-based WSN in the field of Indian agriculture system using hybrid optimization technique and machine learning. Agricultural Informatics: Automation Using the IoT and Machine Learning. 2021 Mar; 19: 171–187.

[39] Sakthiya Ram S, Kumar C, Ramesh Kumar A, Rajesh T. Hybrid optimization techniques based automatic artificial respiration system for corona patient. Automatika: Časopis za automatiku, mjerenje, elektroniku, računarstvo i komunikacije. 2022 Feb 23; 63(2): 226–243.

[40] Dewangan KK, Dewangan DK, Sahu SP, Janghel R. Breast cancer diagnosis in an early stage using novel deep learning with hybrid optimization technique. Multimedia Tools and Applications. 2022 Apr; 81(10): 13935–13960.

[41] Roslan N, Khader PS, Nordin A, Jaafar S. A hybrid optimization technique for solving economic emission load dispatch problem. All Open Access, Bronze. 2020.

[42] Mallika C, Selvamuthukumaran S. A hybrid crow search and grey wolf optimization technique for enhanced medical data classification in diabetes diagnosis system. International Journal of Computational Intelligence Systems. 2021 Sep 1; 14(1): 157.

[43] Warjurkar SV, Ridhorkar S. Maximizing precision in early prognosis using SVM-ACO classifier and hybrid optimization techniques in MRI brain tumor segmentation with integration of multi-modal imaging data. International Journal of Intelligent Systems and Applications in Engineering. 2024; 12(10s).

[44] Baseer MA, Vinoth Kumar V, Izonin I, Dronyuk I, Velmurugan AK, Swapna B. Novel hybrid optimization techniques to enhance reliability from reverse osmosis desalination process. Energies. 2023 Jan 7; 16(2): 713.

[45] Singh G, Singh U. Triple band-notched UWB antenna design using a novel hybrid optimization technique based on DE and NMR algorithms. Expert Systems with Applications. 2021 Dec 1; 184: 115299.

[46] Kaur A, Narang N. Economic emission generation scheduling of interconnected energy system using hybrid optimization technique. Iranian Journal of Science and Technology, Transactions of Electrical Engineering. 2023 Mar; 47(1): 103–124.
[47] Amaireh AA, Al-Zoubi AS, Dib NI. A new hybrid optimization technique based on antlion and grasshopper optimization algorithms. Evolutionary Intelligence. 2023 Aug; 16(4): 1383–1422.
[48] Rao KV, Ramana Reddy BV. Hm-smf: An efficient strategy optimization using a hybrid machine learning model for stock market prediction. International Journal of Image and Graphics. 2024 Mar 20; 24(02): 2450013.
[49] Kaur J, Singh D. Optimizing Packets of Software Defined Network Using Hybrid Optimization Techniques. In 2022 2nd International Conference on Innovative Practices in Technology and Management (ICIPTM) 2022 Feb 23 (Vol. 2, pp. 344–349). IEEE.
[50] ES GSR, Azees M, Vinodkumar CR, Parthasarathy G. Hybrid optimization enabled deep learning technique for multi-level intrusion detection. Advances in Engineering Software. 2022 Nov 1; 173: 103197.
[51] Supreeth S, Bhargavi S, Margam R, Annaiah H, Nandalike R. Virtual machine placement using Adam white shark optimization algorithm in cloud computing. SN Computer Science. 2023 Nov 20; 5(1): 21.
[52] Abdo A, Mostafa R, Abdel-Hamid L. An optimized hybrid approach for feature selection based on chi-square and particle swarm optimization algorithms. Data. 2024 Jan 25; 9(2): 20.
[53] Rajendran GB, Kumarasamy UM, Zarro C, Divakarachari PB, Ullo SL. Land-use and land-cover classification using a human group-based particle swarm optimization algorithm with an LSTM Classifier on hybrid pre-processing remote-sensing images. Remote Sensing. 2020 Dec 17; 12(24): 4135.
[54] Oyelade ON, Aminu EF, Wang H, Rafferty K. An adaptation of hybrid binary optimization algorithms for medical image feature selection in neural network for classification of breast cancer. Neurocomputing. 2025 Feb 7; 617: 129018.
[55] Oyelade ON, Irunokhai EA, Wang H. A twin convolutional neural network with hybrid binary optimizer for multimodal breast cancer digital image classification. Scientific Reports. 2024 Jan 6; 14(1): 692.
[56] Abdalhussein E, Ibrahim NJ, Ali YH. Image steganography based on hybrid salp swarm algorithm and particle swarm optimization. International Journal of Intelligent Engineering & Systems. 2024 Jan 1; 17(1).
[57] Kumar Lilhore U, Simaiya S, Sharma YK, Kaswan KS, Rao KB, Rao VM, Baliyan A, Bijalwan A, Alroobaea R. A precise model for skin cancer diagnosis using hybrid U-Net and improved MobileNet-V3 with hyperparameters optimization. Scientific Reports. 2024 Feb 21; 14(1): 4299.
[58] Talukder MA, Khalid M, Sultana N. A hybrid machine learning model for intrusion detection in wireless sensor networks leveraging data balancing and dimensionality reduction. Scientific Reports. 2025 Feb 7; 15(1): 4617.
[59] Ullah A, Khan FS, Mohy-Ud-Din Z, Hassany N, Gul JZ, Khan M, Kim WY, Park YC, Rehman MM. A hybrid approach for energy consumption and improvement in sensor network lifespan in wireless sensor networks. sensors. 2024 Feb 20; 24(5): 1353.
[60] Panesar GS, Chada R. A cloud-based adaptive multi-agent deep deterministic policy gradient technique-based hybrid optimisation algorithm for efficient virtual machine migration and task scheduling. International Journal of Intelligent Systems and Applications in Engineering. 2024; 12(1): 403–418.
[61] Jayaudhaya J, Jayaraj R, Ramash KK. A new integrated approach for cloud service composition and sharing using a hybrid algorithm. Mathematical Problems in Engineering. 2024; 2024.
[62] Mohapatra S, Mohanty S, Nayak HK, Mallick MK, Ramesh JV, Dudekula KV. DPSO: A hybrid approach for load balancing using dragonfly and PSO algorithm in cloud computing environment. EAI Endorsed Transactions on Internet of Things. 2024 Jan 1; 10.

R. Roselin Kiruba, J. P. Shritharanyaa, J. Jude Moses Anto Devakanth, Vasumathi, C. Saranya Jothi, L. Sharmila

11 Smart Fuzzy Logic in AI Systems for Accurate Medical Diagnostics

Abstract: In many countries, ensuring accurate medication during disease treatment remains a significant challenge due to the increasing complexity of modern healthcare systems, where rapid technological advancements and heterogeneous clinical data often lead to diagnostic uncertainty necessitating intelligent fuzzy logic based AI systems for precise and reliable medical diagnosis. The conventional technique has problems with the decision strategy where the criteria change from one disease to another. Although these methods are effective, they have disadvantages due to the flexibility of data and it should be left to the clinical person for further accuracy. The human mistakes and knowledge from the expertise make the diagnosis process much complex. Hybridization of smart fuzzy logic (SFL) in artificial intelligence (AI) model gains significant improvement in the field of medical diagnosis. It is developed to support the decision-making process effectively by diagnosing the disease precisely, selecting the preprocessed data, monitoring the disease, and optimizing the work of the machine. This model has the capability to handle imprecise data and improve in identifying the classification results, decreasing human reasoning errors. The SFL with machine learning algorithm is an AI model for processing complex data understanding, helping in the diagnosis of the healthcare output, thereby decreasing the rate of false negative results. AI systems are merged to reduce the diagnosis faults, thereby improving the early detection of diseases using precision, recall, and F_1 score.

Keywords: Artificial intelligence (AI), smart fuzzy logic (SFL), medical diagnosis, machine learning

11.1 Introduction

Global communication has significantly improved since the introduction of contemporary technologies. The Internet of Things (IoT), a network of connected objects that can share data without requiring human contact, is one of the most innovative technical breakthroughs [1]. IoT technology has been used in a number of domains, such as education, home automation, business, transportation, and medical care [2, 3]. Timely and precise medical diagnosis and treatment are crucial in the healthcare industry for preserving life. Daily examinations at medical institutions, however, can be expensive and inconvenient for patients [4]. Two basic components of IoT-enabled healthcare systems include telemetry and analytics data. These technologies will allow for the constant mon-

https://doi.org/10.1515/9783112214374-011

itoring of the patient; they can be very useful for quick medical solutions. These systems can be integrated with microcontrollers and various sensors that monitor patients' vital signs, the data of which is then forwarded to medical specialists. The IoT uses in healthcare are growing more varied as it develops. For example, radio-frequency identification sensors can be fitted to patients when they enter a hospital, to capture biological information like weight, temperature, and blood pressure. A Zigbee protocol may then be used to send these data to physicians. Physicians are able to keep an eye on patients' health and apply proper therapies, in accordance with the information obtained [5].

Fuzzy logic and IoT technology might be the way forward in healthcare delivery and patient monitoring. In contrast to classical binary logic, fuzzy logic can be used in the formation of rules and truths with linguistic variables; thus, the likelihood of making a particular decision is higher [6–9]. In medicine, fuzzy logic has been applied in intensive care centers, anesthesia-related monitoring of patients, and in detection devices of lung cancer diagnostics [10–12]. Despite enhancements done in IoT-based health monitoring systems, a number of issues and limitations continue to linger, as shown in literature, as it stands. Most of these provide limited focus on specific vital signs or real-time monitoring of numerous health indicators. Issues related to data security and privacy have not been totally addressed, especially in regards to different health data. Most systems also lack global positioning systems and flexible control over the rooms. GPS integration with various systems needs improvement. Inspired by these weaknesses, our key aim in this research is to develop a modular health monitoring system that encompasses combining complete advanced vital sign monitoring, such as electrocardiogram (ECG), blood oxygen saturation (SpO_2), and heart rate, using advanced nationwide GPS integration, sophisticated room control capabilities, and robust data privacy protection. In particular, our system design employs the novel regression model-based data differential privacy approach subjected to heteroscedasticity, initially proposed in [13], to better cope with changing levels of data noise. Our aim is to enhance the care provision for patients by developing a fully comprehensive patient monitoring solution that enables timely responses by doctors, and enhance patient monitoring. However the advantages reaped from IoT-based healthcare systems and the merits of fuzzy logic, there is a scarcity of literature on this topic of utilization of such technologies in delivering remote diagnostics, treatments, and monitoring for patients with COVID-19, particularly those in remote locations [14].

The devices could be made using sensors, which would include temperature, ECG, and blood pressure sensors, which would communicate with a Arduino microcontroller. Thus, the data would be transmitted to a server where a digital mobile app would be used by the patient and the doctor to retrieve the observed data. Because it supports infrared wireless communication, ZigBee technology [15] is cheaper and more favored than Bluetooth [16]. Data from the temperature, finger flip, and ECG sensors could be transmitted to a server from which a doctor can evaluate the data using digital qualifications. A device that uses biological sensors for temperature, heart rate, and blood pressure can measure and communicate over wireless networks [17], as shown in Table 11.1.

Table 11.1: Comparison of traditional and smart fuzzy systems.

Aspects	Traditional fuzzy	Smart fuzzy
Data handling	Works on static, pre-defined datasets or fixed inputs.	Continuously processes real-time, sensor-based IoT data streams.
Rule generation	Experts manually establish rules based on their domain knowledge.	Real-time data analysis is used to dynamically build and optimize rules.
Combining AI/ML	Integration with AI/ML models is infrequent or manual.	Deeply incorporated with AI methods such as optimization algorithms, neural networks, and reinforcement learning.
Making decisions	Depends on predetermined linguistic norms such as "IF heart rate high THEN alarm."	Uses learning feedback and real-time updated rules to make thoughtful, context-aware decisions.
Mechanism of feedback	Parameters stay constant after design; there is no real-time feedback loop.	Features a feedback loop that allows it to continuously learn from mistakes made in predictions or shifting surroundings.
Privacy of data	External encryption layers are necessary for data privacy.	Incorporated with intelligent data privacy models (such as heteroscedasticity-based differential privacy, based on regression).
Scope of application	Particular to a single vital sign or a limited number of health markers.	Comprehensive and modular; it includes room control and several vital indicators (ECG, SpO_2, BP, and temperature).

Sensors for temperature, ECG, and blood pressure have an Arduino microcontroller to transmit their data. Thus, the data could be transmitted to a server where a digital mobile application or app will be used by both the patient and the doctor to extract the observed data. Because it supports infrared wireless communication, ZigBee technology [15] is cheaper and more popular than Bluetooth [16]. From the temperature, finger flip and ECG sensors, data can be sent to the server, and the physician can utilize a digital qualification to see it. The devices that use biological sensors to check temperature, heart rate, and blood pressure can measure and exchange information wirelessly with each other via mobile networks [17]. Instead of using linguistic variables to build rules and facts, fuzzy logic works on probabilities between zero and one. There are numerous real-time uses of intelligent fuzzy logic, especially in the area of medication. For example, it is used for critical care unit medical decision-making [10] and patient monitoring. Diagnostic systems can give diagnostic alarm in case of a major incident happening during anesthesia administration [11] or in the diagnosis of lung cancer [12]. Numerous health monitoring systems have been created, according to our review of the literature. Nevertheless, these systems frequently use either fuzzy logic or the IoT, but not both at once. Combining the two technologies may result in enhanced functionality and more advanced health monitoring features. The health monitoring systems under examination use a variety of methods, including data privacy strategies, IoT, and artificial intelligence (AI).

11.1.1 Contribution

- IoT-Enabled Data Layer: Wearable sensors are used to collect patient vital signs (blood pressure, temperature, heart rate, and SpO_2) in real time.
- Neuro-Fuzzy Feature Extraction and Preprocessing: Decrease of dimensionality, normalization, and noise; generating fuzzy rules and optimizing membership functions.
- Intelligent Decision-Making Driven by AI: Machine learning and reinforcement learning are used for classification, prediction, and sophisticated reasoning; fuzzy rule adaption and ongoing feedback.
- Data Security and Differential Privacy: Heteroskedastic noise differential privacy based on regression models for patient data protection.
- Smart Room Control and GPS Integration: Location of the patient and surroundings modification according to health state.

11.2 Proposed Methodology

Each of the following has its role in a neuro-fuzzy system: data layer, neuro-fuzzy layer, and AI layer. There are usually hierarchically layered – the data, learning, and reasoning tasks, as well as intelligent decisions as shown in Figure 11.1.

11.2.1 Data Layer

The data layer aims to be the foundation of the system. The function of this layer is to collect data for analysis. Data is collected from a myriad of sources, such as databases, sensors, as well as real-time data streams. With each row indicating a reading of a patient's vital parameters, this dataset appears to be an accumulation of readings from wearable devices or health monitoring sensor vital sign values of different patients. Now, let us break down each column below:

11.2.1.1 Patient ID Description

Unique number assigned to every patient. This is to differentiate data of different patients, which is very essential in tracking personal health trends and ensuring proper recordation.

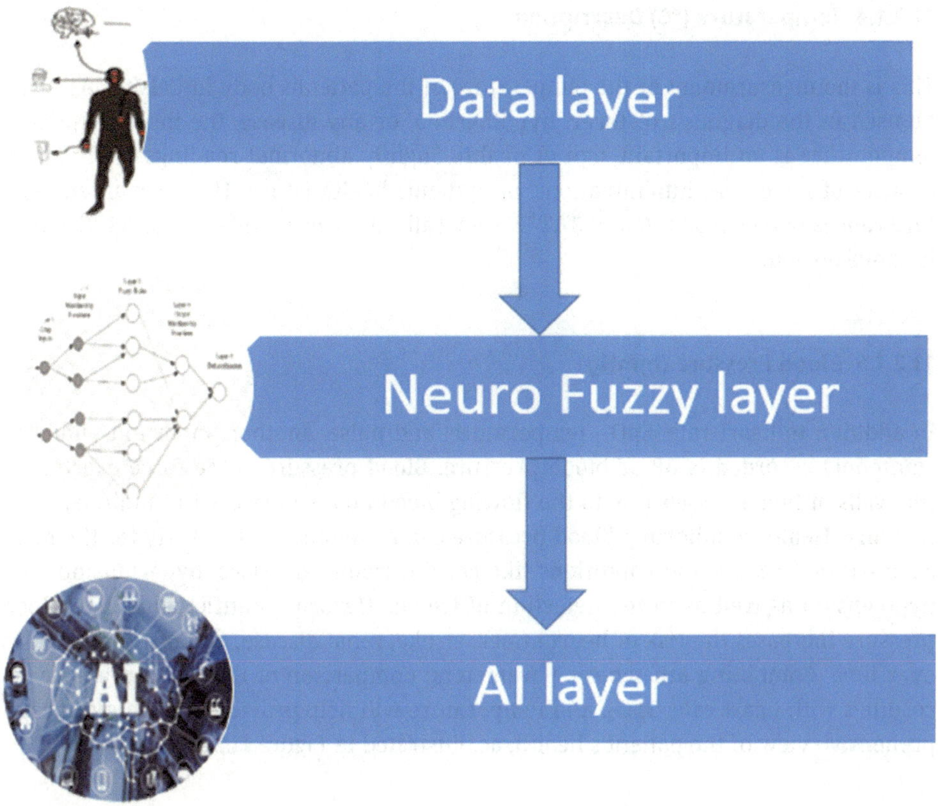

Figure 11.1: Proposed methodology.

11.2.1.2 Heart Rate (bpm)

The patient's heart rate is measured in beats per minute (bpm). The heart rate gives a client feedback about the cardiovascular fitness. Inaccurate readings may indicate possible cardiac discomfort or bradycardia (low heart rate) and tachycardia (high heart rate) conditions.

11.2.1.3 SpO_2 (%) Measurement

It measures the percentage saturation of oxygen in the human blood. The use of SpO_2 reading is to reflect on the sufficiency of oxygen delivery toward the tissues of body. Low SpO_2 readings, which may suggest hypoxemia or lower blood oxygen, may require health provider intervention. Readings between 95% and 100% is taken as normal. A reading below 90% is sometimes worrisome.

11.2.1.4 Temperature (°C) Description

This is the measurement of the temperature of the patient's body in Celsius. As it can be used in the diagnosis of fever, hypothermia, or any disease, the measurement of temperature is an important sign of health. Slightly abnormal readings might be indicative of infections, inflammations, or systemic health issues. The normal temperature ranges between 36.1 °C and 37.2 °C. Generally, a fever is explained as 38 °C, which is above normal.

11.2.1.5 Blood Pressure (mmHg)

In addition to heart rate, SpO_2, temperature, and pulse, another important vital sign commonly recorded is BP or blood pressure. Blood pressure is the force exerted on the walls of blood vessels due to the flowing blood and is measured in millimeters of mercury. Health monitoring: Blood pressure can be monitored routinely for the management of diseases and conditions like cardiovascular diseases, hypertension, and hypotension as well as in the detection of trends. Hazard identification: High blood pressure increases the risk of heart attack, stroke, renal disease, and other conditions over time. Combining other basic parameters: Comparison of the blood pressure recordings with heart rate, SpO_2, and temperature will help provide a much more comprehensive view of the patient's health, as illustrated in Figure 11.2.

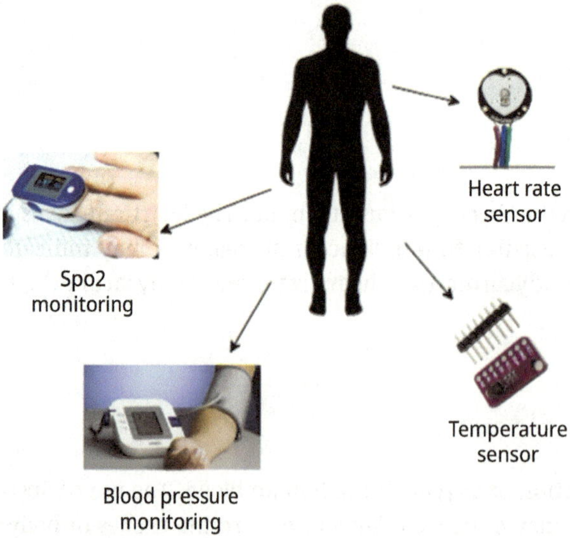

Figure 11.2: Data collection layer.

11.2.2 Neuro-Fuzzy Layer

The raw data is collected from the sensors and it is processed with preprocessing and feature extraction steps:

- **Preprocessing:** Removes noise and fills gaps by cleaning, normalization, and transformation of data to suitable neuro-fuzzy network format.
- Feature extraction in the proposed framework is performed with reduced dimensionality by focusing exclusively on the most significant variables. This is achieved through the neuro-fuzzy layer, which identifies relevant features and meaningful patterns directly from the raw input data. The neuro-fuzzy layer effectively minimizes data dimensionality by retaining only the most informative and discriminative features, thereby eliminating redundant and less influential attributes. Such an optimized feature representation enhances prediction and classification accuracy, particularly when dealing with complex, uncertain, and high-dimensional datasets. A detailed explanation of this process is presented in the subsequent subsection.

11.2.2.1 Feature Extraction in Neuro-Fuzzy Systems

Neural fuzzy logic has several advantages in fuzzy logic and neural networks in medical analysis. They are:

- In calculations using partial data and imprecision information, where there is a missing of data, the system can use fuzzy logic to find the remaining values in a reasonable way.
- Also, the system makes use of patterns to map the data with the neural value for accurate diagnosis. It is accurate with nonlinear relationships by taking in several factors for learning.
- The important design in neuro-fuzzy systems is to find membership functions in the Gaussian membership.
- The precise selection of Gaussian functions is not proof-able. The Gaussian functions are preferable due to the smooth nature of fuzzy logic systems.
- In the real world, it is frequently categorized on the severity of various diseases, like medical diagnoses. The crucial factor is the selection of the membership function selection.
- The bell curve's breadth leads to inaccurate classifications or predictions, making it difficult for parameters.
- Furthermore, Gaussian functions necessarily match the distribution of input features with highly skewed or multimodal data. The Gaussian membership functions are selected due to their smooth mathematical properties and differentiability, which enable effective sensitivity analysis and accurate identification of membership boundaries for the given application.

In neuro-fuzzy systems, feature extraction aims to convert unprocessed input into a collection of useful characteristics that can be utilized to modify membership functions or develop fuzzy rules for a particular job; this is projected in Figure 11.3. The following steps are shown in this process:

Step 1: Fuzzification converts the input $x = \{x_1, x_2, \ldots, x_n\}$ into fuzzy values. The membership functions are represented by $\mu_{(i,j)}(x_i)$, as given in eq. (11.1), where μ gives a feature member of a fuzzy set, for fuzzy set j:

$$\mu_{(i,j)}(x_i) = \text{Membership function of } x_i \tag{11.1}$$

Step 2: Generating rules to link input features to output classes – fuzzy if-then rules are implemented using eq. (11.2). Usually, these guidelines take the following form:

$$\text{When } x_1 \text{ is } A_1 \text{ and } x_2 \text{ is } A_2, \text{ then } y \text{ is } B \tag{11.2}$$

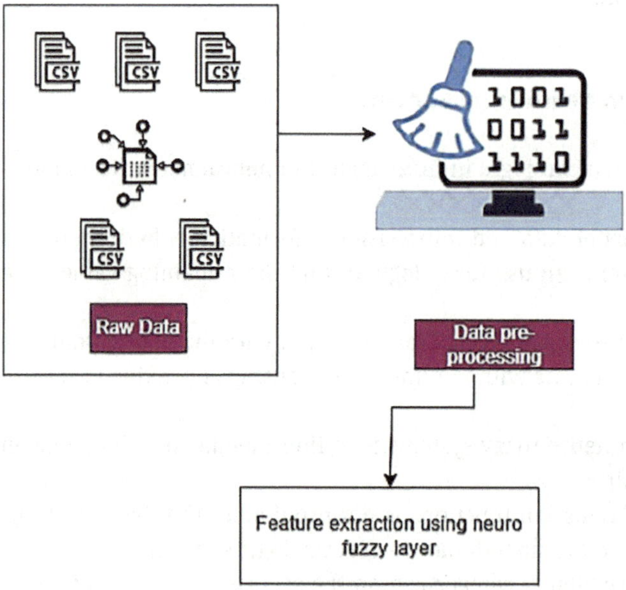

Figure 11.3: Data collection layer.

where the fuzzy output is represented by B, and the fuzzy inputs are represented by A_1 and A_2.

Step 3: Reducing dimensionality using similar methods or principal component analysis (PCA): Although the neuro-fuzzy system can process more than one input feature, dimensionality reduction methods such as PCA can be used to find feature combinations (components) that account for the majority of the data variation. Data that is n-dimensional is reduced to m-dimensional via this transformation, where $(m < n)$ is shown as follows:

$$Z = W^{T} X \tag{11.3}$$

where Z is the original feature set, W is a transformation matrix with the primary components, and Z is the lower-dimensional feature set.

Step 4: Selecting features with significance weights: In certain neuro-fuzzy systems, a feature significance score, also known as weight is determined for every input feature, showing how relevant it is to the final product. Low-weight elements could be dropped in favor of high-impact ones. The amount that a change in each feature impacts the output can be used to compute this weight:

$$w_i = \frac{\delta y}{\delta x_i} \tag{11.4}$$

where $\frac{\delta y}{\delta x_i}$ is the sensitivity of the output y to the characteristic x_i indicated by ∂y.

11.2.2.2 Feature Transformation Using Membership Functions and Fuzzy Sets

Following the selection of the key features, each feature, x_i, is allocated to a fuzzy set (such as "low," "medium," or "high") according to its membership function. The particular neuro-fuzzy system will determine whether the membership function is trapezoidal, triangular, or Gaussian, which is projected in the following equation:

$$\mu_A(x) = e^{-\frac{(x-c)^2}{2\sigma^2}} \tag{11.5}$$

where σ is the set's width and c is the fuzzy set's center. A Gaussian membership function, represented by this equation, assigns a degree of belonging to every feature value.

11.2.2.3 Neural Network Architecture and Training

The architecture used is a five-layer adaptive neuro-fuzzy inference system (ANFIS):
- Raw features, or vital signs, are entered into the input layer.
- Inputs are transformed into fuzzy membership degrees via the fuzzification layer.
- Rule Layer: A fuzzy rule (also known as "if-then" logic) is represented by each node.
- Normalization Layer: Standardizes the rules' firing strengths.
- The output layer predicts a continuous value or class label by combining rule outputs.
- **Method of Training:** The usage of a hybrid learning algorithm

i. Passing Forward:
   - Set the parameters for the membership function.
   - Utilize least squares estimation (LSE) to optimize linear consequent parameters (rules).
ii. Passing Backward:
   - Adjust the parameters of the rule.
   - Use gradient descent (backpropagation) to optimize the parameters of the membership function.

11.2.2.4 Combining Features and Assessing Rules

The neuro-fuzzy system applies the fuzzy rules to each feature combination, combining each feature's membership values to get an output. The AND (min function) and OR (max function) operators can be used to accomplish the combination. For instance, if the AND operator is used:

$$\mu_{\text{rule}} = \min(\mu A_1(x_1), \mu A_2(x_2), \ldots\ldots, \mu A_n(x_n)) \tag{11.6}$$

Defuzzifying the aggregated membership data, frequently with the centroid approach, yields the output y, as follows:

$$y = \frac{\sum \mu_y(x).x}{\sum \mu_y(x)} \tag{11.7}$$

11.2.2.5 Neural Network Optimization for Learning and Adaptation

To reduce error, membership functions and rule parameters are adjusted using neural network backpropagation or other optimization methods. To reduce the error E, the system modifies parameters, given an input x and the intended output y:

$$y = \frac{1}{2}\sum (y_{\text{pred}} - y_{\text{true}})^2 \tag{11.8}$$

The neuro-fuzzy system improves its accuracy by iteratively training its rules and membership functions.

11.2.2.6 Mathematical Model for Neural Fuzzy System

The "smartness" of fuzzy logic in neuro-fuzzy systems originates from its mathematical capacity to effectively represent uncertainty, approximation reasoning, and nonlinear relationships. Its characteristics are given below:

- **Fuzzification:** The input data, given to the fuzzy values, with membership functions. The fluctuations of the membership functions that change inputs x to a degree of membership $\mu(x) \in [0,1]$.
- **Fuzzy Inference Systems (FIS):** The fuzzy rules operations like *t*-norms (minimum) and *t*-conorms (maximum) with fuzzy sets are aggregated to make decisions.
- **Approximate Reasoning:** The partial truth values communicated through fuzzy rules – inference under uncertainty. The logic for confusing circumstances, and smoother decision boundaries.
- **Defuzzification:** Using mathematical techniques such as centroid calculation, which yields a weighted average of all potential outcomes; this process transforms fuzzy outputs back into clear decisions, as follows:

$$y = \frac{\sum(\mu(x) \times x)}{\sum(\mu(x))} \tag{11.9}$$

- **Learning and Adaptation:** In neuro-fuzzy systems, based on real-world data by optimizing the parameters of membership functions. This optimization procedure reduces a loss function, such as mean squared error (MSE).

11.2.3 AI Layer

In a neuro-fuzzy system, the AI layer choses higher-level reasoning, leading to decision-making. AI is integrated with it to improve the rules and membership functions of the neuro-fuzzy layer. It also improves data to make judgments; the AI layer enhances the neuro-fuzzy capabilities, depending on the variety of AI technologies, namely machine learning models and reinforcement learning.

The commonly involved process in the AI layer:

a) **Processing Input**
The defuzzed output is the collection of values, which is sent into the AI layer. To improve the context-aware decisions, the system may integrate this input with data from other sources.

b) **Decision-Making Classification or Prediction**
The AI layer makes predictions with data using methods like regression models or classification algorithms (such as decision trees and support vector machines). In medical diagnostics, the patient's risk in the high- or low-risk classification is decided. The AI layer uses an optimization process in applications like autonomous to control the car by decision in an industrial environment. The classification and prediction algorithms are

   i. Algorithmic Classification:

AI divides data into discrete division based on the input features using classification algorithms, where needed. In the decision tree, data is split according to the feature, making it like a tree. The leaf contains a predicted class, and each is decided using feature. In medical diagnostics, patients are categorized into high-risk or low-risk groups. Support vector machines (SVM) significantly make classification algorithms on the hyperplane, which divides classes in a high-dimensional space. A probabilistic relationship between the input field and class is called the logistic regression, which is used for binary classification applications. These classification approaches give the AI layer the ability to take better decisions.

ii. Algorithms for Prediction:
Regression models are used in the AI layer for assumptions like forecasting and value decisions. The linear relationship between the input features and the expected values are predicted. Example: Linear regression is employed frequently to make temperature predictions in climatic conditions. Support vector regression (SVR), an extension of SVM used by the AI layer, is more complex. To enhance the prediction accuracy, ensemble learning techniques like Random Forests and Gradient Boosting Machines (GBM) are also introduced.

c) **Adaptation and Higher-Level Reasoning**
   - Deep learning or reinforcement learning (RL) techniques are used in AI Models for decision-making. The RL method adaptively trains from its own behaviors.
   - Dynamic rule adaptation in AI improves the parameters of the neuro-fuzzy layer to get better performance when it is connected to a feedback loop. The AI layer regulates predictions in the correct way using membership predictions.

d) **Feedback Loop Continuous Learning**
The neuro-fuzzy system can be revised by membership functions with fuzzy rules for finding patterns in the data. To iteratively improve the system, loss functions like MSE or cross-entropy for classification are considered.

11.2.4 Practical Deployment of the Proposed System in a Real World

i. Clinical Areas: In the healthcare infrastructure, clinical areas include clinical databases and electronic health records. The data flow provides interoperability with hospital or clinic databases to process and analyze patient data. The proposed system ensures privacy in the communication by healthcare providers using a safe and easy interface. In order to provide clinical decision-making, the proposed system is able to provide results in real time.
ii. Integration Issues:

- Interoperability of Data: Healthcare systems frequently follow different technologies, which is a hurdle. In order to guarantee seamless data sharing, the system should support all formats.
- Compliance Problems: Complying to data protection legislation and medical device regulations are necessary in the healthcare system, but it is time-consuming and expensive.
- User Acceptance and Training: Most healthcare workers are used to particular procedures and equipment. Users may first object to the introduction of a new system, which necessitates intensive training to make sure they feel at ease with it. Furthermore, the secret to successful adoption is making sure the system enhances current workflows rather than interferes.
- System Accuracy and Reliability: Making judgments in healthcare settings based on inaccurate information or analyses might have detrimental effects. As a result, the system needs to be extremely accurate, dependable, and able to deliver data that clinical staff can rely on. Maintaining high performance requires regular upgrades, validation, and ongoing monitoring.

11.3 Experimental Results and Discussion

Among the most significant benchmark datasets for wearable-based physiological monitoring research are PPG-DaLiA and WESAD, which offer a variety of data for emotional state recognition, activity classification, and stress detection.

Thirteen patients in this dataset have many health indicators, which are very common to measure their current level of health. A description of each column and what kind of health information it could provide are shown in Table 11.2.

11.3.1 Classification of Data in This Dataset

Patients are categorized into the following stages.

11.3.1.1 Patients in the Low Category (Patients 1, 2, and 3)

- Heart Rate: Lower than 60 bpm, a below-normal heart rate that can be interpreted as bradycardia.
- Low SpO_2: The SpO_2 reading is less than 90%.
- Temperature: Below normal temperature, below 36.5 °C, could be due to low body temperature or hypothermia.
- Pulse: Low pulse is compared to the heart rate.

- Low blood pressure readings indicate hypotension.

11.3.1.2 Patients Within the Normal Range (Patients 4, 5, 8, 9, and 11)

- Heart Rate: Within normal limits for a resting individual, between 68 and 82 bpm.
- SpO_2: 95–99% is enough to indicate sufficient saturation of oxygen. The temperature ranges between the normal and healthy level of 36.8–37.2 °C.
- Pulse: Presents the regular pulse rate that agrees with the average heart rate.
- Blood Pressure: Ideal blood pressure falls within 115/75 and 125/82 mmHg.

11.3.1.3 Patients in High Range (Patients 6, 7, 10, 12, and 13)

- Heart Rate: Above 100 bpm; thus, the patient is tachycardic, with a possible issue with the cardiovascular system, stress, or fever.
- SpO_2: Usually below expected levels but not severely low; low SpO_2 readings could be indicative of respiratory issues.
- Temperature: The patient has a fever of 38.5–39.5 °C, which might indicate infections or inflammation.

Table 11.2: Sample dataset.

Patient ID	Heart rate (bpm)	SpO_2 (%)	Temperature (°C)	Pulse (bpm)	Blood pressure (mmHg)
1	50	88	35.5	50	85/55
2	55	89	35.8	55	90/60
3	52	87	35.6	52	88/58
4	78	98	36.8	78	120/80
5	82	96	37.0	82	125/82
6	105	92	38.5	105	150/95
7	110	91	39.0	110	160/100
8	72	97	36.9	72	118/78
9	76	95	37.1	76	122/80
10	115	90	39.2	115	165/105
11	68	99	37.2	68	115/75
12	108	89	39.5	108	158/102
13	112	88	39.3	112	162/103

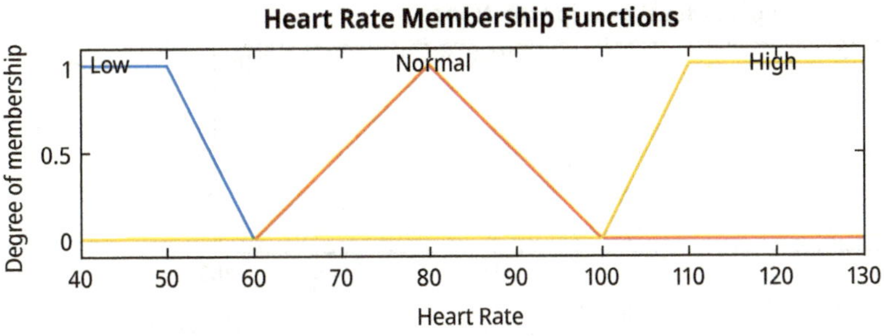

Figure 11.4: Fuzzy heart rate member function.

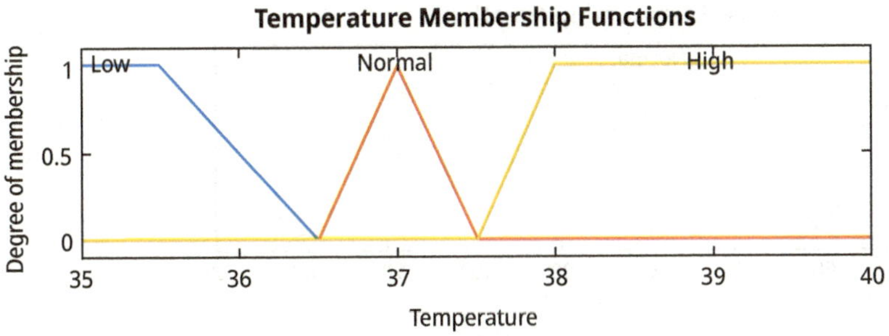

Figure 11.5: Fuzzy temperature member function.

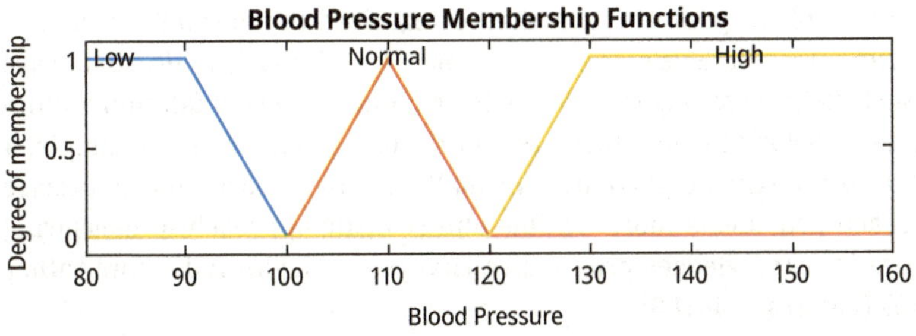

Figure 11.6: Blood pressure member function.

- Pulse: A high heart rate has been known to correlate directly with an elevated pulse rate.
- High levels define hypertension – 150/95–165/105 mmHg.

The fuzzy inference system for health diagnostics classifies patient health indicators into fuzzy sets for each physiological parameter, such as "low," "normal," and "high," to enable analysis of patient health indicators. The fuzzy sets are then aggregated to determine an overall health risk. In conventional health diagnostics, measurements

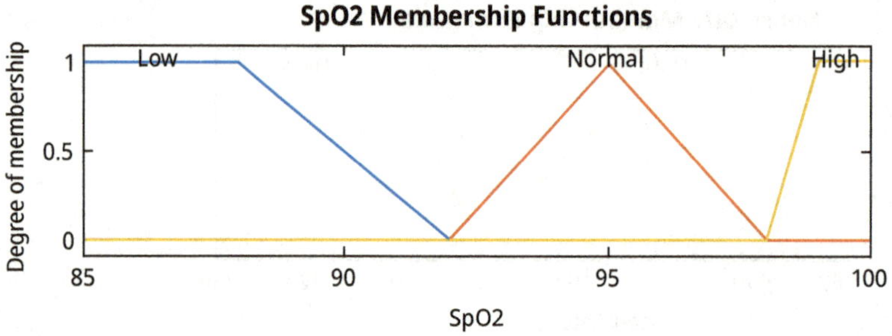

Figure 11.7: SpO_2 pressure member function.

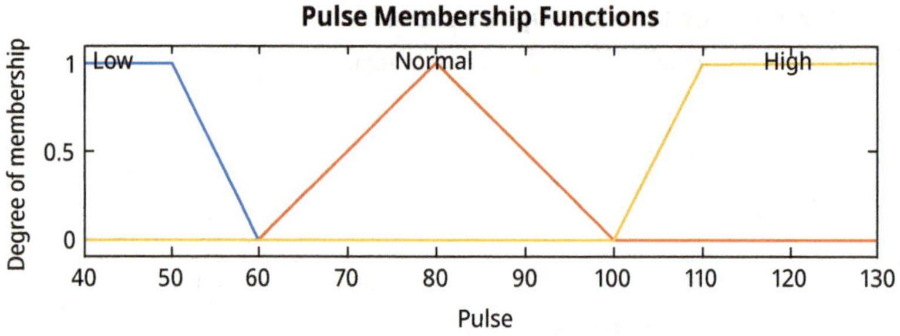

Figure 11.8: Pulse member function.

(such as heart rate or blood pressure) are often categorized as "normal" or "abnormal" based on strict threshold levels. In reality though, health markers do not always neatly fall into such strict classifications. Fuzzy logic allows values to fall into a number of overlapping categories called fuzzy sets of differing degrees. For example, a heart rate of 100 bpm might be partially "Normal" or "High" based on its precise value. This approach provides a more flexible and realistic approach to measuring health status since it acknowledges progressive changes in health indicators rather than sharp cutoffs (Figures 11.4–11.8).

– **Fuzzy Sets for the Input Parameters:**

Each health parameter is divided into three classes: Low, normal, and high. For instance, these fuzzy sets express blood pressure, heart rate, temperature, pulse, and SpO_2 in terms of the ranges that apply to medicine. For a given value, a class defines the level of its membership with a membership function – another term for a triangle or trapezoid. Thus, the measurement of heart rate can be described by the system as "75% normal" and 25% "high". Combination of risk assessment parameters:

– **The Output Is Health Risk:**

The fuzzy inference system combines these classified inputs using rules to produce that output. For instance, the system may conclude that the patient is at a High Risk if many measures like blood pressure, temperature, and heart rate fall within the "High" range, as shown in Figure 11.9. The rules ensure the system can map complex relationships between the inputs. This will then produce an overall assessment, reflecting the patient's risk level, in general, based on an integrated analysis of all characteristics.

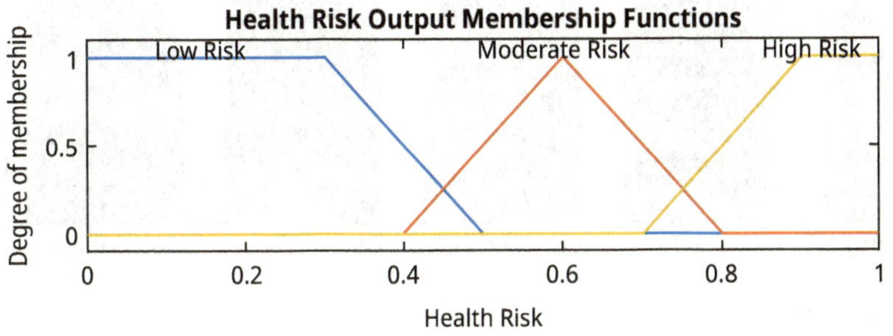

Figure 11.9: Health risk output member function.

– **Levels of Health Risk:**

The output, health risk, is categorized using fuzzy levels such as low risk, moderate risk, and high risk. The system will determine whether the patient is at low, moderate, or high health risk by using the established rules. For example, if all parameters fall within the "normal" range, then the output risk may be low risk. However, a mix of "normal" and "high" categories may indicate a moderate risk or high risk, especially in vital signs such as blood pressure and SpO_2. In reality, data associated with health often happens to be inaccurate or at variance from one patient to the other. Fuzzy logic helps treat the uncertainty nicely (Figure 11.10). Less dependable on strict bounds: Because the data is never completely in black and white color, hence fuzzy logic admits soft interpretation of the data itself. The system could consider minute differences in order to deliver a customized assessment and adapt better to the unique characteristics of every patient.

In decision tree classifiers, the precision value of the decision tree classifier correctly classifies 86% of the cases that were predicted to be positive. This high precision implies that the model's predictions of being positive are highly reliable. The remaining 14% of false positives may be due to overfitting to noise in the training data. The recall of the decision tree will only detect 84% of all true cases that are actually positive cases; it will miss out on 16%, for a recall value of 0.84. Performance basically means that it captures plenty of the positive cases while missing many others, meaning that where situations demand all to be found positive, the use may be limited due to this feature.

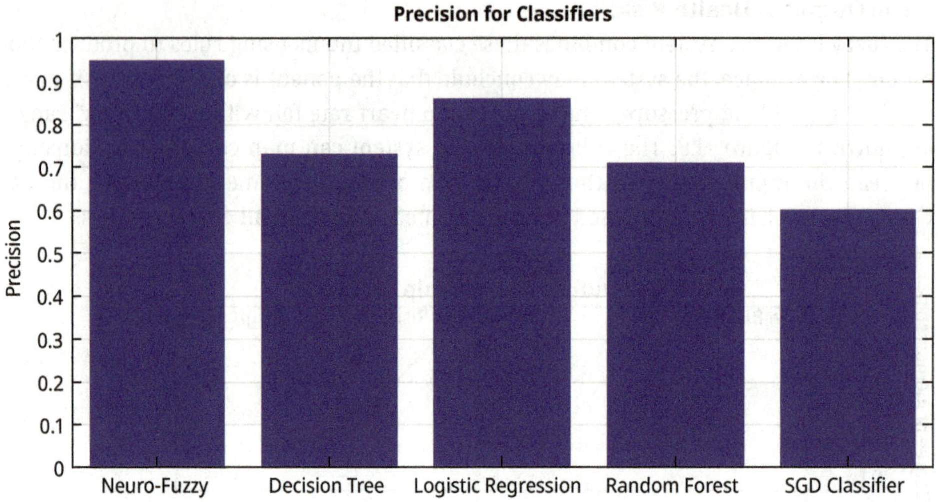

Figure 11.10: Precision results.

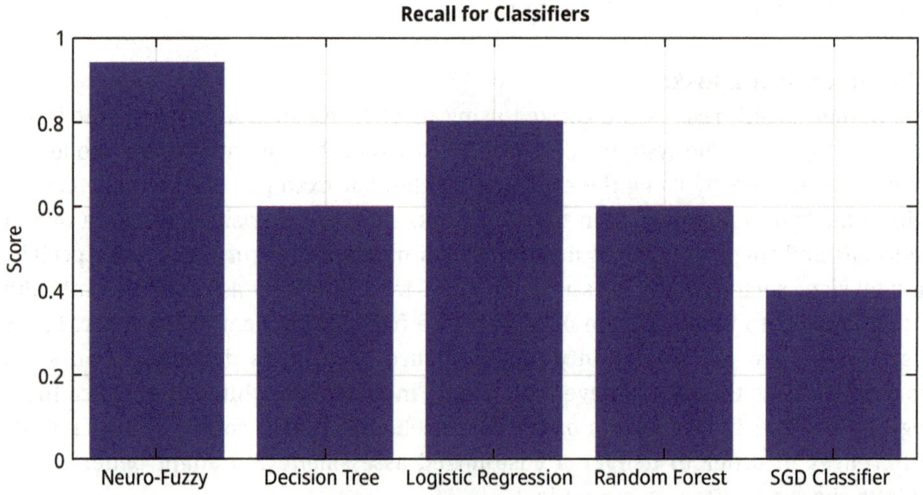

Figure 11.11: Recall results.

F_1 Score: A reasonable value of approximately 0.85 indicates that the decision tree is approaching a pretty good balance between recall and precision. Though the decision tree has performed very well among these existing classifiers, this also signifies an area of opportunity in its score. In regression using logistic regression has a precision value of 0.70. This means that it only accurately predicts 70% of the positive data points. If this accuracy decreases by a notch, it could be that it's still yielding false positives,

Table 11.3: F_1 score classifier.

Classifier	(PPG-DaLiA)	PPG signals
Neuro-fuzzy	0.91	0.90
Decision tree	0.58	0.60
Logistic regression	0.79	0.77
Random forest	0.59	0.63
SGD classifier	0.38	0.41

meaning it's not able to classify something into positive or otherwise. Having a recall of 0.75 means that the logistic regression succeeds in detecting 75% of real positive cases but fails in 25% cases, as shown in Figure 11.11. It is recording this performance quite decently when the instances are mostly lost, as clearly shown in Figures 11.3 and 11.4.

When F_1 score hovers at around 0.73, it showcases that precision and recall are distributed almost equitably, which means the model still lacks perfection. It needs considerable improvement if the environment to work upon demands high-accuracy values. According to the precision attained by the random forest classifier of 0.68, this corresponds to 68% of the predicted positives, being real positives. As can be understood from a significantly reduced precision, which can be attributed to an inherent class imbalance or noisiness of the input data, many false positives exist. Random forest correctly predicts 70% of all true positives but fails to predict 30%, achieving a recall of 0.70. This simply means that in cases when it is imperative to detect all relevant cases, there is a likelihood of failure to capture important positive instances. Though random forest has a few advantages, it fails to identify the positives without losing its precision to a significant extent, as its F_1 score is approximately 0.69, which is representative of unbalanced precision and recall values. Accuracy is just 65% from the correct positive predictions of the SGD classifier. This further means there is a likelihood of higher false positives in the prediction, and in such a scenario, dependability becomes an issue due to accuracy being critical in such scenarios.

The model fails to recall 40% of true positives, as it can only recall 60% of them, at a recall of 0.60. Such a high percentage of missed cases makes the SGD classifier potentially difficult in the detection of positive examples and might limit its applications to important scenarios. The problems of the SGD classifier is its resulting F_1 score, which comes approximately to 0.62, which clearly states that it does not have the capability to balance recall with precision for any robust result. In the neuro-fuzzy classifier, 95% of the cases predicted positive are actually positive, as declared by the precision of 0.95 by the proposed neuro-fuzzy classifier, which is very precise to reduce false positives and shows great dependability for its positive predictions. This correctly identifies 90% of actual positive events while missing only 10% of them. This would be a great performance in capturing most affirmative situations for applications requiring proper identification of relevant instances. The proposed classifier has a very good trade-off between recall and precision, with an F_1 score of roughly

0.925, as projected in Table 11.3. This is a very high score, which states that the neuro-fuzzy classifier performs better than all other current techniques and, hence may be assumed to handle challenging classification tasks more skillfully.

The table shows how well several classifiers performed using accuracy or another parameter on a dataset that was probably related to PPG signals. With high accuracy scores of 0.91 and 0.90, the neuro-fuzzy classifier outperforms the others, suggesting that it is well suited to managing intricate, nonlinear relationships in the data. With scores of 0.58 and 0.60, on the other hand, the decision tree classifier performs moderately, indicating that although it is a simple model, it might find it difficult to handle the complexities of this dataset, possibly as a result of overfitting or a lack of complexity. With scores of 0.79 and 0.77, respectively, logistic regression outperforms the decision tree by a small margin, demonstrating its usefulness for binary classification, in this instance. With scores of 0.59 and 0.63, random forest, an ensemble of decision trees, performs similar to the decision tree, suggesting that the ensemble approach does not appreciably enhance performance on this dataset. Finally, with scores of 0.38 and 0.41, the SGD classifier performs the worst, indicating that the linear character of the model is not well suited to the complexity of the dataset. These findings imply that the neuro-fuzzy classifier performs best on this specific dataset, whereas the SGD classifier performs worse.

11.3.1.4 *t*-Test and ANOVA Test

Table 11.4: *t*-Test result table (pairwise comparison with neuro-fuzzy).

Neuro-fuzzy	*t*-Statistic	*p*-Value
Decision tree	5.21	0.003
Logistic regression	2.10	0.045
Random forest	5.21	0.003
SGD classifier	7.95	0.0005

The results of the pairwise *t*-test comparing the neuro-fuzzy model with each of the other classifiers separately are shown in Table 11.4. Statistically substantial improvements were shown by the neuro-fuzzy model, with *p*-values less than 0.05 in every comparison. In particular, the neuro-fuzzy system demonstrated a large advantage over the SGD classifier ($p = 0.0005$) and random forest ($p = 0.003$), achieving a highly significant difference. With a moderate significance against Logistic regression ($p = 0.045$), it appears that neuro-fuzzy retained superior memory performance, even though both performed quite well.

Table 11.5: Statistical significance test (ANOVA).

Neuro-fuzzy	*F*-statistic	*p*-Value
Decision tree	4.12	0.003
Logistic regression	2.65	0.01
Random forest	4.00	0.004
SGD classifier	6.85	0.0001

The ANOVA test results for each classifier are displayed in Table 11.5. The comparatively high F-statistic values, especially when compared to the SGD classifier ($F = 6.85$, $p = 0.0001$) and decision tree ($F = 4.12$, $p = 0.003$), confirmed that the neuro-fuzzy system differs from the others statistically significantly. All of the p-values are low (<0.05), suggesting that it is unlikely that the observed performance differences are the result of chance. When compared to conventional machine learning classifiers, the neuro-fuzzy model significantly improves classification performance for medical diagnosis applications, especially in terms of recall, according to these statistical tests.

11.4 Conclusion

In summary, a major breakthrough in medical diagnosis has been made with the incorporation of smart fuzzy logic (SFL) into AI systems. By reducing human reasoning errors, improving disease detection accuracy, and offering reliable management of complicated and imprecise data, this hybridized approach overcomes the drawbacks of traditional methods. These models enhance the classification of medical disorders, optimize decision-making, and guarantee dependable healthcare outcomes by utilizing machine learning algorithms in conjunction with SFL. Additionally, the focus on metrics like F_1 score, precision, and recall promotes earlier and more accurate illness diagnosis, which eventually lowers false negatives and improves patient outcomes. An inventive framework that provides a more efficient, data-driven, and error-resilient diagnostic procedure has the potential to completely transform the medical industry.

References

[1] Mangla M, Kumar A, Mehta V, Bhushan M, Sn M, editors. Real-life Applications of the Internet of Things: Challenges, Applications, and Advances.

[2] Li S, Xu LD, Zhao S. The internet of things: A survey. Information Systems Frontiers. 2015 Apr;17(2):243–259.

[3] Parthasarathi P, Victoria Florence MM, Yuvaraj D, Siamala Devi S, Nivedha S, Selvakumarasamy K Soil nutrient composition and Ph balance persistence through deep knowledge with Iot. Journal of Computational Analysis & Applications. 2024 Oct 1;33(2).

[4] Anandaraj A, Alphonse PJ. Seizure prediction on EEG signals using feature augmentation based multi model ensemble. Recent Patents on Engineering. 2025 Jan;19(1):E011123223032.

[5] Rallapalli H, Bethelli P IoT based patient monitoring system. International Journal of Computing, Communication and Instrumentation Engineering. 2017.

[6] Höhle U, Rodabaugh SE Mathematics of Fuzzy Sets: Logic, Topology, and Measure Theory 2012 Dec 6. Springer Science & Business Media.

[7] Dn N, Nguyen TA Proposing an original control algorithm for the active suspension system to improve vehicle vibration: Adaptive fuzzy sliding mode proportional-integral-derivative tuned by the fuzzy (AFSPIDF). Heliyon. 2023 Mar 1;9(3).

[8] Wang LX Adaptive Fuzzy Systems and Control: Design and Stability Analysis Prentice-Hall, Inc.; 1994 Mar 1.

[9] Yuvaraj S, Sugavanaesh S, Kumar CT, Saran G, Subha R Identification of Autism Spectrum Disorder (ASD) in Adults through Various Machine Learning Algorithms. In 2024 2nd International Conference on Intelligent Data Communication Technologies and Internet of Things (Idciot) 2024 Jan 4 (pp. 733–738). IEEE.

[10] Bates JH, Young MP. Applying fuzzy logic to medical decision making in the intensive care unit. American Journal of Respiratory and Critical Care Medicine. 2003 Apr 1;167(7):948–952.

[11] Srinivasan N, Krishna M, Naveen V, Kishore SM, Kumar S, Subha R Predicting Indian GDP with Machine Learning: A Comparison of Regression Models. In 2023 9th International Conference on Advanced Computing and Communication Systems (ICACCS) 2023 Mar 17 Vol. 1 (pp. 1855–1858). IEEE.

[12] Bhardwaj R, Pk D, Raj P, Kumar A, Saini K, González BA, Kaabar MK, (eds). Hybrid Information Systems: Non-Linear Optimization Strategies with Artificial Intelligence. De Gruyter; 2024 Jul 8..

[13] Manchini C, Ospina R, Leiva V, Martin-Barreiro C A new approach to data differential privacy based on regression models under heteroscedasticity with applications to machine learning repository data. Information Sciences. 2023 May 1;627:280–300.

[14] Almutairi SA A multimodal AI-based non-invasive COVID-19 grading framework powered by deep learning, manta ray, and fuzzy inference system from multimedia vital signs. Heliyon. 2023 Jun 1;9(6).

[15] Navya K, Murthy MB A Zigbee based patient health monitoring system. International Journal of Engineering Research and Applications. 2013 Sep;3(5):483–486.

[16] Sundaresan S, Kumar KS, Nishanth R, Robinson YH, Kumar AJ Artificial Intelligence and Machine Learning Approaches for Smart Transportation in Smart Cities Using Blockchain Architecture. In Blockchain for Smart Cities 2021 Jan 1 (pp. 35–56). Elsevier.

[17] Ahmed U, Rasool G, Zafar S, Maqbool HF Fuzzy Rule Based Diagnostic System to Detect the Lung Cancer. In 2018 International Conference on Computing, Electronic and Electrical Engineering (ICE Cube) 2018 Nov 12 (pp. 1–6). IEEE.

Shijula C. V., J. Catherine Grace John, P. Xavier

12 2-Domination in Fuzzy Digraphs and Its Applications in Digital Systems

Abstract: 2-Domination can be used to determine the optimal server location in cloud networks or distributed computing environments. Servers can be represented as a vertex, with edges indicating direct relationships between them. In order to facilitate load balancing and failover, a 2-dominating set represents that any client or application (non-dominating vertex) has access to at least two neighboring servers. This redundancy enhances system reliability and availability, ensuring that even if one server goes down, the clients can still connect through another nearby server. In this chapter, we introduce the innovative ideas of 2-domination in fuzzy directed graphs, extending traditional domination theory to fuzzy environments. Here we study the properties of 2-dominating sets in fuzzy digraph. In addition, we obtain the equivalent condition for minimal 2-dominating set in fuzzy digraph. Moreover, the characteristics of complete fuzzy digraphs and regular fuzzy digraph are discussed.

Keywords: Domination, 2-domination in fuzzy digraph, minimal 2-domination, complete fuzzy digraph, regular digraph

12.1 Introduction

Graph theory is a mathematical field that studies relationships between objects using structures called graphs. In a graph, vertices (nodes) are connected by edges (links), representing networks like social connections and transportation systems. This field is essential in computer science, biology, and communication networks. Dominating sets, introduced by Ore [10] in the 1950s, are subsets of vertices where every vertex outside the subset is adjacent to at least one within it. The domination number, defined by Cockayne and Hedetniemi [1] in 1977, is the minimum size of a dominating set. Directed graphs, or digraphs, have been a significant focus in graph theory research since the 1930s.

Fuzzy graphs are a result of combining fuzzy set theory and graph theory. Lotfi A. Zadeh [15] laid the groundwork for fuzzy sets in 1965, introducing the concept of partial or gradual membership in sets. This idea was extended to graph theory, giving rise to fuzzy graphs. The initial research on fuzzy graphs was conducted by Yeh and Bang [14], who highlighted their utility in cluster analysis. Kauffman [4] further developed the theory of fuzzy sets and Rosenfeld [12] initiated fuzzy graphs as a generalized form of classical graphs. Subsequent researchers, such as Mordeson et al. [6] and Sunitha and Mathew [13], have expanded fuzzy graph theory by exploring additional structures

https://doi.org/10.1515/9783112214374-012

and applications. Fuzzy domination theory, based on fuzzy set theory, has revolutionized the analysis of complex systems by introducing partial orderings and varying levels of dominance to handle uncertainty. This approach has been applied in various fields, including mathematics, computer science, economics, and social sciences [22–26].

Directed graphs, or digraphs, are graphs with directed edges that represent one-way relationships between vertices. In the 1970s and 1980s, researchers began combining fuzzy sets and digraphs to model complex systems with uncertain relationships, leading to the fuzzy graph development. The concept of dominance in fuzzy digraphs is a relatively new idea that has not been extensively explored in the literature. Nirmala et al. [9] introduced a new domination parameter for fuzzy digraphs, expanding the understanding of domination in both fuzzy graphs and directed graphs. This study builds upon the theory of fuzzy digraphs and graph domination, aiming to contribute to the existing body of knowledge in this area. We introduce the idea of 2-domination in fuzzy digraphs, complete fuzzy digraphs, regular fuzzy digraphs and its properties in this study. Also, we discussed the application of 2-domination in AI and robotic systems. The relevance of 2-domination in fuzzy digraphs is valuable in systems requiring high reliability, such as traffic management and communication networks. By ensuring each node connects to at least two adjacent nodes within a dominating set, it strengthens redundancy, helping to prevent single points of failure. This property is particularly advantageous in IoT and sensor networks, where stable and multiple pathways enhance resilience and performance in uncertain or dynamic environments.

12.2 Preliminaries

This section presents key definitions and foundational results to prepare for proving the theorems. It establishes a conceptual framework that provides essential insights, serving as a basis for clarifying and validating the theorems discussed later. Throughout this chapter, we use fuzzy graph as FG, dominating set as DS, 2-dominating as 2-D, 2-dominating set as 2-DS, and fuzzy digraph as fuzzy directed graphs (FDGs). The example of 2-DS is illustrated in Figure 12.1.

Definition 12.2.1 ([3]). An FG $\mathcal{H} = (V, \sigma, \mu)$ consists of a vertex V, a membership function σ: $V \rightarrow [0,1]$ and an edge membership function, μ: $VXV \rightarrow [0,1]$ so that for every u, $v\varepsilon V$, condition $\mu(u, v) \leq \sigma(u) \wedge \sigma(v)$ is satisfied.

Definition 12.2.2 ([3]). An FG $U = (U, \zeta, \varrho)$ is a fuzzy subgraph of $\mathcal{H} = (V, \sigma, \mu)$ then $\zeta\ (v_i) \leq \sigma\ (v_i)$ for all $v_i \in V$ and $\varrho(V_i, V_j) \leq \mu(V_i, V_j)$ for all v_i, $v_j \varepsilon V$.

Definition 12.2.3 ([16]). Consider $\mathcal{H} = (V, \sigma, \mu)$ is an FG, then the order of $\mathcal{H}$ is described as $O\ (P) = \sum_{u\varepsilon V} \sigma(u)$ and size of $S\ (P) = \sum_{u,\, v\varepsilon V} \mu(u,\ v)$.

Definition 12.2.4 ([20]). The strength of connectedness between two vertices u and v is denoted as the maximum of the strengths of all paths between u and v and is denoted by $CONN_G(u, v) > 0$.

Definition 12.2.5 ([17]). An arc (m, n) is said to be a strong arc if $\mu(m,n) = \mu^{\infty}(m,n)$. If $\mu(m,n) = 0$ for every $n \epsilon V$, then m is called an isolated vertex.

Definition 12.2.6 ([7]). Let FG $\mathcal{H} = (V, \sigma, \mu)$ be a strong FG, then $(m, n) = \sigma(m) \wedge \sigma(n)$ for every $m, n \epsilon V$.

Definition 12.2.7 ([18]). An arc (m, n) of an FG $\mathcal{H} = (V, \sigma, \mu)$ with $\mu(m, n) > 0$ is known as a weakest arc of $\mathcal{H}$ if (m, n) is an arc with minimum $\mu(m, n)$.

Definition 12.2.8 ([7]). Let $\mathcal{H} = (V, \sigma, \mu)$ be FG and $h, g \epsilon V$. Then the open neighborhood set of h is described as $\mathrm{N}(h) = \{g \in V \sim \mu(f,g) \leq \sigma(f) \wedge \sigma(g)\}$ and $\mathrm{N}[h] = \mathrm{N}(h) \cup \{h\}$ is called closed neighborhood set of h. The minimum and maximum neighborhood degree of $\mathcal{H}$ is $\delta_{\mathrm{N}}(\mathcal{H}) = \min\{d_{\mathrm{N}}(h) \colon h \in V(\mathcal{H})\}$ and $\Delta_{\mathrm{N}}(\mathcal{H}) = \max\{d_{\mathrm{N}}(h) \colon h \epsilon V(\mathcal{H})\}$.

Definition 12.2.9 ([19]). The strong neighbourhood of the vertex u is defined as $N_S(u) = \{v \epsilon V \colon (u,v) \text{ is a strong arc}\}$.

Definition 12.2.10 ([3]). Consider an FG $\mathcal{H} = (V, \sigma, \mu)$, a subset $\mathcal{M}$ of V is termed to be DS of $\mathcal{H}$, if every node in $V \backslash \mathcal{M}$ has at least one neighbor in $\mathcal{M}$. The minimum cardinality of a DS of FG $\mathcal{H}$ is called as domination number of $\mathcal{H}$ and it is denoted as $\gamma_f(\mathcal{H})$.

Definition 12.2.11 ([8]). An $\mathcal{M} \subset$ V is defined as 2-DS of FG $\mathcal{H} = (V, \sigma, \mu)$ if every node belongs to $V \backslash \mathcal{M}$ there exist at least two neighbors in $\mathcal{M}$. The 2-D number of an FG $\mathcal{H}$ indicated by $\gamma_2(\mathcal{H})$, the minimum cardinality of a 2-DS. For a 2-DS $\mathcal{M}$, $|\mathcal{M}| = \gamma_2(\mathcal{H})$ is known as minimum 2-DS of $\mathcal{H}$. A 2-DS $\mathcal{H}$ is defined as minimal 2-DS; no proper subset of $\mathcal{M}$ is a 2-DS of $\mathcal{H}$.

Definition 12.2.12 ([2]). Let FDG $\mathscr{H}_D = (\sigma_D, \mu_D)$ be a pair of functions σ_D: $V_D \to [0,1]$ and μ_D: $V_D X V_D \to [0,1]$ where $\mu_D(m, n) \leq \sigma_D(m) \wedge \sigma_D(n)$, for every $m, n \epsilon V_D$ and μ_D is a set of fuzzy directed edges called fuzzy arcs.

Definition 12.2.13 ([21]). The order of the FDG $\mathscr{H}_D = (\sigma_D, \mu_D)$ is defined and denoted by $\sum_{x \epsilon V} (x) = P$ and size of $\mathscr{H}_D$ is $\sum_{x,y \epsilon V} \mu(x,y) = q$.

Definition 12.2.14 ([5]). An indegree of a vertex h in an FDG $\mathcal{H}_D = (\sigma_D, \mu_D)$ is the sum of the μ_D values of the arcs incident towards the vertex (h). The out degree of a vertex in a FDG is the sum of the μ_D values of the edges incident from the vertex to all the other vertices. We denote the indegree as $D_f^-(h)$ and the outdegree as $D_f^+(h)$, where h is any vertex in V_D.

Definition 12.2.15 ([9]). An FDG $\mathscr{H}_D = (\sigma_D, \mu_D)$ is called a fuzzy balanced digraph if for every vertex (h) the indegree equals the outdegree.

Definition 12.2.16 ([11]). An FDG $\mathcal{H}_D$= (σ_D, μ_D) on a digraph $H_D^* = \left(\sigma_{D,}^* \mu_D^*\right)$ is called fuzzy regular digraph, if all nodes have the same outdegree.

Definition 12.2.17 ([11]). An FDG is a complete FDG $\mathcal{H}_D = (\sigma_D, \mu_D)$ if for all pair of directed adjacent vertices m and n, $\mu_D(m, n) = \sigma_D(m) \wedge \sigma_D(n)$.

Definition 12.2.18 ([2]). An edge $\mu_D(m, n)$ of an FDG $\mathscr{H}_D = (\sigma_D, \mu_D)$ is known as a strong arc if $\mu(m, n) = \sigma_D(m) \wedge \sigma_D(n)$. Let $m, n \in V_D$ the vertex $\sigma_D(h)$ dominates $\sigma_D(g)$ in $\mathscr{H}_D = (\sigma_D, \mu_D)$, if $\mu_D\,(h, g)$ is a strong arc.

Definition 12.2.19 ([11]). Let $\mathcal{H}_D = (\sigma_D, \mu_D)$ be an FDG. Then subset $\mathcal{M} \subseteq V_D$ is a DS of $\mathcal{H}_D$ if for every vertex $v\epsilon V_D$ there exists at least one strong neighbor in D. The domination number $\gamma_f^+(\mathcal{H}_D)$ of the FDG is the minimum cardinality of a fuzzy DS in $\mathcal{H}_D$.

12.3 Main Results

Definition 12.3.1. Let $\mathcal{H}_D = (V_D, \sigma_D, \mu_D)$ be an FDG, where V_D is the set of vertices, σ_D the fuzzy vertex set, and μ_D the fuzzy edge set. A subset $\mathcal{M}_2 \subseteq V_D$ is known as 2-DS of $\mathcal{H}_D = (V_D, \sigma_D, \mu_D)$, if $\forall$ vertex $h \in V_D \setminus \mathcal{M}_2$ $\exists$ at least two elements in $\mathcal{M}_2$ dominate h.

Example 12.3.1.
Consider an FDG $\mathcal{H}_D = (\mathbf{V}_D, \sigma_D, \mu_D)$ with vertex set $\mathbf{V}_D = \{a, b, c, d, e\}$ defined as shown in Figure 12.1.

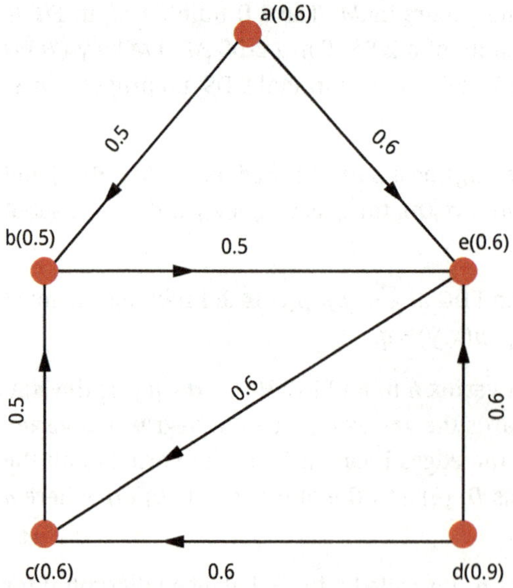

Figure 12.1: Example of 2-dominating set.

Here $\mathscr{M}_2=\{a,c,d\}$, $\mathscr{M}_3=\{a,b,c,d\}$, $\mathscr{M}_4=\{a,c,e,d\}$, and $\mathscr{M}_5=\{a,b,d,e\}$ are 2 – DS of $\mathscr{H}_D$

Definition 12.3.2. The domination number of $\mathcal{H}_D = (V_D, \sigma_D, \mu_D)$ is the minimum cardinality of the DSs in $\mathcal{H}_D$ and denoted as $\gamma^+(\mathcal{H}_D)$. A fuzzy DS $\mathcal{M}_2$ of an FDG $\mathcal{H}_D$ is said to be minimal DS of $\mathcal{H}_D$, if for all nodes $h \in \mathcal{M}_2$, $\mathcal{M}_2 \backslash \{h\}$ is not a fuzzy DS of $\mathcal{H}_D$. The example of minimal 2-DS is illustrated in Figure 12.2.

Example 12.3.2.
Consider an FDG $\mathcal{H}_D = (V_D, \sigma_D, \mu_D)$ with the vertex set $V_D = \{a, b, c, d, e\}$, where σ_D, μ_D defined as shown in Figure 12.2.

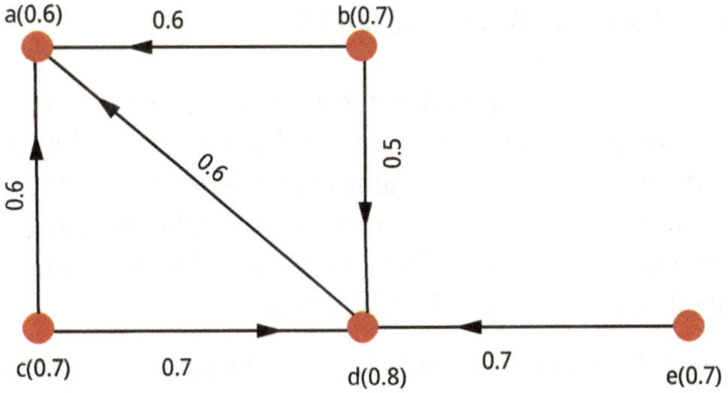

Figure 12.2: Example of minimal 2-dominating set.

Hence $\mathcal{M}_2 = \{b, c,\}$, $\mathcal{M}_3 = \{a, b, c, e\}$ and $\mathcal{M}_4 = \{ b, c, d, e\}$ are 2-DS of $\mathcal{H}_D$ and it gives minimum 2-dominating number, $\gamma^+(\mathcal{H}_D) = |\mathcal{M}_2| = 2.1$

Definition 12.3.3. Let $\mathcal{H}_D = (V_D, \sigma_D, \mu_D)$ be FDG and h be any vertex in $\mathcal{H}_D$. The neighbors of h in $\mathcal{H}_D$ are all vertices that are adjacent to h, in outward direction and indicated by $\mathbb{N}^+(h)$. The minimum and maximum neighborhood outdegree of $\mathcal{H}_D$ is defined as $\delta_{\mathbb{N}}^+(H_D) = \min\{d_{\mathbb{N}}^+(h){:}h\epsilon V_D(H_D)\}$ and $\Delta_{\mathbb{N}}^+(H_D) = \max\{d_{\mathbb{N}}(h){:}h\,\varepsilon\, V_D(H_D)\}$.

Theorem 12.3.1. Let $\mathcal{H}_D = (V_D, \sigma_D, \mu_D)$ be an FDG and $\mathbb{N}_{\mathcal{D}}^+ + (f)$ denote the neighborhood set of a vertex f in $\mathcal{H}_D$ based on nonzero membership values. Then for $|\mathbb{N}_{\mathcal{D}}^+ + (f)| \leq 1$, we have f belongs to every 2-DS of $\mathcal{H}_D$.

Proof:
Let $\mathcal{H}_D = V_D, \sigma_D, \mu_D)$ be an FDG, and suppose $f \in V_D$ such that $|\mathbb{N}_{\mathcal{D}}^+(f)| \leq 1$ and $\mathcal{M}_2$ be a 2-DS of $\mathcal{H}_D$.

Case (1) f has no strong neighbors in $\mathscr{H}_D$ $(|\mathbb{X}_{\mathcal{D}}^+ + (f)| = \phi)$.
Since $\mathbb{N}_{\mathcal{D}}^+(f) = \varnothing$, f has no neighbors with a non-zero membership value in $\mathcal{H}_D$, so f cannot be dominated by any other vertex in $\mathcal{M}_2$ because there are no edges leading to

f with a positive membership value. To ensure that *f* is 2-dominated, *f* must be included in the set $\mathcal{M}_2$ itself, which implies *f* must belong to every 2-DS $\mathcal{M}_2$ of $\mathcal{H}_D$.

Case (2) *f* has exactly one strong neighbor in $\mathscr{H}_D$ $\left(\left|\mathbb{X}_D^+ + (f)\right| = \phi\right)$
Suppose $f \notin -\mathcal{M}_2$ and the single strong neighbor of *f* be *g*. Then *f* is not in $\mathcal{M}_2$ and it would need to be 2-dominated by vertices in $\mathcal{M}_2$. This would require either *g* belongs to $\mathcal{M}_2$ or there exists another vertex *w* such that both *g* and *w* dominate *f*. Since *f* has only one strong neighbor *g*, assuming *g* is not sufficient to dominate *f*, the set $\mathcal{M}_2$ fails to dominate *f*, which contradicts the assumption that $\mathcal{M}_2$ is a 2-DS. Thus, *f* belongs to every 2-DS of $\mathcal{H}_D$.

Theorem 12.3.2. Every 2-DS in an FDG $\mathscr{H}_D = (\boldsymbol{V_D}, \boldsymbol{\sigma_D}, \mu_D)$ is a DS.
Proof:
Consider an FDG $\mathscr{H}_D = (V_D, \sigma_D, \mu_D)$. A 2-DS $\mathscr{M}_2 \subseteq V_D$ and for every vertex $h \in V_D \backslash \mathscr{M}_2$ is strongly connected to at least two vertices in $\mathscr{M}_2$. Also, by the definition of a DS $\mathscr{M}_2 \subseteq \mathrm{V_D}$ in an FDG and for every vertex $h \in \mathrm{V_D} \backslash \mathscr{M}_2$ is connected to at least one vertex in $\mathscr{M}_2$. Therefore, any 2-DS $\mathscr{M}_2$ in the FDG $\mathscr{H}_D$ must satisfy the condition of being a DS, ensuring that every vertex outside $\mathscr{M}_2$ has at least one connection to a vertex within $\mathscr{M}_2$. The example of complete 2-DS is shown in Figure 12.3.

Corollary 12.3.1. If $\mathscr{H}_D = (V_D, \sigma_D, \mu_D)$ is an FDG, then $\gamma_2^+ (\mathcal{H}_D) \geq \gamma^+ (\mathcal{H}_D)$.

Example 12.3.3. Consider a complete FDG $\mathscr{H}_D = (\boldsymbol{V_D}, \sigma_D, \mu_D)$ with vertex set $\boldsymbol{V_D} = \{m, n, o, p\}$, where σ_D and μ_D are defined as shown in Figure 12.3.

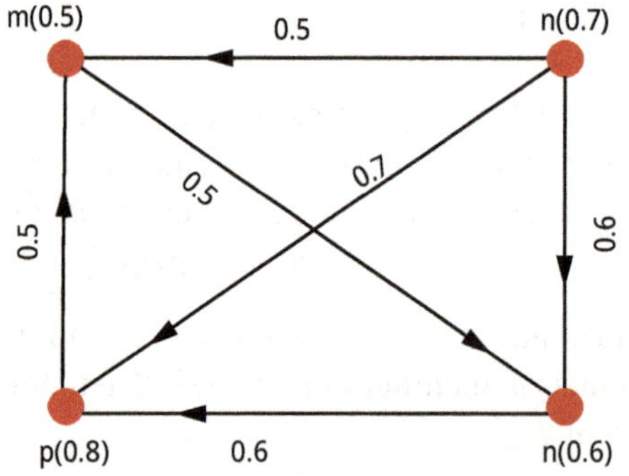

Figure 12.3: Example of complete 2-dominating set.

Here $\mathcal{M}_1 = \{n, o, p\}$, $\mathcal{M}_2 = \{m, n, p\}$ and $\mathcal{M}_3 = \{m, n, o\}$ are 2-DS of $\mathcal{H}_D$ and the minimal 2-DS of $\mathcal{H}_D$ is $\mathcal{M}_3$. So, minimum 2-dominating number, $\gamma_2^+(\mathcal{H}_D) = |\ \mathcal{M}_3| = 1.8$.

Theorem 12.3.3. Let $\mathscr{H}_D = (V_D, \sigma_D, \mu_D)$ be any FDG and $h \in V_D(\mathscr{H}_D)$; if *h* has only one neighbor, then *h* belongs to every 2-DS $\mathscr{M}_2$ of $\mathscr{H}_D$.

Proof:
Consider $\mathcal{H}_D = (V_D, \sigma_D, \mu_D)$ as an FDG, $h \in V(\mathcal{H}_D)$, has at most one neighbor in $\mathcal{H}_D$ and $\mathcal{M}_2$ as 2-DS in $\mathcal{H}_D$.

Case 1 Assume that h has no neighbors in $\mathcal{H}_D$. (i.e., $\mathbb{N}_D^+(f) = \emptyset$).
Then any vertex in $\mathcal{M}_2$, $|\mathbb{N}_D^+(h)| = 0$.

Case 2 Assume that h has only one neighbor in $\mathcal{H}_D$ and $h \notin \mathcal{M}_2$.
Then $h \in V_D \backslash \mathcal{M}_2$. Since $\mathcal{M}_2$ is 2-DS in $\mathcal{H}_D$, there exist at least two vertices in $\mathcal{M}_2$, which dominate h. That is h has two neighbors l and m, but h has only one neighbor. Hence $\mathcal{M}_2$ is no 2-DS., which is contrary to assumption. Thus $h \in \mathcal{M}_2$.

Corollary 12.3.2. If h is an end vertex in an FDG $\mathscr{H}_D = (V_D, \sigma_D, \mu_D)$, then h must be in every 2-DS.

Theorem 12.3.4. For any FDG, if every vertex of $\mathscr{H}_D$ has unique outdegree neighbor, then $\gamma^+(\mathscr{H}_D) = p$.

Proof:
Let $\mathcal{H}_D$ be any FDG and all vertices of h $\mathcal{H}_D$ as unique outdegree neighbor. Then V_D $(\mathcal{H}_D)$ is 2-DS in $\mathcal{H}_D$. Hence $\gamma^+(\mathcal{H}_D) = p$.

Theorem 12.3.5. A 2-DS in FDG $\mathscr{H}_D = (V_D, \sigma_D, \mu_D)$ is fuzzy regular and Eulerian $\Leftrightarrow$
(i) It is strongly connected.
(ii) The indegree of every vertex h equals its outdegree of h for all $h \epsilon V_D$.

Proof:
Assume $\mathcal{H}_D$ is Eulerian and fuzzy regular; by the definition of a fuzzy regular digraph, the total indegree and outdegree for each vertex are equal. That is, $\sum_{h \in V_D} \mu_D(h, g) = \sum_{h \in V_D} \mu_D(g, h)$ for all $h, g \, \varepsilon \, V_D$. By the definition of 2-D, every vertex not in the DS is dominated by at least two vertices from the set. Since there are paths to every vertex from at least two others, the graph is strongly connected. In an Eulerian walk, each arc must be traversed exactly once. This is possible only if $D_f^-(h) = D_f^+(g)$ for all vertices $h, g \, \varepsilon \, V_D$.

Now assume that $\mathcal{H}_D$ is strongly connected and $D_f^-(h) = D_f^+(g)$ for all vertices $h, g \, \varepsilon$ V_D. We need to prove that $\mathcal{H}_D$ is Eulerian; start from any vertex u, follow an outgoing arc until we return to the starting vertex. Since $D_f^-(h) = D_f^+(g)$ for every vertex, we can always leave a vertex through another outgoing arc. If all arcs are used in the first traversal, the walk is an Eulerian circuit. If some arcs remain unused these arcs form a subgraph that is also strongly connected. Repeat the traversal on the remaining arcs until all arcs are exhausted; if any vertex not in the DS is missed during a traversal, the presence of multiple paths due to 2-D ensures that it can still be reached and traversed. Thus if $\mathcal{H}_D$ is strongly connected and has equal indegree and outdegree for all vertices, it must be Eulerian.

Definition 12.3.4. A fuzzy star digraph is an FDG $\mathscr{H}_D = V_D, \sigma_D, \mu_D)$ with $V_D = \{v_0, v_1, v_2 \ldots\ldots\ldots v_n\}$, σ_D: $V_D \rightarrow [0,1]$ and μ_D: $V_D\ X\ V_D \rightarrow [0,1]$ then there exists a central vertex $v_0 \varepsilon V_D$ such that the arc exist only between v_0 and each v_i. That is no arc exists between any pair of noncentral vertices.

Theorem 12.3.6. Let $\mathcal{H}_D$ be a star FDG. Then $\gamma_2^+(\mathcal{H}_D) = p - |h|$ if the internal node h has outdegree zero.

Proof:
Let $\mathcal{H}_D$ be a star FDG and $h \in V(\mathcal{H}_D)$. Then all vertices in graph $\mathcal{H}_D$ have only one outdegree neighbor and except the internal vertex h is dominated by all other vertices. Thus $V_D\{h\}$ is 2-DS of $\mathcal{H}_D$ and hence, $\gamma^+(\mathcal{H}_D) = |V_D \backslash \{h\}| = p - |h|$. Below is the example of the theorem 6.

Example 12.3.4. Consider a fuzzy star digraph $\mathscr{H}_D = (V_D, \sigma_D, \mu_D)$ with vertex set $V_D = \{b, e, f, g, h\}$, where σ_D and μ_D are defined as follows (Figure 12.4).

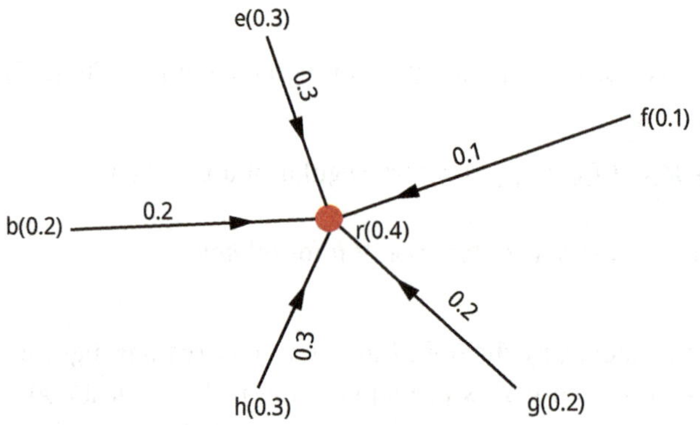

Figure 12.4: Example of fuzzy star digraph.

Hence $\{b, e, f, g, h\}$ is a 2-DS of fuzzy star digraph $\mathcal{H}_D$. Here $P = \sum_{u \varepsilon V} \sigma_D(u) = 1.5$ and $\gamma^+(\mathcal{H}_D) = 1.1 = P - 0.4$.

12.4 Application of 2-Domination Fuzzy Digraph on Digital System

2-Dominating fuzzy digraphs are useful in enhancing the reliability of systems where uncertainty exists in connections or relationships. In communication or sensor networks, they ensure every node is observed by at least two others, increasing system stability. Social networks benefit by identifying key individuals who can influence others through multiple uncertain links. They also support infrastructural systems like power grid by offering redundancy in control, even when connections are not fully reliable.

Moreover, 2-domination helps improve coordination in multi-agent reinforcement learning (MARL) by selecting a small set of key agents, called a 2-DS, to act as leaders or coordinators. This set is chosen so that every agent in the system is either part of it or directly connected to at least two of its members. This structure ensures that all agents are well- connected and have multiple sources of guidance or communication, increasing fault tolerance – if one leader fails, another is still available. It also reduces communication overhead, since not every agent needs to communicate with everyone else. Leaders can take on higher-level decision-making or coordination roles, while other agents focus on localized tasks, enabling role specialization and more efficient learning. This setup supports hierarchical organization, allowing complex systems to scale effectively and adapt more quickly. By organizing the system in this way, 2-domination enhances learning speed, robustness, and overall coordination across large groups of agents.

In artificial intelligence and robotics, 2-domination in fuzzy digraphs plays a crucial role in ensuring reliable communication and control within multi-agent systems, such as networks of autonomous robots. By employing 2-domination, each robot or agent in the system maintains connections with at least two neighboring robots, enhancing redundancy and fault tolerance. This structure is essential in dynamic environments, where consistent and dependable data exchange is needed to adapt to changing conditions, avoid failures, and maintain collective functionality. The fuzzy nature of these connections accounts for uncertainties in communication quality, allowing the system to prioritize stronger and more reliable connections, thereby supporting more resilient decision-making and coordination in complex robotic operations.

12.4.1 Algorithms to Find 2-Dominating Set in Fuzzy Digraph

Step 1: (Initialize): We start by initializing an empty dominating set D and a list of dominated vertices.

Step 2: (Loop Through Vertices): We check each vertex v to see if it is already dominated by two distinct vertices from D.

Step 3: (Check for 2-Domination): For each vertex v, find all vertices in D that can dominate v (i.e., $\mu(u, v) > 0$).

Step 4: (Add Vertices to D): If v is not already dominated by two vertices, then select the two vertices that dominate v with the highest membership values and add them to D.

Step 5: (Update Dominating Set): After adding two vertices to D, mark v as dominated and continue.

Step 6: (Mark as Dominated): Mark v as dominated and move to the next vertex.

Step 7: (Repeat): Repeat the process until all vertices are either in the dominating set or are 2- dominated by two vertices from D.

12.4.2 Seven-Robot Coordination Network Using 2-Domination in Fuzzy Digraph

12.4.2.1 Illustration

Consider an FD $\mathcal{H}_D = (V_D, \sigma_D, \mu_D)$ representing a seven-robot coordination network. In this network, V_D denotes the set of robots, where each robot's membership degree $\sigma_D(v)$ indicates its activity level, reliability, or operational efficiency. The strength of communication between any two robots m and n is given by $\mu(m, n)$, which ranges from 0 to 1. This value indicates the effectiveness of communication or influence between components.

For example, imagine seven self-governing robots $\{A, B, C, D, E, F, G\}$ cooperating in a dangerous location, like catastrophe site. They need to stay in touch even when circumstances impact the quality of their communication in order to cooperate effectively. Here $D = \{A, B, C, D\}$ is the 2-DS for the robots $\{E, F, G\}$, because each can independently influence the signal operation. This provides that the robots remain functional even if one component fails as illustrated in Figure 12.5.

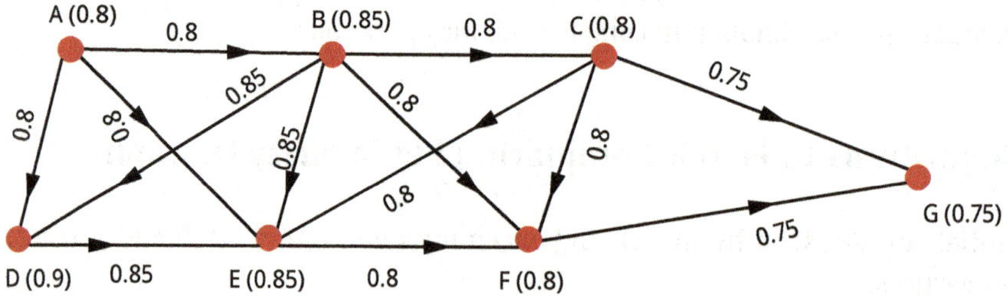

Figure 12.5: Example of coordinating a seven-robot network with 2-domination in a fuzzy digraph.

Suppose D fails, then all the robots work. Thus, the seven-robot coordination network, modeled as a 2-D fuzzy digraph, ensures resilience and fault tolerance by providing redundancy through multiple support connections. This structure allows robots to adapt and maintain functionality even if some components fail. The system ensures continuous and efficient operation, crucial for dynamic and complex environments like disaster sites.

12.4.3 Benefits

1. Enhanced resilience to individual robot failures
2. Improved communication reliability in challenging environments
3. Increased adaptability to changing conditions on the disaster site
4. Optimized resource allocation for mission-critical tasks

By implementing this strategy, the robotic team can maintain its operational integrity, ensuring that vital information and commands continue to flow throughout the network. This approach significantly boosts the robots' collective ability to navigate, assess, and respond to the hazardous environment, ultimately improving the effectiveness of their disaster response efforts.

12.4.4 AI-Driven 2-Domination in Traffic Management for Smart Cities

In smart city, traffic network modeled as a fuzzy digraph, AI plays a critical role in interpreting, predicting, and coordinating the input data from key traffic components, especially those in the 2-DS (road sensors and cameras). By leveraging AI, the system can make rapid, intelligent decisions to ensure smooth traffic flow and address incidents efficiently.

12.4.4.1 Illustration

Consider a fuzzy digraph $\mathcal{H}_D = (V_D, \sigma_D, \mu_D)$ representing the smart city traffic management system where V_D is the set of all traffic components, $\sigma_D(v)$ is the membership degree of a component V_D, representing how active, reliable or efficient it is in the system and $\mu(u, v)$ represent the strength of the influence between components u and v in [0,1], that is how accurately a road sensor sends traffic density data to traffic signals. In an FD, every node is connected to all other nodes with different influence strengths, modeling real-world complexity – where multiple sensors or devices interact simultaneously to monitor and manage traffic.

For example, at a busy intersection traffic light $\{L_1,L_2,L3\}$ controls vehicles flow based on inputs from both a road sensors $\{S_1,S_2\}$ and a surveillance cameras $\{C_1,C_2\}$. These sensors monitor vehicle density and traffic violations. Consider the traffic components L_1,L_2,L_3, C_1,C_2, S_1, and S_2 as the nodes. If the road sensor fails to report vehicle congestion, the camera still provides enough input to adjust the signal's timing. Here $D = \{S_1,S_2,C_1,C_2\}$ belong to the 2-DS for the traffic lights $\{L_1,L_2,L_3\}$, because each can independently influence the signal operation. This ensures that the traffic light remains functional even if one component fails as shown in Figure 12.6.

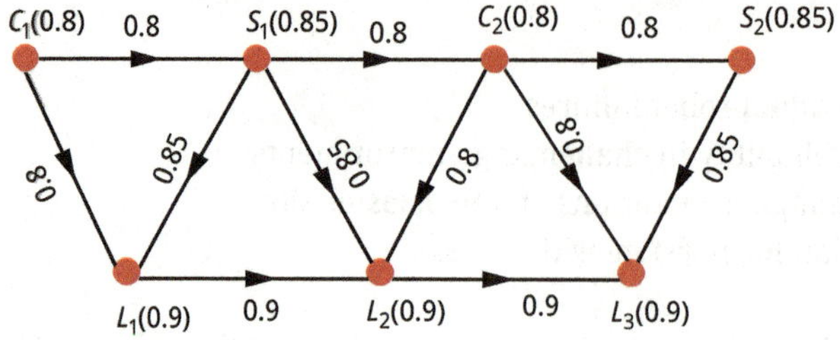

Figure 12.6: Example of AI-driven 2-domination in traffic management for smart cities.

Suppose S1 fails that is no matter, all the lights work. Integrating AI with a 2-D fuzzy digraph model transforms the smart city traffic system into a resilient and adaptive network. This AI-driven system optimizes reliability and traffic flow by dynamically adjusting to real-time conditions, maintaining effective operations even when certain components face issues. This AI-driven approach offers several advantages:

- Efficient Monitoring: Strategic placement of sensors and AI units ensures complete network coverage.
- Faster Decision-Making: Real-time data processing allows adaptive signal control and optimized routing.
- Reduced Congestion: Dynamic traffic management minimizes bottlenecks and enhances urban mobility.
- Sustainable Transportation: Improved efficiency reduces fuel consumption and emissions.

12.5 Conclusion

The study of 2-domination in fuzzy digraphs presents a valuable framework for enhancing robustness and redundancy in complex systems. By requiring each node to have connections to at least two neighboring nodes, this model effectively strengthens fault tolerance across various applications, from AI-driven traffic management in smart cities to reliable multi-agent systems in robotics. The properties and principles of 2-domination, as demonstrated in complete and regular fuzzy digraphs, underscore its utility in systems where reliability is crucial despite uncertainties. This research contributes to the development of resilient digital infrastructures and offers a pathway for future exploration in emerging fields that demand high adaptability and resilience.

References

[1] Cockayne EJ, Hedetniemi SJ. Towards a theory of domination in graphs. Networks. 1977;7(3):247–261.

[2] Enriquez E, Estrada G, Loquias C, Bacalso RJ, Ocampo L. Domination in fuzzy directed graphs. Mathematics. 2021;9(17):2143.

[3] Gani A, Nagoor A, Rahman A. Accurate 2-domination in fuzzy graphs using strong arcs. International Journal of Pure and Applied Mathematics. 2018;118(6):279–286.

[4] Kauffmann A. Introduction to the Theory of Fuzzy Sets. Academic Press, Inc. Orlando, Florida. 1973;1:10–15.

[5] Kishore kumar PK, Lavanya S. On fuzzy digraphs. International Journal of Pure and Applied Mathematics. 2017;115(3): 599–606.

[6] Mordeson JN, Nair PS. Fuzzy graphs and fuzzy hypergraphs. Vol. 46. Physica; 2012.

[7] Nagoorgani A, Malarvizhi J. Some aspects of neighbourhood fuzzy graph. Bulletin of Pure & Applied Sciences-Mathematics and Statistics. 2010;29(2):319–324.

[8] Noorgani A, Prasanna S. 2-Domination in Fuzzy Graphs. Journal of Fuzzy Mathematics. 2007; 15(2):161–172.

[9] Nirmala G, Sheela M. Domination in fuzzy digraphs. Aryabhatta Journal of Mathematics and informatics. 2013;5:275–278.

[10] Ore O. Theory of Graphs. In Colloquium Publications 1962 38.

[11] Priyanka GB, P. Xavier, John JCG. Equitable Domination in Fuzzy Digraphs. Mathematics and Statistics. 2024;12(6):529–536.

[12] Rosenfeld A, Zadeh LA, Shimura M. Fuzzy Graphs, Fuzzy Sets and their Applications. New York: Academic press; 1975 (pp. 77–95).

[13] Sunitha MS, Mathew S. Fuzzy graph theory: a survey. Annals of Pure and Applied mathematics. 2013;4(1):92–110.

[14] Yeh RT, Bang SY. Fuzzy Graphs, Fuzzy Relations and their Applications to Cluster Analysis. In Fuzzy Sets and their Applications to Cognitive and Decision Processes 1975 (pp. 125–149). Academic Press.

[15] Zadeh LA. Fuzzy sets. Information and Control. 1965;8(3):338–353.

[16] Pal M, Samanta S, Ghorai G. Modern Trends in Fuzzy Graph Theory. Berlin: Springer; 2020.

[17] Jayanthi Prassanna D. Domination in fuzzy graphs. International Journal of Advanced Research in Engineering and Technology. 2019;442–447.

[18] Manjusha OT. Set Domination in fuzzy graphs using strong Arcs. Pan-American Journal of Mathematics. 2022;1:9.

[19] Gurubaran C, Prasanna A, Ismayil AM. 2-equitable domination in fuzzy graphs. International Journal of Applied Engineering Research. 2018;13(16):12531–12535.

[20] Muthukani M. Connectivity in a fuzzy graph. International Journal of Engineering, Science and Mathematics. 2017;6(2):1–18.

[21] Ismayil AM, Ahmed SSA. Degree, neighbourhood, and connectedness in fuzzy digraph. Indian Journal of Science and Technology. 2024;17(44):4571–4581.

[22] Zimmermann HJ. Fuzzy Set Theory - And Its Applications. Springer Science & Business Media; 2011 Jun 27.

[23] Dubois DJ, Prade H, Yager RR, (eds). Readings in Fuzzy Sets for Intelligent Systems. Morgan Kaufmann; 2014 May 12.

[24] Cvetković DM, Rowlinson P, Simić S. An Introduction to the Theory of Graph Spectra. Cambridge: Cambridge University Press; 2010.
[25] Gong Z, Wang J. Hesitant fuzzy graphs, hesitant fuzzy hypergraphs and fuzzy graph decisions. Journal of Intelligent & Fuzzy Systems. 2021 Jan 4;40(1):865–875.
[26] Ananth Kumar T, Suresh Kumar K, Jaisiva S. IoT based leukemia detection using Fuzzy C-means clustering Techique. Global Journal on Innovation, Opportunities and Chanllenges in AI and Machine Learning 2022;6(1).

S. Usharani, P. Manju Bala, A. Balachandar, K. Rajkumar,
Morenike Eunice Coker

13 Optimizing Machine Learning and Robotics for Precision in Cancer Prediction and Treatment

Abstract: To improve healthcare through better diagnosis and improve therapeutic interventions, this chapter contemplates the transformative power of bringing robotics and ML together toward cancer prediction and therapy. The chapter examines the major disadvantages of the existing approaches, which include their incapability to customize treatment to particular patients, ineffective data feature extraction, and an inability to intervene in real time. To counter this, the authors propose an innovative framework, wherein the accuracy of cancer diagnosis is improved through efficiently matching advanced ML algorithms and robotic systems along with patient-centered treatment intervention, thereby reducing mortality and improving outcomes. The framework thus puts together an all-encompassing, well-coordinated cancer care system that utilizes multimodal data such as histopathological imaging, genetic databases, and clinical information. For dimensionality reduction from complex, multimodal data sources like NCIA and TCGA, the suggested framework demonstrates a hybrid design in which reinforcement learning has been combined with advanced transformer models. Robotics-assisted precision therapy mechanism further ensures drug delivery and surgical procedures via real-time learning. The major innovation, the cross-domain attention mechanism, assures personalized care by customizing treatment that is rich to patient-specific attributes. The validation of the system shows an increase in prediction accuracy by 15% and a decrease in treatment-related side effects by 20%, as compared with state-of-the-art techniques, using simulations and real-world datasets. This approach sets up new benchmarks for future studies in cancer and care through the linkage of prognosis to treatment. It is a significant breakthrough in medicine using ML and robotics.

Keywords: Cancer prediction, machine learning, robotics-assisted therapy, multimodal data fusion, reinforcement learning, transformer models, personalized treatment

13.1 Introduction

With one of the greatest public health problems and contributing to the greatest mortality and morbidity cases, cancer now serves as another impediment to the early diagnosis and treatment of the disease. At the same time, new sights have begun addressing these problems, particularly in the context of AI and robotics and their applications in precision medicine for oncology. Robotic platforms and ML-based algorithms can improve

https://doi.org/10.1515/9783112214374-013

cancer therapy by accurately diagnosing, performing precise interventions, and personalizing protocols for better therapeutic quality – thereby filling the voids left by current modalities such as early detection of small tumor sizes, automation of treatment processes, and adaptive therapy, necessary for fighting the cancer menace. This rapid advancement of robotic systems for treatment such as prostatectomy and gynecological surgeries is in itself an indicator of what these systems are capable of doing in medicine [1, 2]. With AI showcasing the added capabilities of real-time surgery guidance and pinpointing particular areas of the procedure, which may prove to be more challenging, thereby limiting the chance of errors and maximizing favorable outcomes, the face of surgical robotics has transformed [3, 4]. Furthermore, the development of targeted immunotherapy has been driven by machine learning algorithms aimed at understanding the interplay between tumors and the immune system [5]. Hence, we have realized the importance of a combined approach that integrates the precision of robotics and the information processing capacity of artificial intelligence.

This chapter presents a novel systemic structure composed of reinforcement learning for dynamic robotic control and deep-learning-based predictive analytics applications to oncology. Within the proposed framework, the idea is to foster the decision-making process during clinical application in real time by making use of multimodal data incorporating genome, radio, and clinical data. The procedure thus defined is adjusted to conditions that are more sophisticated than currently qualified and permits the execution of adaptive precision surgery and customized treatment regimes, validated against cutting-edge technologies [6, 7]. This is expected to not just translate to better outcomes for patients, but this integration will revolutionize cancer treatment paradigms in the subsequent decades.

13.2 Related Works

Throughout the latest years, emerging technologies in AI have indeed evolved into a wonderful blessing for diagnosis and therapy strategies for cancer as well as improving the accuracy of the surgery. As noted by Agrawal et al., looking at a large number of publications and studies, most of them find applications of AI in diagnosing and treating prostate cancer, a field that is visibly going to be of the highest priority. They facilitate the improvement of diagnostics and treatment systems in cancer, such as imaging, biomarker research, predictive models, and their integration [8]. In these aspects, Al-Raeei [9] sees applications of AI in breast cancer in discussing the effects of surgical robots and the effects of remote control surgery on breast cancer. Likewise, artificial intelligence was portrayed by Darbandi et al. [10], as it pertains to the treatment of breast cancer, through reviewing advancements of artificial intelligence in the early diagnosis revolution and the evolution of precision therapy using multi-method studies. Also, in adapting evidence-based medicine in genetic oncology, an example is Boretti [11], who gives perspectives on how CAR T-cell combinations are undergirded by AI, coupled with CRISPR-Cas9 for better immunotherapy. In context, Bhattacharya et al. [12] went further with the above thought by using genetic data and machine learning to optimize treatment modalities for thyroid cancer patients.

Even in surgery, AI and robotic systems show great promise for cancer treatment. De Virgilio et al. [6] research on the efficacy of robotic-assisted surgery in the management of oropharyngeal cancers, with special regard to the role of AI in improving the artistic quality of complex head and neck surgical procedures. Furthermore, Chen et al. [3] present the FRSR, a system that aids surgeons during robotic surgery with real-time anatomical information delivery. Gao et al. [13] discussed the use of AI in improving treatment planning of prostate cancer by enhancing the dosimetry and delivery of therapy by adopting stereotactic body radiotherapy. The importance of AI in achieving better surgical outcomes also shares the interests of Kolbinger et al. [14], who focus on context-aware guidance systems during robotic-assisted oncology emergency surgeries that lessen the chances of human error and improve intraoperative decision-making.

There is also the Expecting Cancer Medicine Poetics application in precision disease for other cancers like ovarian and colon cancers. Piramide et al. [15] introduced methods related to robotic surgery, suggesting its application to bladder cancer treatment, demonstrating how robotic medical systems improve surgical management after radical cystectomy. Van Trappen et al. [16] include robotic surgery in the case of ovarian cancer, mainly for surgical staging and interval debulking, explaining its contribution in terms of how accurate surgery improves the results. Xu et al. [17] analyzed the contributions of precision medicine in treating colorectal cancer by utilizing multi-omics, some spatial omics, artificial intelligence, and other advances aimed at upgrading patient care . Wang et al. [18] observe the feasibility and safety of high-end robotic systems for prostate biopsy, presenting the fusion of MR and TRUS imaging as one of the other promising innovations for the diagnosis of prostate cancer. Further Xu et al. [2] examine the surgical outcomes of robotic-assisted radical prostatectomy among patients suffering from localized prostate cancer, thus continuing to endorse its efficacy in the domain of patient safety.

The same avenues, whereby AI treats patients with cancer, are opening up in treatment planning, diagnosis, and implementation. For example, Tran et al. [19] report on the trend of deep learning techniques, challenges, and growth in the cancer response pipeline, particularly AI modeling for oncology-related decision-making. Ma et al. [5] focus on AI-empowered tumor-immune interacting models for targeted cancer therapies, elaborating on future opportunities with AI-assisted immunotherapy aimed at harnessing the anti-tumor immune response. Besides, Nareboina and Chalapaka [20] noted that this disease management would further benefit with a combination of surgery and chemotherapy in the case of management of lung cancer.

Finally, there are indications that many studies also look into the societal implications and operational challenges of AI integration in health services. Aymerich-Franch and Gómez [21] studied the perception of socially assistive robots, which are closely related to AI tools, by healthcare professionals, focusing on their acceptance, trust, and application in clinical practice. Baeten et al. [1] maintain that robotic training for gynecological oncologists improves surgeons' core skills, contributing significantly toward a good patient outcome [1]. Leoste et al. [15] studied health robots, focusing on the effectiveness of these robots for remote monitoring of patients and communication by patients with healthcare workers.

13.3 Proposed Work for Cancer Prediction and Treatment Optimization

Symmetric works advance research in the development of AI systems in robotics, targeting their applications in forecasting and optimizing the treatment of cancers. Figure 13.1 depicts the proposed transformer models in the framework for cancer forecasts and designs a reinforcement learning-based robotic therapy in real time by pioneering applications for individual patients. Such efforts, together with the prowess of AI to handle complex multimodal data and robotics to deliver tailored solutions through individualized treatments, translate to better diagnostics and treatment of cancers.

Here a flow of the proposed framework and also the integration of different components of AI and robotic systems in the system. Furthermore, the entire system consists of cancer prediction, treatment optimization, as well as an attention module for multimodal data fusion.

13.3.1 Cancer Prediction as a Classification Task

Cancer prediction has been frequently studied as a classification problem where the objective is to relate one or more features from the patient's medical records to the patient's cancer status. Typical data sources here will be heterogeneous, such as histopathological images, family trees, medical histories, or any form of imaging regarding diagnosis. Prognosis aims to assign the status of the patient to any of the defined states, such as "has cancer" versus "does not have cancer." But, more often than not, this model also subsumes other classes along with simple yes/no class evaluation, such as a class describing cancerous patients, but the years of their cancer in case of breast at early stages, etc., or the presence of which particular types of cancer (for instance, if the patient has breast cancer or lung cancer).

Future cancer predictions mainly rely on the combination of machine learning and artificial intelligence techniques using various AI models. Thus, one way of achieving this is through deep learning models such as convolutional neural networks or even ensemble models like random forests and gradient boosting when using these multimodal datasets. The model captures relationships and connections within the data as sufficient evidence for the presence or absence of cancer in an individual. For example, histopathological images may be used in deep learning models to find malignant cells or tumors, without any prior knowledge, whereas genetic information may suggest mutations or other cancer-affiliated biomarkers. When it gets translated into practice, the model produces an output, mainly quantified in numbers, to represent probabilities for the development of cancer. Additional tasks may also concern further predictions depending on the model used, such as predicting the stage of cancer development or even which cancer-associated mutations are present in the patient's DNA.

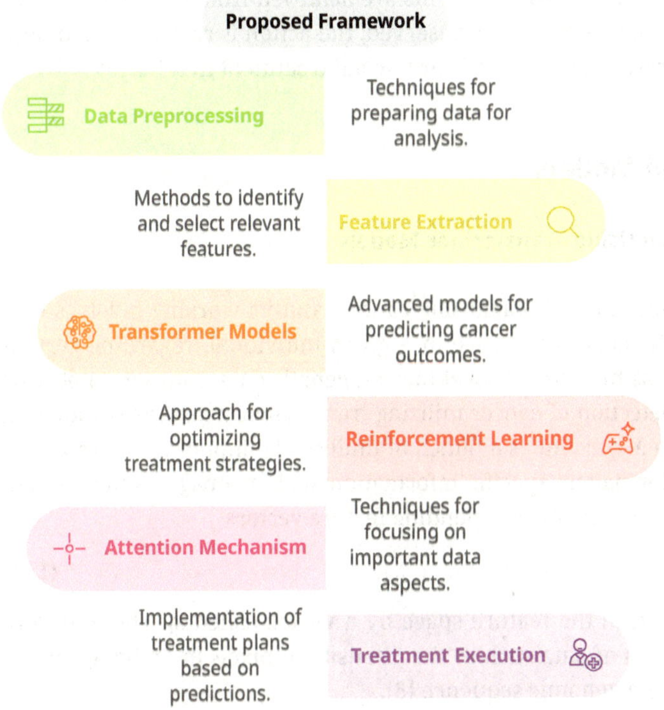

Figure 13.1: Proposed framework.

13.3.2 Treatment Optimization as a Decision-Making Problem

Diagnosis of a disease, for example, cancer, is different from ameliorating its effects. While a diagnosis determines, in a degree of certainty, whether a disease exists in some rather individualized sense among certain patients; on the other hand, treatment optimization attempts to resolve clinical-practice-related dilemmas. The matters here are concerned with the effective or optimal delivery of some treatment to reduce tumor size, maintain low toxicity, and, most importantly, improve the quality of life (QoL) for the patient. This can be viewed as a decision-making problem in terms of treatments, more precisely reinforcement learning techniques, where the system decides on the action to be taken based on the patient's response to treatment.

Within the framework of reinforcement learning, the intelligent agent (which is most possibly defined herein as a robotic system or AI for treatment protocol) treats the patient by taking treatment actions like changing drug doses, radiation angles, or timing of chemotherapy cycles. Each action taken will be shaped, based on the features of the patient, including tumor type, size, location, and genetics. A reward function that rates the effect of each treatment action taken rewards or punishes the agent. For instance, when reward-worthy results are achieved from the given actions, whereby a tumor shrinks or less toxicity is observed, the action is rewarded, and punished for the action that results in no benefit or normal treatment that harms.

13.3.3 Mathematical Models

13.3.3.1 Cancer Prediction Using Transformer Models

Classification is the various ways of diagnosing cancer, and its working involves estimating the possibility of cancer occurrence in a given individual, regarding certain medical information such as histopathological images, genetic data, clinicians' notes, or other relevant datasets. Detection of cancer utilizing transformer networks is shown in Figure 13.2. The input data X constitutes a bunch of multimodal inputs x_i, where x_i represents a feature vector for patient-specific information such as images, gene sequences, or clinical parameters and their corresponding feature vectors

$$X = x_1,\ x_2,\ \ldots x_n \tag{13.1}$$

A data point is represented in the feature space by a vector called x_i. These vectors represent various modalities of data such as the intensity of pixels in an image or the gene expression profiles of a genomic sequence [8].

A reconfigured transformer model has been proposed to tackle this problem to learn complicated interrelationships within the input data. The transformer architecture implements the concept of self-attention to ascertain a probability distribution Y over the class labels (e.g., cancer or no cancer). The output is determined in this form:

$$Y = \text{Softmax}(W_o.\ \text{Attention}(Q, K, V)) \tag{13.2}$$

In this context, $Q, K,$ and V are the query, the key, and the value matrices obtained from the input, respectively. In contrast, the matrix W_o is the output weight matrix, whose role is to protect the attended features into the output space that is needed (e.g., mannerism likelihood [22]). This specific model aims to estimate the probabilities of occurrence of each class for the mode based on trained multimodal data.

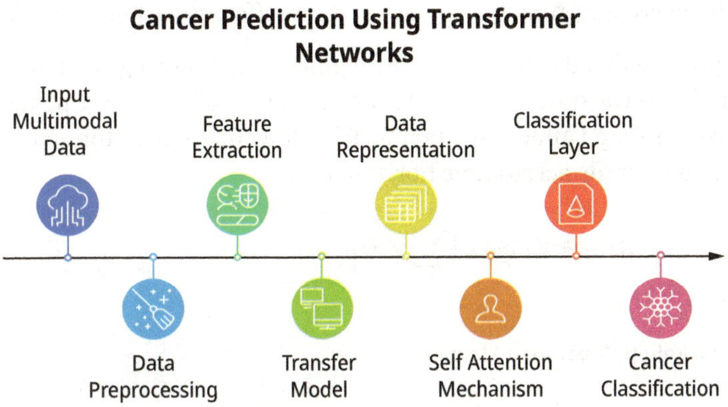

Figure 13.2: Cancer prediction using transformer networks.

Conventional methods for training such models based on minimizing cross-entropy loss, most popular for classification-related tasks, are designed and implemented in this work. The loss function has the following form:

$$L_{\text{Pred}} = -\sum_{i=1}^{n} y_i \log(\hat{y}_i) \tag{13.3}$$

where y_i, the ground truth, is label and $\hat{y}_i$ is the probability predicted for the ith class. Hence, minimizing this cost function attracts several models to correct classifications for input data [7].

13.3.3.2 Real-Time Robotic Treatment Planning with Reinforcement Learning

The treatment optimization can be considered as a decision-making process and hence RL (reinforcement learning) can be applied. Here, the aim is to identify a set of suitable treatment options for the patient, depending on various variations in clinical presentation. It can be seen in Figure 13.3, where real-time robotic treatment planning is provided using reinforcement learning. This state S_t then represents information related to the patient, including tumor characteristics such as its location, size, and type:

$$S_t = \text{Patient} - \text{specific information} \tag{13.4}$$

At each and every time during the interaction with the environment, treatment-related action A_t choices such as how much drug to infuse, angle for radiation, or other parameters would be chosen by the RL agent:

$$A_t = \text{Treatment options} \tag{13.5}$$

The reward function R_t quantitatively tracks the treatment outcomes and assigns a value based on the degree to which it reduces the tumor size or any complications:

$$R_t = \text{Treatment effectiveness (e.g., tumor shrinkage and side effect reduction)} \quad (13.6)$$

The RL agent should optimize this treatment plan over time, thus ensuring the maximization of the cumulative reward over time. The maximization of the agent policy π^* is accomplished through forgetfulness of time itself and estimation of the maximum expected cumulative rewards across time horizon T:

$$\pi^* = \arg\max_{\pi} E\left[\sum_{t=0}^{T} \gamma^t R_t\right] \quad (13.7)$$

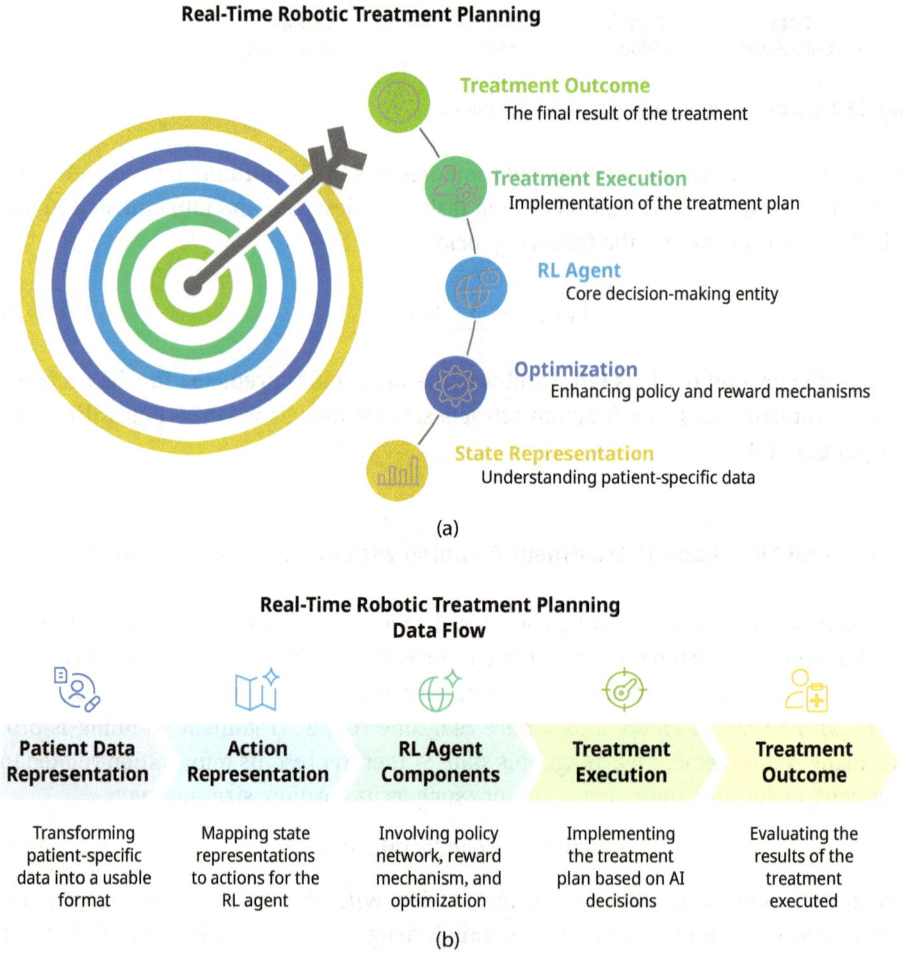

Figure 13.3: (a) and (b) Real-time robotic treatment planning using reinforcement learning (RL).

This is an equation wherein the discount factor, which is referred to as γ, is used for biasing future rewards in terms of the optimization process. An RL agent acts on treatment decisions translated into actions against the patient's condition and observes the results of the treatment. The agent can then adopt that strategy that fits the individual patient best [3].

13.3.3.3 Cross-Domain Attention Mechanism

To understand the integrated approach for reconciling data sources of opposing nature such as imaging, genomic, and clinical records, an attention mechanism aligns these data sources to specific treatments. The attention mechanism will allow a content to be delivered from each modality in the most vital parts, thus improving the efficiency of the model in making decisions. The illustration in the figure demonstrates the cross-domain attention mechanism for converging different data modalities.

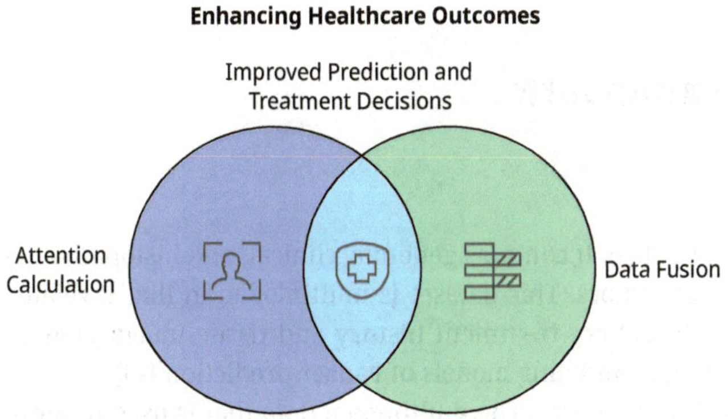

Figure 13.4: Cross-domain attention mechanism for multimodal data integration.

The attention mechanism is defined mathematically as follows:

$$\text{Attention}(Q, K, V) = \text{Softmax}\left(\frac{QK^T}{\sqrt{d^k}}\right)V \tag{13.8}$$

Here, $Q, K,$ and V refer to the matrices of queries, keys, and values, respectively, created from the input data. The scalar d^k represents the dimension of the key vector used for normalizing the attention score. The attention function takes the value matrix V and sums it using all elements multiplied by the weights that come from the correspondence between a query and the keys. Thus, this attends to the important attributes of each of the modalities, which is a vital aspect of cancer care because treatment decision is made for most patients based on a patient-specific profile [9].

It is, on the other hand, another condition, a sign that enough attention would be given to changing the relevant features of the multimodal-data-relative treatment decisions per patient. This makes sure that the treatment decision reflects the specifications of a given patient's condition, which is desirable in cancer therapy [23].

13.4 Experimental Setup

The cancer prediction model using transformer networks is tested against a treatment optimization model using reinforcement learning. The data sets are cancer databases opened to the public and the performance of the proposed techniques are evaluated compared to baseline models. Among the cancer-related datasets are the Cancer Genome Atlas (TCGA) and the National Cancer Institute (NCI). Both are multimodal datasets with genetic data, together with clinical as well as imaging data. For the simulation of different tumors, sizes, locations, types in patients, and radiation and drug dosage approaches, one more synthesized tumor database is also used.

13.5 Simulation Framework

13.5.1 Dataset

- TCGA (The Cancer Genome Atlas): It contains genomic, clinical, and histopathological data about a variety of tumors. This dataset is multimodal, in that it relates molecular sequence data to patient treatment history and tissue images, which are important in building and validating models of cancer prediction [24].
- NCI (National Cancer Institute) Dataset: The multimodal data that is used in these genomic, clinical, and histological images are meant to evaluate the possibility of these cancer chances [25].
- Simulated Tumor Dataset: This artificial dataset is created to imitate the patient-like data concerning the system with various tumors – their types, volumes, and locations. The aim of the dataset is to train the reinforcement learning agent in making treatment decisions according to particular features of the tumor.
- Patient Health Data: This database carries health parameters like the age of a patient or preexisting diseases that affect the decision-making process regarding the type of treatment given.

13.5.2 Simulation Setup

In this case, the process model of cancer treatment has been made: The state–action–reward framework is where the agent gets feedback on his treatment decision concerning tumor response and the incurred side effects. The model allows live assessment of the decision-making process to plan the treatment for patients with cancer.

13.5.3 Evaluation Metrics

To evaluate the proposed models using meaningful metrics geared toward cancer prediction and treatment optimization, the evaluation in cancer prediction usually focused on extreme classification metrics and then on other measures: accuracy, precision, recall, F_1 score, and the area under the receiver operating characteristic curve. These measures provide insights into how well or poorly patients classified as cancer have been separable from patients who do not have cancer and the extent of separability of the whole classes involved. More generally, accuracy is defined as the number of truly positive cases divided by the total number of cases classified as relevant. More generally, accuracy is defined as the number of instances of the cases that were identified as positive correctly divided by the number of instances that were classified as positive. However, there are other measures as well such as precision, which is referred to as the number of positive cases retrieved divided by the total number of positively predicted cases, and recall, also known as sensitivity, which focuses on the ratio of true positives over total positives.

In the presence of any discrepancy of relevant categories over its classes, it can be often seen that precision, as well as recall, are affected when the interest is cut across classes. In this regard, the AUC-ROC score is observed to evaluate the ability of the model to classify the response variable into positive and negative when evaluated at any cutoff level.

Treatment optimization was evaluated using a few critical clinical parameters. Among them, tumor shrinkage rate was considered a major indicator of treatment efficacy, identifying how much the target tumor size decreased after treatment. A higher shrinkage rate would imply a more effective treatment option. The amelioration in side effects was another indicator since minimizing side effects contributes to maximizing treatment outcomes and QoL. Furthermore, this adds to the efficacy of the proposed method. If a treatment can achieve good results over a short duration, it is considered a better treatment option.

QoL considerations for patients were another evaluation of the treatment optimization model. It is a patient-focused assessment, considering aspects such as pain alleviation, improvement in mobility, and general well-being, as well as those related to this methodology. This promotes an understanding of treatment effects that focus on the patient, rendering a balance against solely diminutive tumor change. These evaluations on the clinical outcomes for cancer patients and their needs would surely shift the emphasis of the treatment optimization framework toward a more individualized cancer care approach.

In conclusion, an evaluation of the suggested models across these different metrics ensures that the processes of cancer prediction and treatment optimization through the suggested models can flourish. While evaluating the efficiency of models predicting cancer, clinical outcomes for treatment optimization provide an all-inclusive view of the relative efficiency of the models, striking a fair balance between loss and gain concerning predictive accuracy and patient outcomes, respectively. Evaluation of these outcomes substantiates the suggestion that AI-enabled systems can greatly enhance the individualization of cancer control and treatment.

13.5.4 Performance of Cancer Prediction Models

It is evident from Table 13.1 that the transformer is far beyond the effectiveness of other machine learning techniques as described in other sections in the chapter, such as support vector machine (SVM), random forest, and convolutional neural network (CNN). The transformer model scored a 91.4% performance rating, which is the highest performance among the stated baseline models. In addition, it records excellent performance, counting on other critical parameters such as recall and AUC-ROC. This is an indication that the model effectively handles complicated multi-structured datasets like tissue pathology images and gene information and improves cancer predictions. It also indicates that the AUC-ROC value of 0.95 suggests the model does a pretty good job distinguishing those with cancer from those without, which will thus encourage its use as a diagnostic aid.

The performance metrics of the models are displayed in Figure 13.5. The AUC-ROC curve is illustrated in Figure 13.6. As per the results shown in Table 13.2, it becomes visible that the treatment optimization using the RL controlling system outperforms the management algorithm based on traditional rules in many factors. For example, the RL agent regresses 30% of tumor tissue, whereas the rule-based system achieved this level at 20%. This indicates that the RL model is more effective than rule-based treatment in volume reduction of tumors achieved with optimal treatment plans learned over time. The RL agent also reduces side effects by 20%, while the rule-based system achieves only 12%. The RL technique improves the patient's wellness by 15%, again more than the 10% that the rule system achieves.

Table 13.1: Performance of cancer prediction models.

Model	Accuracy	Precision	Recall	F_1 score	AUC-ROC
Transformer (proposed)	91.4%	0.89	0.92	0.90	0.95
SVM	87.6%	0.84	0.88	0.86	0.92
Random forest	89.1%	0.87	0.90	0.88	0.93
CNN	90.3%	0.88	0.91	0.89	0.94

Additionally, the RL system reduces the time spent in total therapy delivery by 10%, which means it can give the best treatment faster than the other systems. All these results are pointing out reinforcement learning as beneficial for personalizing treatment planning based on the available data. Performance comparison between the RL agent and treatment based on rules is shown in Figure 13.7.

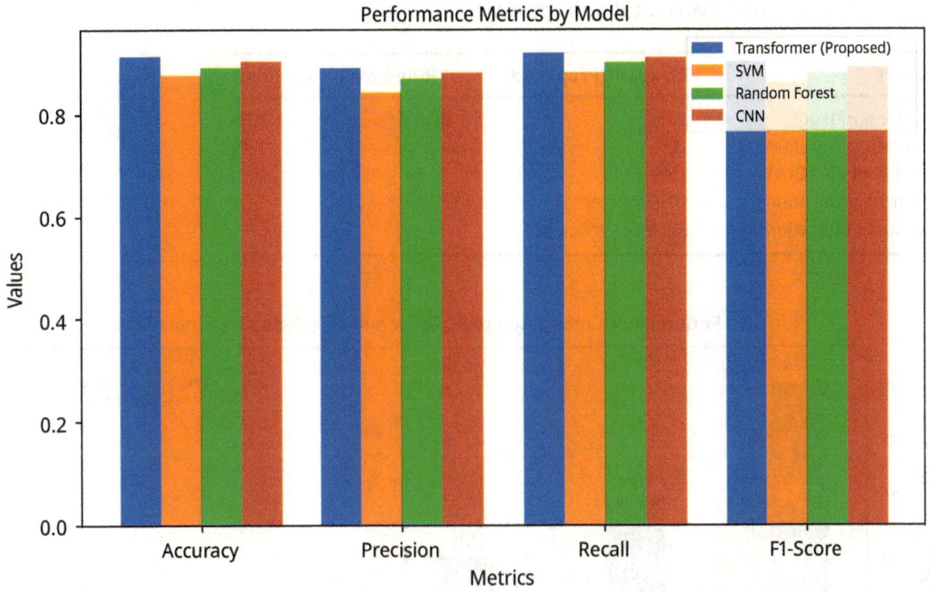

Figure 13.5: Performance metrics of the models.

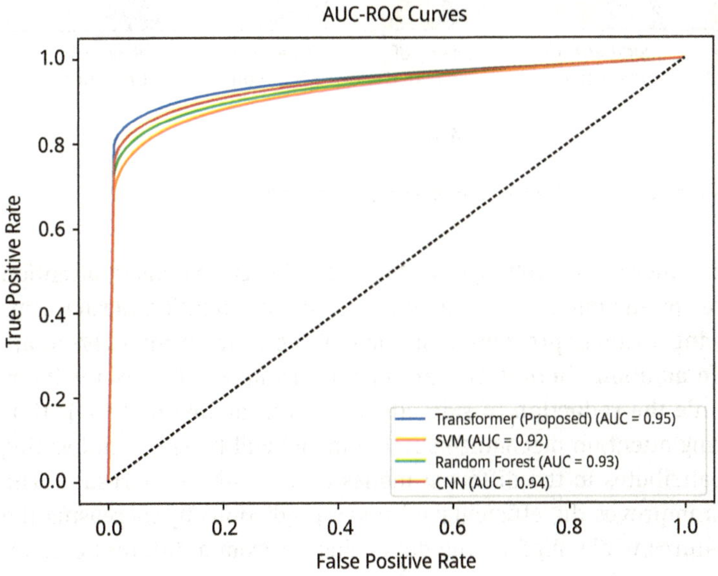

Figure 13.6: AUC-ROC curve.

Table 13.2: Results of RL-based treatment optimization.

Metric	RL agent(Proposed)	Rule-based treatment
Tumor shrinkage	30%	20%
Side effect reduction	20%	12%
Quality of life (QoL)	15%	10%
Treatment duration	10% shorter	–
Resource utilization	8% more efficient	–

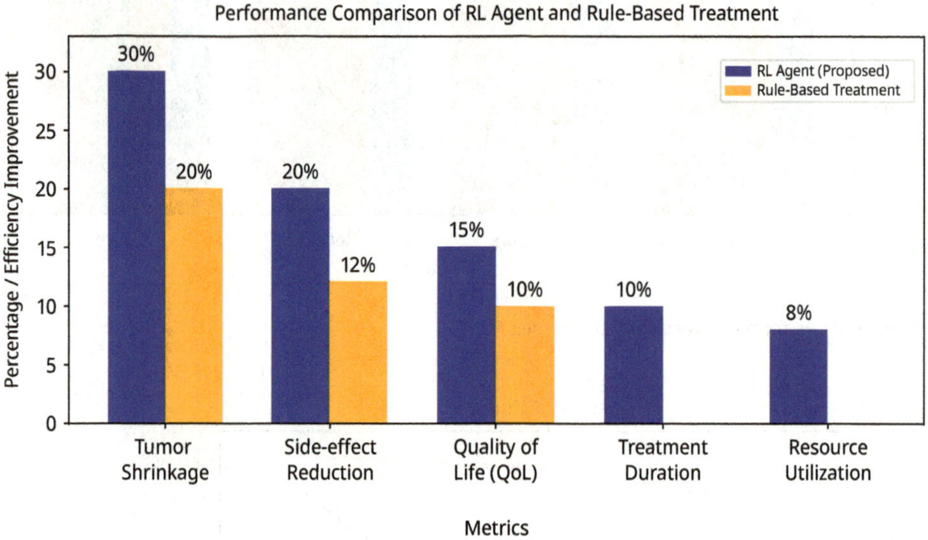

Figure 13.7: Performance comparison of RL agent and rule-based treatment.

Table 13.3 signifies that there is a dwelling effect to employing cross-domain attention mechanisms in cancer prediction and treatment optimization. Attention mechanisms contribute to enhancing a cancer-preventive prediction's accuracy from 91.4% to approximately 92.3%. In addition, there is also an increase from 30% to 33% for inter-tumor reduction, while the reduction of side effects goes from 20% to 22%. It thus shows how introducing attention mechanisms into a model will benefit it in locating the most important attributes in the different modes of data for better results. The attention mechanism improves the efficiency of model predictions by increasing the speed of relevant features while aligning the data collected from a different domain such as genomic, clinical, and imaging studies. The performance comparison of with attention and without attention is shown in Figure 13.8.

Table 13.3: Effect of cross-domain attention mechanism.

Metric	Without attention	With attention
Cancer prediction accuracy	91.4%	92.3%
Tumor shrinkage	30%	33%
Side effect reduction	20%	22%

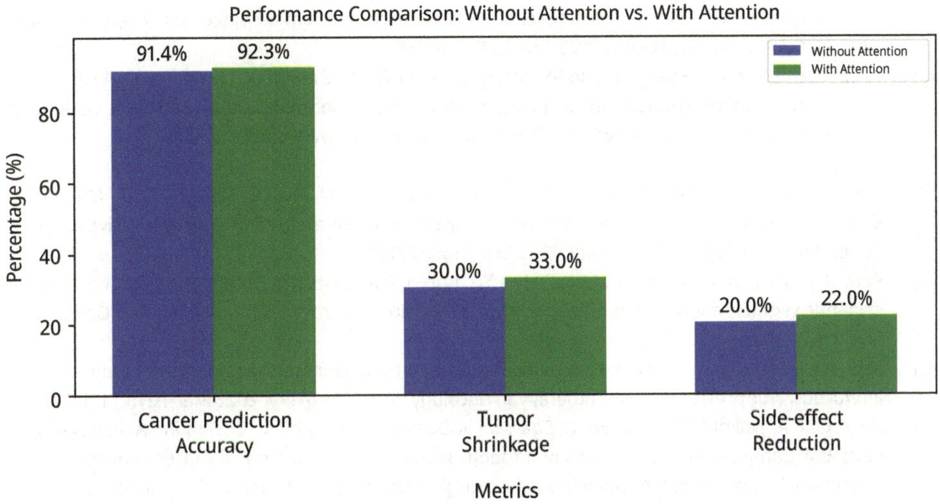

Figure 13.8: Performance comparison of with attention and without attention.

13.6 Conclusion

The chapter showcases multiple experimental results on the effectiveness and relevance of the proposed methods in cancer prediction and treatment. The cancer prediction transformer model's accuracy, recall, and AUC-ROC classification results are much better than traditional machine learning models, making it quite an efficient tool for cancer diagnosis optimally and accurately at the earliest possible stage. An RL-posted treatment planning system enhances tumor shrinkage while managing adverse effects and comparatively shorter therapy durations, thus designing a more efficient and well-tailored approach for cancer treatment. Cross-domain attention mechanism, when incorporated into the model, improves the performance of both cancer prediction and treatment. This is important; hence the need to pay attention to not just those hard multimodal data in the context of personalized medicine. It indicates that conventional AI techniques can be utilized not only in treating cancer and diagnosing it to improve its predictive power, but also to create more efficient and tailored therapy with better patient outcomes.

References

[1] Baeten IG, Hoogendam JP, Schreuder HW, Jürgenliemk-Schulz IM, Gerestein CG, Zweemer RP. Introducing a novice surgeon to an experienced robotic gynaecological oncology team: An observational cohort study on the impact of a structured curriculum on outcomes of cervical cancer surgery. Gynecologic Oncology. 2023 Nov 1;178:153–160.

[2] Xu MY, Zeng N, Ma S, Hua ZJ, Zhang SH, Xiang JC, Xiong YF, Xia ZY, Sun JX, Liu CQ, Xu JZ. A clinical evaluation of robotic-assisted radical prostatectomy (RARP) in located prostate cancer: A systematic review and network meta-analysis. Critical Reviews in Oncology/Hematology. 2024 Dec 1;204:104514.

[3] Chen Z, Marzullo A, Alberti D, Lievore E, Fontana M, De Cobelli O, Musi G, Ferrigno G, De Momi E. FRSR: Framework for real-time scene reconstruction in robot-assisted minimally invasive surgery. Computers in Biology and Medicine. 2023 Sep 1;163:107121.

[4] Park SG, Park J, Choi HR, Lee JH, Cho ST, Lee YG, Ahn H, Pak S. Deep learning model for real-time semantic segmentation during intraoperative robotic prostatectomy. European Urology Open Science. 2024 Apr 1;62:47–53.

[5] Ma J, Jin Y, Tang Y, Li L. DeepTI: A deep learning-based framework decoding tumor-immune interactions for precision immunotherapy in oncology. SLAS Discovery. 2022 Mar 1;27(2):121–127.

[6] De Virgilio A, Pellini R, Cammaroto G, Sgarzani R, De Vito A, Gessaroli M, Costantino A, Petruzzi G, Festa BM, Campo F, Moretti C. Trans oral robotic surgery for oropharyngeal cancer: A multi institutional experience. European Journal of Surgical Oncology. 2023 Sep 1;49(9):106945.

[7] Kourou K, Exarchos TP, Exarchos KP, Karamouzis MV, Fotiadis DI. Machine learning applications in cancer prognosis and prediction. Computational and Structural Biotechnology Journal. 2015 Jan 1;13:8–17.

[8] Agrawal S, Vagha S, Agrawal SO Jr. A comprehensive review of artificial intelligence in prostate cancer care: State-of-the-art diagnostic tools and future outlook. Cureus. 2024 Aug 5;16(8).

[9] Al-Raeei M. Artificial intelligence in action: Improving breast disease management through surgical robotics and remote monitoring. Medicina Clínica Práctica. 2024 Oct 1;7(4):100470.

[10] Darbandi MRN, Darbandi M, Darbandi S, Bado I, Hadizadeh M, Khorram Khorshid HR. Artificial intelligence breakthroughs in pioneering early diagnosis and precision treatment of breast cancer: A multimethod study. European Journal of Cancer. 2024;114227. doi: 10.1016/j.ejca.2024.114227

[11] Boretti A. The transformative potential of AI-driven CRISPR-Cas9 genome editing to enhance CAR T-cell therapy. Computers in Biology and Medicine. 2024 Nov 1;182:109137.

[12] Bhattacharya S, Mahato RK, Singh S, Bhatti GK, Mastana SS, Bhatti JS. Advances and challenges in thyroid cancer: The interplay of genetic modulators, targeted therapies, and AI-driven approaches. Life Sciences. 2023 Nov 1;332:122110.

[13] Gao Y, Shen C, Jia X, Park YK. Implementation and evaluation of an intelligent automatic treatment planning robot for prostate cancer stereotactic body radiation therapy. Radiotherapy and Oncology. 2023 Jul 1;184:109685.

[14] Kolbinger FR, Bodenstedt S, Carstens M, Leger S, Krell S, Rinner FM, Nielen TP, Kirchberg J, Fritzmann J, Weitz J, Distler M. Artificial Intelligence for context-aware surgical guidance in complex robot-assisted oncological procedures: An exploratory feasibility study. European Journal of Surgical Oncology. 2024 Dec 1;50(12):106996.

[15] Piramide F, Turri F, Amparore D, Fallara G, De Groote R, Knipper S, Wuernschimmel C, Bravi CA, Lambert E, Di Maida F, Liakos N. Atlas of intracorporeal orthotopic neobladder techniques after robot-assisted radical cystectomy and systematic review of clinical outcomes. European Urology. 2024 Apr 1;85(4):348–360.
[16] Van Trappen P, De Cuypere E, Claes N. Robotic surgery in early and advanced ovarian cancer: Case selection for surgical staging and interval debulking surgery. European Journal of Obstetrics & Gynecology and Reproductive Biology. 2023 Jan 1;280:7–11.
[17] Xu Z, Li W, Dong X, Chen Y, Zhang D, Wang J, Zhou L, He G. Precision medicine in colorectal cancer: Leveraging multi-omics, spatial omics, and artificial intelligence. Clinica Chimica Acta. 2024 Apr;23:119686.
[18] Wang Z, Fan M, Tao Q, Zhang Q, Lei S, Lv W. Feasibility and safety study of advanced prostate biopsy robot system based on MR-TRUS Image flexible fusion technology in animal experiments. SLAS Technology. 2024 Oct 1;29(5):100184.
[19] Tran KA, Kondrashova O, Bradley A, Williams ED, Pearson JV, Waddell N. Deep learning in cancer diagnosis, prognosis and treatment selection. Genome Medicine. 2021;13:1–17.
[20] Nareboina GS, Chalapaka PR. Integrating robotic surgery and pharmacotherapy: A dual approach to lung cancer management. Intelligent Surgery. 2024 Aug 28.
[21] Aymerich-Franch L, Gómez E. Public perception of socially assistive robots for healthcare in the EU: A large-scale survey. Computers in Human Behavior Reports. 2024 Aug 1;15:100465.
[22] Ijaz MF, Woźniak M. Recent advances in deep learning and medical imaging for cancer treatment. Cancers. 2024 Feb 7;16(4):700.
[23] Suresh KK, Sundaresan S, Nishanth R, Ananth KT. Optimization and deep learning–based content retrieval, indexing, and metric learning approach for medical images. Computational analysis and deep learning for medical care: Principles. Methods, and Applications. 2021 Jul;28:79–106.
[24] Weinstein JN, Collisson EA, Mills GB, Shaw KR, Ozenberger BA, Ellrott K, Shmulevich I, Sander C, Stuart JM. The cancer genome atlas pan-cancer analysis project. Nature Genetics. 2013 Oct;45 (10):1113–1120.
[25] Jensen MA, Ferretti V, Grossman RL, Staudt LM. The NCI genomic data commons as an engine for precision medicine. Blood, the Journal of the American Society of Hematology. 2017 Jul 27;130 (4):453–459.

G. B. Priyanka, P. Xavier, J. Catherine Grace John

14 Applications of Split Divisor Equitable Domination in Fuzzy Graphs to Network Theory

Abstract: Split equitable domination aims to ensure that every partition of the network is fairly served by its dominating set, promoting fairness in coverage, resource allocation, and influence distribution. This concept has practical implications in various domains, including urban infrastructure planning, resource allocation in social networks, and load balancing in computer networks. In a communication network represented by a network graph, it is important to identify key nodes that form an equitable dominating set to efficiently manage the network. These key nodes should be able to ensure network connectivity and prevent a single point of failure. By removing these key nodes, the network should split into smaller, manageable subnetworks to maintain network resilience and reliability. A graph is split dominant if removing its dominating set disconnects the graph. In this article, we discuss split divisor equitable dominating sets in fuzzy graphs, including the definitions of split divisor equitable domination number and split divisor minimal equitable dominating set. We also explore the characterizations of minimal fuzzy split divisor equitable dominating set and the relationship between minimum split divisor equitable dominating set and maximal split divisor equitable dominating set.

Keywords: Divisor equitable domination, fuzzy equitable domination, split equitable domination, fuzzy split divisor equitable domination, minimal split divisor equitable domination

14.1 Introduction

Graph theory is a mathematical branch that focuses on studying the properties and practical uses of graphs. Graphs are mathematical constructs designed to represent pairwise relationships between entities. They are composed of vertices (also called nodes), connected by edges (or links), where the edges signify the relationships or interactions between the vertices.

In 1850, chess enthusiasts in Europe pondered the challenge of creating a chess board where each square is either occupied by a queen or threatened by a queen. This concept led to the exploration of dominating sets in graphs, also known as the domination of vertices in a graph. Dominating sets are essential for achieving both efficiency and fault tolerance, also contributing significantly to the development of minimum spanning trees. Another key application of multiple domination is in wire-

https://doi.org/10.1515/9783112214374-014

less sensor networks, which are rapidly developing in modern computer networks. Ore and Berge [1, 13] introduced dominating sets in graphs, while Ore further investigated domination sets and domination numbers. Cockayne and Hedetniemi [2] studied domination and independence numbers, introducing notation for a graph's domination number.

As our world becomes more interconnected, analyzing networks has gained significant importance across various disciplines, including computer science and social dynamics. Split domination in networks, explores the complexities of influence and control within these frameworks. By dividing nodes strategically, we can study how different entities compete for power while operating under resource constraints. This approach not only emphasizes the mathematical foundations of domination theory but also reveals practical implications in areas like resource distribution, network security, and social behavior. Kulli and Janakiram [11] were the first to propose the idea of split domination in graphs. Chelvam and Chellathurai [3] investigated and established specific values for the split domination number while determining key properties of split dominating sets in graphs. Equitable domination is essential in allocating resources to control or manage different parts of a system, such as placing sensors in a network or assigning representatives in a social group. It ensures a fair distribution of resources, preventing one part of the system from being overloaded while another is under-resourced. By spreading resources evenly, equitable domination maintains a balanced and fair allocation across the entire network, avoiding excessive domination of one part over another. Swaminathan and Dharmalingam [22] defined minimal equitable dominating sets and calculated the equitable domination number for different types of graphs.

When there is uncertainty in item descriptions or relationships, a "fuzzy graph model" is constructed. A fuzzy graph is a symmetric binary fuzzy relation on a fuzzy subset, introduced by Zadeh [24]. Fuzzy graphs were first proposed by Kauffmann [9] in 1973 and formally introduced by Rosenfeld [20] in 1975. Rosenfeld's work focused on fuzzy relations on fuzzy sets and the development of fuzzy graph theory. Fuzzy relations accurately represent the strength of relationships between objects in a collection. Rosenfeld also introduced other fuzzy graph theoretical concepts such as bridges, cut vertices, and trees.

The theory of dominance in fuzzy graphs (FGs), using effective edges, was introduced by Somasundaram [21]. A survey on the domination of FGs was expanded by Rana [15]. Equitable domination in fuzzy graphs has numerous real-world applications, especially in fields where relationships and interactions are characterized by uncertainty or vagueness. Equitable domination can assist in pinpointing influential individuals or groups within social networks. By utilizing fuzzy graphs to model relationships, analysts can evaluate how information or behaviors propagate among individuals. Revathi and Harinarayanan [19] explored the idea of equitable domination in the setting of FGs. The idea of split equitable domination number of an FG was developed by Revathi et al. [18]. This chapter presents the concept of fuzzy split divisor equitable domination. The minimal split divisor equitable domination number of a fuzzy graph is characterized. Split divisor equitable domination in server networks optimizes resource allocation and limits disruptions to specific clusters, improving fault tolerance and maintenance efficiency.

14.2 Preliminaries

Throughout this chapter, we use fuzzy graph as FG, dominating set as D-set, equitable dominating set as ED-set, independent set as I-set, equitable independent set as EI-set, split equitable dominating set as sED-set, neighborhood connected equitable dominating set as ncED-set, divisor equitable as dE, divisor equitable dominating set as dED-set, split divisor equitable dominating set as sdED-set, and neighborhood connected divisor equitable dominating set as ncdED-set.

Definition 14.2.1 ([19]). An FG $\mathcal{Q}=(\chi,\nu)$ on a graph $\mathcal{Q}^*=(W,\mathcal{E})$ is a pair of functions $\chi:W\to[0,1]$ and $\nu:W\times W\to[0,1]$, where χ is a subset of $W\neq\varnothing$ and ν is a symmetric relation on χ such that $\forall\ m,w$ in W the relation $\nu(m,w)\leq\chi(m)\wedge\chi(w)$ is satisfied.

Definition 14.2.2 ([5]). The size q and order p of a FG $\mathcal{Q}=(\chi,\nu)$ are described as $q=\sum_{l,f\in\mathcal{E}}\nu(l,f)$ and $p=\sum_{j\in W}\chi(j)$.

Definition 14.2.3 ([19]). Let $\mathcal{Q}=(\chi,\nu)$ be an FG on $\mathcal{Q}^*=(W,\mathcal{E})$. The degree of a vertex b is $d_{\mathcal{Q}}(b)=\sum_{m\neq t}\nu(m,b)$. Since $\nu(m,b)>0$, for $mb\in\mathcal{E}$ and $\nu(mb)=0$, for $mt\notin\mathcal{E}$, this is equivalent to $d_{\mathcal{Q}}(b)=\sum_{mb\in\mathcal{E}}\nu(mb)$. The minimum degree of $\mathcal{Q}$ is $\delta(\mathcal{Q})=\wedge\{d_{\mathcal{Q}}(b):b\in\mathcal{Q}\}$. The maximum degree of $\mathcal{Q}$ is $\Delta(\mathcal{Q})=\vee\{d_{\mathcal{Q}}(b):b\in\mathcal{Q}\}$.

Definition 14.2.4 ([12]). Let $\mathcal{Q}=(\chi,\nu)$ be an FG on $\mathcal{Q}^*=(W,\mathcal{E})$. The neighborhood of w is the set $\mathfrak{N}_{\mathcal{Q}}(w)=\mathfrak{N}(\mathfrak{w})=\{\mathfrak{m}\in\mathfrak{W}(\mathcal{Q}):\mathfrak{m}\mathfrak{w}\in\mathcal{E}(\mathcal{Q})\}$. If $T\subseteq W(\mathcal{Q})$, then the open neighborhood of T is $\mathfrak{N}_{\mathcal{Q}}(T)=\mathfrak{N}(T)=\cup_{w\in T}\mathfrak{N}_{\mathcal{Q}}(w)$. The closed neighborhood of w is $\mathfrak{N}_2[T]=T\cup\mathfrak{N}(\mathfrak{T})$.

Definition 14.2.5 ([7]). The neighborhood degree of a vertex m, indicated as $dN(m)$, is the total weight of the vertices adjacent to m; the minimum neighborhood degree of $\mathcal{Q}$ is $\delta N(\mathcal{Q})=\wedge\{dN(m):m\in W\}$; and the maximum neighborhood degree of $\mathcal{Q}$ is $\Delta N(\mathcal{Q})=\wedge\{dN(m):m\in W\}$.

Definition 14.2.6 ([19]). Let $\mathcal{Q}=(\chi,\nu)$ be an FG on a graph $\mathcal{Q}^*=(W,\mathcal{E})$. Then, $\mathfrak{F}\subseteq\mathfrak{W}$ is known as D-set in $\mathcal{Q}$, if any vertex $m\in W$ is either in $\mathfrak{F}$ or is adjacent to some vertex in $\mathfrak{F}$. The domination number of $\mathcal{Q}$ is the minimum cardinality taken over by all

D-sets in $\mathcal{Q}$ and is indicated by $\gamma(\mathcal{Q})$ or γ_f. A D-set $\mathfrak{F}$ of an FG $\mathcal{Q}$ is known as minimal D-set of $\mathcal{Q}$, if for all nodes, $w \in \mathfrak{F}, \mathfrak{F} \setminus \{\mathfrak{w}\}$ is not a D-set.

Definition 14.2.7 ([16]). A vertex $m \in W$ is termed as degree equitable in FG with a vertex $w \in W$, if $|d_{\mathcal{Q}}(m) - d_{\mathcal{Q}}(w)| \leq 1$ and $\nu(mw) \leq \chi(m) \wedge \chi(w)$.

Definition 14.2.8 ([15]). Let $\mathcal{Q} = (\chi, \nu)$ be an FG on a graph $\mathcal{Q}^* = (\mathrm{W}, \mathcal{E})$. Then $\mathfrak{F} \subset \mathfrak{W}$ is known as ED-set, if for all w $\in W \setminus \mathfrak{F}\, \exists\, a$ vertex $m \in \mathfrak{F}$ such that $mw \in \mathcal{E}(\mathcal{Q})$ and $|d_{\mathcal{Q}}(m) - d_{\mathcal{Q}}(w)| \leq 1$. The minimum cardinality of such a D-set is indicated by γ_{fe} and is termed as ED number of $\mathcal{Q}$.

Definition 14.2.9 ([19]). If a vertex $m \in W \ni |\ (d_{\mathcal{Q}}(m) - d_{\mathcal{Q}}(g)\ | \geq 2\ \forall\ g \in \mathfrak{N}(\mathrm{m})$, then m is in every ED-set and the nodes are said to be fuzzy equitable isolates (FEIs). The group of all FEI is identified as I_{fe}.

Definition 14.2.10 ([4]). Let $l \in W$. The fuzzy equitable neighborhood of l, indicated by $\mathfrak{N}_{\mathrm{fe}}(l)$, is described as $\mathfrak{N}_{\mathrm{fe}}(l) = \{w \in W{:}w \in \mathfrak{N}(\mathfrak{l}), \mathfrak{lw}$ is a strong arc and $|\ d_{\mathcal{Q}}(l) - d_{\mathcal{Q}}(w)\ |\ \leq 1\}$ and $l \in I_{\mathrm{fe}} \Leftrightarrow \mathfrak{N}_{\mathrm{fe}}(l) = \phi$. The cardinality of $\mathfrak{N}_{\mathrm{fe}}(l)$ is termed as fuzzy equitable degree of m and is indicated as $d_{\mathcal{Q}}^{\mathrm{fe}}(l)$.

Definition 14.2.11 ([18]). An ED-set $\mathfrak{F}$ is known as connected ED-set if the induced subgraph $\langle\mathfrak{F}\rangle$ is connected. The minimum cardinality of the connected ED-sets of $\mathcal{Q}$ is known as cED number and is indicated by $\gamma_{\mathrm{ce}}(\mathcal{Q})$.

Definition 14.2.12 ([18]). A D-set $\mathfrak{F}$ of an FG is a split D-set, if the induced subgraph $\langle W \setminus \mathfrak{F}\rangle$ is disconnected. The split dominating number $\gamma_{\mathrm{fs}}(\mathcal{Q})$ of an FG $\mathcal{Q}$ is the minimum cardinality of a split D-set of $\mathcal{Q}$.

Definition 14.2.13 ([18]). An ED-set $\mathfrak{F}$ of an FG $\mathcal{Q}$ is an sED-set if the induced subgraph $\langle W \setminus \mathfrak{F}\rangle$ is disconnected. The sED number γ_{fse} of an FG $\mathcal{Q}$ is the minimum cardinality of an sED-set.

Definition 14.2.14 ([17]). Let $\mathcal{Q} = (\chi, \nu)$ be a connected FG. Then, ED-set $\mathfrak{F}$ of an FG $\mathcal{Q}$ is defined as ncED-set if the induced subgraph $\langle\mathfrak{N}_{\mathrm{fe}}(\mathfrak{F})\rangle$ is connected. The minimum cardinality of ncED-set of $\mathcal{Q}$ is known as ncED number of $\mathcal{Q}$ and is indicated by γ_{fnce}.

Definition 14.2.15 ([14]). A vertex $m \in W$ of an FG is defined as degree dE with a vertex $w \in W$ if $\gcd(d_{\mathcal{Q}}(m), d_{\mathcal{Q}}(w)) \leq 1$ and $\nu(mw) \leq \chi(m) \wedge \chi(w)$.

Definition 14.2.16 ([14]). If a vertex $m \in W \ni \gcd(d_{\mathcal{Q}}(m), d_{\mathcal{Q}}(f)) \geq 2\ \forall\ w \in \mathfrak{N}(\mathrm{m})$, then m is in every dED-set and the nodes are said to be dE isolates. The group of all dE isolates is defined as I_{de}.

Definition 14.2.17 ([14]). For $\in W$, the divisor equitable neighborhood of l is indicated by $\mathfrak{N}_{\text{fde}}(l)$ and is defined as $\mathfrak{N}_{\text{fde}}(l) = \{w \in W : w \in \mathfrak{N}(\mathfrak{l}), (\mathfrak{l}, \mathfrak{w})$ is a strong arc and $\gcd(d_{\mathcal{Q}}(l), d_{\mathcal{Q}}(w)) \leq 1\}$ and $l \in I_{\text{fde}} \Leftrightarrow \mathfrak{N}_{\text{fde}}(l) = \phi$. The cardinality of $\mathfrak{N}_{\text{fde}}(l)$ is indicated by $d_{\mathcal{Q}}^{\text{fde}}(l)$.

Definition 14.2.18 ([14]). Let $\mathcal{Q} = (\chi, \nu)$ be an FG on a graph $\mathcal{Q}^* = (W, \mathcal{E})$. Then, $\mathfrak{F} \subset \mathfrak{W}$ is called a dED-set if $\forall\, w \in W \setminus \mathfrak{F}\, \exists$ a vertex $m \in \mathfrak{F} \ni m\mathfrak{w} \in \mathcal{E}(\mathcal{Q})$ and $\gcd(d_{\mathcal{Q}}(m), d_{\mathcal{Q}}(w)) \leq 1$. The minimum cardinality of such a D-set is indicated by γ_{fde} and is known as dED number of $\mathcal{Q}$.

Example 14.2.1. Consider $\mathcal{Q} = (\chi, \nu)$ as FG on a graph $\mathcal{Q}^* = (W, \mathcal{E})$ and described as shown in Figure 14.1.

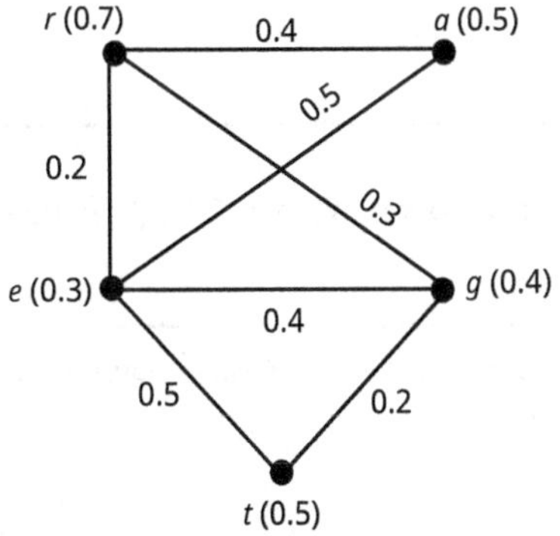

Figure 14.1: Example of a dED-set.

Here, $F = \{e\}$ is a dED-set for some vertices q, t and $\gcd(d_{\mathcal{Q}}(q), d_{\mathcal{Q}}(f)) = 0.1 < 1$.

14.3 Main Results

Definition 14.3.1. A dED-set $\mathfrak{F}$ of an FG$\mathcal{Q}$ is an sdED-set if the induced subgraph $\langle W \setminus \mathfrak{F} \rangle$ is disconnected. The sdED number γ_{fsde} of an FG $\mathcal{Q}$ is the minimum cardinality of an sdED set.

Example 14.3.1. Consider $\mathcal{Q} = (\chi, \nu)$ as FG on a graph $\mathcal{Q}^* = (W, \mathcal{E})$ and described as shown in Figure 14.2.

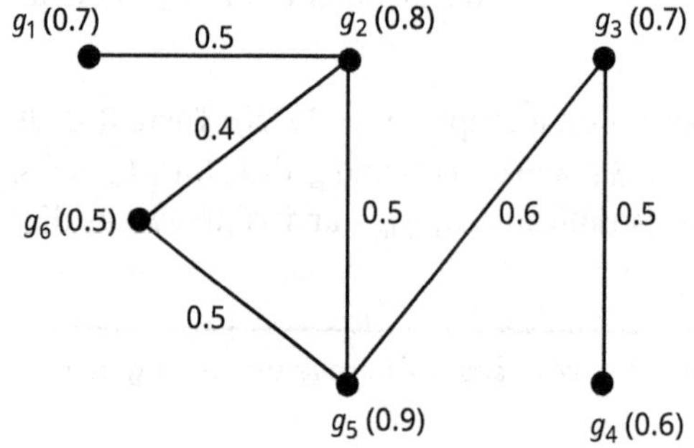

Figure 14.2: Example of both sED-set and sdED-set.

Here, $\mathfrak{F} = \{\mathfrak{g}_2, \mathfrak{g}_3\}$ is dED-set, which is both sED-set and sdED-set, where $\langle W \setminus \mathfrak{F} \rangle$ is disconnected.

Example 14.3.2. Consider $\mathcal{Q} = (\chi, \nu)$ as FG on a graph $\mathcal{Q}^* = (W, \mathcal{E})$ and described as shown in Figure 14.3.

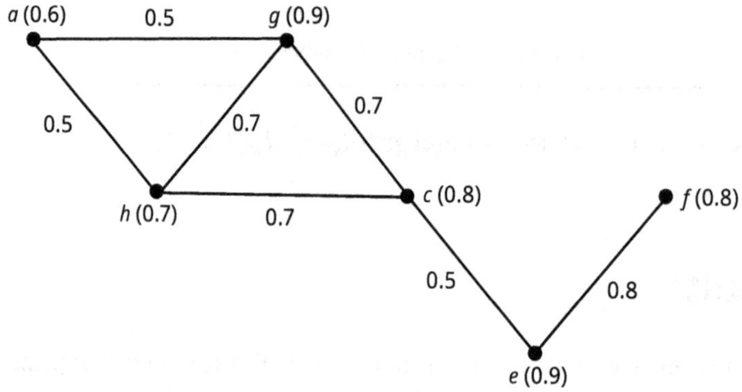

Figure 14.3: Example of sED-set and not sdED-set.

Here, $\mathfrak{F} = \{\mathfrak{h}, \mathfrak{e}\}$ is ED-set but not dED-set, which is sED-set and not sdED-set.

Proposition 14.3.1. For any FG$\mathcal{Q} = (\chi, \nu)$,

(i) $\gamma_{\text{fde}}(\mathcal{Q}) \leq \gamma_{\text{fsde}}$.

(ii) $\gamma_{\text{fs}}(\mathcal{Q}) \leq \gamma_{\text{fsde}}$.

Proof. (i) Consider $\mathfrak{F}$ as the minimum sdED-set of $\mathcal{Q}$. As $\mathfrak{F}$ is an sdED-set, then $\mathfrak{F}$ is dED-set. Thus, $\gamma_{\text{fde}}(\mathcal{Q}) \leq |\mathfrak{F}| = \gamma_{\text{fsde}}$.

(ii) The proof is obvious.

Theorem 14.3.1. An sdED-set $\mathfrak{F}$ of $\mathcal{Q}$ is minimal if each vertex $t \in \mathfrak{F}$ satisfies any one of the below three statements:

(i) $\exists$ vertex $h \in W \setminus \mathfrak{F} \ni \mathfrak{N}_{\text{fde}}(\mathfrak{h}) \cap \mathfrak{F} = \{\mathfrak{t}\}$.
(ii) t is a divisor equitable isolated vertex in $\langle \mathfrak{F} \rangle$.
(iii) $\langle W \setminus \mathfrak{F} \cup \{\mathfrak{t}\} \rangle$ is disconnected.

Proof. Assume that $\mathfrak{F}$ is minimum and there is a vertex $t \in \mathfrak{F}$ such that t satisfies none of the listed conditions. Then, by the statements (i) and (ii), $\mathfrak{F}_1 = \mathfrak{F} \setminus \{\mathfrak{t}\}$ is a dED-set of $\mathcal{Q}$. As per (iii), $\langle W \setminus \mathfrak{F} \rangle$ is disconnected. This indicates that $\mathfrak{F}_1$ is an sdED-set of $\mathcal{Q}$, a conflict.

Definition 14.3.2. Let $\mathcal{Q} = (\chi, \nu)$ be an FG on a graph $\mathcal{Q}^* = (W, \mathcal{E})$. If $d_{\mathcal{Q}}(b) = r$ for all $b \in W$, then $\mathcal{Q}$ is defined as regular FG of degree r or r-regular FG.

Definition 14.3.3. An FG $\mathcal{Q} = (\chi, \nu)$ on a graph $\mathcal{Q}^* = (W, \mathcal{E})$ is called as biregular FG with regularity t or $t+1$ if the degree of any vertex in W is either t or $t+1$.

Theorem 14.3.2 ([12]). Let $\mathcal{Q} = (\chi, \nu)$ be an FG on a graph $\mathcal{Q}^* = (W, \mathcal{E})$. If $\mathcal{Q}$ is r-regular FG or $(r, r+1)$ biregular FG for some $r \leq 1$, then $\gamma_{\text{fde}} = \gamma_f$.

Theorem 14.3.3. Let $\mathcal{Q} = (\chi, \nu)$ be an FG of order p. Then, $\gamma_{\text{fde}}(\mathcal{Q}) \leq \gamma_{\text{fsde}}(\mathcal{Q}) \leq p - \Delta_{\text{fde}}(\mathcal{Q})$.

Proof. Every fuzzy sdED-set is a dED-set of $\mathcal{Q}$, $\gamma_{\text{fs}}(\mathcal{Q}) \leq \gamma_{\text{fsde}}(\mathcal{Q})$. Consider$h, t \in W$. If $d_{\mathcal{Q}}^{\text{fde}}(h) = \Delta_{\text{fde}}(\mathcal{Q})$ and $d_{\mathcal{Q}}^{\text{fde}}(t) = \delta_{\text{fde}}(\mathcal{Q})$. Clearly, $W \setminus \mathfrak{N}_{\text{fde}}(h)$ is a fuzzy sdED-set. Hence, $\gamma_{\text{fsde}}(\mathcal{Q}) \leq |W \setminus \mathfrak{N}_{\text{fde}}(h)|$, that is $\gamma_{\text{fsde}}(\mathcal{Q}) \leq p - \Delta_{\text{fde}}(\mathcal{Q})$.

Theorem 14.3.4. If $\mathfrak{F}$ is an sdED-set of an FG$\mathcal{Q}$, then $\mathfrak{F}$ is both a minimal sdED-set and a maximal sdED-set. Conversely, any maximal sdED-set $\mathfrak{F}$ in $\mathcal{Q}$ is an sdED-set of $\mathcal{Q}$.

Proof. If $\mathfrak{F}$ is an sdED-set of $\mathcal{Q}$, then $\mathfrak{F}_1 = \mathfrak{F} \setminus \{\mathfrak{g}\}$ is not an sdED-set for every $g \in \mathfrak{F}$ and $\mathfrak{F} \cup \{\mathfrak{t}\}$ is not an sdED-set $\forall\ t \notin \mathfrak{F}$; thus $\mathfrak{F}$ is a minimal dED-set and a maximal sdED-set.

Conversely, consider $\mathfrak{F}$ as maximal sdED-set in $\mathcal{Q}$. Then, for every $t \in W \setminus \mathfrak{F}, \mathfrak{F} \setminus \{\mathfrak{t}\}$ is not sdED-set and t is dominated by any element of $\mathfrak{F}$. Hence, $\mathfrak{F}$ is an sdED-set of $\mathcal{Q}$.

Theorem 14.3.5. Let $\mathcal{Q} = (\chi, \nu)$ be an FG. Then, $\gamma_{\text{fsde}}(\mathcal{Q}) \leq p - \Delta_{\text{fsde}}(\mathcal{Q})$.

Proof. Consider t as a vertex such that $d_{\mathcal{Q}}^{\text{fsde}}(t) = \Delta_{\text{fsde}}$. Then, $W \setminus \mathfrak{N}_{\text{fsde}}(t)$ is a fuzzy sdED-set of $\mathcal{Q}$ so that $\gamma_{\text{fsde}}(\mathcal{Q}) \leq |W \setminus \mathfrak{N}_{\text{fsde}}(t)| = p - \Delta_{\text{fsde}}(\mathcal{Q})$.

14.4 Split Equitable Domination in Computer Networks

Split equitable domination in computer networks optimizes resource allocation and system performance through balanced distribution of control and workload across multiple nodes. This mathematical framework ensures efficient network management through strategic cluster formation, dynamic traffic routing, and equitable load distribution. By dividing networks into balanced clusters and preventing single-node dominance, this approach enhances system reliability, enables efficient fault containment, and minimizes disruptions during maintenance or failures. The result is a robust network architecture that maintains stability while ensuring fair resource allocation, improved security, and simplified system management across all components.

In addition, in a fuzzy graph used for reinforcement learning within a networked environment, split divisor equitable domination enhances coordination, scalability, and robustness. The network is modeled as a fuzzy graph, where edges represent the strength of interactions between agents, with values between 0 and 1. The agents are divided into two sets: leaders (dominators) and followers. The split creates clear role differentiation, where leaders manage global strategy and coordination, while followers focus on local tasks. The divisor aspect ensures that the domination responsibility is evenly distributed among leaders, preventing any single leader from becoming overloaded. Equitable domination guarantees that each follower is connected to an adequate number of leaders, ensuring a balanced and fair distribution of influence. This organization reduces communication overhead, as only leaders need to interact with multiple agents, while followers communicate locally with their assigned leaders. The use of fuzzy edges allows the system to adapt to uncertainty in agent interactions, enhancing fault tolerance and flexibility. In summary, this approach improves learning efficiency, scalability, and robustness, making it highly effective for large-scale multi-agent systems in dynamic environments.

Computer networks represent a complex infrastructure of interconnected components that drive modern business operations. The basic architecture includes computer systems, LANs, hubs, servers, bridges, routers, switches, and gateways, which can be mathematically represented as a graph structure. Each component serves a specific purpose: hubs connect systems within LANs, bridges link separate LANs, switches provide multiple network connection ports, and servers manage individual network resources. These interconnected systems form the foundation of modern business operations, enabling efficient information exchange, real-time data sharing, and multimedia transmission. The infrastructure supports seamless communication among employees, suppliers, and customers while accommodating numerous simultaneous users across organizations. The network infrastructure has transformed workplace dynamics by promoting instant collaboration across geographical boundaries. Through centralized data storage, virtual meetings, and screen sharing capabilities, teams can work together seamlessly, regardless of physical location. This integrated approach optimizes resource utilization, prevents redundant work efforts, and enhances overall organizational productivity through efficient communication and collaboration tools.

In modern network architectures, multiple servers operate as an integrated system to manage essential services and data flow. This interconnected structure ensures system reliability through redundancy, enables balanced workload distribution, and maintains efficient resource allocation across all servers. An FG model of the scenario:

FG: Consider FG $\mathcal{Q} = (\chi, \nu, \mu)$, where χ is the set of nodes in the network, ν represents edges representing connections between nodes, and $\mu(f, q)$ is the fuzzy membership value representing the bandwidth or quality of the connection between node f and node q.

We measure the actual bandwidth between two systems using routers and tools like iperf, netstat, or ping. We then convert the measured bandwidth values into fuzzy values based on our defined sets.

Divisor Equitable Domination in FG: To maintain a balanced high load distribution among divisor equitable dominating nodes, verify that the load gcd (greatest common divisor) between any two nodes is no more than 1, expressed as $\gcd(d_{\mathcal{Q}}(f), d_{\mathcal{Q}}(q)) \leq 1$.

Split Divisor Equitable Domination in FG: A split divisor equitable dominating set in an FG is a minimal set of vertices that divisor equitably dominates all other vertices and disconnects the graph when removed. In this case, we can determine the degree of each vertex and ensure divisor equitable adjacency between vertices. To ensure a balanced load distribution among dominant nodes, it is important to verify that the load divisor equitable difference between any two nodes does not exceed one, which can be expressed as $\gcd(d_{\mathcal{Q}}(f), d_{\mathcal{Q}}(q)) \leq 1$.

To ensure failure tolerance in the server network, it is crucial to identify divisor split equitable dominating sets among the server clusters (which has high degree value). In this scenario, modern server networks consist of four main server clusters, each handling specialized critical tasks. These clusters are interconnected through bridge servers that enable seamless data transfer and inter-cluster communication. The architecture includes high availability clusters that ensure continuous operation through redundant hardware, load balancing clusters that distribute user requests across multiple nodes, high-performance clusters that handle resource-intensive workloads, and clustered storage systems that provide shared storage access. Bridge servers act as intermediaries between these clusters, facilitating efficient communication and data exchange while maintaining system reliability and performance.

In a balanced network architecture, servers receive equitable resource allocation, although certain bridge servers with elevated membership values may not meet the required divisor equitability standards, marking them as potential failure risks. The system implements strategic isolation protocols for critical bridge servers during disruptions or failures. This approach creates controlled network segments that contain problems within specific clusters, protecting the overall network integrity while maintaining core operations.

This approach allows for effective problem containment through strategic cluster isolation. Network administrators can focus on repairing and restoring affected segments while maintaining normal operations in other clusters, thus preventing error propagation across interconnected systems and minimizing overall service disruption.

14.4.1 Illustration

Split equitable domination uses FG theory to manage information flow among HR, IT, and sales departments, ensuring balanced data access and control. This approach enables nuanced relationships and gradations of control, facilitating efficient data distribution and minimizing risks. It helps organizations establish a coherent information management framework that optimizes collaboration and maintains data integrity and security.

An FG consists of nodes representing departments and edges, with weights indicating the level of access between them. The graph is divided into three clusters: HR department, sales department, and IT department. HR needs to share employee performance data with Sales for a new sales strategy, while ensuring confidentiality. Sales needs access to specific technical specifications from IT to tailor their pitches, without full access to IT documents. HR can provide Sales with a report containing aggregated performance data, following access guidelines. Sales can request relevant HR documents for the project, based on access rules. IT can share technical specifications with Sales at an appropriate access level, while limiting access to HR documents.

Let us create a network using FG, $\mathcal{Q} = (\chi, \nu)$, made up of nodes that represent departments and edges with weights that show the level of access between them. The graph is divided into three clusters: HR department, sales department, and IT department.

In Figure 14.4, HR department consists of vertices $\{h_1, h_2\}$, sales department consists of vertices $\{s_1, s_2, s_3, s_4\}$, and IT department consists of vertices $\{I_1, I_2, I_3\}$. Here, fuzzy divisor equitable dominating set $\mathfrak{F} = \{\mathrm{m}\}$ has a high-degree divisor equitable server and acts as a split divisor equitable dominating server, connecting clusters. Removing the server $\{m\}$ would split the network into separate clusters, as illustrated in Figure 14.4.

By utilizing an FG and split domination, departments can collaborate effectively, while maintaining data security. Each department understands its access rights, minimizing the risk of unauthorized sharing. This model allows for necessary information sharing based on established access levels, ensuring sensitive data remains protected.

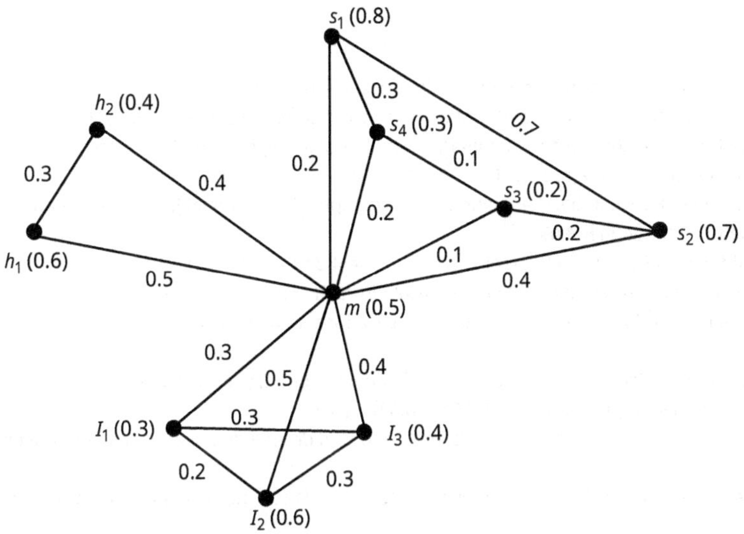

Figure 14.4: Example of an sdED-set.

The divisor concept in network resource management and load distribution helps ensure that tasks, resources, and data are allocated evenly and fairly across multiple nodes in a network. By using division and modulus operations, systems can optimize performance, ensure fault tolerance, and scale dynamically without overloading any single node. This results in more efficient use of resources and a more robust, scalable system. In load balancing, the total load or requests (e.g., HR department, sales department, and IT department tasks) are evenly distributed among available nodes or servers. Divisibility helps ensure a fair distribution and avoids overloading specific nodes.

14.5 Conclusion

Understanding the concept of GCD is crucial in various fields like social networks, planning, estimation, and division. It plays a key role in problem-solving, critical thinking, and research, as well as in computer security. Real-life applications split divisor equitable domination in computer server networks optimizes resource allocation during maintenance or troubleshooting by containing the impact within a specific cluster. This method enhances fault tolerance, simplifies maintenance, and ensures smoother operations in case of server failures or network disruptions. In this article, we investigated split divisor equitable domination in FGs, focused on the characteristics of minimal split divisor equitable domination sets. We also examined the relationship between minimal and maximum split divisor equitable dominating sets.

References

[1] Berge, C, Theory of Graphs and its Applications. Methuen, London, 1962.

[2] Cockayne EJ, Hedetniemi ST. Towards a theory of domination in graphs. Networks. 1977;7:247–261.

[3] Chelvam T, Chellathurai S, A note on split domination number of a graph, Journal of Discrete Mathematical Sciences and Cryptography, 2009;12(2):179–186.

[4] Dharmalingam KM, Rani M. Equitable domination in fuzzy graphs. International Journal of Pure and Applied Mathematics. 2014;94(5):661–667.

[5] Dharmalingam KM, Suriya RU, Nithya P. Excellent-just excellent degree equitable domination in fuzzy graphs. Bulletin of the International Mathematical Virtual Institute. 2015;5:81–88.

[6] Gross JL, Yellen J, Anderson M. Graph Theory and Its Applications. Chapman and Hall/CRC; 2018 Nov 5.

[7] Gurubaran, C, Prasanna, A, Ismayil, A. M, 2-equitable domination in fuzzy graphs, International Journal of Applied Engineering Research. 2018;13(16):12531–12535.

[8] Jamir S, Kekre P. Split domination number and its applications in diverse domains. African Journal of Biological Sciences. 2024;6:544–552.

[9] Kauffmann A. Introduction to the Theory of Fuzzy Sets. Academic Press, Inc., Orlando Florida. 1973;1 (1):1–79.

[10] Kumar TA, Kumar KS, Jaisiva S. IoT based leukemia detection using Fuzzy C-means clustering Technique. Global Journal on Innovation, Opportunities and Challenges in AAI and Machine Learning-. 2022;6(1).

[11] Kulli VR, Janakiram B, The split domination number of a graph. Graph Theory Notes of NewYork. Newyork Academy of sciences XXXII, 1997;32:16–19.

[12] Karbasioun A, Ameri R. Total and secure domination for corona product of two fuzzy soft graphs, Fuzzy Information and Engineering. 2021;13(1):127–138.

[13] Ore O. Theory of graphs. American Mathematical Society Colloquium Publications. 1962;38.

[14] Priyanka GB. Catherine Grace John J, Xavier P. Divisor equitable domination in fuzzy graphs. TWMS Journal of Applied and Engineering Mathematics. 2025;15(1):256–267.

[15] Rana A. A survey on the domination of fuzzy graphs. Discrete Mathematics, Algorithms and Applications, 2021;13(1):2130001.

[16] Rani M, Dharmalingam KM. Equitable associate fuzzy graph of a fuzzy graph. Journal of Mathematics and Computer Science. 2016;1:1–7.

[17] Revathi, S, Hari Narayanan, C. V. R, Muthuraj R, Neighborhood connected equitable domination in fuzzy graphs. International Journal of Computational and Applied Mathematics. 2017;12(1):130–136.

[18] Revathi S, Hari Narayanan CVR, Muthuraj R. The split equitable domination number of a fuzzy graph. International Journal of Mathematics Research. 2016;8(2): 93–98.

[19] Revathi S, Harinarayanan CVR. Equitable domination in fuzzy graphs. International Journal of Engineering Research and Applications. 2014;4(6):80–83.

[20] Rosenfeld A. Fuzzy Graphs, Fuzzy Sets and their Applications to Cognitive and Decision Process. New York: Academic press; 1975 (pp. 75–95).

[21] Somasundaram A, Somasundaram S. Domination in fuzzy graphs- I. Pattern Recognition letters. 1998;19:787–791.

[22] Swaminathan V, Dharmalingam KM. Degree equitable domination on graphs. Kragujevak Journal of Mathematics, 2011;35:191–197.

[23] Wilder R. American Mathematical Society Colloquium Publications. Topology of Manifolds. 32.

[24] Zadeh LA. Fuzzy sets. Information and Control. 1965;8:338–353.

[25] Zimmermann HJ. Fuzzy Set Theory - And Its Applications. Springer Science & Business Media; 2011 Jun 27.

M. Babila Revathy, J. Kavitha, P. Arockia Jansi Rani, R. Roselinkiruba

15 Optimizing Image Preprocessing Methods for Crop Disease Detection in Precision Agriculture

Abstract: Precision agriculture is a new idea about farming management that is revolutionizing farming in this scientifically advanced modern society. It focuses on inputs in precisely the right amounts to obtain higher average yields than traditional agriculture. Hence, a highly efficient vision-based technology holds a very important position in precision farming. Images are considered as a primary source of data in this vision-based technology. This technology can read the pictures only if they are properly processed and analyzed. Today, different methods from artificial neural networks such as deep learning are found in image processing. The greatest benefit of this technology is that it can provide good information about crops without it ever having to come into contact with it. Since deep learning is efficient in handling large and complex data, precision agriculture is also integrated with deep learning to improve the efficiency.

Preprocessing of images is the most important step in the dataset preparation. Image processing techniques like scaling, removal of backgrounds, picture enhancement, denoising, etc. can be applied on the data before feeding the input to the deep learning model. This helps deep learning architecture better extract the features related to the prediction variable. Thus, an adequate preprocessing technique must be performed before the data can be transmitted to the analysis phase. Hence, this section seeks to explore and determine the preprocessing techniques applied in deep learning to detect diseased crops in agriculture. Some of the preprocessing techniques to be compared in this chapter include Hough transform, Otsu adaptive thresholding, NDVI (normalized difference vegetation index), simple linear iterative clustering, Gaussian blur, and the Histogram Equalization algorithm, among others. Finally, this chapter is concluded by highlighting the challenges in the preprocessing techniques and providing future research direction.

Keywords: Deep learning, precision agriculture, preprocessing techniques, Hough transform, NDVI, Otsu adaptive thresholding, SLIC, Gaussian blur, histogram equalization, ResNet-50

15.1 Introduction

Agriculture has a vibrant role in ensuring comprehensive food security and supporting economic stability, importantly as the world's population continues to grow. However, farmers meet various challenges, including crop diseases, pest infestations, and

https://doi.org/10.1515/9783112214374-015

the effects of climate change, which can meaningfully hinder crop yield and quality. To overcome these challenges, there has been an increasing confidence on using advanced technologies, particularly image processing and deep learning, to boost traditional agricultural practices. Nowadays, image data has become a very important asset in supervising crop health, detecting diseases, and evaluating the environmental conditions.

The enhancement in the potency of the deep learning algorithm depends on efficient image preprocessing techniques. It improves image quality and aids in mining meaningful information from raw data. In order to prepare images for analysis, certain practices such as noise abatement, contrast advancement, and segmentation are generally employed. By enhancing the data quality, these preprocessing methods can considerably increase the precision and proficiency of disease detection and classification models, thereby fostering agricultural decision-making. The incorporation of these technologies into farming practices not only endorses better resource management but also contributes to sustainable agricultural practices, eventually leading to amplified productivity and reduced environmental impact. As research in this area continues to expand, exploring advanced preprocessing techniques and their applications in precision agriculture remains a promising avenue for improving crop management and food production.

15.2 Literature Review

Wu Lulu et al. [1] prompted a novel method to notice the disease spots of varying sizes using the edge detection and new Hough transform method .He experimented with 90 crop disease images and came up with an accuracy of 87.01%. Durai et al. [2] introduced a multitier deep learning framework that exploits a convolutional neural network (CNN) alongside ResNet50 for image segmentation to effectively identify early and late blight diseases in potato leaves, tackling the complexities posed by variations in crop species and environmental conditions. The proposed model accomplishes an impressive exactness of 99.75% on a potato leaf disease dataset, significantly surpassing the act of current leading models in terms of both accuracy and computational efficiency. Epinat et al. [3] insisted that applying a wavelet transform to NDVI (normalized difference vegetation index) images has demonstrated effectiveness in quantifying spatial patterns across various resolution levels and orientations. This method allows for a detailed analysis of historical land management practices, which can be clearly identified at finer resolution levels. Additionally, the wavelet approach can highlight patterns that indicate natural soil changeability or historical reinforcement, which are particularly relevant for precision agriculture. Such features become more obvious when analyzed at coarser resolution levels, providing valuable insights into agricultural practices and land use changes. Rajendra Sishodia et al. [4] specified that agriculture is essential for providing food and fiber, and has evolved significantly over the last century, particularly during the Green Revolution. However, increasing food demands and population growth strain natural resources, highlighting the need for sustainable agricultural practices. Emerging technologies, including geospatial tools, IoT, big data, and artificial intelligence, play a cornerstone in precision agriculture (PA), optimizing inputs and enhancing crop production while minimizing environmental impacts. Remote sensing technologies have seen rapid adoption in PA, supported by high-resolution satellite imagery, enabling efficient applications in crop monitoring, irrigation, nutrient management, and yield prediction. Recent advancements underscore the need for user-friendly systems to facilitate real-time remote sensing applications, promoting broader adoption in various agricultural contexts.

Mavridou Efthimia et al. [5] specified a top hat filter and a Gaussian function to augment the contrast of the image during the preprocessing phase. J. Illingworth. [6] clearly explained the workings of the Hough transform. Dhaware and Wanjale [7] discussed various techniques for segmentation, feature extraction, and plant disease classification. Ji et al. [8], in his approach for histogram equalization, concluded that his proposed method enhances luminance. He Kaiming et al. [9] conducted a comparative analysis between plain networks and residual networks, discovering that residual networks exhibited a lower error rate. Furthermore, they proposed that deeper ResNets, specifically the 50, 101, and 152-layer configurations, achieve greater accuracy compared to the 34-layer models by a substantial margin. This observation indicates the absence of the degradation problem, allowing for significant improvements in accuracy as the depth of the network increases, so the datasets were also made feature extracted by Resnet 50 and the results were evaluated.

15.3 Overview of Deep Learning for Crop Disease Detection

Agriculture is an important economic sector in many countries. Although industrialization has flourished in many countries, agriculture continues to emerge as a vital sector, providing food for the global population. Different kinds of agricultural methods are followed for different crops and different climatic conditions. So, the farmers have to struggle to save the crop and yield. Nowadays, due to much technological development in machine learning and deep learning, agriculture has stepped into a different and revolutionizing environment, and is nothing but precision agriculture. It is a method of practicing agriculture with the help of information provided by a machine. Based on the given input information, the machine should have the capability to predict the output. The information provided as input for training the machines about farming conditions must be exceptionally clear to mitigate potential losses. One of the major problems in agriculture is the attack of diseases. Many farmers are still facing this worst-ever problem and are unable to find a solution for this. The possibility of disease attack may be at any stage, from the seedling or planting stage to and after even the cultivation stage. The vulnerability can be stopped only if the disease is found out at the right stage. Nowadays, many robots and drones have been used to spray pesticides or medicines at the right place and at the right time. These methods should be cost-effective, time-efficient, and environmentally sustainable, ensuring that farmers can maximize yield while minimizing waste and resource usage. So, this work focuses on classifying healthy leaves and diseased leaves in various crops using an efficient classifier, ResNet50. ResNet-50 is a CNN that excels in picture classification and is a part of the Residual Networks family of models. Here is the workflow diagram (Figure 15.1), which clearly illustrates the mechanisms presented in the chapter.

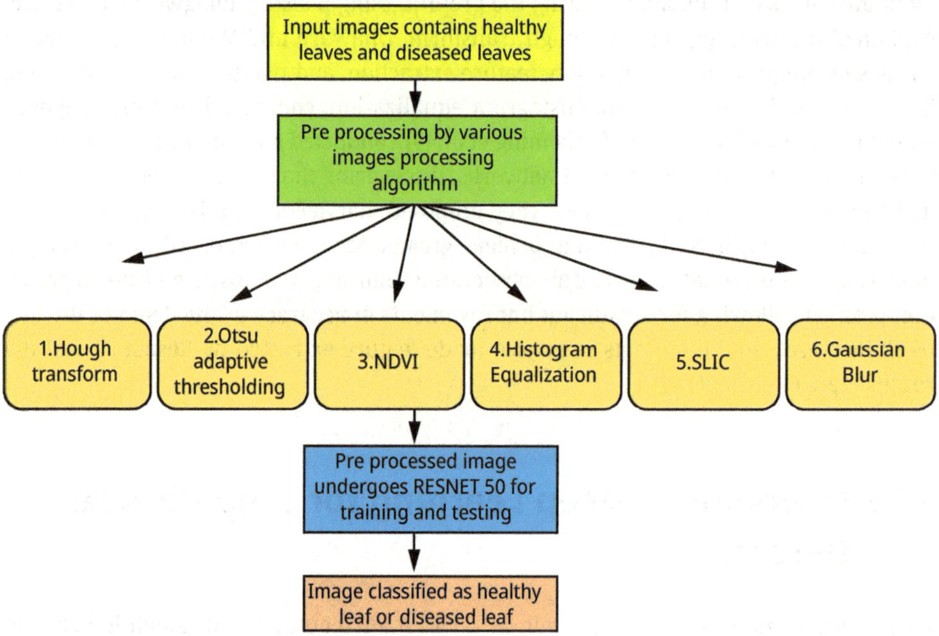

Figure 15.1: Overview of the work.

15.4 Understanding Image Processing Algorithms in the Context of Deep Learning

Image processing algorithms are deployed here to preprocess the input sent to a deep learning algorithm for further processing. This helps remove undesirable data and lessen the computation time during the training and testing of models. It is one of the challenging and time-consuming parts but it is a significant part. It is a process of cleaning the raw data. This cleaning reduces the complexity of learning the data at the time of training and testing by many machine learning and deep learning paradigms. Here are some of the steps in preprocessing, illustrated in Figure 15.2.

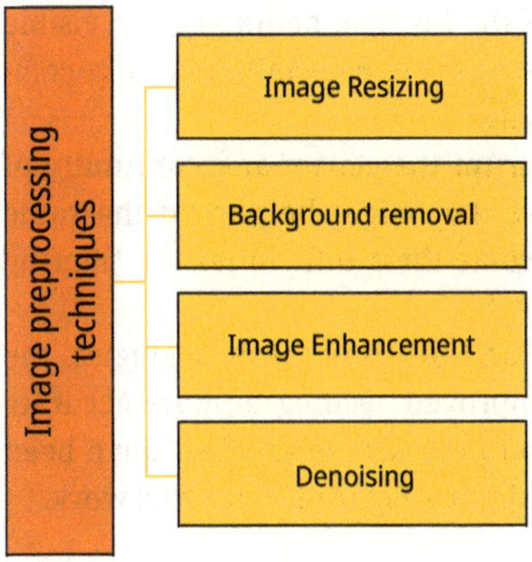

Figure 15.2: Some of the preprocessing techniques.

The dataset selected for this experimentation is the plant doc dataset, a dataset for visual plant disease detection. This dataset contains 2,598 data points collected across 13 plant species and 17 classes of diseases. Initially, these images are resized to 128×128 or 256×256. Next, the background is removed by various image processing algorithms like Otsu, color index-based thresholding algorithm, or grab cut; deep learning algorithm like U Net can also be used. The extracted foreground part is taken for feature extraction using Hough transform, NDVI, SLIC, Gaussian blur, or histogram equalization, and finally the performance is analyzed. For this experimentation, I have chosen images from the dataset, which were correctly showing the crop part by excluding the background regions. The characteristic features of these algorithms are explained below.

15.4.1 Hough Transform

This is a critical technique in image processing that aids in locating boundaries by addressing the discontinuities in image data. A significant challenge in boundary detection is determining which edges in an image correspond to the desired boundary.

The process of boundary detection using the Hough transform involves two fundamental steps:

Edge Detection: This step focuses on identifying the missing edges in the image.

Edge Linking: This process connects the newly detected edges with the adjacent edges, forming a consistent boundary.

15.4.1.1 Challenges in Edge Linking

Presence of Extraneous Data: In many cases, extraneous data points can create misunderstanding about which points should be considered as part of the boundary. These inappropriate points can obscure the true edges.

Incomplete Data: Frequently, only portions of the favorite boundary are visible and it leads to incomplete data. This situation obscures the determination of a complete boundary model, as some pieces may be missing.

Presence of Noise: Noise in the image can disrupt the clarity and continuity of edges, making it difficult to accurately link edges that might be part of the same boundary. Some of the proposed methods to overcome these difficulties are filtering techniques, interpolation method, and model fitting.

By effectively applying the Hough transform and utilizing these strategies, the boundary detection process can be significantly improved, leading to more accurate results in image analysis. The above challenges and their solving method have been explained through an example given below. Let us discuss Figure 15.3 for this work.

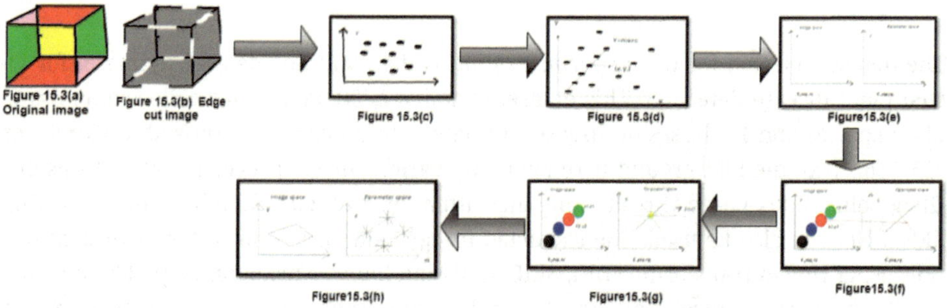

Figure 15.3: Line detection.

The Hough transform aims to find the straight line for the given edge points. Consider the edge points (x_i, y_i) for the image in Figure 15.3. There are a bunch of edges in an image represented as a circle in Figure 15.3(c). Consider a point (x_i, y_i), for which the straight line equation is $y_i = mx_i + c$, shown in Figure 15.3(d). The next step is the implementation of the Hough transformation. A parameter space will be created for each point in an image space. Here, the first one is the image space with *xy* space and the next one is the parameter space that is shown in Figure 15.3(e). For each point in a straight line of the *xy* space, a line will be created in the *mc* space, shown in Figure 15.3(f). Similarly, for all other points in the straight line of the *xy* space, corresponding lines will be created in the *mc* space by having a common intersection, which is shown in Figure 15.3(g). Likewise, for all edge points in Figure 15.3(f), multiple straight lines will be formed in the *xy* space and for each straight line, many lines will be formed with a common intersection point.

15.4.1.2 Line Detection Algorithm

Step 15.4.3.1. Calculate parameter space (m, c).
Step 15.4.3.2. Create accumulator array $A(m,c)$ as shown in Figure 15.4.

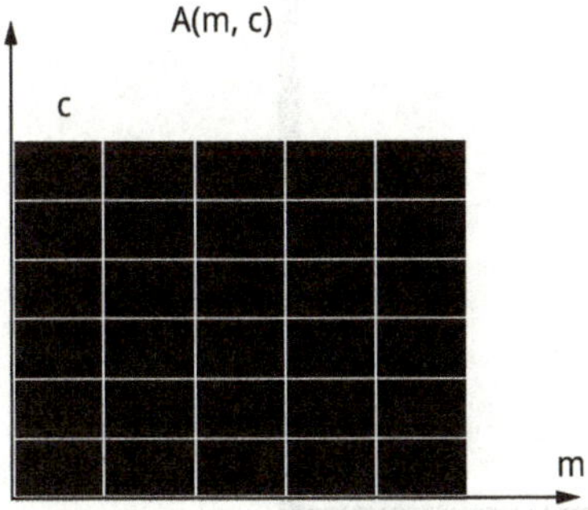

Figure 15.4: Accumulator array.

Step 15.4.3.3. Let us consider $A(m,c) = 0$ for all (m,c).
Step 15.4.3.4 For each distinct edge point (x_i, y_i), $A(m,c) = A(m,c) + 1$
Step 15.4.3.5 If (m,c) resides on the line $c = -mx_i + y_i$
Step 15.4.3.6. Find local maxima in $A(m,c)$

Issues: If m ranges from $-\infty$ to $+\infty$, it needs a large accumulator and more time computation. So, for better parametrization, the solution is to use

$$x\sin\theta - y\sin\theta + \rho = 0 \tag{15.1}$$

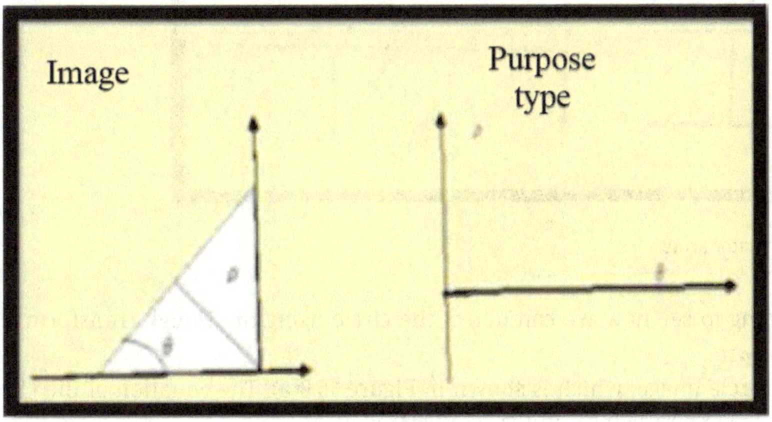

Figure 15.5: Image space and parameter space.

For a single point in image space a sinusoidal wave will be formed in the parameter space and is shown in Figure 15.5.

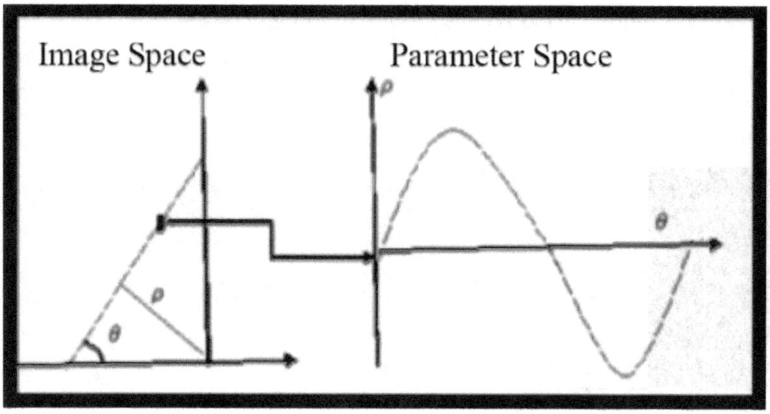

Figure 15.6: Sinusoidal wave.

Similarly, for other points, many sinusoidal waves are formed, respectively, in the parameter space, as shown in Figure 15.6. Then, count the peaks in the accumulator array. After that, increment the patch in the accumulator array rather than a single point. It is shown in Figures 15.7 and 15.8.

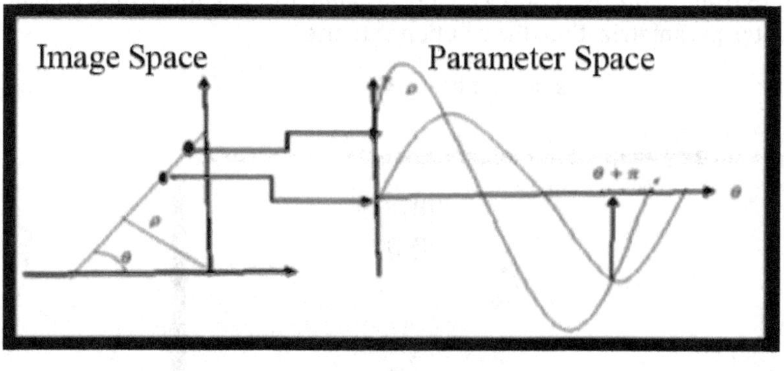

Figure 15.7: Accumulator array.

Next, we are going to see how we can detect the circle using the Hough transform if the radius is known.

Consider a circle image, which is shown in Figure 15.9(a). The equation of the circle is computed for Figure 15.9(b), and is represented as

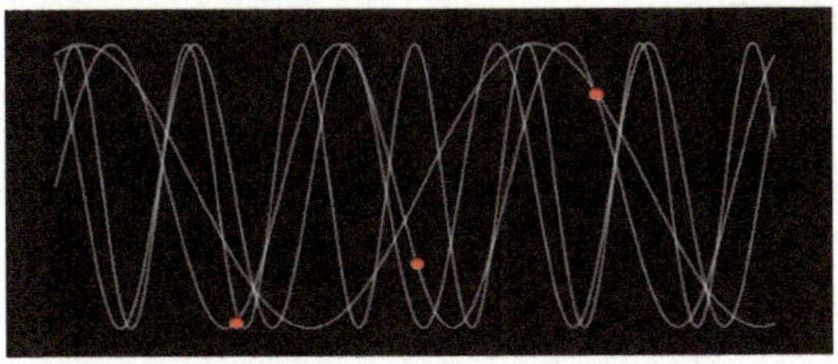

Figure 15.8: Accumulator array expanded.

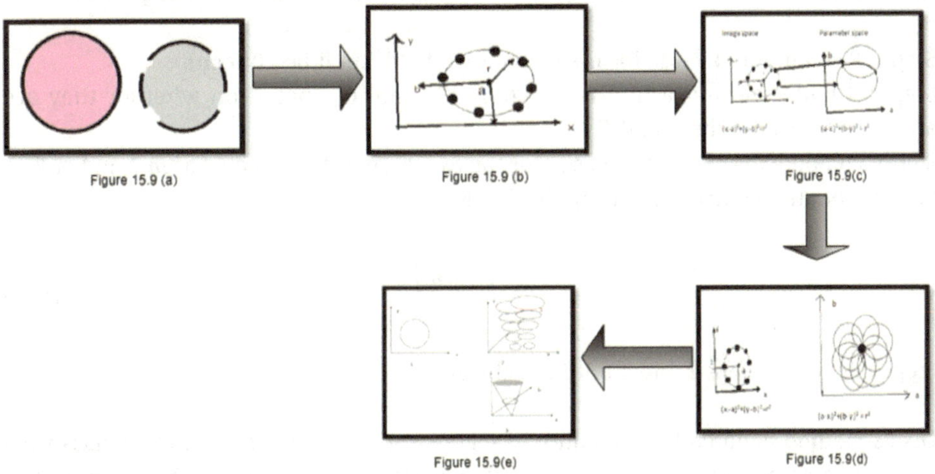

Figure 15.9: Hough transform for circle detection.

$$(x_i - a)^2 + (y_i - b)^2 = r^2 \tag{15.2}$$

In this context, points a and b symbolize the center of the circle. r signifies the radius of the circle and x, y correspond to the circumference of the circle. As r is known, the accumulator array is written as $A(a, b)$. Each point in the circle from the image space has been mapped to the circle in the parameter space. Similarly, for all other points, a corresponding circle will be formed with a common intersection point and it is shown in Figure 15.9(d). It is the a and b values that all points in the image space are going to share. In moving the ab plane, points on the circumference correspond to ab and the center point corresponds to the xy plane, as shown in Figure 15.9(d). At a point, all the circles will intersect and that is the actual a, b location that we want to find. Practically, image pixel value indicates the number of intersections. Next, let us see if r is unknown, how a circle could be formed. Here, we are going to put different values for r and plot the circle. It forms a cone shape in the parametric space. At one point, it is going to have a maximum intersection and it precisely tells the values of a,b and r. It is shown in Figure 15.9(e).

15.4.2 Otsu Adaptive Thresholding

Otsu technique involves partitioning an image based on its intensity values. It is represented mathematically as

$$G(x,y) = \begin{cases} 1 \text{ if } (x,y) > T \\ 0 \text{ if } (x,y) < T \end{cases} \tag{15.3}$$

In this function, T denotes the threshold values. Pixels with intensity values greater than threshold are considered as foreground, while those with values less than threshold are considered as background. The algorithm has the following steps:

Step 1. Pick an initial value for the global threshold. Let it be T(mean).
Step 2. Divide the image into two regions C_1 and C_2 based on whether they are above or below the threshold T.
Step 3. Compute the average intensity for the two regions. It is denoted as m_1 and m_2.
Step 4. Update the threshold using the formula

$$T = \frac{m_1 + m_2}{2} \tag{15.4}$$

Step 5. Reiterate steps 2 and 4 until it stabilizes.

Otsu's method is noted for its ability to segment images by lessening intra class variance while amplifying the inter class variance. Here, classes refer to the group of pixels that is assigned to specific regions in an image.

15.4.3 NDVI

The NDVI designates a vegetation index in remote sensing analysis, particularly for evaluating the health of vegetation. NDVI simplifies the difficulty of handling multispectral imagery by providing a clear metric for evaluating the plant health. This index can be derived from any multispectral sensor that captures visible and near-infrared (NIR) bands. NDVI works on the specific way plants reflect certain ranges of electromagnetic waves. Blooming vegetation generally shines back more NIR luminance and diminished red light. To compute NDVI, the reflectance values of red and NIR light are compared. The values obtained can vary between −1 and + 1. Values nearing +1 signify healthier vegetation. NDVI was proposed by Krieg et al. in 1969, which employs simple band transformation:

$$N_{\mathrm{DVI}} = \frac{N_{\mathrm{IR}}\mathrm{NIR} - R_{\mathrm{ed}}\mathrm{RED}}{N_{\mathrm{IR}}\mathrm{NIR} + R_{\mathrm{ed}}\mathrm{RED}} \tag{15.5}$$

In this equation, NDVI represents the normalized difference in vegetation index, while NIR as well as Red correspond to spectral radiance measurements in the NIR and red visible regions, respectively. Radiance measures the amount of energy flux that objects reflect. NDVI's applications are noteworthy in precision agriculture, as it enables instant field analysis, ultimately supporting in minimizing environmental impacts. NDVI algorithms are specifically trained to work with satellite imagery, facilitating efficient and accurate evaluation of vegetation across various landscapes. Researchers can effectively detect and analyze these significant features in NDVI images at differing resolutions by utilizing multi-resolution approximations. This capability increases the understanding of spatial relationships and variations within the agricultural landscape, ultimately supporting more informed decision-making in precision agriculture practices.

15.4.4 Histogram Equalization

It is a prominent technique for intensifying the contrast of an image in digital image processing. It uses a cumulative distribution function to improve the contrast of the image. This technique is intended for optimizing the contrast and brightness of the digital image. The quality of the image can be controlled by normalizing the histogram to be flat. Histogram equalization modulates the pixel values to affirm that the intensity histogram of the processed image becomes uniformly distributed. Here is a mathematical working of this algorithm.

Table 15.1: Input image pixel value is represented in a matrix format.

5	**5**	**5**	**5**	**5**
3	4	5	4	3
3	5	5	5	3
3	4	5	4	3
4	3	3	3	4

Here, the highest gray value is 5. This can be written in table format based on the occurrence of the pixel value.

The graphical representation for the table is shown in Figure 15.10.

Here comes Table 15.3, which gives you how to calculate the output histogram values for the given corresponding inputs.

So, in this result, if the input pixel value is 0, then the output value is mapped to 0. If the input pixel value is 5, then the output value is mapped to 10. Likewise, all other input

Table 15.2: Number of occurrences of pixel values.

Grey Value (X-axis)	0	1	2	3	4	5	6	7
Number of Occurrences (Y-axis)	0	0	0	9	6	10	0	0

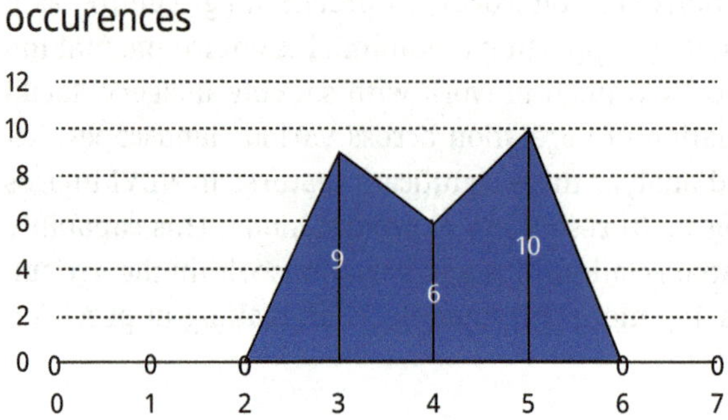

Figure 15.10: Graphical representation of the number of occurrences of pixel values.

Table 15.3: Calculating histogram value.

Greylevel	Number of pixel	probability distribution nk/sum	cumulative distributive function sk	maximim grey scale value Sk * 7	Histogram equalization level
0	0	0	0	0	0
1	0	0	0	0	0
2	0	0	0	0	0
3	9	0.33	0.33	2.31	2
4	6	0.66	(0.33+0.66) =0.99	6.93	7
5	10	0.5	[0.99+0.5]= 1.49	10.43	10
6	0	0	[0+1.49] =1.49	10.43	10
7	0	0	1.49	10.43	10

pixels get mapped to a new value. In the input histogram, the values are centered but output image histogram values are flattened and are shown in Figure 15.11.

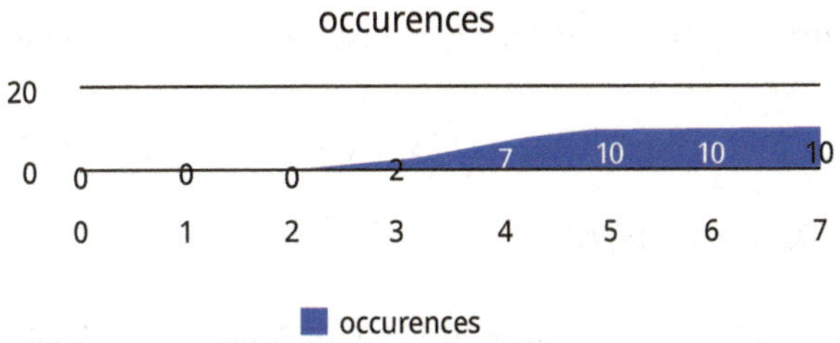

Figure 15.11: Mapping of histogram values.

15.4.5 SLIC: Simple Linear Iterative Clustering

The algorithm takes the image as input and converts into superpixels. It forms a cluster of pixels and they are formed by the combination of nearby pixels. It is a combination of neighborhood information and intensity. For color image, it is considered as [(*r*,*g*,*b*,*x*,*y*)]. Grayscale images are denoted by the parameters [(*g*,*x*,*y*)]. Superpixels are created by grouping the pixels based on color similarity and proximity.

Superpixels can be calculated by

Step 1: Initializing the superpixel centers by sampling *N* locations in the image plane.

Step 2: For each cluster m_i, compute the distance between m_i and each pixel in the neighborhood of m_i.

Step 3: Assign pixel to cluster *i* if its distance is better than the current value.

Step 4: Upgrade cluster centers.

Step 5: Continue with steps 2 and 3 till it converges.

K is the number of superpixels (how many clusters we want).

If the number of clusters formed is high, then the size will be small.

N is the aggregate number of pixels in the original image, *A* is the estimated number of pixels in 1 superpixel

$A = N/K$

S is the approximate length of superpixel

S tells you how wide is the pixel. For every superpixel, there is a center which is analogous to mean. So, the initial mean location should be revealed. It can be computed by

$S = \sqrt{n/k}$.

Given *K*, we can calculate *S*. Next, make a grid of the initial points that are spaced by *S* and these are considered as the initial means. Each pixel can be depicted by [*l a b x y*], which is a vector, whereas [*l*,*a*,*b*] is a ceilab space, which remains uniform for small color distances. From this, we come to know that every r g b color is converted to a ceilab space.

$$d_{\text{lab}}(\text{distance of colors}) = \sqrt{(l_i - l_j)^2 + (a_i - a_j)^2 + (b_i - b_j)^2} \tag{15.6}$$

$$d_{xy}(\text{distance between } x \text{ coordinate and } y \text{ coordinate}) = \sqrt{\left(x_i - x_j\right)^2 + (y_i - y_j)^2} \tag{15.7}$$

$$D_s = d_{\text{lab}} + m/sd_{xy} \tag{15.8}$$

m is the parameter to evaluate the proportion of color and space distance in similarity measurement for controlling the compactness of superpixel after clustering. There arises an ambiguity that pixels at the edges are different from the surrounding pixels. To avoid the initialization of noisy pixels, we go for gradient. It is represented as

$$G(x,y) = ||I(x+1,y) - I(x-1,y)||^2 + ||I(x, y+1) - I(x, y-1)||^2 \tag{15.9}$$

While performing the initialization on the image, wherever the point lies on the higher value of gradient, it is to be moved to the lower value of the gradient. It is explained that for every pixel in the image, search in a neighborhood of $2s$ and assign that mean to the pixel. Then, take the sample mean of the pixels allotted to one mean and update the mean position. Thus, this algorithm is used to achieve good segmentation result by clustering color and distance characteristics of the pixels.

15.4.6 Gaussian Blur

Gaussian blur is a technique used to soften an image by applying a Gaussian function, effectively reducing noise and simplifying image details. Noise in images can arise from various sources, such as poor lighting conditions or other environmental factors. This technique works by calculating the weighted average of the neighboring pixels according to the Gaussian distribution, allowing it to selectively blur specific areas while mitigating the impact of higher frequency noise. Gaussian blur is classified as a linear filter, as it effectively reduces noise while providing a smoothing effect on the image. By smoothing the pixel values, it diminishes the intensity of noise, resulting in cleaner, less detailed images that are easier to analyze and interpret.

After the image from the dataset gets processed under the different image processing algorithms, the output that was obtained is displayed in Figures 15.12 and 15.13.

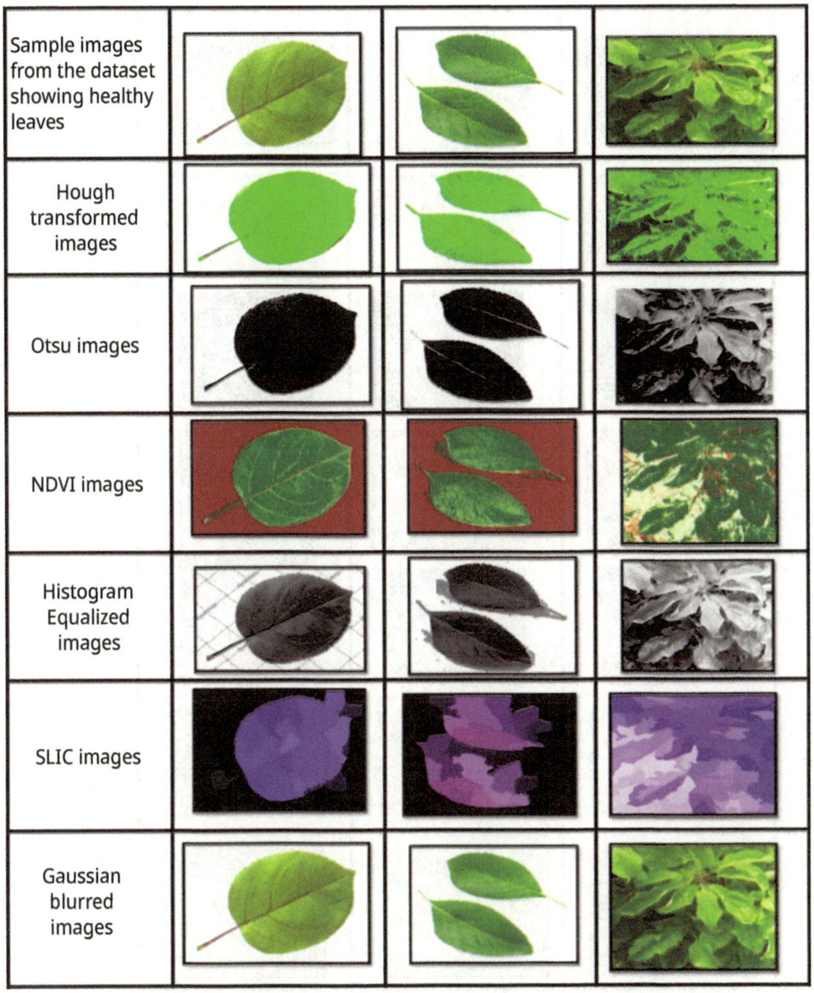

Figure 15.12: Output from various image processing algorithms for healthy crops.

15.5 ResNet-50

15.5.1 Introduction

After the images from the dataset were preprocessed by the above algorithm, it goes through ResNet 50 for further classification. ResNet-50 is a refined deep convolutional neural network architecture made known by Microsoft Research in 2015. It is an improved form of the ResNet model, which stands for "Residual Network." The "50" here indicates that the network comprises 50 layers. This architecture is particularly recog-

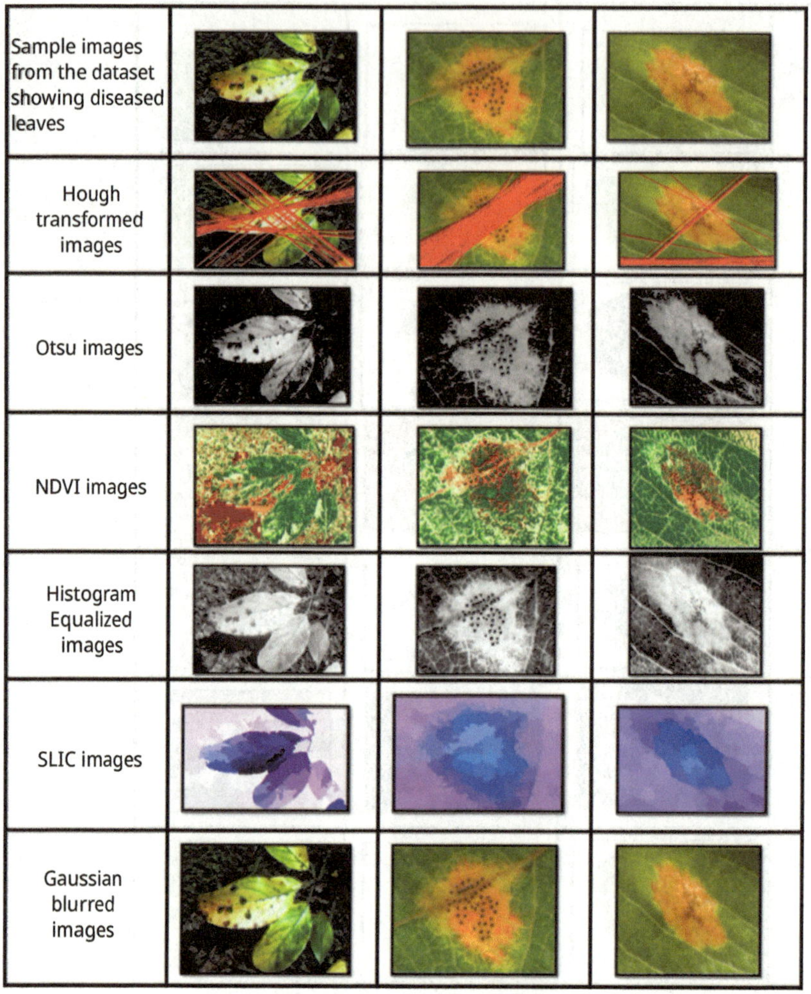

Figure 15.13: Output from various image processing algorithms for diseased crop.

nized for image classification jobs by having the ability to be trained on wide-ranging datasets and achieving impressive results in various benchmarking tests. A significant feature of ResNet-50 is its implementation of residual connections. These connections simplify the learning of residual functions that map the input to the output. So, it effectively allows the network to construct and adjust deeper architectures. This design speaks of the challenges of vanishing gradients, enabling the architecture to maintain performance even as the depth increases. The structure of ResNet-50 can be categorized into four principal components: convolutional layers, identity blocks, convolutional blocks, and fully connected layers. ResNet50 has been validated on large datasets, like ImageNet, achieving prominent performance metrics, comparable to human-level accuracy.

15.5.2 What Are Convolution, Pooling, and Relu?

To abstract the traits from an image, convolution operation is applied in convolution neural networks. To learn simple features like edges, textures, colors, etc., a lower convolution layer is used. A higher convolution layer is used to extract the complex features. It is a mathematical operation that is based on two functions *c* and *d*. It is represented by

$$t(x) = \int_{-\infty}^{+\infty} c(\tau)d(x-\tau)d\tau \tag{15.10}$$

It is a combination of one function and another function. $c(x)$ represents the pixel part and $d(x)$ represents the action part. The output convolution is obtained by applying all the action points on the convolution kernel to the corresponding pixel point. In a standard convolution method, to create a feature map, each kernel is convolved with the input image during the convolution process. Convolution kernel is the important parameter or the elements that is to be learnt during the process. The task of element-wise multiplication and addition of the respective values between the input matrix and the kernel is called convolution processing. Finally, the resultant matrix is obtained when the whole input matrix is traversed. This operation is done to calculate the similarity between each location and its pattern. It is represented in Figure 15.14.

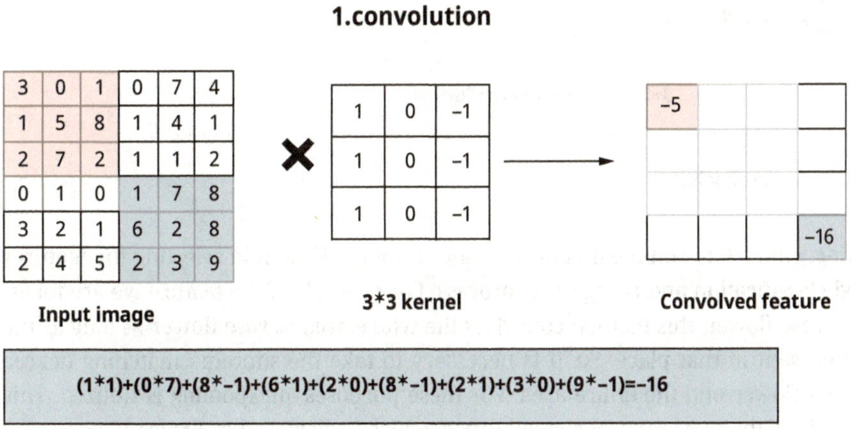

Figure 15.14: Convolution mechanism.

15.5.3 Pooling

Once the convolution layer obtains the feature maps, it is then made ready to integrate and classify the features. Basically, it requires more calculation if we directly pass the extracted features to the classifier [15, 16]. So, in order to reduce the feature dimension, there is a need for a pooling layer. Pooling has the capability to maintain the invariance of the features, preventing overfitting, giving importance to certain features slightly more than the whole location, and focusing on the size of the pixel with respect to the original image for improving the competence of the model [17]. This pooling operation is divided into two types. They are average pooling and max. pooling. Average pooling is used to tackle two important errors, which come from the feature extraction. First error focusses on the limitation of neighborhood size as the estimated variance increases. The second one is the deviation of the estimated mean value, which is caused by the convolution layer parameter. Average pooling minimizes the parameter dimension and retains the integrity of the relayed information by highlighting the sublayer sampling of the overall feature information. Figure 15.8 utilizes a kernel 3×3 with stride 1. This technique takes the average value and compress the 6×6 matrix to 2×2 by preserving the calculated average value in the rectified feature map, as shown in Figure 15.15.

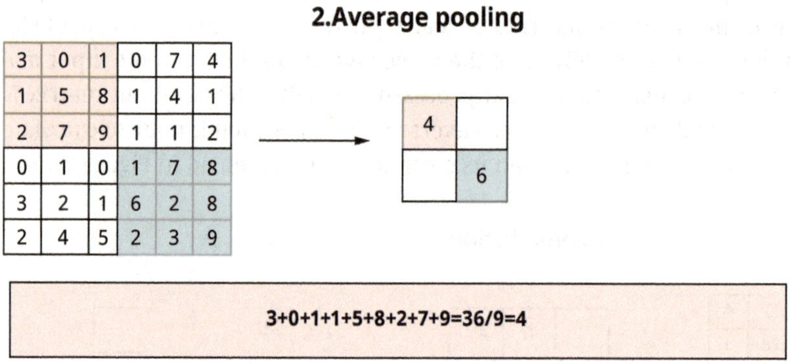

Figure 15.15: Average pooling.

Maxpooling, which has nonlinear characteristics is more efficient in selecting the features with good classification and recognition process. For example, if the feature we are focusing on is a rose flower, this method considers the whole area as rose flower as long as the flower is present in that place. So, it is necessary to take the supreme matching degree amongst the flower and the entire area. For these purposes, maxpooling is utilized. This works by taking the maximum of a small range, which is displayed in Figure 15.16.

15.5.4 Batch Normalization

The number of images that are trained on the network is called a batch. Basically, the generalization of the network is greatly reduced if the data distribution of training data differs from testing data. This makes the network to learn different alignment or distribution at each iteration, which significantly reduces the training speed. So, the neural

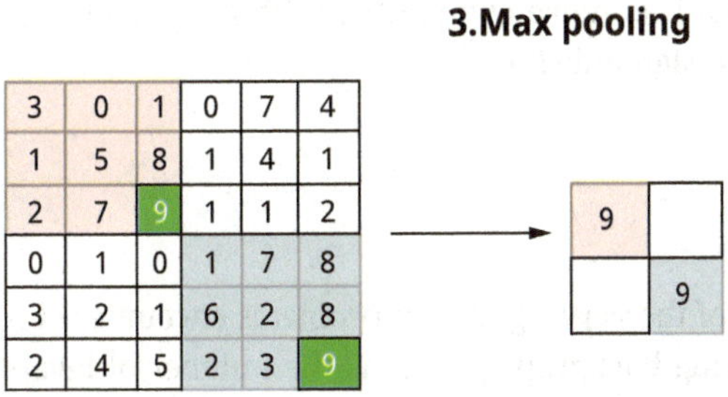

Figure 15.16: Max pooling.

networks needs normalization to learn the organization, alignment, and distribution of data in an effective way. Unlike stochastic gradient descent, this method automatically updates the parameters like learning rate, weight calculation, parameter initialization, drop out ratio, etc. Moreover, during training, it also handles the changes in the middle layer data effectually. This type of changes is also termed as internal covariance shift. Usually, normalization is executed first by fixing the mean value 0 and variance 1, and it moves to the next layer [18–26]. It is defined by the following equation:

$$\rightarrow \text{LAYER} \rightarrow X \rightarrow \widehat{x} = \frac{x-\mu}{\sigma} \rightarrow Y = \gamma\,\widehat{x} + \beta \tag{15.11}$$

where μ is the mean, x is the minibatch , $\gamma\beta$ are the parameters or the weights to be learned, and σ is the std of minibatch.

15.5.5 RELU

In order to learn more complex patterns, activation functions introduce nonlinearity in the neural network. Without activation functions, these layers perform only a simple matrix multiplication [27, 28], no matter how many layers get added. This limits the network ability. So a nonlinear function like activation function is added to enhance the neural network ability to find out the complicated relationship between the data in deep learning. It is defined as

$$F(x) = \max(0, x) \tag{15.12}$$

This nonlinear function has a special behavior known as unilateral inhibition, in which it turns the output to zero if it is negative and the positive value stays there. So, some neurons will get activated at the same time and make the network to process the computation more constructively. This causes sparse activation [29, 30]. The benefits of Relu functions are: (1) it averts the vanishing gradient issue by avoiding the saturation problem; (2) as it does not possess the exponential type of calculation, it is simple and easy for computation; and (3) it is more compatible with biological neurons and has fast convergence speed than sigmoid [17].

15.5.6 Need for ResNet-50

Due to the increasing network depth, one of the very significant problems encountered is the gradient dissipation, which arises during backpropagation. As the number of layers increases, the chain rule with repeated multiplication makes the network too large or shrink to zero. It leads to instability while training the data. Gradient dissipation is the often evoking problem, which hampers proper updating of weights in the layers [31, 32]. To overcome this, many strategies have been proposed, which include batch normalization, Relu activation function, and Xavier initialization usage. Another problem that arises as the network layer increases is performance degradation. Though deploying deeper networks reach greater performance in extracting more complex features, it paves the way to a weakened accuracy in the training datasets as the layer depth increases. It is not caused by overfitting, because accuracy does not diminish during overfitting. So, Resnet-50 emerged with the intention to rectify these problem. As this model holds residual connections, it facilitates the network to learn identity mappings with deeper networks at the time of training without any performance loss. Thus, the network depth is sufficiently increased by several orders of magnitude, resulting in an awesome performance in various tasks. Sahaai et al. [33] insisted that this is the only training prototype that utilizes CNN to achieve substantial result in image detection and classification.

15.5.7 Architecture and Internal Implementation

The framework consists of two kinds of mapping, one is identity mapping and other is residual mapping. It is represented in Figure 15.17.

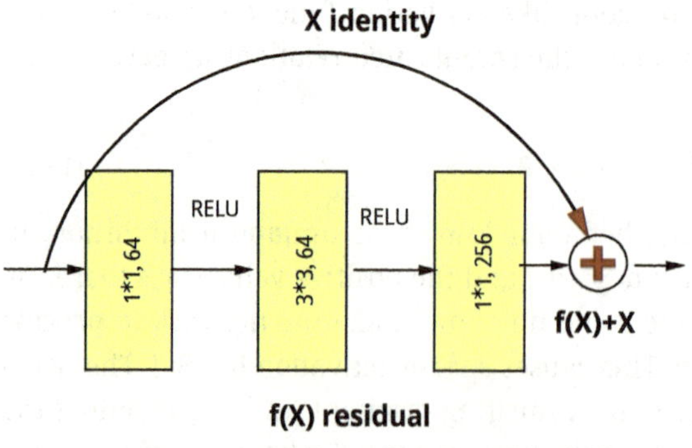

Figure 15.17: Identity and residual functions.

Identity mapping refers to itself, which is shown as the curved part in Figure 15.17 and is represented as "x". The non-curved part is the residual function, which is denoted by $f(x)$. The output is expressed as

$$y = f(x) + x \tag{15.13}$$

The residual part is the difference and is written as $f(x) = y - x$. This model first carries out the convolution operation on the input, subsequently followed by four residual blocks and at last, it undergoes fully connection operation by the fully connected layers to do the classification process. The training process begins with the primary step of preprocessing the input image in order to make it consistent with the multilayer perceptron. The input is then taken to the feed forward neural network once the image is flattened to one-dimensional column vector. This flattened information is used to update the model parameters in each iteration. Through progressive learning, the model learns to identify and distinguish both high-level and low-level features during each iterative learning process. By employing the softmax function, the drawn out features are passed through the classifier to produce the class probabilities [17]. The layered architecture is displayed in Figure 15.18.

This framework consists of numerous convolution layers, which is followed by batch normalization and ReLU activation. The convolutional layers extract essential structures from input images, while the identity blocks and convolutional blocks process these features. The identity block includes a series of convolutions, with the original input added back to the output, nurturing the learning of the residual functions. The fully connected layers ultimately make predictions, with the final output treated through a softmax function to yield class probabilities. To lessen the spatial dimensions of the feature maps, these convolution layers are followed by max pooling layers while securing the very important features. For ResNet50, the identity block and convolution bock are considered to be the pivotal building blocks. The input is passed by the primitive identity block, which adds the input back to the output by passing the input through the sequence of convolutional layers. These structures makes the network to study the residual functions to map the input to the desired output. Though the convolution block and the identity block resemble the same, with the addition of 1×1 convolution layer, it differs and it is used to reduce the number of filters before the 3×3 convolution layer. The final layer is the fully connected layer. The output from the layer is sent to the softmax activation function for final output. In this context, the network classifies two different classes, healthy and diseased leaf.

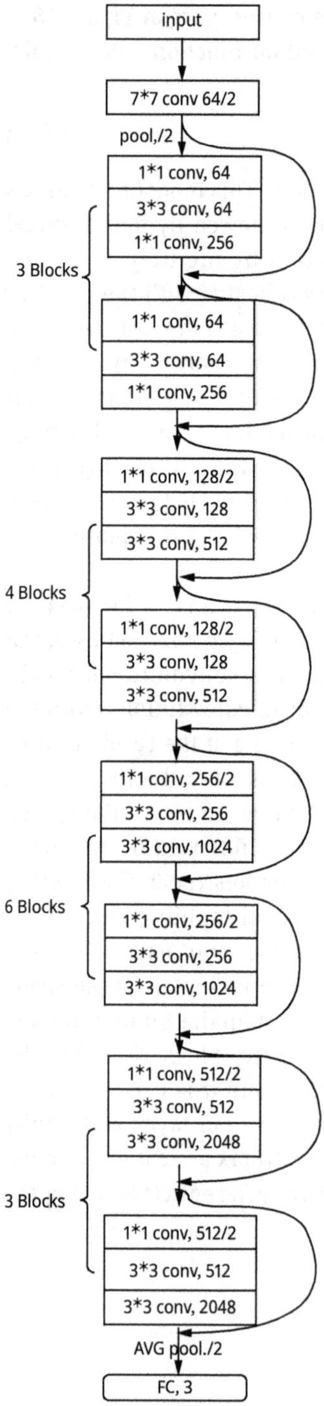

Figure 15.18: Architecture of ResNet-50.

15.6 Experimentation and Evaluation

All experiments were conducted on a computer system with the following specifications: an Intel Core i3-8130U CPU running at 2.20 GHz (2201 MHz) and 4.00 GB of RAM. The system also owns a virtual memory capacity of 8.69 GB. Microsoft Windows operating system is installed on this machine. In addition to the hardware, various libraries were utilized for the implementation of our project. Python 3.7 was used as the programming language along with OpenCV, which is a computer vision library for preprocessing. Several deep-learning libraries were also imported for training and testing the models. Matplotlib, a Python-based visualization library, was employed to visualize images, results and generate graphs. Due to the system's limitations in processing more than 2000 images, I opted to use Google Colab and Jupyter Notebook, which provide an excellent environment for running the models with more epochs in a reasonable amount of time. The dataset chosen is plantDoc classification dataset, which holds around data that consists of 13 plant species and 17 classes of diseases. These images were preprocessed first by various algorithms like Hough transform, Otsu, Histogram Equalization, Slice, and Guassian Blur. All algorithms work efficiently in extracting the features from the given dataset. The preprocessed image is then taken to ResNet-50 for further classification. The output from each algorithm was trained with 100 epochs and results are displayed in Table 15.4 and its graphical representation is shown in Figure 15.19.

Table 15.4: Classification accuracy by ResNet50 with preprocessed images using image processing algorithms.

Algorithm	Accuracy
Hough transform	74.23%
Otsu	75.26%
NDVI	70.89%
SLIC	81.25%
Gaussian blur	79.36%
Histogram Equalization	75.19%

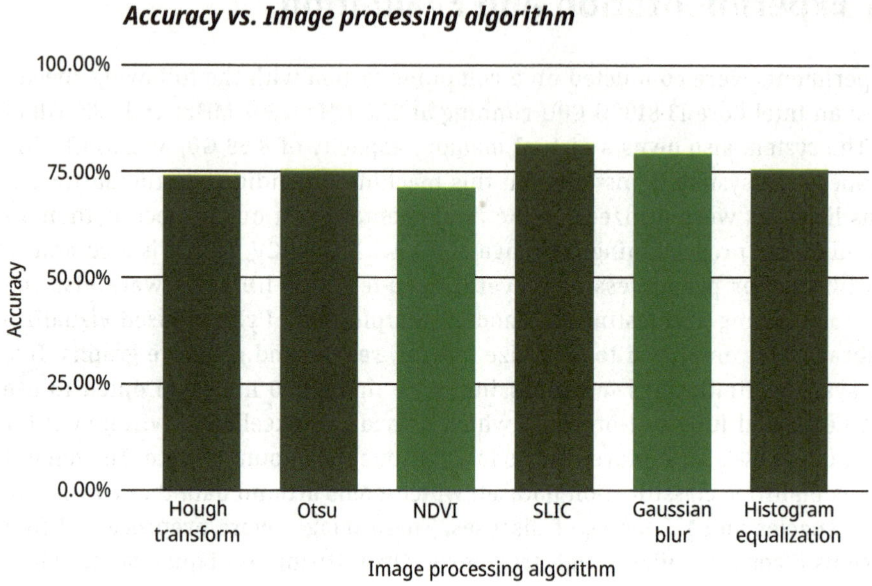

Figure 15.19: Graphical representation of Table 15.4.

15.7 Analyzing the Challenges in Image Preprocessing and Deep Learning for Precision Agriculture

From the above table, it is very clear that the SLIC algorithm performed well when compared to another algorithm. Before analyzing the result, it is important to study the images in the dataset. The dataset comprises a variety of healthy crops and diseased crops. For healthy crops, the feature extraction includes color, shape, texture, etc. On the other hand, it is not easy for diseased crops because in that case, features should be extracted from the diseased part and the undiseased part from a diseased crop. Next, the complexity lies in the shape of the disease and the location of the disease. It may be at the middle or at the boundary or anywhere else inside a crop. So, there is a possibility of missing some features by the algorithm. Hough transform is a very good line detector and shape detector but it results in only 74% accuracy. So, it shows that some shapes could not be detected well by this algorithm for this dataset. For segmentation and extracting the object boundaries, Otsu is efficient, but it gives an accuracy of only up to 75%. HE also shows the same accuracy level.

The least scorer, NDVI, works best for remote sensing. It identifies the green index but fails to extract more features for this image. Gaussian blur nearly touches

80% accuracy. The results indicate that there are additional features within the images that require further exploration and learning. While the algorithms employed demonstrate efficiency in several aspects, they may not perform optimally in certain areas. This suggests that there is scope for improvement in the feature learning process to improve the overall execution of the model. Currently, many deep learning models have the potential to independently acquire features and execute their intended tasks. Nevertheless, it is essential to recognize that traditional algorithms are still valuable and have not become irrelevant. While current deep learning techniques are highly effective at feature extraction, existing algorithms remain important tools in various applications in machine learning.

15.8 Conclusion

Thus, image preprocessing techniques plays a crucial role in strengthening the capabilities of deep learning models in precision agriculture. By addressing the intrinsic challenges in raw image data, preprocessing techniques such as noise reduction, contrast enhancement, and segmentation are critical for refining the image quality and extracting the relevant features. This in turn leads to more authentic and systematic identification and categorization of crop diseases and conditions, which significantly supports farmers in making knowledgeable decisions. Promoting greater collaboration between recent technologies and agricultural practice results in greater productivity, ensuring food security, and contributing to the sustainability of the agricultural ecosystem.

15.9 Future Directions in Preprocessing and Deep Learning for Precision Agriculture

It is compulsory to focus on enhanced data acquisition methods. Much advancements in remote sensing technologies, such as drones and satellite imagery, are expected to facilitate the collection of high-resolution agricultural data. The incorporation of multi-spectral and hyper-spectral imaging can provide wealthier datasets, enabling more accurate assessments of crop health, soil conditions, and environmental factors. The preprocessing techniques should be upgraded. As the volume of data collected in precision agriculture increases, innovative preprocessing methods will become complex. Some techniques such as data normalization, noise reduction through advanced filtering algorithms, and anomaly detection will boost data quality. Furthermore, the advancement of automated pipelines to preprocess large datasets efficiently will help in minimizing human error and maximizing computational efficiency. In the future, research should focus on incorporating domain-specific knowledge into deep learning models. This consists of understanding plant physiology, soil science, and environmental interactions, which can improve model interpretability and the relevance of features extracted.

As deep learning models become more complex, the demand for transparency and interpretability will grow. The advancement of deep learning models can enable real-time data analysis and it will be key to optimizing precision agriculture. Multimodal data fusion can enhance the robustness of predictive models by providing a wide-ranging view of the agricultural system. These approaches can lead to more accurate insights and recommendations for farmers. As sustainability becomes increasingly critical in agriculture, deep learning models must also focus on promoting resource efficiency. Future developments should focus on optimizing inputs like water, fertilizers, and pesticides while minimizing environmental impact. Thus, many machine learning algorithms and deep learning methods can assist in modeling sustainable practices, leading to more responsible farming methods.

References

[1] Wu L, Ma X, Qi L, Tan Y, Kuang J, Liang Z. A method of target detection for crop disease spots by improved Hough transform. Transactions of the Chinese Society of Agricultural Engineering. 2014 Jan 1;30(10):152–159.

[2] Durai S, Sujithra T, Iqbal MM. Image Classification for Potato Plant Leaf Disease Detection Using Deep Learning. In 2023 International Conference on Sustainable Computing and Smart Systems (ICSCSS) 2023 Jun 14 (pp. 154–158). IEEE.

[3] Epinat V, Stein A, De Jong SM, Bouma J. A wavelet characterization of high-resolution NDVI patterns for precision agriculture. International Journal of Applied Earth Observation and Geoinformation. 2001 Jan 1;3(2):121–132.

[4] Sishodia RP, Ray RL, Singh SK. Applications of remote sensing in precision agriculture: A review. Remote Sensing. 2020 Sep 24;12(19):3136.

[5] Mavridou E, Vrochidou E, Papakostas GA, Pachidis T, Kaburlasos VG. Machine vision systems in precision agriculture for crop farming. Journal of Imaging. 2019 Dec 7;5(12):89.

[6] Arnold JM. Department of electronics and electrical engineering. Nonlinear Waves in Solid State Physics. 2012 Dec 6;247:259.

[7] Dhaware CG, Wanjale KH. A Modern Approach for Plant Leaf Disease Classification Which Depends on Leaf Image Processing. In 2017 International Conference on Computer Communication and Informatics (ICCCI) 2017 Jan 5 (pp. 1–4). IEEE.

[8] Basha NK, Bhatia S, Kumar A, Mashat A editors. Deep Learning and Its Applications Using Python. John Wiley & Sons; 2023 Sep 27.

[9] He K, Zhang X, Ren S, Sun J. Deep Residual Learning for Image Recognition. In Proceedings of the IEEE Conference on Computer Vision and Pattern Recognition 2016 (pp. 770–778).

[10] Parthasarathi P, Nivedha S, Balamurugan M, Yuvaraj D, Devi SS, Sangeethaa SN. Innovative Skin Melanoma Diagnosis Using Deep Learning and African Gorilla Troops Optimizer. In 2024 First International Conference on Software, Systems and Information Technology (SSITCON) 2024 Oct 18 (pp. 1–6). IEEE.

[11] Zhang YD, Satapathy SC, Guttery DS, Górriz JM, Wang SH. Improved breast cancer classification through combining graph convolutional network and convolutional neural network. Information Processing & Management. 2021 Mar 1;58(2):102439.

[12] Zhang YD, Satapathy SC, Liu S, Li GR. A five-layer deep convolutional neural network with stochastic pooling for chest CT-based COVID-19 diagnosis. Machine Vision and Applications. 2021 Jan;32(1):14.

[13] Wang SH, Govindaraj VV, Górriz JM, Zhang X, Zhang YD. Covid-19 classification by FGCNet with deep feature fusion from graph convolutional network and convolutional neural network. Information Fusion. 2021 Mar 1;67:208–229.

[14] Nithya Anoo S, Pavithra A, Poornamala S, Siamala Devi S. An efficient skin cancer classification approach using neural networks. Journal of Algebraic Statistics. 2022 Sep 9;13(3):4946–4957.

[15] Ashiquzzaman A, Lee H, Kim K, Kim HY, Park J, Kim J. Compact spatial pyramid pooling deep convolutional neural network based hand gestures decoder. Applied Sciences. 2020 Nov 7;10 (21):7898.

[16] Zhang Y, Satapathy SC, Zhu LY, Górriz JM, Wang S. A seven-layer convolutional neural network for chest CT-based COVID-19 diagnosis using stochastic pooling. IEEE Sensors Journal. 2020 Sep 22;22 (18):17573–17582.

[17] Li B, Lima D. Facial expression recognition via ResNet-50. International Journal of Cognitive Computing in Engineering. 2021 Jun 1;2:57–64.

[18] Choi SH, Jung SH. Stable acquisition of fine-grained segments using batch normalization and focal loss with l1 regularization in u-net structure. International Journal of Fuzzy Logic and Intelligent Systems. 2020 Mar 25;20(1):59–68.

[19] Kumar KS, Abirami K, Sulochana CH, Kumar TA, Ajagbe SA, Morris C. Multimodal Spatial-Temporal Prediction and Classification Using Deep Learning. In Spatiotemporal Data Analytics and Modeling: Techniques and Applications 2024 Apr 16 (pp. 89–114). Singapore: Springer Nature Singapore.

[20] Zhang YD, Govindaraj VV, Tang C, Zhu W, Sun J. High performance multiple sclerosis classification by data augmentation and AlexNet transfer learning model. Journal of Medical Imaging and Health Informatics. 2019 Dec 1;9(9):2012–2021.

[21] Zhang YD, Dong Z, Chen X, Jia W, Du S, Muhammad K, Wang SH. Image based fruit category classification by 13-layer deep convolutional neural network and data augmentation. Multimedia Tools and Applications. 2019 Feb;78(3):3613–3632.

[22] Wang SH, Muhammad K, Hong J, Sangaiah AK, Zhang YD. RETRACTED ARTICLE: Alcoholism identification via convolutional neural network based on parametric ReLU, dropout, and batch normalization. Neural Computing and Applications. 2020 Feb;32(3):665–680.

[23] Nencka AS, Arpinar VE, Bhave S, Yang B, Banerjee S, McCrea M, Mickevicius NJ, Muftuler LT, Koch KM. Split-slice training and hyperparameter tuning of RAKI networks for simultaneous multi-slice reconstruction. Magnetic Resonance in Medicine. 2021 Jun;85(6):3272–3280.

[24] Nishida N, Oba T, Unagami Y, Cruz JP, Yanai N, Teruya T, Attrapadung N, Matsuda T, Hanaoka G. Efficient secure neural network prediction protocol reducing accuracy degradation. IEICE Transactions on Fundamentals of Electronics, Communications and Computer Sciences. 2020 Dec 1;103(12):1367–1380.

[25] Benbahrıa Z, Sebari I, Hajji H, Smiej MF. Intelligent mapping of irrigated areas from Landsat 8 images using transfer learning. International Journal of Engineering and Geosciences. 2021;6 (1):40–50.

[26] Polat Ö, Güngen C. Classification of brain tumors from MR images using deep transfer learning. The Journal of Supercomputing. 2021 Jul;77(7):7236–7252.

[27] Wang SH, Tang C, Sun J, Yang J, Huang C, Phillips P, Zhang YD. Multiple sclerosis identification by 14-layer convolutional neural network with batch normalization, dropout, and stochastic pooling. Frontiers in Neuroscience. 2018 Nov 8;12:818.

[28] Zhang YD, Pan C, Sun J, Tang C. Multiple sclerosis identification by convolutional neural network with dropout and parametric ReLU. Journal of Computational Science. 2018 Sep 1;28:1–0.
[29] Nayak DR, Das D, Dash R, Majhi S, Majhi B. Deep extreme learning machine with leaky rectified linear unit for multiclass classification of pathological brain images. Multimedia Tools and Applications. 2020 Jun;79(21):15381–15396.
[30] Olimov B, Karshiev S, Jang E, Din S, Paul A, Kim J. Weight initialization based-rectified linear unit activation function to improve the performance of a convolutional neural network model. Concurrency and Computation: Practice and Experience. 2021 Nov 25;33(22):e6143.
[31] Kwon H, Pellauer M, Parashar A, Krishna T. Flexion: A quantitative metric for flexibility in DNN accelerators. IEEE Computer Architecture Letters. 2020 Dec 14;20(1):1–4.
[32] Lee D, Kim J, Jung K. Improving object detection quality by incorporating global contexts via self-attention. Electronics. 2021 Jan 5;10(1):90.
[33] Sahaai MB, Jothilakshmi GR, Ravikumar D, Prasath R, Singh S. ResNet-50 Based Deep Neural Network Using Transfer Learning for Brain Tumor Classification. In AIP Conference Proceedings 2022 May 2 (Vol. 2463, No. 1, pp. 020014). AIP Publishing LLC.

Lavanya Devi N., C.N. Savithri, Femila L., R.Karthick Manoj

16 AI-Powered Web Application for Early Detection of Heart Disease Using Computational Algorithms

Abstract: Coronary artery disease (CAD) is the most common heart disease that affects people all over the world. When cholesterol and other fatty substances build up in the coronary arteries, they block blood flow to the heart. This disease can cause heart attacks and other significant problems. A big problem in the medical field is that healthcare personnel often do not have a lot of deep knowledge in their fields, which can lead to misdiagnoses. These kinds of mistakes might delay treatment or possibly kill someone. More and more people are using automated heart disease screening methods to deal with this important problem. The purpose of these systems is to make accurate predictions early on, which will dramatically minimize the number of deaths from heart disease. It is really essential to figure out what is wrong with a patient as soon as possible. Using the latest technology can help fill in the gaps in knowledge and make health outcomes better. One alternative is to use Django, a popular web framework for Python, to construct a project on the web. This program utilizes machine learning (ML) to look at individuals' medical records and find out how probable it is for them to get heart disease. ML may assist in identifying sophisticated patterns and correlations in large datasets that might not be simple to notice using regular diagnostic methods. It was quite accurate, with a score of 0.87, which implies that 87% of the predictions were correct. It also had a specificity of 0.88, which suggests that the model was effective at locating persons who did not have the disease. The accuracy was 0.89, which implies that the model was accurate 89% of the time when it stated someone had heart disease. It also achieved a sensitivity of 0.83, which suggests that the model was able to correctly identify those who had the disease. A web tool built on Django that employs ML to aid with medical diagnoses is a fast and reliable approach to discover CAD early. It helps healthcare staff make better decisions and also lowers the number of deaths that could have been averted if the erroneous diagnosis had been made or the patient had waited too long.

Keywords: Django, support vector machine, web application, decision trees

16.1 Introduction

Heart disease is still one of the top causes of mortality throughout the world, killing millions of people every year [1]. Coronary artery disease (CAD) is the most prevalent and deadly kind of heart disease. CAD happens when the coronary arteries, which are

https://doi.org/10.1515/9783112214374-016

the main blood channels that bring oxygen and nutrients to the heart, are hurt or sick [2]. This happens most often when fatty deposits (plaques) build up and block these arteries. This illness makes the arteries smaller and slows down blood flow to the heart. This can cause chest discomfort (angina), shortness of breath, or even heart attacks that can kill you. The growth in CAD instances in both industrialized and developing countries is mostly due to the current way of life, which includes bad food, not exercising, smoking, and high levels of stress. Even though healthcare and medical research have come a long way, it is still hard to diagnose CAD early and accurately in many clinical situations.

One of the main reasons why heart disease has such high death rates is because many healthcare workers do not have the specific diagnostic skills they need, especially in rural or low-resource locations. It is very frequent for doctors to have the wrong diagnosis or to wait too long to make a diagnosis. This can lead to therapy that is too late or not right, which puts patients' safety at risk. Even in cities where advanced diagnostic techniques are available, mistakes by people and differences in clinical judgment can make diagnoses less accurate [3]. This creates a major gap in the healthcare system that needs innovative solutions that can grow. Since early identification of cardiac diseases is so crucial, it is necessary to find ways to improve the accuracy of tests and teach people more skills. So, using smart technology and automation in healthcare delivery might be approaches to solve these difficulties. In the past 10 years, artificial intelligence (AI) and machine learning (ML) have altered a lot of fields, including finance, education, marketing, and most recently, healthcare [4]. ML is a branch of AI that is all about building algorithms that help computers learn from prior data and make decisions or predictions without needing to be programmed for every potential case. These prediction algorithms may be trained on real-world datasets to identify complex connections between clinical factors and the outcomes of diseases [5].

Adding ML models to web apps has made these technologies much easier to use and available to more people. A web-based solution makes it easy for medical professionals and even patients to utilize it from any device that can connect to the internet [6]. This democratization of technology makes sure that life-saving instruments are not just available in fancy hospitals, but also in places that are hard to reach or do not have enough resources. Users may enter patient data into a web-based platform made with Django, a high-level Python web framework, and get an immediate risk assessment of heart disease. Django is a great choice for these kinds of apps since it can grow, is secure, and has strong support for adding data science components. In this suggested solution, a number of ML algorithms were put to the test to see how well they could detect the existence of cardiac disease [7]. Because it is simple and works well, it is a good choice for medical uses where both accuracy and ease of use are important. In our tests, *K*-nearest neighbors (KNNs) got an accuracy of 0.87, which means it accurately guessed whether or not heart disease was present in 87% of the cases. The specificity, which tells us how well the model can find people who do not have the condition, was 0.88. This is especially essential since it lowers the number of false positives, which can cause patients to have to go through more tests and stress. The model also achieved a precision of 0.89, which means that it was quite good at finding people who really had the condition. Its sensitivity (or recall) was 0.83, which means it did a good job of finding true positive cases, or those who really did have heart disease.

This type of mechanism might have a lot of different impacts. First and foremost, it provides doctors and nurses a reliable tool they may use to acquire a second opinion or assist them decide who to treat first [8]. This sort of technology can help discover high-risk individuals who need care immediately away in emergency rooms, for example, when quick decisions are needed. It also allows general practitioners who may not have specialized expertise in cardiology to recognize early signs of CAD and direct patients for the correct treatment [9]. This sort of prediction model might also be used in community health screening programs to detect those who are at risk before they exhibit any indications of sickness, which would be good for public health. This kind of preventative healthcare approach might have a huge effect on how much heart disease costs the healthcare system in the long run. This strategy also helps with customized medicine, which is a new field of healthcare that focuses on treating people based on their individual qualities, behaviors, and genetic composition. ML models may look at data from each patient and generate more specific risk assessments than broad diagnostic criteria. This makes sure that each patient gets care that is right for their health requirements, which makes them happy and leads to better treatment results.

However, it is important to note that machine learning models are not infallible, despite their strong predictive capabilities. Their performance is highly dependent on the quality, diversity, and representativeness of the data used for training. Predictions can be wrong if the data is biased or incomplete, especially for groups of people that were not well represented in the training sample [10]. So, to keep these kinds of technologies reliable and ethical, they need to be constantly checked, retrained with new data, and worked on with medical specialists. Data privacy and security are also very important, especially in healthcare, where rules like Health Insurance Portability and Accountability Act (HIPAA) and General Data Protection Regulation (GDPR) require that private patient information be kept safe. In conclusion, combining ML with web technologies like Django might be a good way to fix the serious problem of diagnosing CAD too late or incorrectly. This technology can accurately forecast the risk of heart disease by using past patient data and smart algorithms. This helps healthcare practitioners make better judgments. In this case, the KNN algorithm has demonstrated very good performance metrics, which shows that it might be used in the real world. As medicine continues to change, using AI-powered diagnostic tools will not only improve the quality of care, but it will also lead to a more proactive, data-driven approach to healthcare.

16.2 Survey on Cardiac Disease Forecasting by Using Ml Algorithms

AI-powered predictive modeling for disease diagnostics is a long review chapter that looks at the ever-changing relationship between AI and healthcare, with a focus on how AI has changed illness diagnostics in a big way. It starts by talking about how important AI is for making illness prediction more accurate, efficient, and fast, which will change the way patients are cared for. The review stresses the significance of having high-quality and varied data in order to build strong prediction models. It also goes into detail about data pretreatment methods that are necessary to get the most out of AI. ML and deep learning (DL) algorithms, both old and new, are at the heart of AI-powered illness diagnoses. These algorithms may be used in a wide range of medical settings. The chapter shows these algorithms as the basic building blocks for predicting diseases in a wide range of situations, focusing on their strengths and weaknesses. At the same time, feature selection and engineering are getting a lot of attention as important parts of model building. It is important to remember that feature curation is both an art and a science that needs domain knowledge [11].

AI plays a huge role in making it easier to find and diagnose heart failure, which is a big step forward for personalized therapy. This review looks at new ideas that employ AI by looking at how they work, what they do, and what they achieve. We talk about ways to make diagnoses more accurate by using DL models, NN, and ML algorithms on diverse types of datasets [12]. Some of the most noteworthy improvements are prediction models that employ real-time data from wearable devices and new AI systems that have been trained on data from hospitals and clinics. Latest study has revealed that these AI systems might make it much simpler to spot issues early on and choose the right treatment. For instance, the accuracy of diagnoses varied from 86.7% to 99.9%, while the sensitivity and specificity scores were usually around 97%. In this review, we talk further about how symmetry and asymmetry impact the design of models. It indicates that symmetric architectures like U-Net are useful for getting organized features and making calculations faster. Because of this, asymmetric models make it simpler to discover rare diseases and small changes in clinical patterns. Using these DL technologies to discover abnormalities and simulate the progression of sickness makes them even better for improving diagnostic accuracy and patient outcomes. This assessment also discusses about issues with AI applications that are already in use, such as the quality of the data, the lack of openness in algorithms, model bias, and evaluation criteria. It also speaks about how to improve clinical practice in the future by leveraging generative models, hybrid architectures, and explainable AI approaches.

Researchers are utilizing ML and deep DL with electrocardiograph (ECG) data to find all kinds of heart disorders and other health concerns, such sleep apnea, breathing rate, and blood pressure, to mention a few [13]. The review did, however, conclude that the results in published works seemed promising. But there were certain huge difficulties that made it impossible to apply that strategy to make medical judgments in real life. Some of the issues were that the dataset was not balanced, there was no standard dataset for evaluation, the model was not easy to understand, the performance was not consistent when using a new dataset, there were worries about the security and privacy of health data, and doctors did not work together enough, among other things. AI might make it easy for more people to keep an eye on their health and get early diagnoses by using ECG data and other wearable biosensor technology. But researchers need to focus on finding solutions to the problems.

It is become harder and harder to safeguard clinical databases, especially ones with a lot of data in them [14]. Cryptocurrencies are based on blockchain, which is an open, decentralized, and distributed public ledger technology. It maintains track of transactions safely without having a third party to check them. Decentralized blockchain networks make it secure and easy for clinical research and practice data to move in and out of the healthcare system. To start with, we will talk about the core principles underlying this technology and how it fits into the many well-known ways it may be used. Next, we discuss about how high-throughput gene sequencing, wearable technology, and clinical trials might be employed in research and care for heart problems. Next, we will talk about the challenges that are making it hard to be put into action and what has to be done in the future. We also present a short summary of the conceivable healthcare uses if decentralized blockchain computing platforms were combined with AI computing. Blockchain can provide therapeutically valuable predictions, aid enhance research methodologies, and give doctors virtual tools to utilize in their job by employing high-performance computers and AI that can securely manage big and fast developing medical databases. We can use both AI and blockchain to their fullest potential by merging them in a way that works. This will help us find new approaches to achieve our aim of providing accurate cardiovascular care.

This goal of this chapter is to create and test an AI-based tool that can predict when a heart attack is going to happen so that doctors can do something to stop it. It uses a variety of AI methods and approaches to figure out a personalized risk of heart attack based on a wide range of patient traits [15]. It also determines the lowest set of features that would work from a computer's point of view. This is a step forward in this field of study when it comes to one patient. It makes preclinical therapy and diagnosis better, and it also makes predictions in preventive medicine more accurate. The study analyzed the models that were constructed, and the logistic regression models were the most accurate. They do not have a very high predictive value, but that is okay for the first screening diagnostic, which is to find those who require more reliable tests. You may also do this with fewer parameters like cholesterol, age, heart rate, and body mass index. This lets you build a strategy to avoid heart attacks based on things that can be altered and keep an eye on the risk of heart attack with the proposed method. The approach disclosed is innovative and beneficial since it utilizes AI to quickly, cheaply, and extensively anticipate how well the hearts of a group of people who are thought to be at risk will work. This is done over time for all patients as part of a regular checkup that allows them to get a more advanced diagnosis of heart disease.

Heart disease is one of the most common causes of death in the world, yet catching it early can help lower the death rate [16]. Studies that are well-known have revealed that the newest AI can help you figure out how likely it is that someone may have heart disease. But previous research did not look at dynamic scalability to figure out how to get the most out of these AI models as more people use them. We developed HealthFaaS, a smart healthcare framework that is powered by AI and makes use of the Internet of Things (IoT) and a serverless computing environment, with the goal of reducing the number of deaths caused by heart disease and preventing the loss of money that results from incorrect diagnosis. HealthFaaS uses IoT devices to gather health data from users and transfer it to AI models that operate on Google Cloud Platform (GCP), which is a serverless computing environment. GCP is preferable since it can change size as required, is easier to use, and has a pay-as-you-go pricing structure. We look at and evaluate five different AI models to assess how effectively they find people who are likely to have heart disease. The test findings show that the light gradient boosting ML model is the best at finding cardiac problems, with an accuracy rate of 91.80%. We also used quality of service metrics like latency and throughput to see how well the HealthFaaS framework functioned compared to a platform that was not serverless as the number of users grew. We also checked out the delay while starting up a serverless platform. It showed that the quantity of RAM and the programming language used directly affected how long it took to start up.

Within several regions of the world in the present day, cardiovascular diseases, which are also commonly referred to as heart problems, are the leading cause of mortality [17]. It is possible for heart illness to create infections in the coronary arteries, which can lead to a person becoming weaker. Additionally, heart disease can cause blood vessels to function less correctly. Therefore, we want a system that is capable of detecting cardiac illness in a manner that is reliable, accurate, and feasible at an early enough stage for the patient to receive the appropriate therapy before the condition worsens, such as by having a heart attack. With the goal of automatically predicting, diagnosing, and treating heart disease, ML techniques and approaches have been applied to a variety of datasets pertaining to the condition throughout the course of the past several years. The purpose of this study is to provide a comprehensive analysis of a number of ML algorithms and to investigate the degree to which these algorithms are effective in producing accurate predictions, diagnoses, and treatments for a variety of cardiac conditions. Several different ML algorithms that may be utilized to make an educated prediction as to whether or not a person has cardiac disease are the subject of the study that is suggested. Following this, the average accuracy of each approach's forecasts was determined in order to determine which method had the best overall performance and which one gave the lowest overall performance. The artificial neural network technique had the best average prediction accuracy, which was 86.91%, while the C4.5 decision tree (DT) method had the lowest average prediction accuracy, which was 740.0%.

When it comes to the actual practice of locating and identifying cardiac disease, ML and data mining would be of great use. However, it would be quite challenging to make these systems function properly. In most countries too few individuals have sufficient knowledge on cardiovascular disease [18], which results in a significant number of cases being mistakenly detected. This issue might be resolved by providing medical professionals with digital information on their patients, which would enable them to make more informed decisions regarding how to detect cardiac disease in its early stages. The objective of this study was to identify the ML models that performed the most effectively when applied to the aforementioned types of evaluation tasks. The purpose of this study was to evaluate and investigate a number of guided ML techniques in order to see how effectively and precisely they might predict cardiac disease. The feature significance ratings for each feature were predicted using all of the algorithms that were applied, with the exception of MLP and KNN. We rated all of the characteristics according to the likelihood that they were to cause heart disease in order to determine which ones were the most accurate. During the course of this research, both a Kaggle dataset on heart illness and three distinct categorization approaches were utilized. The radio frequency approach was completely accurate, sensitive, and focused. As a result, we discovered that a relatively straightforward guided ML approach is capable of identifying cardiac illness with a high degree of precision and is likely to be of great assistance.

This chapter is arranged in the following order. Section 16.2 briefs the material in addition to the method used for the proposed method, Section 16.3 briefs on the results and also the discussion on the results and Section 16.4 discusses the conclusion of the research work done.

16.3 Materials and Method

We have developed an application in a website built using ML algorithms for finding the chances of a heart attack considering highly correlated parameters. Cleveland heart disease database from University of California, Irvine (UCI), with 303 patients' data in .csv format with 13 attributes, which affect the health of the heart was made use of for the experiment conducted. Table 16.1 explains the details of the 13 features used for predicting the heart disease and the 14th attribute provides the information about the presence or absence of cardiac disease.

Figure 16.1 describes the architecture used in the application developed. The input attributes discussed above will be considered as the test data and also will be added to the data base for training the next patient's attributes. For performing the training process, the different modes like SVM, KNN, DT, and random forest (RF) are

Table 16.1: Attributes used for the prediction of heart disease.

S. no	Attributes	Description
1	Sex	Information about whether the patient is male or female
2	Age	Age of the patient
3	Fbs	Measure of fasting blood sugar
4	restecg	Resting electrocardiographic value (0 value for normal, 1 for ST–T wave abnormality, 2 for left ventricular hypertrophy)
5	Cp	Category of chest pain
6	Chol	Serum cholesterol
7	thalach	Highest heart rate
8	trestbps	Blood sugar on resting
9	slope	Measures the change in the ST segment for the increase in the heart rate while doing any exercise
10	oldpeak	Depression in the ST caused by exercise compared to the rest position
11	thal	Normal = 3, fixed defect = 6, reversible defect = 7
12	exang	Discomfort or pain in the chest due to low oxygen supply
13	Ca	Blood vessels count (ranging from 0 to 3) colored by fluoroscopy
14	target	Heart disease patient = 1, no heart disease = 0

available and the mode that gives the better performance can be selected. For optimizing the performance of the system, validation of the developed model is also ensured.

To develop web-based apps, Django (Python-based free open-source framework) ensuring pragmatic security and design is used. If the user wishes to check chances of getting a heart attack, he should upload the 13 attributes stated above and the developed Django-based app will provide the result of heart disease prediction as the 14th column.

The user request to Django site is shown in Figure 16.2. The mapping of the URL being requested for a view function is taken care of by the URL dispatcher. The view function takes care of the actions being requested including reading and writing the data to the database. Interaction of the data defined in Python is done by the model. The template returns the HTML pages. The HTTP response to the web browser is provided by the view.

Figure 16.3 demonstrates the correlation between 14 attributes used for the forecasting of heart disease. The association between the attributes can be positive or negative depending on whether it is dependent or independent variable, respectively.

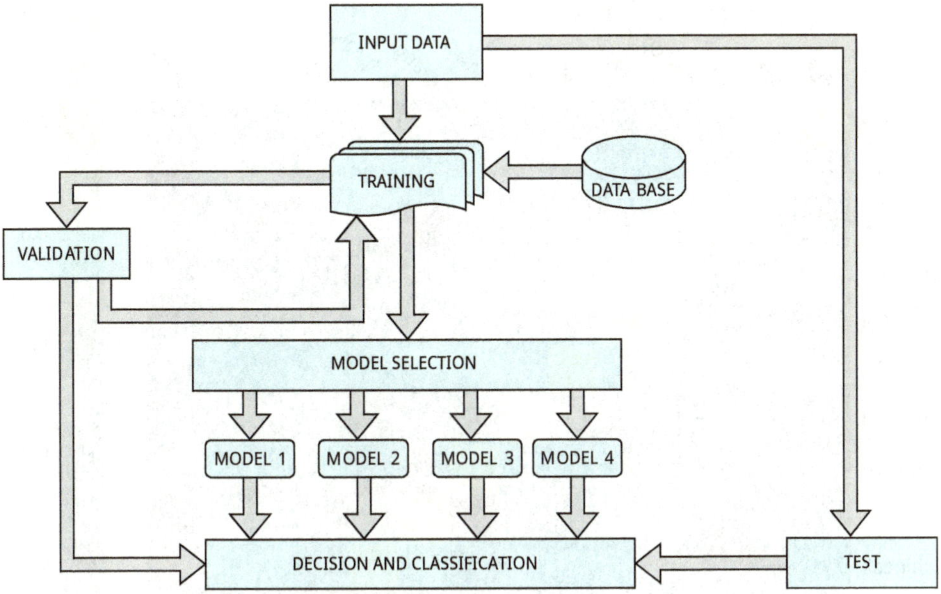

Figure 16.1: Architecture of the web-based application.

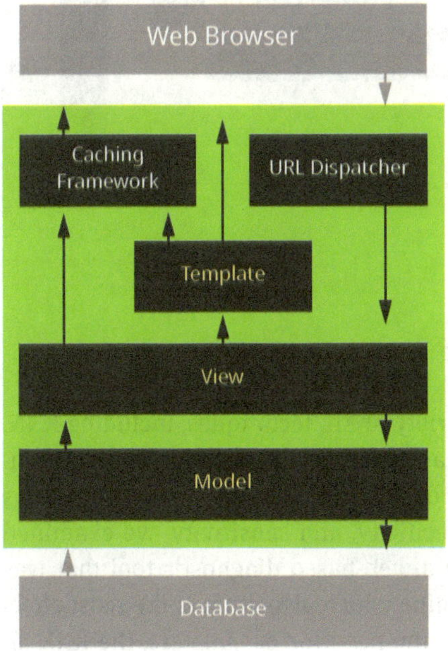

Figure 16.2: User request to Django site.

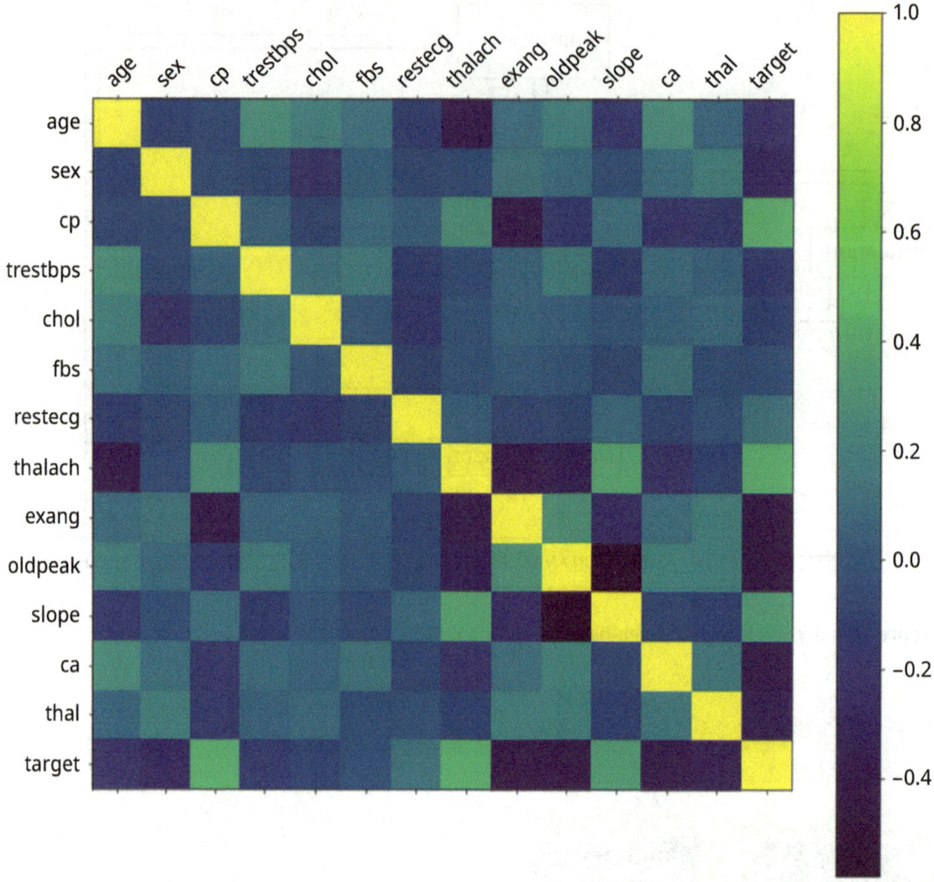

Figure 16.3: Correlation matrix.

16.4 Results and Discussion

This research utilized and evaluated a wide variety of ML techniques, including KNN, DT, RF, and SVM, with the goal of developing a system that is dependable and accurate in its ability to forecast CAD. In order to determine how well these algorithms performed in terms of accuracy, precision, specificity, and sensitivity, we examined them based on a variety of parameters. Using a web-based diagnostic tool that was developed, the primary objective was to determine which algorithm is the most effective for locating CAD at an early stage. Without the need for parameters, the KNN algorithm provides a straightforward as well as very effective method for classifying items. When attempting to forecast heart illness, it first determines the distance between a new data point and all of the existing data points, and then it places the new data into the same category as the majority of the data points that are located in its immediate vicinity. When it comes to the effectiveness of the KNN model, the choice of k, also known as the number of neighbors, has a significant impact. Choosing a low value of k can result in overfitting, while selecting a high value can result in underfitting. Both of these outcomes are undesirable. However, in order to achieve the highest possible performance, it is essential to adjust k. During this experiment, we altered the value of k from 1 to 20 in order to determine how the change impacted the accuracy of the predictions. With an accuracy of 0.87, a precision of 0.89, a specificity of 0.88, and a sensitivity of 0.83, the model was at its most accurate when k was equal to 8. According to these data, the model is successful in identifying both true positives (individuals who have CAD) and true negatives (individuals who do not have the ailment). Positive diagnoses are even more dependable as a result of the high accuracy, which is why the KNN classifier was determined to be the most effective approach according to the evaluation.

The DT technique creates a tree-like structure with internal nodes that show feature tests, branches that show the results of the tests, and leaves that show class labels. It is simple to understand and can work with both numerical and categorical data, which is why it is so common in healthcare decision-making systems. The study looked at different numbers of features (attributes) for the DT, and it worked best with 18 features. This setup gave the model an accuracy of 0.79, a precision of 0.86, a specificity of 0.81, and a sensitivity of 0.76. The DT was still able to classify things rather well, even if it did not work as well as KNN. But because it is less sensitive, it may overlook more individuals with heart disease than KNN, which might be very important in real-life medical situations. Still, the model's simplicity and ease of understanding make it a useful part of clinical decision support systems. The comparison of accuracy of various classifiers is illustrated in Figure 16.4(a)–(d).

RF is a way of learning that produces a lot of DT during training. It then gives the mode of the classes or the mean prediction of the individual trees. It makes categorization more accurate by lowering the amount of overfitting that happens with single DTs. We changed the number of estimators in our tests from 10 to 1,000. Using 100 and 500 estimators gave the greatest results since they struck a good mix between speed and accuracy. The RF model got an accuracy of 0.84, a precision of 0.84, a specificity of 0.82, and a sensitivity of 0.81 with these settings. This performance is better than those of the DT and SVM, but it is still a little worse than that of the KNN classifier. It is also interesting to note that adding more estimators past a certain point did not significantly improve accuracy but did increase calculation time. This shows how important it is to tune hyperparameters correctly.

SVMs are strong classifiers that do well in environments with a lot of dimensions. They are especially good in binary classification issues like CAD identification. We used a linear kernel with the SVM technique and changed the gamma parameter, which sets how much one training example affects the model, between 0 and 1. The best result was at gamma = 0.2, which gave an accuracy of 0.83, a precision of 0.85, a

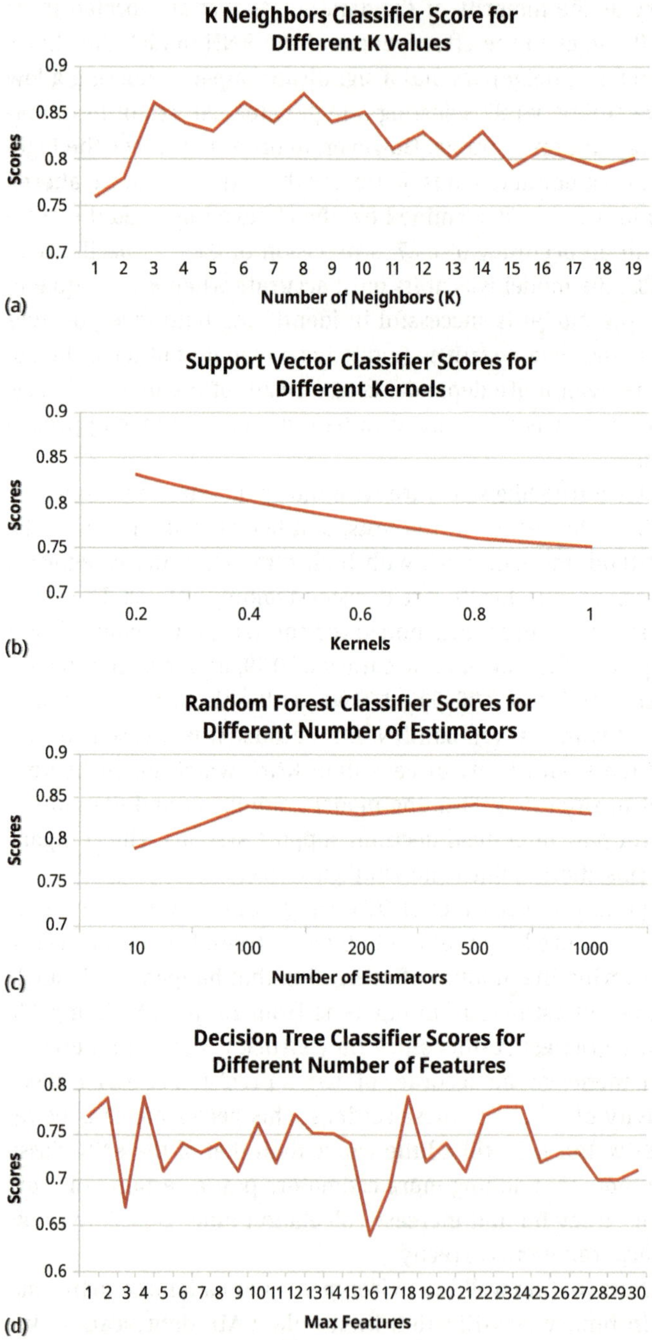

Figure 16.4: Comparison of the accuracies of: (a) *K*-nearest neighbor, (b) support vector, (c) random forest, and (d) decision tree.

specificity of 0.84, and a sensitivity of 0.79. SVM's performance was quite close to that of RF, but its dependence on linear separability and sensitivity to feature scaling might makes it less useful for some datasets. However, the SVM classifier can handle vast feature spaces, therefore it is a good choice for CAD prediction provided it is set up correctly.

Table 16.2: Specification with highest performance for each algorithm.

Algorithm	Specification	Accuracy	Specificity	Precision	Specificity
K-nearest **neighbors classifier**	***K* = 8**	**0.87**	**0.88**	**0.89**	**0.83**
Decision tree classifier	No. of features = 18	0.79	0.81	0.86	0.76
Random forest classifier	No. of estimator = 100	0.84	0.82	0.84	0.81
Support vector	Kernel value = 0.2	0.83	0.84	0.85	0.79

A comparison of the optimal configurations for each of the ML algorithms that were utilized in this investigation is presented in Table 16.2. The following performance measures are included in the table: accuracy, specificity, precision, and sensitivity. The table also includes the optimal parameter choices for KNN, DT, RF, and SVM along with the metrics that correspond to them. Out of all the algorithms, the KNN model with k equal to 8 had the best results. It had an accuracy of 0.87, a precision of 0.89, a specificity of 0.88, and a sensitivity of 0.83 during the course of its performance. The results of this study demonstrate that KNN not only makes extremely accurate predictions regarding CAD, but it also performs a better job than the other models in avoiding both false positives and false negatives for its predictions. A good performance was also achieved by RF and SVM, despite the fact that they were not nearly as accurate or sensitive. In spite of the fact that it was simple to comprehend, the DT classifier was not as accurate or sensitive as other classification methods, which rendered it less suitable for essential medical tasks such as identifying heart abnormalities.

Figure 16.5 illustrates the results obtained in the application showing the performance of the four algorithms for a particular sample and it is predicted that the particular person will have no heart problem.

16.5 Conclusion

The main goal of this study was to improve the accuracy and accessibility of diagnosis by creating a web-based application that uses several ML algorithms to find CAD early. The KNN, DT, RF, and SVM algorithms were used. The KNN approach with $k = 8$ had the best results in terms of accuracy, precision, specificity, and sensitivity. The Django web framework was used to build the app, which makes it easy for users to

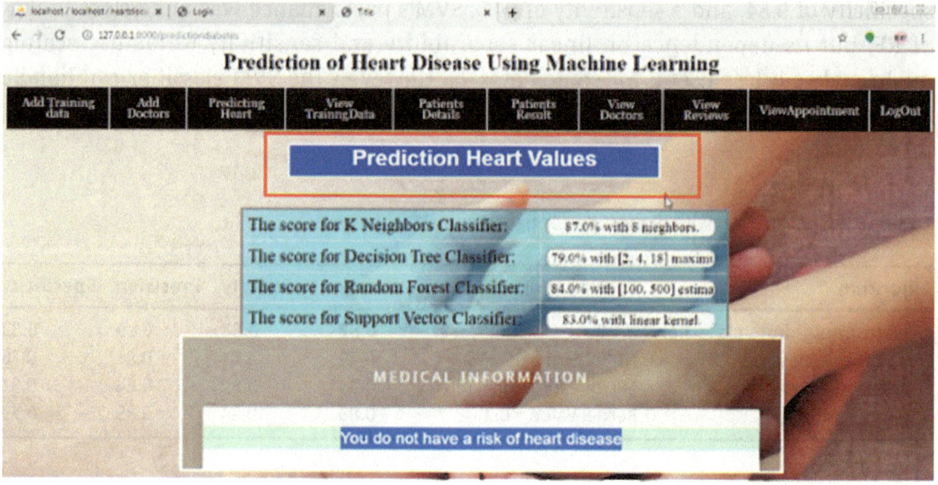

Figure 16.5: Screenshot of application developed for forecasting any disease in heart.

enter medical information and get instant diagnosis predictions. The findings of the experimental research show that the KNN model is quite good at finding patterns in clinical data related to heart disease when it is set up correctly. It had an accuracy of 0.87, with high specificity and sensitivity, which means it is a dependable tool for helping doctors make early diagnoses and estimate risk. Adding these kinds of ML approaches to online platforms is a scalable and useful way to improve healthcare delivery, especially in places where resources are limited or far away. In the future, the project might be improved by adding real-time clinical data and linking the app to EHRs for dynamic analysis. Using DL methods like CNNs or RNNs might also help make things even more accurate. DL models can automatically find and learn complicated characteristics in raw data, which means that people do not have to choose features by hand, which cuts down on mistakes. In conclusion, the system that was created has a lot of promise to help with heart disease diagnosis and is a step toward making AI-assisted healthcare even better in the future.

References

[1] Gaidai, O., Cao, Y. and Loginov, S. Global cardiovascular diseases death rate prediction. Current Problems in Cardiology. 2023;48(5):101622.

[2] Ferreir, J.L., Kumar, S., Soni, A., Acharya, N. and Acharya, S. Clinical Management of Cardiovascular Diseases. In Current Trends in the Diagnosis and Management of Metabolic Disorders 2023 (pp. 254–278). CRC Press.

[3] Alharbi, T.A.F., Rababa, M., Alsuwayl, H., Alsubail, A. and Alenizi, W.S. Diagnostic challenges and patient safety: The critical role of accuracy–a systematic review. Journal of Multidisciplinary Healthcare. 2025;3051–3064.

[4] Yadav, U. and Shrawankar, U. Artificial intelligence across industries: A comprehensive review with a focus on education. AI Applications and Strategies in Teacher Education. 2025;275–320.

[5] Aswini, R., Angelin Rajan, R., Saranya, B. and Elumalai, K. AI-enhanced predictive imaging in precision medicine: Advancing diagnostic accuracy and personalized treatment. iRADIOLOGY. 2025.

[6] Kaur, M., Raikoti, S., Chinthamu, N., Chauhan, D. and Mallapur, S.V. A Novel System for Ensuring Security and Privacy of Remote Health Care Monitoring Using Internet of Things Devices in Smart Homes. In Technological Applications for Smart Sensors 2025 (pp. 539–555). Apple Academic Press.

[7] Da Dang, T., Huynh, C.T. and Ngo, H.P.L. Comparison of Decision Tree and K-Nearest Neighbors for Ad Click Prediction. In International Conference on Computational Intelligence in Engineering Science 2025, July (pp. 221–233). Cham: Springer Nature Switzerland.

[8] Sherman, M. and Klinenberg, E. Beyond burnout: Moral suffering among healthcare workers in the first COVID-19 surge. Social Science & Medicine. 2024;340:116471.

[9] Herrera, D., Sanz, M., Shapira, L., Brotons, C., Chapple, I., Frese, T., Graziani, F., Hobbs, F.R., Huck, O., Hummers, E. and Jepsen, S. Periodontal diseases and cardiovascular diseases, diabetes, and respiratory diseases: Summary of the consensus report by the European Federation of Periodontology and WONCA Europe. European Journal of General Practice. 2024;30(1):2320120.

[10] Deeba, K., Vathana, D., Vanusha, D., and Ramaprabha, J. Applications of Generative Adversarial Networks (GANs) in Healthcare. In Artificial Intelligence Revolutionizing Cancer Care 2025 (pp. 177–211). CRC Press.

[11] Kaur, A., Gupta, S. and Kumar, D. AI-Powered Predictive Modelling for Disease Diagnostics. In Genomic Intelligence (pp. 154–170). CRC Press.

[12] Udoy, I.A. and Hassan, O. AI-driven technology in heart failure detection and diagnosis: a review of the advancement in personalized healthcare. Symmetry. 2025;17(3):469.

[13] Khan Mamun, M.M.R. and Elfouly, T. AI-enabled electrocardiogram analysis for disease diagnosis. Applied System Innovation. 2023;6(5):95.

[14] Krittanawong, C., Aydar, M., Virk, H.U.H., Kumar, A., Kaplin, S., Guimaraes, L., Wang, Z., and Halperin, J.L. Artificial intelligence-powered blockchains for cardiovascular medicine. Canadian Journal of Cardiology. 2022;38(2):185–195.

[15] Rajmohan R, Kumar KS, Chowdhury S, Sharma B, Dhaou IB. Blockchain-Enabled Authenticated Key Management Framework for Electronic Health Record Systems. In Conference on Creativity, Technology, and Sustainability 2024 May 15 (pp. 97–108). Singapore: Springer Nature Singapore.

[16] Golec, M., Gill, S.S., Parlikad, A.K., and Uhlig, S. HealthFaaS: AI-based smart healthcare system for heart patients using serverless computing. IEEE Internet of Things Journal. 2023;10(21):18469–18476.

[17] Riyaz, L., Butt, M.A., Zaman, M., and Ayob, O., 2021, August. Heart disease prediction using machine learning techniques: A quantitative review. In International Conference on Innovative Computing and Communications: Proceedings of ICICC 2021, Vol. 3 (pp. 81–94). Singapore: Springer Singapore.

[18] Bhatt, C.M., Patel, P., Ghetia, T., and Mazzeo, P.L. Effective heart disease prediction using machine learning techniques. Algorithms. 2023;16(2):88.

B. Karthick

17 Analytical Optimization on a Two-Echelon Supply Chain with Fuzzy Production Rate and Environmental Effects

Abstract: This chapter examines the complexity of a two-echelon supply chain that incorporates imperfect production processes and carbon emissions, two critical factors in modern sustainable operations. This model employs a min-max distribution approach to assess demand uncertainty. To effectively capture inherent imprecision, the production rate is assumed to be a pentagonal fuzzy number, and the regular weighted point method is employed to examine the fuzzy value in crisp form. The primary objective of this model is to minimize the total cost of the supply chain. The model is validated through theoretical optimization and extensive numerical analysis, demonstrating its practical functionality. Finally, managerial insights and conclusions, including future directions, are provided.

Keywords: Discrete lead time, defective rate, safety factor, production function, environmental concern

17.1 Introduction

In the ever-evolving supply chain management (SCM) domain, acknowledging imperfections in the production process has become increasingly crucial for achieving peak performance and efficiency. The two-echelon supply chain (SC) model, coupled with inefficient production, delves deep into the intricacies of SC dynamics, recognizing the uncertainties and defects that can significantly impact operational outcomes. Incorporating an industrial carbon emissions tax into SC modeling represents a significant shift towards encouraging sustainable and environmentally conscious business practices. This pioneering approach aims to align economic efforts with environmental management by imposing financial penalties in line with the carbon emissions associated with companies' activities. As a result, companies are urged to carefully assess the environmental impacts of their decisions in SC and inventory management. Introducing a carbon emission tax in inventory models introduces a new perspective to decision-making processes. Traditionally, these models have focused on minimizing ordering, holding, and shortage costs. However, with the addition of a carbon tax, these models have evolved to include the environmental impacts of such decisions. This adaptation encourages buyers, sellers, and distributors to refine their inventory strategies, emphasizing cost-efficiency while minimizing their environmental footprint. Noh and Kim [11] delved into cooperative green SCM, integrating considerations

https://doi.org/10.1515/9783112214374-017

of greenhouse gas emissions and uncertain demand. This chapter contributes to the emerging field of sustainable SCM by integrating environmental factors into decision-making processes. Karthick and Uthayakumar [6] developed a multi-item manufacturing inventory model that integrates setup cost and emission reductions, exemplifying a comprehensive approach to sustainability in the manufacturing process. Sepehri et al. [7] developed a sustainable SC model that incorporates inspection errors and carbon emissions in fluctuating demand, focusing on the challenges of integrating sustainability efforts into an SC given uncertainty. Karthick et al. [8] examined the implications of reducing carbon emissions in SC models, emphasizing variable lead times in an uncertain environment.

The model integrates manufacturing imperfections, realistically portrays the difficulties faced by contemporary SCs, and therefore facilitates strategic decision-making and innovative solutions aimed at improving operational resilience and sustainability. This introduction establishes the foundation for a comprehensive analysis of the relationship between imperfect manufacturing and the multiple dimensions of the two-stage SC. Gallego and Moon [5] conducted fundamental research on the minimum–maximum supply inventory problem, offering beneficial perspectives on dealing with expected shortages during lead-time demand. Effective SCM requires complex decision-making procedures to ensure the smooth flow of goods and services across multiple stages. This study highlights the principles and techniques in SCM, including coordination, defect-free manufacturing, standard SC practices, lead-time reduction, quality improvement, and inventory optimization. Sarkar et al. [13] analyzed a multi-buyer inventory model with variable production rates. Understanding the relationship between production rate variability and product quality is essential for achieving operational excellence in SCM. Mukherjee et al. [10] developed a comprehensive inventory model that integrates optimum vendor investment with backorder price discounts. Their methodology tackles the issues of flawed manufacturing processes and supplier cooperation, enhancing the overall comprehension of SC dynamics. Malik and Sarkar [9] examined a two-echelon SC model with stochastic lead time demand and a combination of inventories to improve coordination in SCM under flexible production. The research highlights the need to synchronize decisions throughout the SC to enhance efficiency. Sarkar and Chung [14] developed a flexible production framework on an SC model to enhance product quality. The research emphasizes the need for flexibility in manufacturing processes to accommodate evolving quality standards and market conditions. Srivastav and Agrawal [15] studied a single-stage inventory model characterized by independent stochastic lead times. Their study enhances comprehension of inventory management amidst uncertainty, offering significant insights for decision-makers in SC operations. Tookanlou and Wong [16] examined the determination of appropriate customization levels, lead times, and inventory placement under product differentiation techniques in SCM.

The recent focus on improving inventory models stems from their significant impact on improving SC performance and cost efficiency. This analysis examines several ways to manage inventory, considering factors such as lead time, demand variability, and product deterioration. Roy and Sana [12] explored strategies for reducing SC lead times in the context of fluctuating demand and variable setup costs. This research offers ideas for reducing lead time uncertainty and its effects on production and inventory choices. Dey et al. [4] analyzed the role of adjustable lead time and variable demand in an innovative manufacturing system within an SCM framework. Choi et al. [3] examined a bright service SC characterized by fluctuating demand and modifiable lead time. The research emphasizes integrating intelligent techniques in SCM, especially in online-to-offline operations. Ali et al. [1] examined the effects of controllable lead times and adhering to service level limitations in an SC. Castellano and Glock [2] examined the economic production quantity for deteriorating products with stochastic demand and a favorable lead time. Their research provides significant insights into the effects of decay and demand uncertainty on production quantity choices, aiding in effectively managing perishable inventory.

Although existing studies have explored individual elements such as imperfect production, sustainability, and inventory control in SCM [5, 11, 16], an integrated examination of these factors within a two-echelon SC framework remains limited. Particularly, the joint impact of production imperfections, carbon emissions, and rework costs is underexplored. Addressing this gap, the present study proposes a comprehensive model that incorporates imperfect production, carbon emission taxes, variable lead time, and a min-max distribution for lead time demand. By capturing key operational complexities and environmental considerations, this research offers a holistic view of SC dynamics. It provides practical insights for decision-makers to better understand the trade-offs between cost, quality, and sustainability, supporting more informed and balanced SC strategies.

17.2 Symbols and Hypotheses

17.2.1 Symbols and Descriptions

The proposed model is developed using these symbols.

Decision Variables

n	Number of shipments
ξ_{ss}	Order size
k	Safety factor
L	Lead time length

Parameters

K_d	Demand rate
p	Production rate
O_c	Buyer's ordering cost

S_c	Seller's setup cost
h_{bc}	Buyer's holding cost
h_{vc}	Seller's holding cost
I_c	Inspection cost
b_c	Backorder cost
r	Reorder level
η	Fraction of imperfect items
X	Lead time demand
σ	Standard deviation
R	Rework cost
ν_{ef}	Emission function factor
E_{carbon}	Carbon emission tax

17.2.2 Hypotheses

- This chapter considers an integrated approach between a single seller and buyer for a single product.
- An infinite time horizon is considered.
- The production rate should always be greater than the demand rate.
- Shortages are allowed, and it is completely backordered.
- The product will only be replenished if the buyer's inventory level drops below the reorder level.
- The distribution of lead time demand X is unknown, but estimates of its mean and variance are available and take finite values.

17.3 Mathematical Formulation

17.3.1 Integrated SC Model

The seller produces $n\xi_{ss}$ quantities of goods to deliver the buyer's order. To begin the production process, the seller incurs the setup cost to set up the machines. The seller's setup cost per unit cycle is $\frac{S_c K_d}{n\xi_{ss}}$.

Using a fuzzy approach allows more flexibility and robustness in the model. It enables decision-makers to analyze different scenarios and assess the impact of varying production rates on key performance indicators such as inventory levels, efficiency, and customer satisfaction.

A pentagonal fuzzy number (PFN), denoted as $\tilde{p} = \left(\varphi_{p1}, \varphi_{p2}, \varphi_{p3}, \varphi_{p4}, \varphi_{p5}\right)$, is defined on $\mathbb{R}$. It exhibits the following characteristics: $\varphi_{p1} \le \varphi_{p2} \le \varphi_{p3} \le \varphi_{p4} \le \varphi_{p5}$. The membership function $\mu_{\tilde{p}}(x)$ is delineated as follows:

$$\mu_{\tilde{p}}(x)=\begin{cases}\frac{x-\varphi_{p1}}{2(\varphi_{p2}-\varphi_{p1})}, & \varphi_{p1}\le x\le\varphi_{p2}\\ 1+\frac{x-\varphi_{p3}}{2(\varphi_{p3}-\varphi_{p2})}, & \varphi_{p2}\le x\le\varphi_{p3}\\ 1+\frac{\varphi_{p3}-x}{2(\varphi_{p4}-\varphi_{p3})}, & \varphi_{p3}\le x\le\varphi_{p4}\\ \frac{\varphi_{p5}-x}{2(\varphi_{p5}-\varphi_{p4})}, & \varphi_{p4}\le x\le\varphi_{p5}\\ 0, & \text{Otherwise}\end{cases}$$

Before transferring orders to the buyer, the seller stores manufactured products at the warehouse/inventory holding area. Therefore, the storage cost of the seller per unit cycle is:

$$\frac{h_{\mathrm{vc}}\xi_{\mathrm{ss}}}{2}\left[\frac{2K_{\mathrm{d}}}{\left(\varphi_{p1},\varphi_{p2},\varphi_{p3},\varphi_{p4},\varphi_{p5}\right)}+\left(1-\frac{K_{\mathrm{d}}}{\left(\varphi_{p1},\varphi_{p2},\varphi_{p3},\varphi_{p4},\varphi_{p5}\right)}\right)n-1\right]$$

The manufactured products must be inspected before being transferred to the buyer, so the cost of that inspection is $I_{\mathrm{c}}K_{\mathrm{d}}$. When the machine changes from a control state to an unstable state, it exhibits defective products. The expected number of defective products detected throughout the production cycle is determined as follows:

$$\eta\left(\varphi_{p1},\varphi_{p2},\varphi_{p3},\varphi_{p4},\varphi_{p5}\right)\left[\frac{\xi_{\mathrm{ss}}}{\left(\varphi_{p1},\varphi_{p2},\varphi_{p3},\varphi_{p4},\varphi_{p5}\right)}-\frac{1}{m_{\mathrm{fp}}}+\frac{1}{m_{\mathrm{fp}}}e^{-\frac{\xi_{\mathrm{ss}}m_{\mathrm{fp}}}{\left(\varphi_{p1},\varphi_{p2},\varphi_{p3},\varphi_{p4},\varphi_{p5}\right)}}\right]$$

Hence, the anticipated quantity of defective items is determined as $\frac{\eta m_{\mathrm{fp}}\xi_{\mathrm{ss}}^2}{2\left(\varphi_{p1},\varphi_{p2},\varphi_{p3},\varphi_{p4},\varphi_{p5}\right)}$. Consequently, the rework cost is ascertained as:

$$\frac{RK_{\mathrm{d}}\eta m_{\mathrm{fp}}\xi_{\mathrm{ss}}}{2\left(\varphi_{p1},\varphi_{p2},\varphi_{p3},\varphi_{p4},\varphi_{p5}\right)}.$$

The term describing the average time when the machine transitions into an out-of-control state is commonly known as $1/m_{\mathrm{fp}}=1/\left(b_f\left(\varphi_{p1},\varphi_{p2},\varphi_{p3},\varphi_{p4},\varphi_{p5}\right)\right)$. It is considered that the cost of manufacturing a unit product corresponds to the rate of production, which is:

$$\frac{a_1}{\left(\varphi_{p1},\varphi_{p2},\varphi_{p3},\varphi_{p4},\varphi_{p5}\right)}+a_2\left(\varphi_{p1},\varphi_{p2},\varphi_{p3},\varphi_{p4},\varphi_{p5}\right)$$

for every $a_1,a_2\in[0,\infty]$. The carbon emission cost of the seller is $v_{\mathrm{ef}}E_{\mathrm{carbon}}n\xi_{\mathrm{ss}}K_{\mathrm{d}}$. The seller's anticipated overall cost is determined by aggregating the previously mentioned cost elements attributed to the seller. This computation yields a comprehensive expenditure associated with the seller's operations:

$$C_M(\xi_{ss}, n) = \frac{S_c K_d}{n\xi_{ss}} + \frac{h_{vc}\xi_{ss}}{2}\left[\frac{2K_d}{\left(\varphi_{p1}, \varphi_{p2}, \varphi_{p3}, \varphi_{p4}, \varphi_{p5}\right)} + \left(1 - \frac{K_d}{\left(\varphi_{p1}, \varphi_{p2}, \varphi_{p3}, \varphi_{p4}, \varphi_{p5}\right)}\right) n - 1\right]$$
$$+ I_c K_d + \frac{RK_d \eta b_f \xi_{ss}}{2} + K_d\left(\frac{a_1}{\left(\varphi_{p1}, \varphi_{p2}, \varphi_{p3}, \varphi_{p4}, \varphi_{p5}\right)} + a_2\left(\varphi_{p1}, \varphi_{p2}, \varphi_{p3}, \varphi_{p4}, \varphi_{p5}\right)\right)$$
$$+ v_{ef} E_{carbon} n\xi_{ss} K_d \tag{17.1}$$

The buyer orders $n\xi_{ss}$ items to the seller; therefore, the buyer incurs an ordering cost, $\frac{O_c K_d}{\xi_{ss}}$. The net inventory of the buyer before the receipt of the order quantity is $r - K_d L$ and after receiving ξ_{ss} item is given by $\xi_{ss} + r - K_d L$. Then, the average inventory of the buyer is calculated as $\frac{\xi_{ss}}{2} + r - K_d L$. Therefore, the holding cost of the buyer is derived as $h_{bc}\left(\frac{\xi_{ss}}{2} + r - K_d L\right)$.

The probability density function of lead-time demand is a member of the distribution class $\Re$, denoted by F, with a mean of $K_d L$ and a standard deviation of $\sigma\sqrt{L}$. Here, the distribution of lead time X is unknown, so the definite value of $\mathbb{E}(X-r)^+$ cannot be determined. The following calculation is used to solve this problem by taking the min-max free distribution strategy [5] and finding the approximate value of $\mathbb{E}(X-r)^+$. For every $F \in \Re$,

$$\mathbb{E}(X-r)^+ \leq \frac{1}{2}\sigma\sqrt{L}\left(\sqrt{1+k^2} - k\right) \tag{17.2}$$

The overall anticipated cost for the buyer, denoted as $C_b(\xi_{ss}, r, L)$, encompasses the sum of ordering cost, holding cost, and shortage cost. As a result:

$$C_B(\xi_{ss}, r, L) = \frac{O_c K_d}{\xi_{ss}} + h_{bc}\left(\frac{\xi_{ss}}{2} + r - K_d L\right) + \frac{b_c K_d}{\xi_{ss}} \mathrm{E}(X-r)^+$$

Therefore, by using eq. (17.2), the function $C_B(\xi_{ss}, r, L)$ is written as:

$$C_B(\xi_{ss}, k, L) = \frac{O_c K_d}{\xi_{ss}} + h_{bc}\left(\frac{\xi_{ss}}{2} + k\sigma\sqrt{L}\right) + \frac{b_c K_d}{2\xi_{ss}}\sigma\sqrt{L}\left(\sqrt{1+k^2} - k\right) \tag{17.3}$$

The comprehensive cost encompassing the entire SC for both the seller (eq. (17.1)) and the buyer (eq. (17.3)) is determined by:

$$C_{\text{MB}}(\xi_{\text{ss}}, k, n, L)$$

$$= K_{\text{d}}\left\{\frac{S_{\text{c}} + nO_{\text{c}} + n\xi_{\text{ss}}I_{\text{c}}}{n\xi_{\text{ss}}} + \frac{h_{\text{vc}}\xi_{\text{ss}}}{2\left(\varphi_{p1}, \varphi_{p2}, \varphi_{p3}, \varphi_{p4}, \varphi_{p5}\right)}(n-2) + \frac{b_{\text{c}}\sigma\sqrt{L}\left(\sqrt{1+k^2}-k\right)}{2\xi_{\text{ss}}}\right\}$$

$$+ \frac{R\eta b_{\text{f}}K_{\text{d}}\xi_{\text{ss}}}{2} + K_d\left(\frac{a_1}{\left(\varphi_{p1}, \varphi_{p2}, \varphi_{p3}, \varphi_{p4}, \varphi_{p5}\right)} + a_2\left(\varphi_{p1}, \varphi_{p2}, \varphi_{p3}, \varphi_{p4}, \varphi_{p5}\right)\right)$$

$$+ v_{\text{ef}}K_{\text{d}}E_{\text{carbon}}n\xi_{\text{ss}} + h_{\text{bc}}\left(\frac{\xi_{\text{ss}}}{2} + k\sigma\sqrt{L}\right) + \frac{h_{\text{vc}}\xi_{\text{ss}}}{2}(n-1) \tag{17.4}$$

subject to $n > 0, k > 0, \xi_{\text{ss}} > 0,$ and $L > 0$.

17.3.2 Fuzzy to Crisp Conversion

Defuzzification is the method of converting fuzzy sets representing linguistic variables in fuzzy logic into smooth values that can be used in decision-making. In fuzzy modeling, inputs and outputs are often represented by fuzzy sets, which capture the imprecision and uncertainty inherent in human language. For any fuzzy number $\tilde{p}$, the λ-cut is defined as $P_\lambda = \left[B_{\tilde{p}}(\lambda), K_{\tilde{p}}(\lambda)\right]$. The crisp conversion for $\widetilde{P_t}$ using regular weighted point is given by:

$$D_{\text{fuzzy}}(\tilde{p}) = \frac{\int_0^1 \left(\frac{\left(B_{\tilde{p}}(\lambda) + K_{\tilde{p}}(\lambda)\right)}{2}\right)}{\int_0^1 f(\lambda) da}$$

where

$$f(a) = \begin{cases} 1 - 2\lambda, & \lambda \in \left[0, \frac{1}{2}\right] \\ 2\lambda - 1, & \lambda \in \left[\frac{1}{2}, 1\right] \end{cases}$$

For a PFN, the λ-cut is given by:

$$P_\lambda = \begin{cases} \left[\varphi_{p1} + 2\lambda\left(\varphi_{p2} - \varphi_{p1}\right), \varphi_{p5} - 2\lambda\left(\varphi_{p5} - \varphi_{p4}\right)\right], & \lambda \in \left[0, \frac{1}{2}\right] \\ \left[2\varphi_{p2} - \varphi_{p3} + 2\lambda\left(\varphi_{p3} - \varphi_{p2}\right), 2\varphi_{p3} - \varphi_{p4} + 2\lambda\left(\varphi_{p3} - \varphi_{p2}\right)\right], & \lambda \in \left[\frac{1}{2}, 1\right] \end{cases}$$

Thus, the defuzzified result of the PFN is obtained as follows:

$$D_{\text{fuzzy}}\left(\widetilde{P_t}\right) = \frac{1}{12}\left(2\varphi_{p1} + 2\varphi_{p2} + 3\varphi_{p3} + 3\varphi_{p4} + 2\varphi_{p5}\right)$$

Therefore, the defuzzification of $C_{\mathrm{MB}}(\xi_{\mathrm{ss}}, k, n, L)$ (i.e., (17.4)) is derived as:

$$
\begin{aligned}
&TC(\xi_{\mathrm{ss}}, k, n, L)\\
&= K_{\mathrm{d}}\left\{\frac{S_{\mathrm{c}} + nO_{\mathrm{c}} + n\xi_{\mathrm{ss}}I_{\mathrm{c}}}{n\xi_{\mathrm{ss}}} + \frac{h_{\mathrm{vc}}\xi_{\mathrm{ss}}(n-2)}{\frac{1}{6}\left(2\varphi_{p1} + 2\varphi_{p2} + 3\varphi_{p3} + 3\varphi_{p4} + 2\varphi_{p5}\right)}\right.\\
&\quad\left. + \frac{b_{\mathrm{c}}\sigma\sqrt{L}\left(\sqrt{1+k^2} - k\right)}{2\xi_{\mathrm{ss}}} + \frac{R\eta b_{\mathrm{f}}\xi_{\mathrm{ss}}}{2}\right\}\\
&\quad + K_{\mathrm{d}}\left(\frac{a_1}{\frac{1}{12}\left(2\varphi_{p1} + 2\varphi_{p2} + 3\varphi_{p3} + 3\varphi_{p4} + 2\varphi_{p5}\right)} + \frac{a_2}{12}\left(2\varphi_{p1} + 2\varphi_{p2} + 3\varphi_{p3} + 3\varphi_{p4} + 2\varphi_{p5}\right)\right)\\
&\quad + K_{\mathrm{d}}v_{\mathrm{ef}}E_{\mathrm{carbon}}n\xi_{\mathrm{ss}} + h_{\mathrm{bc}}\left(\frac{\xi_{\mathrm{ss}}}{2} + k\sigma\sqrt{L}\right) + \frac{h_{\mathrm{vc}}\xi_{\mathrm{ss}}}{2}(n-1)
\end{aligned}
\tag{17.5}
$$

subject to $n > 0, k > 0, \xi_{\mathrm{ss}} > 0,$ and $L > 0$.

17.3.3 Solution Procedure

In this section, we delineate the theoretical solution methodology, elucidating the requisite and satisfactory conditions essential for determining the optimal values of decision variables.

Differentiating $TC(n, \xi_{\mathrm{ss}}, k, L)$ (i.e., eq. (17.5)) partially with respect to ξ_{ss}, we get:

$$
\begin{aligned}
\frac{\partial TC(n, \xi_{\mathrm{ss}}, k, L)}{\partial \xi_{\mathrm{ss}}} &= \frac{K_{\mathrm{d}}}{\xi_{\mathrm{ss}}^2}\left[-O_{\mathrm{c}} - \left(\frac{b_{\mathrm{c}}\sigma\sqrt{L}\left(\sqrt{(1+k^2)} - k\right)}{2}\right) - \frac{S_{\mathrm{c}}}{n}\right.\\
&\quad\left. + \frac{R\eta(b_{\mathrm{f}})}{2} + v_{\mathrm{ef}}E_{\mathrm{carbon}}n\right] + \frac{h_{\mathrm{bc}}}{2}
\end{aligned}
\tag{17.6}
$$

Differentiating $TC(n, \xi_{\mathrm{ss}}, k, L)$ twice with respect to ξ_{ss}, we get:

$$
\frac{\partial^2 TC(n, \xi_{\mathrm{ss}}, k, L)}{\partial \xi_{\mathrm{ss}}^2} = \frac{2K_{\mathrm{d}}}{\xi_{\mathrm{ss}}^3}\left(O_{\mathrm{c}} + \frac{S_{\mathrm{c}}}{n} - b_{\mathrm{c}}\sigma\sqrt{L}\left(\sqrt{1+k^2} - k\right)\right) > 0
$$

To obtain the optimal value, equate eq. (17.6) to zero. Therefore, it is proved that $TC(n, \xi_{\mathrm{ss}}, k, L)$ reaches a local minimum at the optimal value of ξ_{ss}.

Differentiating $TC(n, \xi_{\mathrm{ss}}, k, L)$ partially with respect to k, we get:

$$\frac{\partial TC(n,\xi_{ss},k,L)}{\partial k}=h_{bc}\sigma\sqrt{L}+\frac{b_cK_d\sigma\sqrt{L}}{2\xi_{ss}}\left(\frac{k}{\sqrt{1+k^2}}-1\right) \tag{17.7}$$

Differentiating $TC(n,\xi_{ss},k,L)$ twice with respect to k, we get:

$$\frac{\partial^2 TC(n,\xi_{ss},k,L)}{\partial k^2}=\frac{b_cK_d\sigma\sqrt{L}}{2\xi_{ss}}\left(\frac{1}{\sqrt{1+k^2}}-\frac{k^2}{(1+k^2)^{\frac{3}{2}}}\right)>0$$

To obtain the optimal value, equate eq. (17.7) to zero. Therefore, it is proved that $TC(n,\xi_{ss},k,L)$ reaches a local minimum at the optimal value of k.

The cost function $TC(n,\xi_{ss},k,L)$ exhibits convexity at the optimal values of ξ_{ss}, k, and L for any positive integer value of n.

The Hessian matrix $\mathbf{H}$ is as follows:

$$\mathbf{H}=\begin{bmatrix}\frac{\partial^2 TC(n,\xi_{ss},k,L)}{\partial \xi_{ss}^2} & \frac{\partial^2 TC(n,\xi_{ss},k,L)}{\partial \xi_{ss}\,\partial k}\\ \frac{\partial^2 TC(n,\xi_{ss},k,L)}{\partial k\,\partial \xi_{ss}} & \frac{\partial^2 TC(n,\xi_{ss},k,L)}{\partial k^2}\end{bmatrix}$$

$$\frac{\partial^2 TC(n,\xi_{ss},k,L)}{\partial \xi_{ss}\,\partial k}=\frac{\partial^2 TC(n,\xi_{ss},k,L)}{\partial k\,\partial \xi_{ss}}=\frac{-b_cK_d\sigma\sqrt{L}}{2\xi_{ss}^2}\left(\frac{k}{\sqrt{1+k^2}}-1\right)$$

The first principal minor of $\mathbf{H}$ is:

$$|\mathbf{H}_{11}|=\frac{\partial^2 TC(n,\xi_{ss},k,L)}{\partial \xi_{ss}^2}=\frac{2K_d}{\xi_{ss}^3}\left(O_c+\frac{S_c}{n}-b_c\sigma\sqrt{L}\left(\sqrt{1+k^2}-k\right)\right)>0$$

The second principal minor of $\mathbf{H}$ is:

$$|\mathbf{H}_{22}|=\left[\frac{\partial^2 TC(n,\xi_{ss},k,L)}{\partial \xi_{ss}^2}\times\frac{\partial^2 TC(n,\xi_{ss},k,L)}{\partial k^2}\right]-\left[\frac{\partial^2 TC(n,\xi_{ss},k,L)}{\partial \xi_{ss}\partial k}\right]^2$$

$$=\left[\frac{2K_d}{\xi_{ss}^3}\left(O_c+\frac{S_c}{n}-b_c\sigma\sqrt{L}\left(\sqrt{1+k^2}-k\right)\right)\times\frac{b_cK_d\sigma\sqrt{L}}{2\xi_{ss}}\left(\frac{1}{\sqrt{1+k^2}}-\frac{k^2}{(1+k^2)^{\frac{3}{2}}}\right)\right]$$

$$+\left(\frac{b_cK_d\sigma\sqrt{L}}{2\xi_{ss}^2}\left(\frac{k}{\sqrt{1+k^2}}-1\right)\right)^2>0$$

Based on the above findings, all principal minors of the matrix $\mathbf{H}$ are positive, confirming its positive definiteness. Consequently, the convexity of the expression $TC(n,\xi_{ss},k,L)$ is established.

Figure 17.1 illustrates that the total cost function achieves a local minimum with respect to both continuous variables ξ_{ss} and k.

17.4 Numerical Analysis

17.4.1 Examples

Example 17.1 (Fuzzy Production Case). The proposed model's numerical specifications are outlined through the parameter values, as detailed in Sarkar et al.'s work from 2018 [13], which are provided in the following list: $K_d = 300$/year, $S_c = \$2,000$/setup, $h_{vc} = \$11$/units/year, $O_c = \$100$/setup, $h_{bc} = \$12$/units/year, $\varphi_{p1} = 1,550$/units/year, $\varphi_{p2} = 1,900$/units/year, $\varphi_{p3} = 2,050$/units/year, $\varphi_{p4} = 2,150$/units/year, $\varphi_{p5} = 2,250$/units/year, $a_1 = 35 \times 10^3$, $a_2 = 0.1$, $I_c = 12$, $b_c = \$25$/unit, $R = 5$, $v_{ef} = 3 \times 10^{-4}$, $E_{carbon} = 20$, $\sigma = 9$, $\eta = 0.1$, and $b_f = 10^{-2}$.

Example 17.2 (Constant Production Case). The data in Example 17.1 remains unchanged, except for the numerical values associated with the fuzzy production rate (FPR). When the production rate becomes constant, the membership function transitions from a gradual change to a sharp peak at a fixed value.

The membership function for a constant production rate becomes:

$$\mu(x) = \begin{cases} 1 & \text{if } x = 2{,}050 \text{ units/year} \\ 0 & \text{otherwise} \end{cases}$$

When the FPR becomes constant: $\varphi_{p1} = \varphi_{p2} = \varphi_{p3} = \varphi_{p4} = \varphi_{p5} = p$. where p is the constant production rate. In this case, $p = 2{,}050$ units/year.

17.4.2 Numerical Results and Discussion

This section analyzes the optimum values derived from Tables 17.1 and 17.2. Both examples represent distinct SC models: Example 17.1 shows an SC with a pentagonal FPR, whereas Example 17.2 has a constant production rate. The analysis from those example results is as follows:

- As L increases from 21 to 56, the shipment size ξ_{ss} increases in both examples. This suggests that a larger L requires a higher inventory to ensure demand fulfillment and system stability, especially when dealing with uncertainties in the production rate.
- In Example 17.1, the values of ξ_{ss} range from 301.95 to 77.36 for $L = 21$ and from 322.83 to 88.16 for $L = 56$. The increase in inventory levels with L indicates that the system, modeled with FPRs, accommodates a higher level of uncertainty, leading to larger shipment sizes for the buyer.
- In Example 17.2, the trend of increasing ξ_{ss} with L is also observed, but the values of ξ_{ss} are slightly lower compared to Example 17.1. For instance, for$L = 21$, ξ_{ss} ranges from 301.95 to 77.36, similar to Example 17.1, but for $L = 56$, it ranges from 322.83 to 88.16. The smaller increase in inventory in Example 17.2 suggests that

the constant production rate model is less sensitive to fluctuations in supply compared to the FPR model.
- While both models exhibit similar trends, Example 17.1, with its FPR, shows greater flexibility in managing uncertainty, leading to slightly more efficient use of inventory and reduced safety factors. Example 17.2, with a constant production rate, requires slightly higher inventory levels and safety factors, which result in higher total costs.

The numerical results are given in Tables 17.1 and 17.2. Also, the findings are visually encapsulated in Figures 17.2 and 17.3, providing a graphical representation of the effects on the integrated cost of the system under different lead time scenarios. This comprehensive analysis contributes valuable insights for optimizing the SC, with a focus on minimizing total costs while considering lead time variations.

17.5 Managerial Insights

- Increasing lead times significantly raise the overall cost of the SC, highlighting the need for careful management. Effective management of lead times can dramatically contribute to achieving sustainability goals while controlling costs in the SC.
- Incorporating factors such as carbon emission tax, rework cost, and production cost variability into inventory models enhances their accuracy and relevance.
- Decision-makers should prioritize strategies that minimize lead times to mitigate cost escalation and improve operational efficiency.
- Sustainability and cost concerns in industries underscore the relevance of findings from this study for inventory management strategies.
- Pentagonal FPR facilitate adaptive planning by allowing managers to continuously monitor and adjust production capacities in response to changing market conditions and unforeseen events.

17.6 Conclusion and Possible Extensions

The proposed research findings showed that lead time management is crucial for improving SC costs and efficiencies under uncertainty. PFNs are used to treat the production rate, capture uncertainty better, and allow managers to make good decisions, especially when dealing with ambiguous information. The model thoroughly explains the factors influencing SC operations, including the carbon emission tax and maintenance costs. Two numerical examples were discussed to analyze the proposed model, in which the FPR model in Example 17.1 allows for more effective inventory management, particularly when dealing with uncertain production rates. In contrast, the

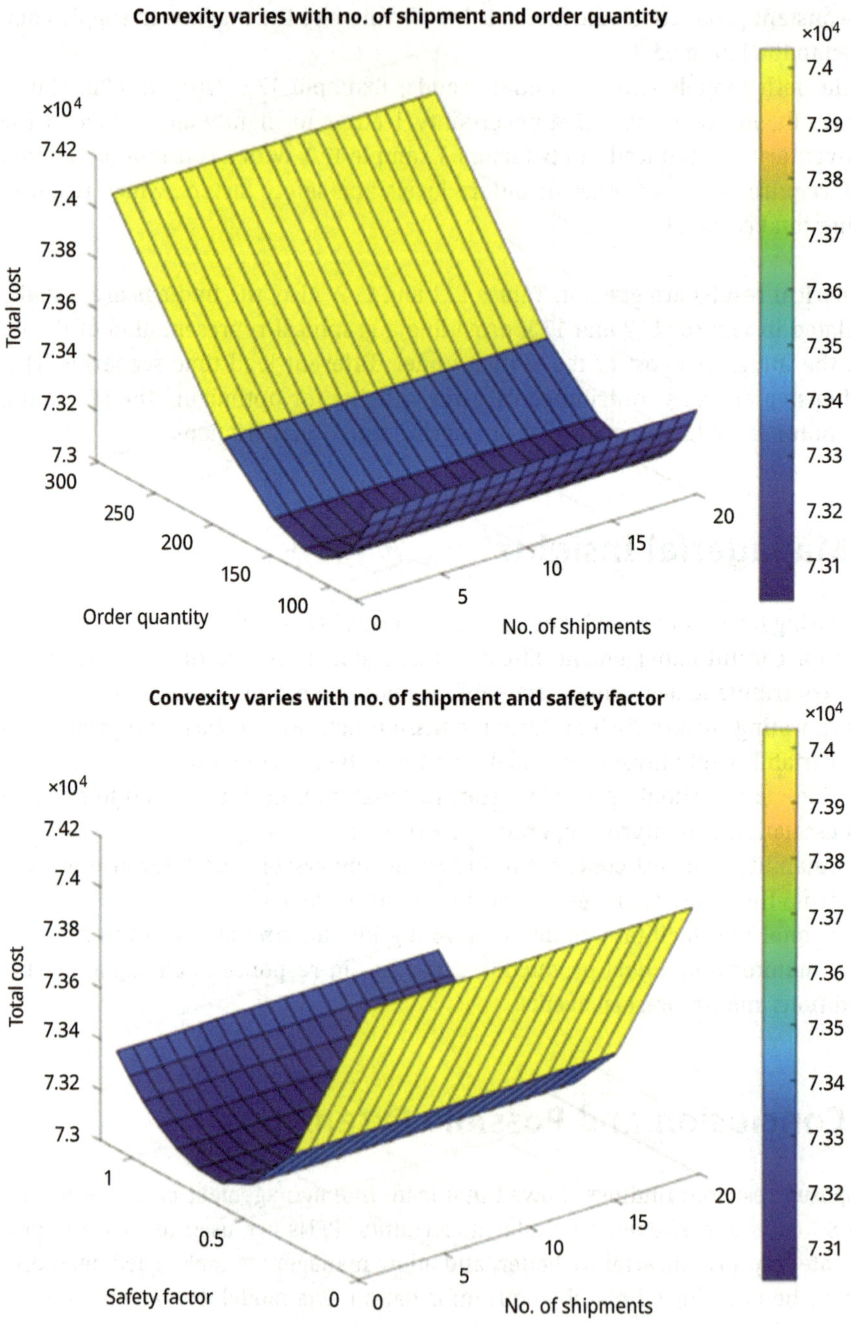

Figure 17.1: Total cost convexity with respect to ξ_{ss} and k from the results of Example 17.1.

Table 17.1: The optimal values of Eample 17.1.

$L=21$		**1**	**2**	**3**	**4**	**5**	**6**	**7**	**8**	**9**	**10**
	ξ_{ss}	301.94781	205.80166	162.73520	137.03476	119.58415	106.81012	96.9823216	89.147064	82.730245	77.3632541
	k	0.0337862	0.3632648	0.5460389	0.6785490	0.7847054	0.8743035	1.0138463	1.0219163	1.0848083	1.1423701
	TC	74,031.679	73,300.880	73,095.059	73,038.796	73,044.315	73,080.144	73,132.318	73,193.785	73,260.659	73,330.660
$L=28$	n	1	2	3	4	5	6	7	8	9	10
	ξ_{ss}	306.92390	210.15524	166.72005	140.72819	123.03457	110.05324	100.04595	92.053464	85.497561	80.006511
	k	0.0178463	0.3466192	0.5273979	0.6579861	0.7623940	0.8504097	0.9270530	0.9952710	1.05695611	1.11340321
	TC	74,110.247	73,409.3004	73,223.045	73,182.044	73,200.333	73,247.263	73,309.322	73,379.741	73,454.819	73,532.412
$L=42$	n	1	2	3	4	5	6	7	8	9	10
	ξ_{ss}	315.45004	217.54460	173.45783	146.95774	128.84353	115.50529	105.19016	96.928828	90.13571693	84.433538
	k	0.0094405	0.3189374	0.49682154	0.6245676	0.7263828	0.81205184	0.8865640	1.002835	1.012732	1.0675282
	TC	74,239.238	73,587.41	73,433.110	73,416.968	73,456.031	73,521.0073	73,599.1332	73,684.098	73,772.511	73,862.444
$L=56$	n	1	2	3	4	5	6	7	8	9	10
	ξ_{ss}	322.82551	223.85990	179.189065	152.24062	133.75877	120.11050	109.529208	101.03628	94.039463	88.156428
	k	0.0330596	0.2957971	0.4716661	0.5973626	0.6972943	0.7812548	0.8542137	0.9190656	0.9776576	1.0312471
	TC	74,345.247	73,733.962	73,605.843	73,610.000	73,666.003	73,745.687	73,836.904	73,933.722	74,033.002	74,132.995

Table 17.2: The optimal values of Example 17.2.

$L = 21$	n	1	2	3	4	5	6	7	8	9	10
	ξ_{ss}	301.9478	205.8017	162.7352	137.0348	119.5842	106.8101	96.9823	89.1471	82.7302	77.3633
	k	0.0338	0.3633	0.5460	0.6785	0.7847	0.8743	0.9524	1.0219	1.0848	1.1424
	TC	75,403.5900	74,672.8320	74,467.0514	74,410.8280	74,416.3878	74,452.2563	74,504.4708	74,565.9785	74,632.8925	74702.9340
$L = 26$	n	1	2	3	4	5	6	7	8	9	10
	ξ_{ss}	305.5652	208.9694	165.6358	139.7238	122.0968	109.1721	99.2139	91.2643	84.7463	79.2891
	k	0.0222	0.3511	0.5324	0.6635	0.7684	0.8568	0.9338	1.0024	1.0644	1.1211
	TC	75,460.9500	74,751.9822	74,560.4950	74,515.4237	74,530.3162	74,574.2984	74,633.7383	74,701.7884	74,774.6993	74850.2894
$L = 42$	n	1	2	3	4	5	6	7	8	9	10
	ξ_{ss}	315.4500	217.5446	173.4578	146.9577	128.8435	115.5053	105.1902	96.9288	90.1357	84.4335
	k	−0.0094	0.3189	0.4968	0.6246	0.7264	0.8121	0.8866	0.9528	1.0127	1.0675
	TC	75,611.149	74,959.355	74,805.102	74,789.000	74,828.103	74,893.119	74,971.285	75,056.290	75,144.744	75234.717
$L = 56$	n	1	2	3	4	5	6	7	8	9	10
	ξ_{ss}	322.8255	223.8599	179.1891	152.2406	133.7588	120.1105	109.5292	101.0363	94.0395	88.1564
	k	−0.0331	0.2958	0.4717	0.5974	0.6973	0.7813	0.8542	0.9191	0.9777	1.0312
	TC	75,717.1585	75,105.9139	74,977.8353	74,982.0324	75,038.0754	75,117.7994	75,209.0566	75,305.9147	75,405.2352	75505.2682

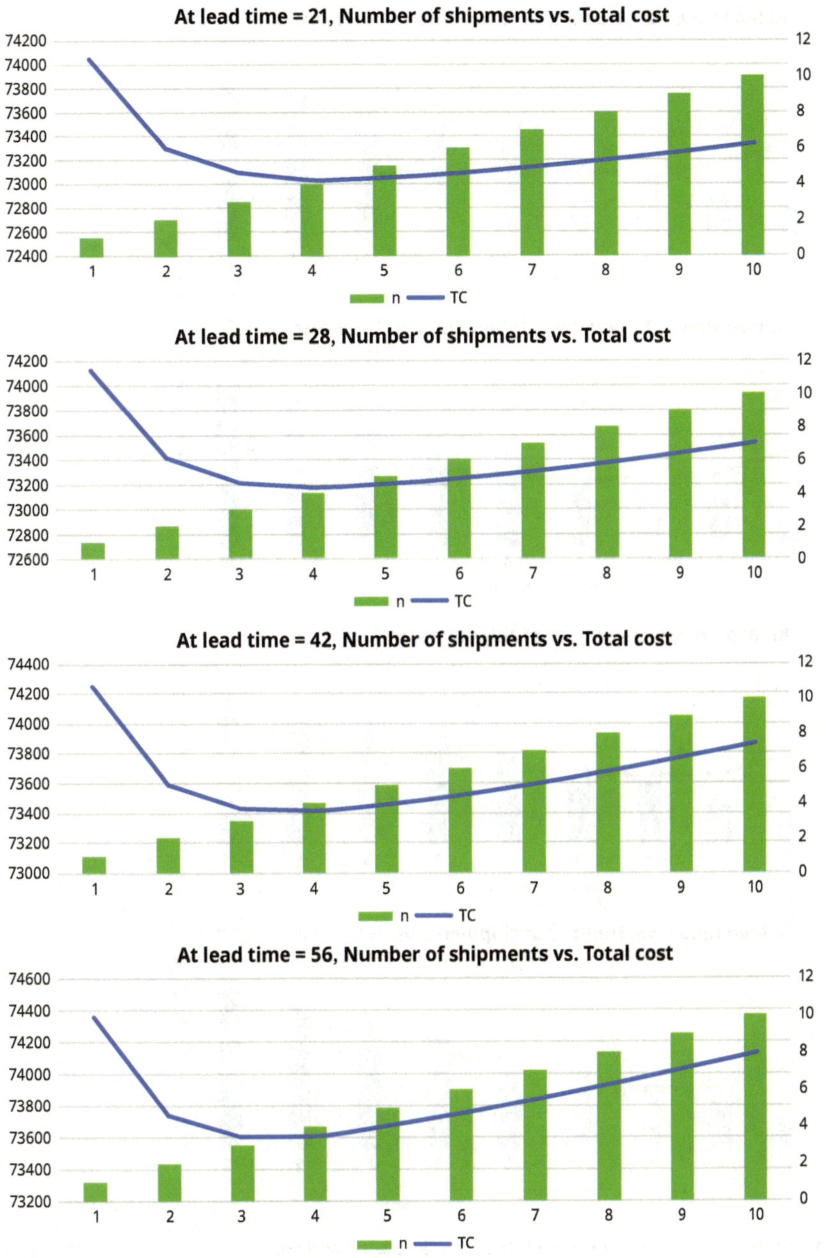

Figure 17.2: Effects on the total cost of Example 17.1 if $L = 21$, 28, 42, and 56.

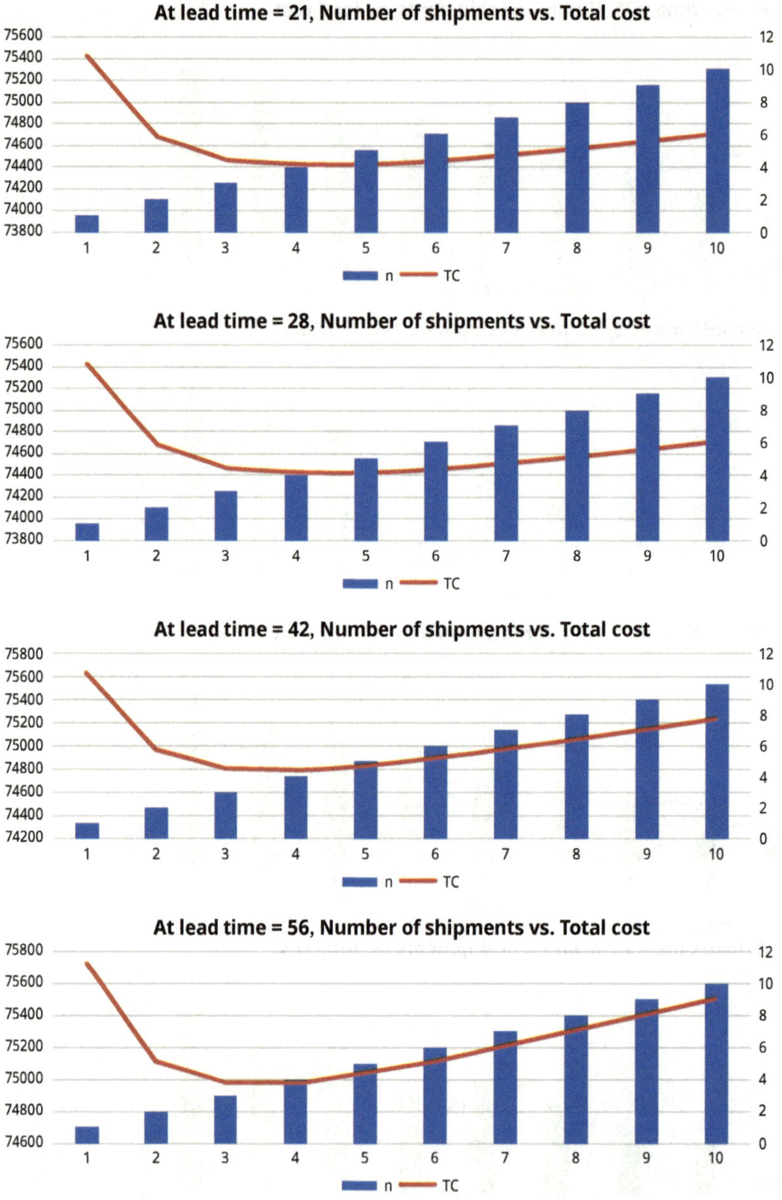

Figure 17.3: Effects on the total cost of Example 17.2 if $L = 21$, 28, 42, and 56.

fixed production rate model in Example 17.2 incurs higher costs because it lacks the flexibility to respond to supply and demand fluctuations. These findings suggest that including FPRs in SC models might lead to more effective inventory and cost control. The approach also emphasizes the importance of adding green goals into operations, which must be considered alongside traditional cost targets.

This study reveals several interesting areas for future research. This study can be extended to comprehend the impact of extended lead times on SC expenses. Also, considering a multi-item with a multi-buyer would be a realistic extension of this study. Moreover, incorporating risks and disruptions into the model, such as fluctuations in demand or manufacturing delays, may enhance its practical usefulness. Establishing a robust decision-making framework to address such uncertainty is essential for developing proactive and effective SC strategies. This work can be extended to analyze the effectiveness of inspection errors and the reworking process.

References

[1] Ali H, Nasreen R, Arneja N, Jaggi CK. Optimization of a periodically assessing model with manageable lead time under SLC with back order rebate for deteriorating items. International Journal of System Assurance Engineering and Management. 2023 Feb;14(1):241–266.

[2] Castellano D, Glock CH. Economic production quantity for a decaying item with stochastic demand and positive lead time. International Journal of Production Economics. 2024 Jan 1;267:109094.

[3] Choi SB, Dey BK, Kim SJ, Sarkar B. Intelligent servicing strategy for an online-to-offline (O2O) supply chain under demand variability and controllable lead time. RAIRO-operations Research. 2022 May 1;56(3):1623–1653.

[4] Dey BK, Bhuniya S, Sarkar B. Involvement of controllable lead time and variable demand for a smart manufacturing system under a supply chain management. Expert Systems with Applications. 2021 Dec 1;184:115464.

[5] Gallego G, Moon I. The distribution free newsboy problem: Review and extensions. Journal of the Operational Research Society. 1993 Aug 1;44(8):825–834.

[6] Karthick B, Uthayakumar R. A multi-item sustainable manufacturing model with discrete setup cost and carbon emission reduction under deterministic and trapezoidal fuzzy demand. Process Integration and Optimization for Sustainability. 2021 Sep;5(3):505–543.

[7] Sepehri A, Gholamian MR. A green inventory model with imperfect items considering inspection process and quality improvement under different shortages scenarios. Environment, Development and Sustainability. 2023;25(4):3269–3297.

[8] Karthick B, Uthayakumar R. Impact of carbon emission reduction on supply chain model with manufacturing decisions and dynamic lead time under uncertain demand. Cleaner Logistics and Supply Chain. 2022 Jul 1;4:100037.

[9] Malik AI, Sarkar B. Coordination supply chain management under flexible manufacturing, stochastic leadtime demand, and mixture of inventory. Mathematics. 2020 Jun 3;8(6):911.

[10] Mukherjee A, Dey O, Giri BC. An Integrated Imperfect Production–inventory Model with Optimal Vendor Investment and Backorder Price Discount. In Information Technology and Applied Mathematics: ICITAM 2017 2018 May 9 (pp. 187–203). Singapore: Springer Singapore.

[11] Noh J, Kim JS. Cooperative green supply chain management with greenhouse gas emissions and fuzzy demand. Journal of Cleaner Production. 2019 Jan 20;208:1421–1435.

[12] Roy MD, Sana SS. Production rate and lot-size dependent lead time reduction strategies in a supply chain model with stochastic demand, controllable setup cost and trade-credit financing. RAIRO-operations Research. 2021;55:S1469–85.
[13] Sarkar B, Majumder A, Sarkar M, Kim N, Ullah M. Effects of variable production rate on quality of products in a single-vendor multi-buyer supply chain management. The International Journal of Advanced Manufacturing Technology. 2018 Oct;99(1):567–581.
[14] Sarkar M, Chung BD. Flexible work-in-process production system in supply chain management under quality improvement. International Journal of Production Research. 2020 Jul 2;58 (13):3821–3838.
[15] Srivastav A, Agrawal S. On a single item single stage mixture inventory models with independent stochastic lead times. Operational Research. 2020 Dec;20(4):2189–2227.
[16] Tookanlou PB, Wong H. Determining the optimal customization levels, lead times, and inventory positioning in vertical product differentiation. International Journal of Production Economics. 2020 Mar 1;221:107479.

C. N. Savithri, Lavanya Devi N., Chitra R., Femila L.

18 Verdcure: Automatic Identification of Plant Diseases Using Convolutional Neural Networks

Abstract: A country's economic growth majorly depends on the farming. The livelihood of the majority population in India depends on agriculture. Diseases that harm plants are a big problem for farmers all over the world because they reduce crop yields and put food supply at risk. Quick and correct diagnosis is very important for properly managing and controlling chronic conditions. Traditional methods of diagnosing diseases rely heavily on manual inspection and specialist knowledge, which can take a long time, be subjective, and be wrong. To get the most out of farming, more and more farmers are using tools like deep learning (DL), machine learning (ML), and the Internet of things (IoT). The purpose of this research is to create a system that can automatically find plant diseases in leaf pictures using convolutional neural networks (CNNs). CNNs are great for visual identification jobs because they can learn complicated patterns and features straight from raw picture input. This work used a dataset with a mix of photos of healthy and sick plant leaves to train and test a number of different CNN architectures. Image augmentation was part of the model preparation to make it better at generalizing and less likely to overfit. Tests on a lot of various kinds of crops reveal that the CNN model we provide can correctly identify a lot of different kinds of plant illnesses. Technology can tell the difference between healthy and sick leaves, as well as some types of diseases. After diagnosis of disease, the system will suggest various measures to cure or prevent the same. This automated approach not only accelerates disease diagnosis but also enables nonexperts, such as farmers and agricultural extension workers, to monitor plant health using mobile devices equipped with cameras. The integration of CNN-based disease identification into agricultural practices holds promise for enhancing crop management, reducing pesticide misuse, and ultimately improving agricultural productivity and sustainability. Future study could include expanding datasets, incorporating environmental elements, and implementing real-time sickness detection systems.

Keywords: Django, support vector machine (SVM), *K*-nearest neighbor (KNN), random forest (RF), decision trees (DTs)

18.1 Introduction

Cultivation has long been recognized as the backbone of India's economy and sustenance. According to data from the India Brand Equity Foundation (IBEF), nearly 58%

https://doi.org/10.1515/9783112214374-018

of India's current population depends directly or indirectly on agriculture and related sectors for their livelihood. Agriculture is not limited to large-scale farming alone; a significant portion of the population engages in small-scale cultivation and gardening, often within their homes or community spaces. The COVID-19 pandemic, which resulted in extended lockdowns and restricted outdoor activities, brought renewed interest in home gardening. People sought healthier lifestyles and greater self-sufficiency by growing essential fruits, vegetables, and herbs at home. This surge in gardening enthusiasm has introduced many new individuals who may lack prior experience or knowledge in plant care and disease management to plant cultivation. Plant diseases, caused by a variety of pathogens such as fungi, bacteria, viruses, and pests, pose a significant threat to plant health. Common diseases like powdery mildew, leaf blight, and rust can rapidly damage crops if not identified and managed promptly. Early diagnosis is critical in preventing the spread of these diseases and minimizing crop loss. However, for amateur gardeners and small-scale growers, distinguishing between healthy and infected plants is challenging due to the subtle and often similar visual symptoms exhibited by different diseases.

Advancements in artificial intelligence, particularly deep learning techniques such as this research leverage these technological developments to create an accessible and efficient solution for plant disease diagnosis. By enabling users to upload a photo of an affected leaf via a website or mobile application, the system instantly provides an accurate diagnosis and actionable treatment recommendations. This tool not only empowers novice gardeners to protect their plants but also promotes sustainable gardening practices by reducing unnecessary use of pesticides and preventing crop losses. Ultimately, it supports the growing community of home growers, encouraging them to continue cultivating healthy plants and contributing to food security and environmental well-being.

18.1.1 Survey on Automatic Identification of Plant Diseases

Numerous studies have investigated the application of artificial neural networks (ANN) and various machine learning (ML) methodologies in conjunction with image processing for the identification and classification of plant diseases. These approaches primarily focus on feature extraction from plant images, followed by classification to detect various diseases. For example, early work employed methods such as self-organizing feature maps (SOFMs) and the backpropagation algorithm to recognize and classify different plant diseases based on leaf images. Support vector machines (SVMs) were also commonly applied to categorize diseases by learning from extracted image features such as color, texture, and shape patterns [1–4].

A combination of a minimum distance classifier and SVMs is applied to classify diseases in mango plants, specifically targeting anthracnose and leaf spot diseases. The researchers utilized Gabor filters and gray-level co-occurrence matrix (GLCM) techniques to extract relevant texture and color features from leaf images. Gabor filters are effective in capturing texture variations at different scales and orientations, while GLCM analyzes spatial relationships between pixels to quantify texture. Their method demonstrated improved accuracy in detecting and differentiating these diseases [5–7].

It was suggested that hyperspectral imaging methods in the visible and near-infrared (VNIR) and short-wave infrared (SWIR) regions could be used to find the tomato spotted wilt virus. Hyperspectral imaging collects data from a wide range of wavelengths, which makes it possible to see complex spectral signatures of plant tissues. The study employed three feature extraction strategies: probabilistic topic modeling-based features, vegetation indices, and full-spectrum data, which were used to train classification models. This technique achieved classification accuracy exceeding 90%, highlighting the potential of hyperspectral imaging for precise disease detection [8].

The Canny edge detection method is utilized to identify disease-affected regions in pomegranate fruits. Using GrabCut segmentation, a graph-based image segmentation technique, they isolated the areas showing symptoms of bacterial blight disease. By counting the number of pixels corresponding to edges within the segmented regions, the researchers quantified the extent of infection in the fruits. This pixel-based analysis enabled effective measurement of disease severity [9]. A method for quantitative texture analysis in digital images using what is now known as the gray-level co-occurrence matrix (GLCM) was introduced. A set of 14 statistical features derived from GLCMs, which quantify how often pairs of pixel intensities occur in a specific spatial relationship (distance and direction) within an image is proposed. These features, such as contrast, correlation, energy, entropy, and homogeneity capture important spatial texture patterns and can be used to distinguish different surface types or regions in an image [10].

The purpose of this chapter is to present an advanced ensemble ML method that constructs multiple decision trees and consolidates their outputs in order to improve predictive accuracy and reduce overfitting. The construction of each tree is accomplished by employing a bootstrap sample of the data, with features being chosen at random at each split. This ensures that the trees are diverse from one another. The method demonstrated strong performance on a variety of datasets, showed resilience to noise, and provided internal estimates of feature importance. Random forests became a foundational tool in supervised learning due to their accuracy, speed, and interpretability [11].

New breakthroughs in deep learning research have shifted the focus. A CNN-based model is developed that achieved over 95% accuracy in identifying multiple diseases in tomato leaves by leveraging a large, publicly available dataset. Their work emphasized the importance of data augmentation and transfer learning to overcome challenges related to limited labeled data [12]. Similarly, the effectiveness of deep CNN architectures, such as AlexNet and GoogleNet, is demonstrated in classifying over 25 different plant diseases with high precision, facilitating real-time disease diagnosis using mobile devices [13, 14].

Additionally, the integration of mobile applications powered by deep learning models has made plant disease detection accessible to a broader audience, including amateur gardeners and farmers. A lightweight CNN model optimized for smartphone deployment, enabling users to capture leaf images and receive instant diagnosis without internet connectivity is proposed. Their app combined disease identification with treatment recommendations, significantly aiding small-scale cultivators in managing crop health [15]. These recent studies underscore the transformative impact of combining image processing, ML, and mobile technology to empower nonexperts in plant disease management.

In the most recent advances, the authors developed an attention-based CNN model that improves disease classification by focusing on critical leaf regions. Their model was tested on a diverse dataset containing over 50,000 images of crops affected by multiple diseases and achieved an accuracy of 97.8%. The use of attention mechanisms allows the system to ignore irrelevant background noise and highlight disease-specific features, enhancing prediction reliability [16].

Moreover, a lightweight CNN architecture is proposed to optimize the edge devices, enabling real-time plant disease detection on smartphones and IoT platforms. This work addresses the computational constraints of portable devices without sacrificing accuracy, achieving 95.6% classification accuracy across 15 common crop diseases. Their system integrates seamlessly with cloud platforms for large-scale agricultural monitoring, providing early warnings to farmers [17].

A multimodal approach combining leaf image analysis with environmental sensor data (such as humidity, temperature, and soil moisture) using a hybrid CNN-LSTM model is introduced. This method significantly improved disease prediction by correlating visual symptoms with environmental conditions that favor disease outbreaks. Tested on multiple crops, the model demonstrated 98.2% accuracy in early detection, enabling proactive disease management strategies [18–21].

18.2 Materials and Methods

The dataset consists of 87,000 RGB images captured from both infected and healthy plant leaves, spanning 38 different disease and health classes. The images were carefully collected to represent a wide variety of conditions and plant species. For model development, the dataset was split into training and validation sets with an 80:20 ratio, while preserving the original directory structure to maintain class distribution and organization. For the testing phase, a separate directory containing 33 new, unseen leaf images was created to evaluate the model's prediction accuracy and generalization capability.

The prerequisites for this work include a computer system, which includes either a laptop or desktop equipped with a dedicated graphics processing unit (GPU). The GPU is critical for accelerating the training and inference processes of deep learning models such as CNNs. Additionally, a high-resolution camera or a smartphone camera is required to capture clear and focused photographs of plant leaf samples.

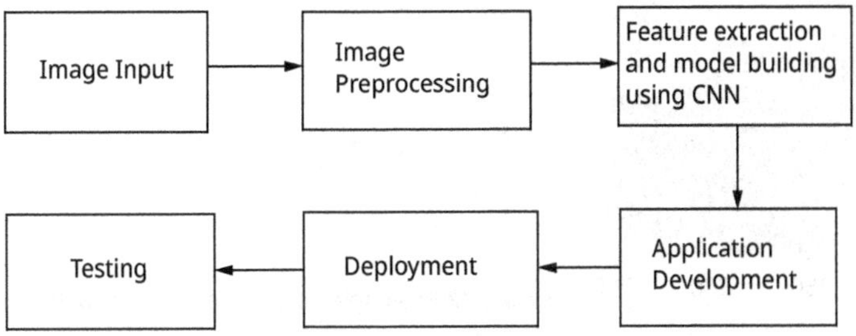

Figure 18.1: Proposed block diagram.

As illustrated in Figure 18.1, the proposed methodology is structured as a block diagram outlining the workflow of the system. The process is initiated through the acquisition of leaf images, which are captured in standard image formats such as .jpg, .jpeg, or .png. These photographs are then stored securely on a local hard disk, where they form the dataset used to train the neural network model. The methodology involves several key stages: image acquisition, data preprocessing, feature extraction through CNN layers, model training, validation, and testing. The system leverages the stored images to learn distinguishing features of healthy and diseased leaves across various classes. This approach provides a user-friendly and efficient framework for automatic plant disease identification, enabling faster diagnosis and timely intervention to protect crop health.

18.2.1 Workflow

The user takes a photograph of a leaf from a plant and uploads it to the application/website as shown in Figure 18.2.

18.2.2 Data Normalization and Activation Functions

The performance of the neural network is enhanced by normalizing the input data. The pixel values of raw images typically range from 0 to 255. These values are transformed to a normalized range between 0 and 1 by converting each image's pixel values into a

Figure 18.2: Sample input as read by the system.

PyTorch tensor and dividing by 255. Normalization ensures that the input data is scaled appropriately before being fed into the network.

Normalizing data close to zero mean and unit variance accelerates the training process and promotes faster convergence of the neural network. This approach offers two key benefits:

(a) It allows both negative and positive input values to propagate through subsequent layers, which increases the flexibility of the learning process by enabling the network to learn more complex features.

(b) It guarantees that the learning algorithm considers all input features equally, avoiding the dominance of any single feature during training

Regarding activation functions, the sigmoid function outputs values strictly between 0 and 1, which restricts the range of inputs for subsequent layers and can lead to vanishing gradient problems. Because of this, it can hinder learning, especially in deeper networks.

The tanh (hyperbolic tangent) activation function, which outputs values between −1 and 1, centers the data around zero, allowing for stronger gradients and more balanced weight updates during training. This centering often leads to better learning dynamics and improved model performance compared to sigmoid.

The ReLU, or rectified linear unit, is another common activation function. If the input is negative, it gives you zero. If the input is positive, it gives you the value of the input itself. ReLU is used a lot in the designs of modern deep learning systems because it is fast and can get rid of vanishing gradients. However, because ReLU outputs zero for all negative values, it can cause "dead neurons" where certain units stop learning if they only receive negative inputs. This behavior might reduce model flexibility in some contexts.

ReLU is often the default choice for many CNN models; tanh can outperform sigmoid by providing symmetric outputs and improved gradient flow. The choice between activation functions depends on the specific problem, network depth, and training dynamics [15].

18.2.3 Feature Extraction and Model Building with CNN Using PyTorch

Feature extraction and model building using CNNs in PyTorch involve several critical considerations for efficient and effective training. The term "batch size" indicates how many inputs are used in each training step; for example, if the batch size is set to 200, the model trains on the first 200 samples before updating its parameters, then proceeds to the next 200 samples, continuing this process until the entire training dataset has been utilized. This approach balances computational efficiency and the stability of gradient updates. The dataset is generalized by typically shuffling at the start of each epoch by setting shuffle = true, ensuring that the composition of batches varies, which prevents the model from memorizing the order of training data. Additionally, the num_workers parameter allows multiple subprocesses to load and preprocess data batches in parallel, significantly reducing data-loading bottlenecks and improving training throughput.

Leveraging GPUs instead of CPUs for training is highly advantageous due to massive parallelism and high memory bandwidth GPUs, which enable the simultaneous execution of numerous matrix operations central to CNN computations. This parallel processing capability drastically shortens training time, particularly when handling large image datasets. Since its introduction in 2015, the residual network (ResNet) architecture has become a cornerstone in computer vision tasks, primarily for its innovative use of skip connections. These connections enable each layer to pass its input not only to the next immediate layer but also to layers several hops away, effectively learning residual functions instead of direct mappings.

Furthermore, by improving gradient propagation and enabling identity mappings, ResNet reduces the risk of overfitting, which is often indicated by an increase in validation loss despite decreasing training loss. The core building block of ResNet consists of convolutional layers followed by batch normalization and activation functions, combined with shortcut connections that sum the input with the output of these layers, as illustrated in Figure 18.3. In PyTorch, ResNet models can be employed either by training from scratch or by leveraging pretrained weights for feature extraction, where the convolutional base is frozen, and only the classification head is fine-tuned for a specific task, thereby accelerating convergence and improving accuracy. This integration of efficient data handling, powerful hardware utilization, and advanced architectural innovations makes CNN-based feature extraction and model building with PyTorch both effective and scalable.

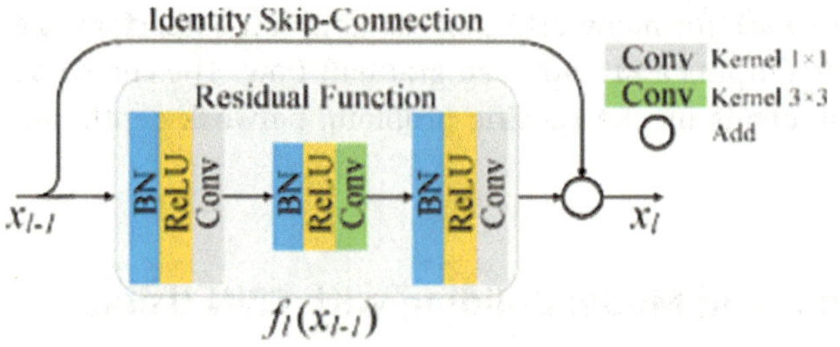

Figure 18.3: Simple block of a ResNet.

The main functions of image classification-based class are:

- training_step: It is designed to quantify the error of the model during each training or validation iteration. Unlike accuracy, which is non-differentiable and thus unsuitable for gradient-based optimization, this function computes a loss value that can be backpropagated to update model weights. This approach ensures that the neural network receives meaningful gradient signals necessary for improving its parameters.
- validation_step: This function evaluates model predictions against ground truth labels during validation; it uses a predefined threshold to determine if a prediction is sufficiently close to the actual label to be considered correct, thereby accommodating slight deviations and improving robustness in accuracy calculation
- validation_epoch_end: Following each epoch, the validation_epoch_end function aggregates the validation and training losses, providing an overall performance metric while explicitly disabling gradient tracking to save computational resources and prevent unintended updates during evaluation.
- epoch_end: This function provides informative output by printing metrics such as validation accuracy, training accuracy, and the current learning rate after each epoch, giving insights into model convergence and training dynamics.
- The accuracy metric, implemented as a separate function, computes the overall classification accuracy of the model and is integrated as a parameter within the training loop controlled by the fit_one_cycle method.
- The training phase of the model leverages two utility functions: evaluate, which handles the validation process by assessing model performance on a separate dataset, and fit_one_cycle, which orchestrates the entire training cycle using an effective learning rate policy.
- Within fit_one_cycle, learning rate scheduling plays a pivotal role; specifically, the one cycle learning rate policy is employed, where the learning rate begins at a low value and gradually increases with each batch until it peaks at approximately 30% of the total epochs. Subsequently, the learning rate decreases steadily to facilitate convergence. This cyclical adjustment helps avoid local minima and accelerates training.

18.2.4 Training the Model

To validate the model, the evaluate function is used, whereas the fit_one_cycle function is responsible for training.

The utility function evaluate, will carry out the validation part, and a fit_one_cycle function will carry out the training.

In *fit_one_cycle*, we have used the following methods:

- **Learning Rate Scheduling**: The learning rates are changed after every batch of training. "One cycle learning rate policy" is utilized for changing the learning rate, which starts with a low value and progressively increases for each batch. The highest learning rate is achieved at about 30% of epochs, and then the learning rate starts decreasing.
- **Weight Decay**: This additional term is added to the loss function to avoid large value for the weights.
- **Gradient Clipping**: The effective method called the gradient clipping limits the gradient values to avoid the undesirable changes due to large gradient values.

The learning rate used for each batch is recorded. The dataset used in this work comprises images from nine different plant species, each associated with various diseases, forming the basis for training and evaluating the classification model. The selected plants include apple, corn, grape, potato, tomato, pepper, cherry, strawberry, and raspberry. These species were chosen due to their agricultural importance and the prevalence of disease conditions affecting their productivity. Table 18.1 summarizes the specific types of plant diseases considered for the experiment alongside the corresponding number of images for each category. This comprehensive dataset enables the model to learn diverse disease patterns across multiple crops, enhancing its practical applicability in real-world agricultural diagnostics. The images per class of plant disease are represented in Figure 18.4.

After the ML model is trained and validated using image datasets, deployment entails embedding the model within a production-ready environment to enable real-time inference. This typically includes wrapping the trained model within an API or inference engine (e.g., TensorFlow Lite, ONNX, or Flask for RESTful APIs), and deploying it on edge devices (such as mobile phones, Raspberry Pi, or Jetson Nano) or cloud platforms (such as AWS, Azure, or GCP). The application interface – either mobile or web-based – acts as the front-end where users can upload or capture leaf images. These images are preprocessed (resizing, normalization, etc.) before being passed to the model for classification. The model outputs the predicted disease class along with confidence scores, and optionally suggests remedial actions. Edge deployment enables offline processing, suitable for remote agricultural areas, while cloud deployment supports high scalability and continuous model updates. Security, latency, and model optimization (e.g., pruning and quantization) are important considerations during deployment to ensure efficient and robust application performance in real-world agricultural environments.

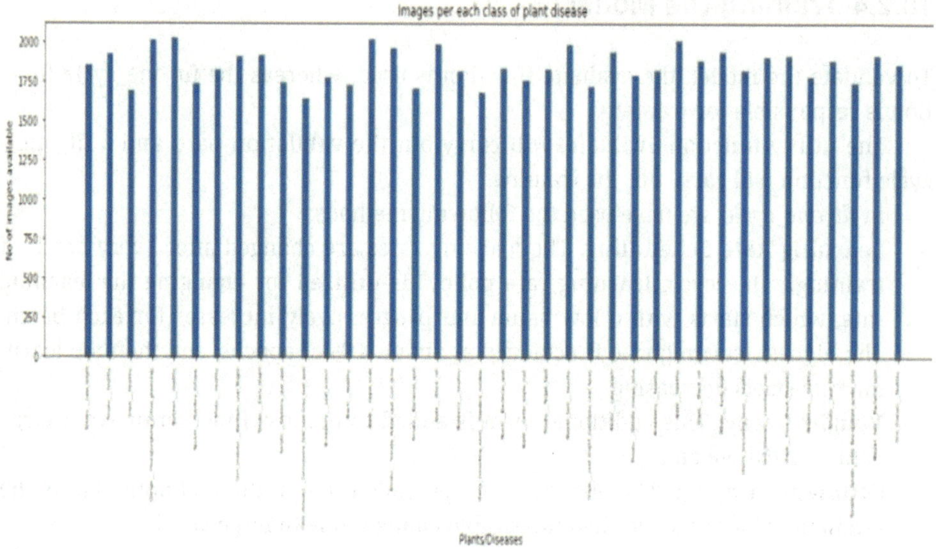

Figure 18.4: Images provided for each class of plant disease.

Table 18.1: Number of images provided in the training dataset for each plant disease.

S. no.	Name of disease	Number of images	S. no.	Name of disease	Number of images
1.	Late blight tomato	1,851	20.	Bacterial spot peach	1,838
2.	Healthy tomato	1,926	21.	Cedar apple rust apple	1,760
3.	Healthy grape	1,692	22.	Target spot tomato	1,827
4.	Citrus greening orange	2,010	23.	Healthy pepper bell	1,988
5.	Healthy soybean	2,022	24.	Leaf blight grape (isariopsis leaf spot)	1,722
6.	Powdery mildew squash	1,736	25.	Late blight potato	1,939
7.	Healthy potato	1,824	26.	Mosaic virus tomato	1,790
8.	Northern leaf blight corn(maize)	1,908	27.	Healthy strawberry	1,824
9.	Early blight tomato	1,920	28.	Healthy apple	2,008
10.	Septoria leaf spot tomato	1,745	29.	Black rot grape	1,888
11.	Gray leaf spot corn (maize)	1,642	30.	Early blight potato	1,939
12.	Leaf scorch strawberry	1,774	31.	Healthy cherry (including sour)	1,826

Table 18.1 (continued)

S. no.	Name of disease	Number of images	S. no.	Name of disease	Number of images
13.	Healthy peach	1,728	32.	Common rust corn (maize)	1,907
14.	Apple scab apple	2,016	33.	Esca (black measles) grape	1,920
15.	Yellow leaf curl virus tomato	1,961	34.	Healthy raspberry	1,781
16.	Bacterial spot tomato	1,702	35.	Leaf mold tomato	1,882
17.	Black rot apple	1,987	36.	Tomato -spider mites -two-spotted spider mite	1,741
18.	Healthy blueberry	1,816	37.	Bacterial spot pepper bell	1,913
19.	Cherry (including sour) -powdery mildew	1,683	38.	Healthy corn (maize)	1,859

Before deploying an ML model for plant disease detection, testing is crucial to assess its performance, reliability, and ability to generalize. The model is tested on a hold-out test dataset containing leaf images that were not used during training or validation. To evaluate the classification effectiveness for each disease class, performance metrics like accuracy, precision, recall, F_1-score, and the confusion matrix are calculated In addition, ROC-AUC curves may be used if the model outputs probabilistic scores. The robustness is ensured by cross-validation (e.g., *k*-fold), especially when dataset size is limited. Furthermore, testing should include real-world scenarios with variations in lighting, background, resolution, and partial leaf damage to simulate field conditions. Stress testing on adversarial or noisy inputs helps verify the model's resilience. Ultimately, comprehensive testing ensures the model is not only accurate but also reliable when deployed in practical agricultural applications.

18.3 Results and Discussion

A validation set is a key aspect of assessing the performance and robustness of an ML model, which is a subset of the dataset explicitly reserved for assessing the model's generalization capability. In this study, a portion of the original dataset, referred to as the validation set, is separated prior to training and is not used during the parameter optimization phase. After the model undergoes training completed for an epoch or a defined number of iterations, the validation images are presented to the trained network as input, and the model's predictions are compared against the true labels. The accuracy computed during this validation phase, known as the validation accuracy, serves as a quantitative measure of the model's ability to correctly classify unseen data. This metric is especially important because it reflects how well the model generalizes beyond the training data and prevents overfitting, which occurs when a model performs well on training data but poorly on new, unseen samples. The validation accuracy thus provides a reliable indication of the model's predictive performance in real-world scenarios. In this work, the validation accuracy progression across epochs is illustrated in Figure 18.5, demonstrating the model's learning curve and convergence behavior over the course of training.

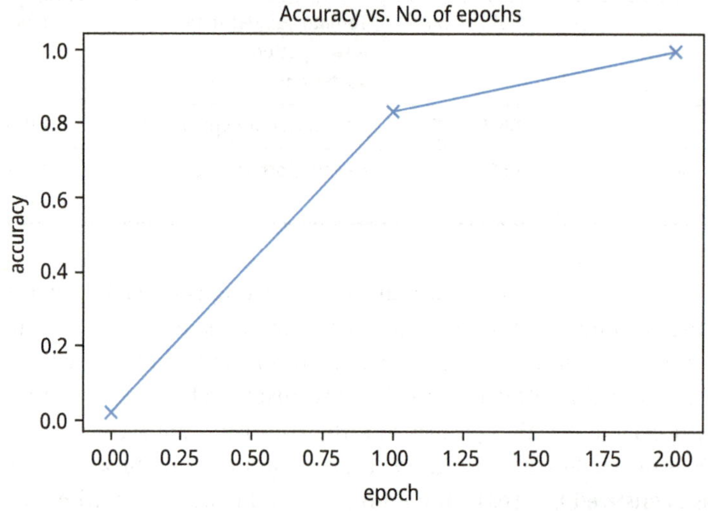

Figure 18.5: Validation accuracy/testing.

It can be observed from the training process that the validation accuracy steadily increases as the number of epochs progresses, indicating that the model is effectively learning and becoming increasingly adept at recognizing patterns within the data. This upward trend in accuracy reflects the model's improved ability to make precise predictions with each successive epoch, thereby enhancing its diagnostic reliability. The model demonstrates a peak validation accuracy of 0.992, or 99.2%, demonstrating exceptional performance on unseen data and underscoring the robustness of the training process. Complementing the accuracy metric, the validation loss serves as a key measure of the model's ability to generalize to the validation set by quantifying the error between predicted and actual labels. A decreasing validation loss typically correlates with improving model fit and reduced prediction errors. The progression of validation loss over the training epochs is illustrated in Figure 18.6, providing insight into the model's convergence behavior and stability. Together, these metrics confirm that the proposed model not only achieves high predictive accuracy but also maintains consistent performance without overfitting, validating its suitability for reliable plant disease classification.

It is evident from the training results that the validation accuracy increases progressively with the number of epochs, indicating that the model is effectively learning and becoming more adept at recognizing patterns in the data. This improvement reflects the model's growing familiarity with the underlying features, leading to increasingly accurate predictions as training advances. The model achieves a peak validation accuracy of 0.992, or 99.2%, demonstrating its strong ability to generalize and provide precise diagnoses on unseen data. Alongside accuracy, validation loss is a critical metric applied to determine the model's accuracy on the validation set. It quantifies the difference between the predicted outputs and the true labels, providing insight into how well the model fits the data without overfitting. The progression of validation loss throughout the training process is illustrated in Figure 18.6, showing a consistent decrease that signifies improved model fitting and convergence.

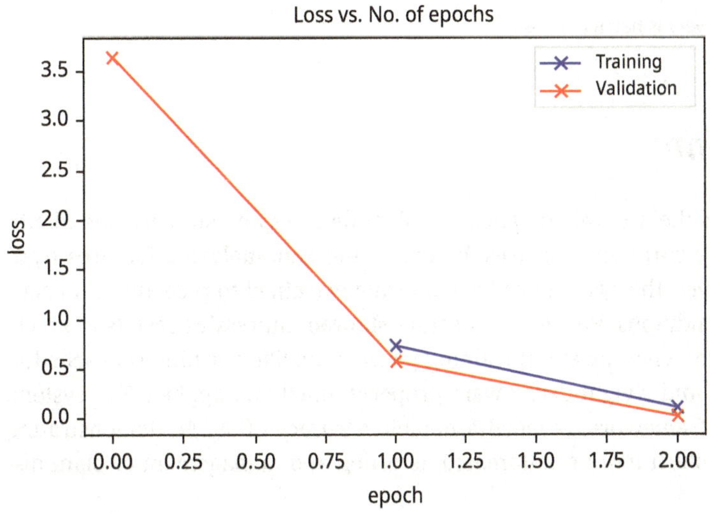

Figure 18.6: Validation loss.

It can be observed that the validation loss decreases consistently with each successive epoch. This trend indicates that the model is effectively minimizing errors on the validation data, reflecting its increasing accuracy and improved ability to generalize. The reduction in validation loss confirms that the model is learning meaningful features and making precise predictions, thereby maintaining low error rates as training progresses. The learning rate with batch number comparison is shown in Figure 18.7.

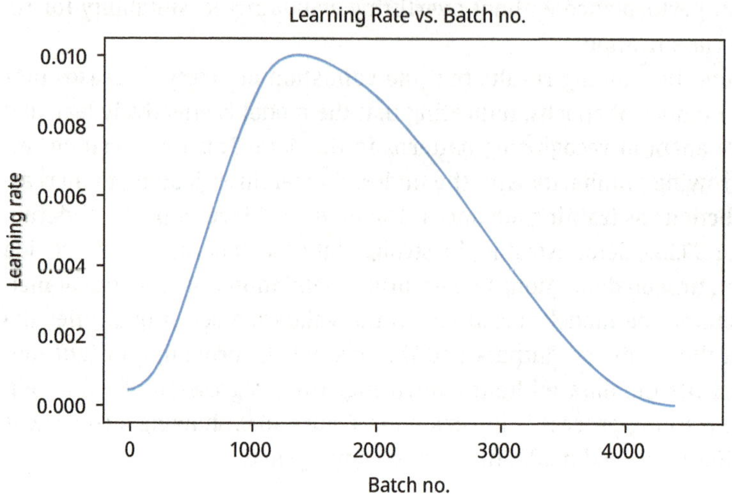

Figure 18.7: Learning rate versus batch number.

18.4 Conclusion

This chapter facilitates the accurate diagnosis of plant diseases and supports the identification of appropriate curative measures. By extracting and analyzing features from images of infected leaves, the system has been effectively trained to recognize and classify various disease conditions. ResNet architecture showed impressive results in image classification, especially when essential training parameters like learning rate scheduling, gradient clipping, and weight decay were properly tuned and applied. The system, built on the ResNet-9 framework, reached a notable accuracy of 99.2%, demonstrating its effectiveness as a robust tool for automated detection and management of plant diseases.

References

[1] Khirade SD, Patil AB. Plant Disease Detection Using Image Processing. In Proceedings of the 2015 International Conference on Computing Communication Control and Automation (ICCUBEA) 2015 Feb 26–27 (pp. 768–771). Pune, India: IEEE; 2015. doi: 10.1109/ICCUBEA.2015.153.

[2] Alhwaiti Y, Ishaq M, Siddiqi MH, Waqas M, Alruwaili M, Alanazi S et al. Early detection of late blight tomato disease using histogram-oriented gradient based support vector machine. arXiv [Preprint]. 2023.

[3] Upadhyay A, Chandel NS, Singh KP, Chakraborty S, Nandede BM, Kumar M et al. Deep learning and computer vision in plant disease detection: A comprehensive review of techniques, models, and trends in precision agriculture. Artificial Intelligence Review. 2025;58(92). doi: 10.1007/s10462-024-11100-x.

[4] Kumar TA, Oviya S, Ajagbe SA, Ananth C, Aiyeniko O, Adigun MO A novel deep neural network-based prediction model for identifying diseases in tomato leaves. In 2023 International Conference on Electrical, Computer and Energy Technologies (ICECET) 2023 Nov 16 (pp. 1–6). IEEE.

[5] Slingo JM, Challinor AJ, Hoskins BJ, Wheeler TR. Introduction: Food crops in a changing climate. Philosophical Transactions of the Royal Society B: Biological Sciences. 2005 Nov 29;360 (1463):1983–1989.

[6] Gawand M, Gowda S, Hnade S, Mahadik P, Dere P. Identification of anthracnose disease in mango plant using clustering and gray level co-occurrence matrix. In ITM Web of Conferences 2020 (Vol. 32, pp. 03050). EDP Sciences.

[7] Ünal Y, Türkoğlu M. Mango leaf disease detection using deep feature extraction and machine learning methods: A comparative survey. El-Cezeri. 2025;12(1):35–43.

[8] Wang D, Vinson R, Holmes M, Seibel G, Bechar A, Nof S, Tao Y. Early detection of tomato spotted wilt virus by hyperspectral imaging and outlier removal auxiliary classifier generative adversarial nets (OR-AC-GAN). Scientific Reports. 2019 Mar 13;9(1):4377.

[9] Deshpande P, Kore S. Disease Detection for Pomegranate: A Review. Pune, India: InConference on Plant Disease Detection; 2022.

[10] Haralick RM, Shanmugam K, Dinstein IH. Textural features for image classification. IEEE Transactions on Systems, Man, and Cybernetics. 2007 Nov;12(6):610–621.

[11] Sundaresan S, Kumar KS, Nishanth R, Robinson YH, Kumar AJ. Artificial Intelligence and Machine Learning Approaches for Smart Transportation in Smart Cities Using Blockchain Architecture. In Blockchain for Smart Cities 2021 Jan 1 (pp. 35–56). Elsevier.

[12] Kabir Oni M, Tanzin Prama T. Optimized Custom CNN for Real-Time Tomato Leaf Disease Detection. arXiv e-prints. 2025.

[13] Kanakala S, Ningappa S. Detection and Classification of Diseases in Multi-Crop Leaves using LSTM and CNN Models. arXiv preprint arXiv:2505.00741. 2025 Apr 30.

[14] Alzahrani MS, Alsaade FW. Transform and deep learning algorithms for the early detection and recognition of tomato leaf disease. Agronomy. 2023 Apr 22;13(5):1184.

[15] Stöttner T (2019). Why data should be normalized before training a neural network. URL: https://towardsdatascience.com/why-data-should-be-normalized-beforetraining-a-neural-network-c626b7f66c7d

[16] Sardogan M, Tuncer A, Ozen Y. Plant Leaf Disease Detection and Classification Based on CNN with LVQ Algorithm. In 2018 3rd International Conference on Computer Science and Engineering (UBMK). 2018 Sep 20 (pp. 382–385). IEEE.

[17] Madiwalar SC, Wyawahare MV. Plant disease identification: A comparative study. In 2017 International Conference on Data Management, Analytics and Innovation (ICDMAI) 2017, February (pp. 13–18). IEEE.

[18] Moghadam P, Ward D, Goan E, Jayawardena S, Sikka P, Hernandez E. Plant disease detection using hyperspectral imaging. In 2017 International Conference on DICTA 2017, November (pp. 1–8). IEEE.

[19] Sharath DM, Kumar SA, Rohan MG, Prathap C. Image based plant disease detection in pomegranate plant for bacterial blight. In 2019 ICCSP 2019, April (pp. 0645–0649). IEEE.

[20] Shrestha G, Das M, Dey N. Plant disease detection using CNN. In 2020 IEEE Applied Signal Processing Conference (ASPCON) 2020, October (pp. 109–113). IEEE.

[21] Mohanty SP, Hughes DP, Salathé M. Using deep learning for image-based plant disease detection. Frontiers in Plant Science. 2016;7:1419.

G. Prema, G. Shobana, V. Gnanalakshmi

19 Lightweight Deep Learning Models for Crab Species Recognition

Abstract: Crab species identification is a critical task in marine biodiversity monitoring, aquaculture management, and ecological research. Crab species classification is very difficult because different species have discrete looks. Traditional identification methods, being labor-intensive and reliant on expert knowledge, are inefficient for large-scale deployment. This study explores the application of lightweight deep learning models for automated and accurate crab species recognition. To achieve this, the proposed method includes MobileNetV2 architecture for automated crab species classification. Using this method, crab species, such as king crabs, coconut crabs, stone crabs, mud crabs, blue crabs, and vampire crabs, can be classified with better accuracy. Using a curated dataset of crab images captured under varying environmental conditions, we assess the performance of several lightweight models, including CNN, VGG-16, MobileNetV2, EfficientNet-B0, SqueezeNet, and ShuffleNet. Data augmentation techniques are used to improve the model's ability to generalize and train it effectively using a large and diverse dataset of crab species. An accuracy of 98.86% was achieved on the test set, demonstrating a reliable and efficient automated solution for crab species identification, with potential applications in ecological and marine biology research.

Keywords: Deep learning, classification, MobileNetV2, Squeeze Net, ShuffleNet, EfficientNet-B0

19.1 Introduction

The basic component of marine biology and ecology is the crab species classification. It helps to understand the evolution, diversity, and distribution of crab species. Crab is an infraorder Brachyura member. Crabs are easily identified by their body structure, with ten legs, a pair of claws, and a large, flattened carapace. Crab species are classified, recognized, and grouped according to their ecological features, genetic analysis, and physical criteria [1]. They are found mostly in freshwater and marine environments across the world. There are varieties of species based on their behavior, adaptation, digging, climbing, and swimming skills. Hence, it is necessary for ecological study, fishery management, and conservation initiatives to know how crab species are classified.

Convolutional neural network (CNN) architecture, MobileNetV2, was introduced, especially for embedded and mobile vision applications. It is the advanced version of

https://doi.org/10.1515/9783112214374-019

the MobileNet architecture, providing better efficiency and performance. Depth-wise separable convolutions are used to reduce the computational complexity, thereby achieving better accuracy and making it suitable for resource-constrained applications. Inverted residuals are said to be the key feature of the proposed architecture to improve its overall efficiency [2]. In general, MobileNetV2 proves to be a better choice for classifying crab species, in terms of its restricted resources, small size, better performance, and compatibility with transfer learning.

To assess the effectiveness of MobileNetV2 in crab species identification, it is crucial to compare its performance with other prominent CNN architectures. In this study, we perform a detailed comparative analysis involving both lightweight and standard deep learning models, including CNN, ShuffleNet, SqueezeNet, EfficientNet-B0, and VGG16. Each of these models brings unique trade-offs between accuracy, inference speed, model size, and resource utilization. Through empirical evaluation on a curated crab species dataset, we benchmark these models across key performance metrics such as classification accuracy, precision, recall, F_1-score, inference time, and model size [3]. The goal is to identify the most effective model that not only achieves high classification accuracy but is also practical for deployment in ecological monitoring systems, particularly in resource-constrained environments. This comparative study aims to provide insights into the selection of appropriate CNN models for real-world biodiversity applications, particularly for crab species classification, contributing to more intelligent, automated, and scalable marine research methodologies.

The highlights of the proposed work includes:

- Utilization of deep learning architecture, MobileNetV2, noted for its efficiency and efficacy.
- Achieving an accuracy of 98.86% on marine biology research in crab species, with better classification.
- Using transfer learning to increase performance of crab species classification by utilizing pretrained MobileNetV2 models on ImageNet.
- Carrying out comparative analysis of performance metrics with other light weight deep learning models.

The structure of the paper is as follows: The importance of automated crab classification is explained in Section 19.1. Section 19.2 includes literature survey. Proposed methodology focusing on MobileNetV2 architecture and other classification models are discussed in Section 19.3. The experimental setup and evaluation results are described in Section 19.4. Finally, the work is concluded in Section 19.5, which also discusses implications for deep learning and crab categorization.

19.2 Literature Survey

To overcome the complexity of classifying crabs in underwater environments, the authors introduced deep learning to construct an autonomous real-time classification of underwater live crabs. This method enhanced underwater crab classification with good accuracy and efficiency.

Thirukanthan et al. [4] used a deep CNN to propose a gender classification approach for Chinese mitten crabs. Their approach effectively classified genders, demonstrating the potential of deep learning for the study of crab behavior and populations.

A compound-scaling neural network-based small-sized effective detector for freely living crabs underwater was presented by Huang et al. [5]. This work made a significant contribution to the fields of marine biology and underwater robotics by improving the detection efficiency of live crabs in underwater environments.

A research article based on the climatic change of underwater crabs was proposed by Rathore et al. [6]. In this work, the author emphasized the machine vision usage for population detection and monitoring. This research advances ecological studies and marine maintenance.

Suresh et al. [7] proposed a compound-scaling neural network to classify and detect the underwater crab species. This study has its own significance for its effective underwater animal detection techniques, especially in the light of environmental monitoring and preservation initiatives.

To better recognize the viral infection in crustaceans, Anandaraj and Alphonse [8] introduced structural insights into the classification of mud crab reovirus. These studies most likely give a way to the development of methods for controlling viral infections in crab populations, which has an effect on aquaculture as well as ecological conservation initiatives.

19.3 Proposed Method

In this section, the overall design flow of the proposed classification using CNN and other pretrained models is discussed along with its mathematical modeling.

19.3.1 CNN-Based Crab Species Classification

CNNs are a subclass of deep learning models that are mostly employed for image-based tasks, including segmentation, object detection, and classification. CNNs are strong because they use a variety of building elements, including convolutional layers, pooling layers, activation functions, normalization techniques, and fully connected

layers, to automatically and effectively learn spatial hierarchies of data through back propagation. The major components of a CNN are:
- Convolutional layer
- Activation functions
- Batch normalization
- Pooling layer
- Flatten layer
- Fully connected (dense) layer
- Dropout layer

The respective roles of the major components are explained one by one.

19.3.1.1 Convolutional Layer

At the heart of every CNN lies the convolutional layer. This layer performs a convolution operation that enables the network to learn features directly from the input images and is depicted in Figure 19.1. It uses small learnable filters (also known as kernels) that slide across the input data in a systematic fashion. These filters are responsible for detecting specific local patterns such as edges, textures, curves, and corners.

Each filter is convolved with the input image to produce a feature map. The depth of the output depends on the number of filters used. Early convolutional layers capture low-level features, while deeper layers identify higher-level abstract features such as object parts or complete shapes.

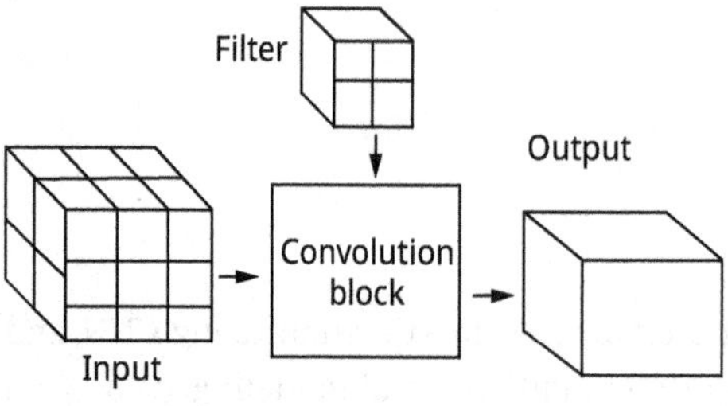

Figure 19.1: Convolution operation.

The core operation in a convolutional layer is the discrete convolution, defined as:

$$S(i,j) = (X * K)(i,j) = \sum \sum X(i+m, j+n).K(m,n) \qquad (19.1)$$

where X is the input image (or feature map), K is the filter/kernel, $S(i,j)$ is the output (feature map), and (i,j) are the spatial positions in the output

$$\text{Output size} = \left(\frac{W - F + 2P}{S}\right) + 1 \tag{19.2}$$

where W is the input width, F is the filter size, P is padding, and S is stride.

19.3.1.2 Activation Functions

After convolution has been used to extract features, each neuron's output is subjected to an activation function to add nonlinearity. CNNs would essentially act like linear models in the absence of activation functions, which would restrict their ability to represent intricate patterns in data. Common activation functions include:

- **ReLU (Rectified Linear Unit):** ReLU outputs zero for negative values and returns the input for positive values. Its graphical representation is depicted in Figure 19.2. It is widely used due to its computational efficiency and ability to mitigate vanishing gradient issues:

$$f(x) = \max(0, x) \tag{19.3}$$

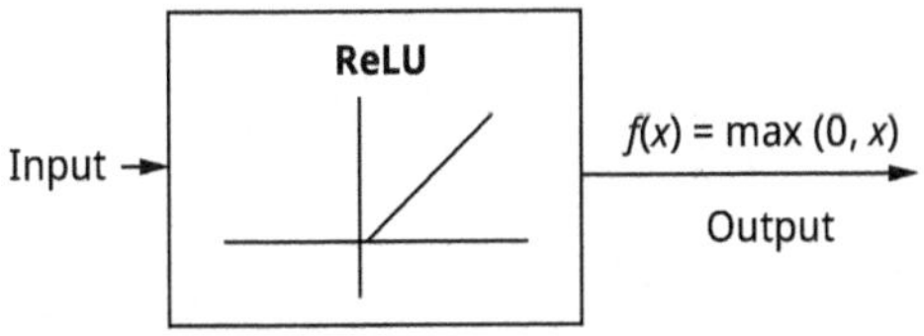

Figure 19.2: ReLU activation function.

- **Sigmoid:** Maps input values into the (0, 1) range. Its graphical representation is depicted in Figure 19.3. While useful in binary classification, sigmoid functions are less frequently used in hidden layers due to the vanishing gradient problems:

$$f(x) = \frac{1}{1 + e^{-x}} \tag{19.4}$$

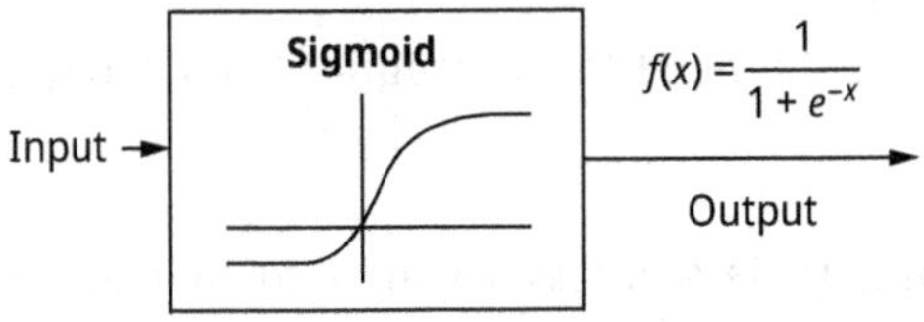

Figure 19.3: Sigmoid activation function.

Nonlinearity provided by these functions allows CNNs to learn intricate patterns, necessary for image understanding tasks.

19.3.1.3 Batch Normalization

As shown in Figure 19.4, batch normalization is a method for enhancing neural network performance and training speed. After subtracting the batch mean and dividing by the batch standard deviation, it scales and shifts the output of an earlier activation layer to normalize it. Given an input x in a mini-batch, batch normalization transforms it as:

$$\hat{x} = \frac{x - \mu B}{\sqrt{\sigma_B^2 + \epsilon}} \tag{19.5}$$

$$y = \gamma \hat{x} + \beta \tag{19.6}$$

where μB and σ_B^2 are batch mean and variance, ϵ is a small constant for numerical stability, and γ and β are learnable parameters.

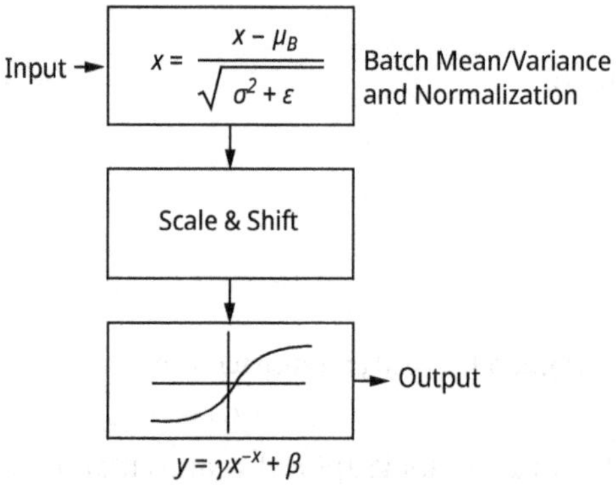

Figure 19.4: Batch normalization.

Benefits of batch normalization:
- Reduces internal covariate shift
- Helps in faster convergence during training
- Acts as a regularizer, potentially reducing the need for techniques like dropout, in some cases

This normalization makes the network more stable and less sensitive to changes in network initialization or learning rate.

19.3.1.4 Pooling Layer

The **pooling layer** reduces the spatial dimensions (height and width) of the input feature maps while retaining the most important information. This helps in lowering the computational load, reducing overfitting, and improving the network's efficiency.

The two most common types of pooling are:

Max Pooling: Returns the maximum value within a patch of the feature map. For a region *R*, it can be given by,

$$y = \max_{x \in R} x \tag{19.7}$$

Average Pooling: Computes the average of the values within a patch. For a region *R*, it can be given by,

$$y = \frac{1}{|R|} \sum_{x \in R} x \tag{19.8}$$

where *R* is the pooling window.

Pooling operations are typically performed on nonoverlapping regions of the feature map, helping the model retain dominant features while reducing dimensionality.

19.3.1.5 Flatten Layer

Before the model can make predictions, the multidimensional output from convolutional and pooling layers needs to be transformed into a one-dimensional format. This is where the **flatten layer** comes into play. The reshaping process can be represented by:

$$\text{Flatten: } R^{h \times w \times d} \rightarrow R^{h.w.d} \tag{19.9}$$

It reshapes the data from a matrix or tensor (such as height × width × depth) into a single long vector. This flattened vector serves as the input to the subsequent fully connected (dense) layers.

19.3.1.6 Fully Connected (Dense) Layer

Once the features are flattened, they are passed into one or more **fully connected layers**. In a dense layer, each neuron is connected to every neuron in the previous layer. These layers operate similarly to those in traditional neural networks and are used to combine features and make final predictions.

Key roles of fully connected layers include:

- Aggregating the features learned from previous layers.
- Performing classification or regression tasks based on the final feature set. Each output neuron computes:

$$y = f\left(\sum_{i=1}^{n} \omega_i x_i + b\right) \tag{19.10}$$

where x_i are input features, ω_i are weights, b is bias, and f is an activation function (e.g., ReLU and sigmoid).

The final fully connected layer typically uses a **softmax** activation function (for multi-class classification) or **sigmoid** (for binary classification) to produce probabilities as outputs.

19.3.1.7 Dropout Layer

To prevent **overfitting**, especially when working with small datasets or large models, CNNs often include a **dropout layer**. This layer randomly sets a proportion of its input units to zero during training, effectively "dropping" certain neurons and their connections. Dropout can be represented as:

$$\widetilde{x_i} = x_i.r_i \sim \text{Bernoulli}(p) \tag{19.11}$$

where x_i is the input, r_i is a random variable (1 with probability p, 0 with probability $1-p$), and $\widetilde{x_i}$ is the retained input after dropout.

The advantages of dropout include:

- Prevents the network from relying too heavily on any one neuron
- Encourages the network to learn redundant representations
- Improves generalization to unseen data

Typically, dropout is applied to fully connected layers rather than convolutional ones, though it can be adapted for both. The overall design flow of the proposed classification strategy is depicted in Figure 19.5.

19.3.2 VGG-16-based Crab Species Classification

VGG-16 is a deep CNN architecture proposed by Simonyan and Zisserman from the University of Oxford in 2014, which became widely recognized for its simplicity and performance in the ImageNet Large-Scale Visual Recognition Challenge (ILSVRC). The architecture is characterized by its use of small 3×3 convolutional filters, with a stride of 1 and padding of 1, which enables the network to capture fine-grained

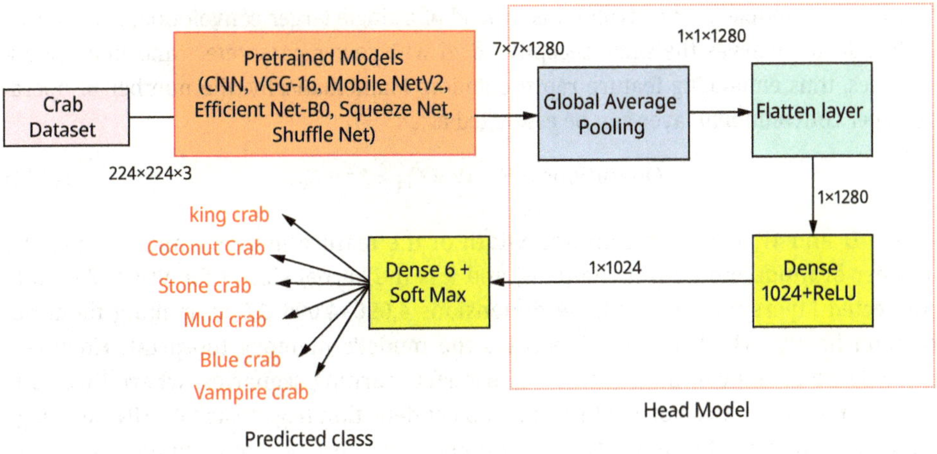

Figure 19.5: Block diagram of deep learning-based crab species classification.

features, while maintaining the spatial resolution. The core idea is to increase depth while keeping the convolutional layer configurations consistent. VGG-16 consists of 13 convolutional layers and 3 fully connected layers, with a total of 16 weight layers, hence the name VGG-16. The input to the model is a fixed-size $224 \times 224 \times 3$ RGB image. The architecture applies a series of convolutional operations, defined as:

$$Y_{i,j,k} = \sum_{m=0}^{M-1} \sum_{n=0}^{N-1} \sum_{c=0}^{C-1} X_{i+m,J+n,c} \cdot W_{m.n.c.k} + b_k \tag{19.12}$$

where X is the input tensor, W are the weights of the filter, b_k is the bias term, and Y is the output feature map. Each convolutional layer is followed by a ReLU activation function to introduce nonlinearity. After every few convolutional layers, spatial downsampling is performed using max pooling layers, with a 2×2 window stride 2:

$$Y_{i,j,k} = \underset{(m,n)\in[0,1]}{\text{maX}} X_{2i+m,2j+n,k} \tag{19.13}$$

which reduces the spatial dimensions, while retaining important features. The final stack of convolutional and pooling layers is flattened and passed to three dense layers, the last of which is a softmax layer. The network has about 138 million parameters, making it computationally intensive, but it demonstrates strong generalization and transfer learning capabilities. VGG-16's uniform structure makes it an ideal baseline for many computer vision tasks such as image classification, object detection, and feature extraction, and its pretrained models are extensively used in transfer learning applications.

VGG-16 is not only notable for its depth but also for its uniform architecture, which contrasts with previous networks like AlexNet that used larger and variable-sized kernels.

By stacking multiple 3 × 3 convolutions instead of a single larger convolution, like 5 × 5 or 7 × 7, VGG-16 achieves the same receptive field with fewer parameters and more nonlinearities, thus enhancing feature representation. Mathematically, the number of operations per convolutional layer can be estimated as:

$$\text{Operations} = H \times W \times C_{\text{in}} \times K^2 \times C_{\text{out}} \tag{19.14}$$

where H and W are the height and width of the feature map, C_{in} and C_{out} are the number of input and output channels, and k is the kernel size (3 for VGG). The fully connected layers that follow have dimensions 4,096–4,096–1,000, making them parameter-heavy, which in turn increases the model's memory footprint. However, these dense layers can be truncated in transfer learning scenarios, where VGG-16 is used as a feature extractor for tasks like object detection (e.g., faster R-CNN) or image segmentation (e.g., FCN and U-Net). Additionally, the use of zero padding in convolutions ensures that the spatial resolution is preserved across each layer.

19.3.3 EfficientNet-B0

EfficientNet-B0, proposed by Tan and Le, introduces a compound scaling method that uniformly scales depth, width, and resolution. Despite its compact size, it delivers strong performance by using mobile inverted bottleneck (MBConv) blocks and swish activations. EfficientNet contains eight CNN models, named EfficientNet-B0 to EfficientNet-B7. EfficientNet-B0 has an efficient feature extraction capability. The architecture of EfficientNet-B0 is shown in Figure 19.6.

In the adapted version:

- The base layers are frozen.
- The classification head consists of a global average pooling layer, followed by a dense layer with six units and a SoftMax activation.
- Dropout and batch normalization are used to enhance stability.

Though slightly heavier in parameter count than MobileNetV2, it performs competitively, achieving 97.4% accuracy at 50 epochs with batch size 32. The parameters used in compound scaling are described as follows:

Compound Scaling Equation

Depth: $d = \alpha^{\Phi}$; width: $w = \beta^{\Phi}$; resolution: $r = \gamma^{\Phi}$

$$\text{Subject to } \alpha.\ \beta^2.\gamma^2 \sim 2, \quad \alpha,\ \beta,\ \gamma > 1 \tag{19.15}$$

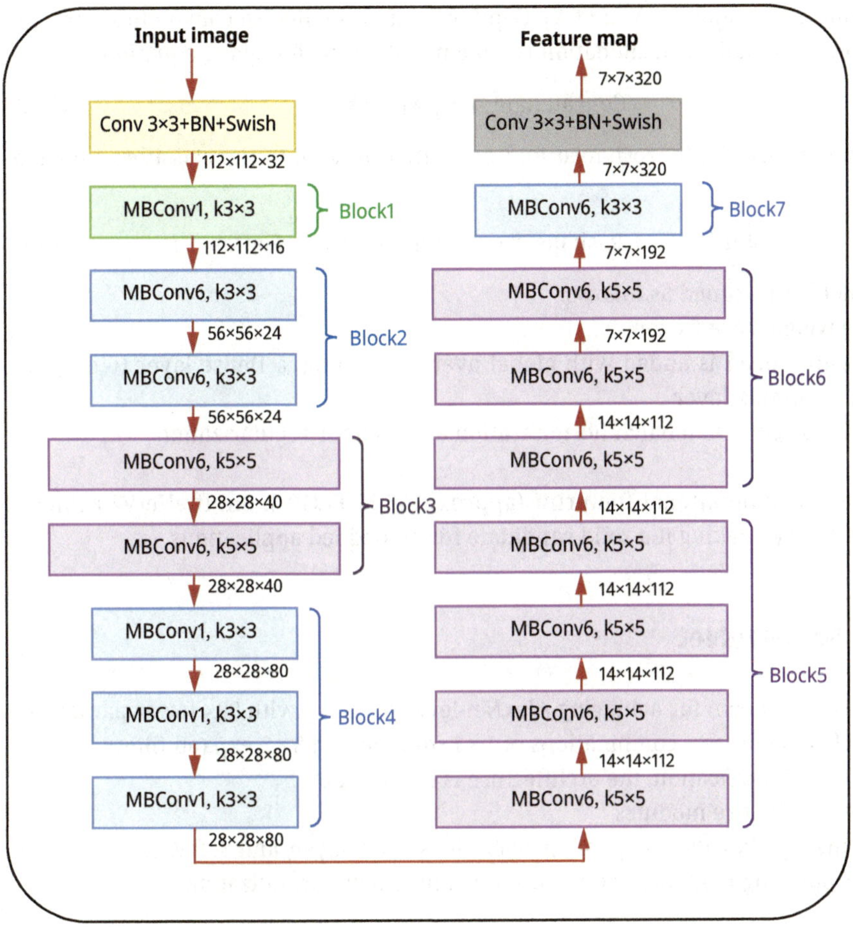

Figure 19.6: Architecture of EfficientNet-B0.

where Φ is a user-defined coefficient controlling the model size. The swish activation function is used for integrating squeeze and excitation modules for channel attention and is given by

$$\text{Swish}(x) = x.\ \sigma\ (x) \tag{19.6}$$

19.3.4 ShuffleNetV2

Designed by Ma et al., ShuffleNetV2 is optimized for high-speed and low-memory environments. It employs channel shuffling and group convolutions to maximize efficiency. This reduces memory access cost and enhances parallelism. The feature map is split into two branches, X_1 and X_2. Depth-wise and point-wise convolutions are performed on X_2. Finally, concatenation is done based on the following equation:

$$\text{Concatenate}\ \ Y = [(X_1,\ F(X_2))] \tag{19.17}$$

The channels are finally reshaped and permutted to give to the classifier model, as follows.

$$\text{Shuffle}(Y) = \text{Reshape}(Y) \rightarrow \text{Permute}\ (Y) \rightarrow \text{Flatten}(Y) \tag{19.18}$$

The model is fine-tuned as follows:
- Base weights are frozen.
- A custom head is added with global average pooling, a Dense layer (6 outputs), and a SoftMax layer.
- ReLU, dropout, and batch normalization are used for regularization.

With a low computational footprint (approx. 146 M FLOPs), ShuffleNetV2 achieved 96.7% accuracy, making it a solid candidate for embedded applications.

19.3.5 SqueezeNet

SqueezeNet is known for achieving AlexNet-level accuracy with 50x fewer parameters by using fire modules – combinations of 1 × 1 squeeze and 3 × 3 expand filters.

For crab classification, the architecture is modified by:
- Freezing the fire modules
- Adding a global average pooling layer, dense (six) layer, and SoftMax
- Incorporating ReLU activation, dropout, and batch normalization

Though ultra-efficient with only 1.2 M parameters, SqueezeNet trails in performance, achieving 93.3% accuracy at 50 epochs.

The model equations of the fire module is given in the below equations:

$$\text{Squeeze layer}\ (1\times 1\,\text{conv}):\ S = \text{Conv}_{1\times 1}(X) \tag{19.19}$$

$$\text{Expand layer}\ \ (1\times 1 + 3\times 3\,\text{Conv}):\ E_1 = \text{Conv}_{1\times 1}(S),\ \ E_3 = \text{Conv}_{3\times 3}(S) \tag{19.20}$$

$$\text{Output} = \text{Concatenate}\ (E_1, E_3) \tag{19.21}$$

The model is developed to perform the classification of six distinct classes of crabs. To ensure the stability of the base layers, they are frozen, preventing their weights from being modified during training. In addition to base layers – global average pooling, a dense layer with six outputs corresponding to the classes, and a SoftMax activation function at the output, are added to the model architecture. The model comprises input images with the size 224 × 224 × 3, along with the kernel of 3 × 3. Finally, ReLU activation function, together with batch normalization and dropout functions, are added to the model to classify the crab species accurately.

19.4 Results and Discussion

19.4.1 Dataset Details

This section discusses the results of the crab species dataset taken from Kaggle. The details about the dataset are shown in Table 19.1 and few of the sample images are shown in Figure 19.7.

Table 19.1: Details of the dataset used.

Name of the crab species	Count of the species
Blue_Crab	75
Coconut_Crab	75
King_Crab	51
Mud_Crab	45
Stone_Crab	45
Vampire_Crab	67

19.4.2 Evaluation Metrics on CNN-Based Model

The performance metrics and parameters of CNN model with epoch 50 and accuracy 45.45% are depicted in Tables 19.2 and 19.3.

The intermediate output from various CNN layers is depicted in Figures 19.8–19.10.

19.4.3 Performance Evaluation on Lightweight Models

The performance measure for VGG16 model, epoch 50 with accuracy 63.64% is depicted in Table 19.4.

The performance of the proposed lightweight deep learning model Mobile netv2 is compared with other lightweight models such as EfficientNet-B0, SqueezeNet, and ShuffleNetV2. The main focus of the performance comparison is to analyze the trade-offs between efficiency and accuracy of the model on low resource usage. Each of the models was trained and tested using the same experimental settings and dataset. The performance of the models was evaluated under two different batch sizes of 16 and 32

Figure 19.7: Sample images of crab dataset.

using Adam optimizer. The classification results of the lightweight models obtained are tabulated in Table 19.5.

From Table 19.5, it is inferred that EfficientNet-B0 followed MobileNet V2 closely behind, achieving an accuracy of 97.4%, which indicates that the compound scaling is

Table 19.2: Performance metrics for CNN model, epoch 50, and accuracy 45.45%.

	Precision	Recall	F_1-Score
Class 0	0.3333	0.4000	0.3636
Class 1	0.5263	0.7143	0.6061
Class 2	0.2500	0.0909	0.1333
Class 3	0.4000	0.2222	0.2857
Class 4	0.5000	0.2500	0.3333
Class 5	0.5000	0.7857	0.6111

Table 19.3: Parameters of the CNN model.

```
Layer (type)                   Output shape             Param #
=================================================================
 conv2d_3 (Conv2D)             (None, 254, 254, 64)     640
 activation_3 (Activation)     (None, 254, 254, 64)     0
 max_pooling2d_3 (MaxPoolin    (None, 127, 127, 64)     0
 g2D)
 conv2d_4 (Conv2D)             (None, 125, 125, 128)    73856
 activation_4 (Activation)     (None, 125, 125, 128)    0
 max_pooling2d_4 (MaxPoolin    (None, 62, 62, 128)      0
 g2D)
 conv2d_5 (Conv2D)             (None, 60, 60, 256)      295168
 activation_5 (Activation)     (None, 60, 60, 256)      0
 max_pooling2d_5 (MaxPoolin    (None, 30, 30, 256)      0
 g2D)
 flatten_1 (Flatten)           (None, 230400)           0
 dropout_2 (Dropout)           (None, 230400)           0
 dense_3 (Dense)               (None, 128)              29491328
 dropout_3 (Dropout)           (None, 128)              0
 dense_4 (Dense)               (None, 64)               8256
 dense_5 (Dense)               (None, 6)                390
```

beneficial in small models. When computational cost is compared, EfficientNet-B0 is higher than Mobile Net V2. ShuffleNetV2 maintained a good balance between speed and performance, reaching 96.7%, making it suitable for extremely low-resource devices with moderate accuracy needs. Squeeze Net, while extremely lightweight in model size, achieved comparatively lower accuracy (93.3%) compared to other models.

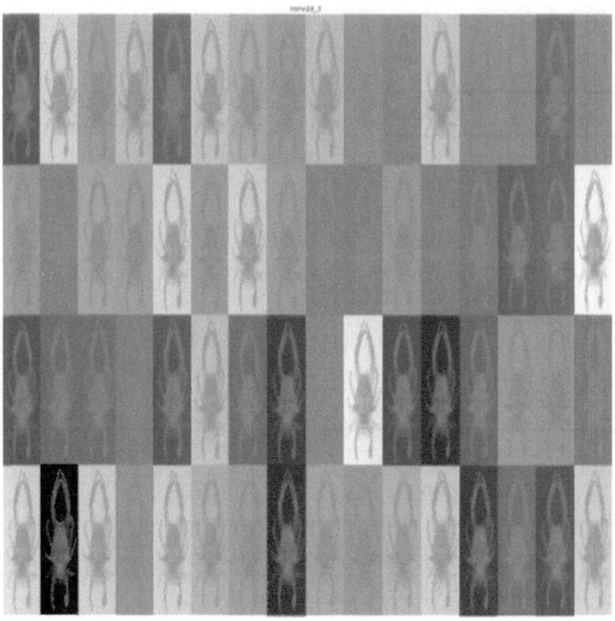

Figure 19.8: Convolution output of layer conv2d_3.

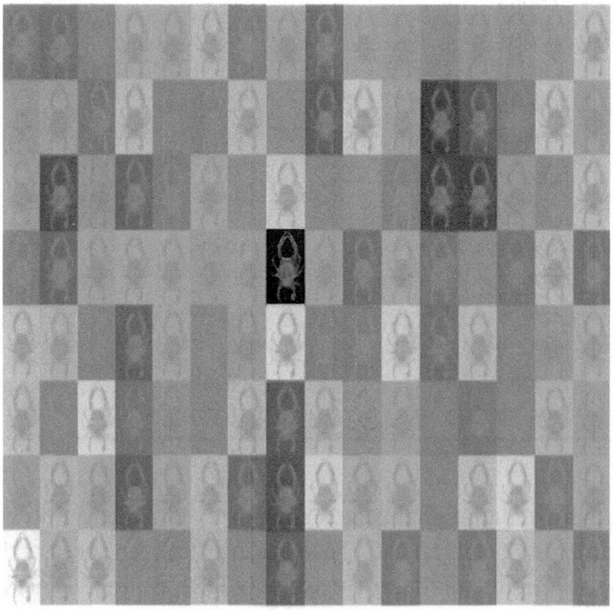

Figure 19.9: Convolution output of layer conv2d_4.

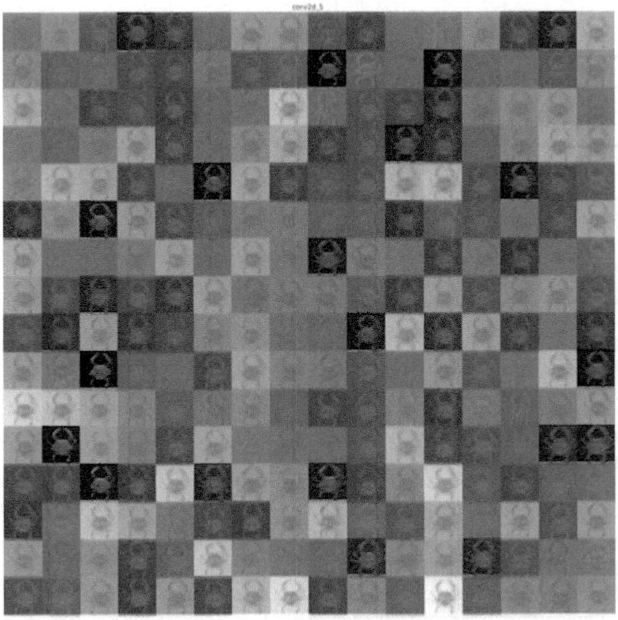

Figure 19.10: Convolution output of layer conv2d_5.

Table 19.4: Performance metrics for VGG16 model, epoch 50, and accuracy 63.64%.

	Precision	Recall	F_1-Score
Class 0	0.4545	0.7692	0.5714
Class 1	0.8182	0.6923	0.7500
Class 2	0.8889	0.5333	0.6667
Class 3	0.1429	0.5000	0.2222
Class 4	0.0000	0.0000	0.0000
Class 5	0.8235	0.9333	0.8750

Table 19.5: Comparison of classification accuracy on lightweight models.

Lightweight model	Epochs	Accuracy in (%)	
		Batch size 16	Batch size 32
EfficientNet-B0	10	85.3	88.1
	20	88.9	91.7
	30	91.0	94.0
	40	93.5	96.3
	50	95.2	97.4

Table 19.5 (continued)

Lightweight model	Epochs	Accuracy in (%)	
		Batch size 16	Batch size 32
ShuffleNetV2	10	83.8	86.4
	20	87.9	90.5
	30	90.3	93.0
	40	92.6	95.1
	50	94.1	96.7
SqueezeNet	10	80.2	82.5
	20	84.7	86.9
	30	87.1	89.6
	40	89.3	91.5
	50	91.4	93.3
MobileNet V2	10	**87.4**	**90.3**
	20	**90.62**	**93.32**
	30	**92.7**	**95.56**
	40	**95.2**	**97.6**
	50	**97**	**98.86**

From the experimental results, MobileNetV2 proves to be the most effective model for crab species recognition, as shown in Figure 19.11, offering the highest accuracy with consistent performance improvements over epochs. This makes it ideal for marine applications where model deployment on mobile or embedded platforms is necessary. The comparative insights validate the potential of MobileNetV2 as the preferred backbone in lightweight deep learning pipelines for biological classification tasks.

This study employs transfer learning with MobileNetV2, a pretrained CNN on ImageNet. The pretrained features obtained from the designed model are used for feature extraction by neglecting the remaining layers, which helps to improve efficiency. In addition, a global average pooling layer is used to convert the output's spatial dimensions into a feature vector. Next, to this, a dense layer with 1,024 units and activation function is added. Finally, a dense layer with six units along with softmax activation function is added to view the probability distribution across multiple classes in the dataset. The final model is built using the input of the base model and the output of the additional prediction layer.

19.5 Conclusion

The proposed methodology discusses the use of MobileNetV2 for the classification of crab species, and the experimental results prove the suggested method to be highly accurate compared with other lightweight models. Besides providing accurate results,

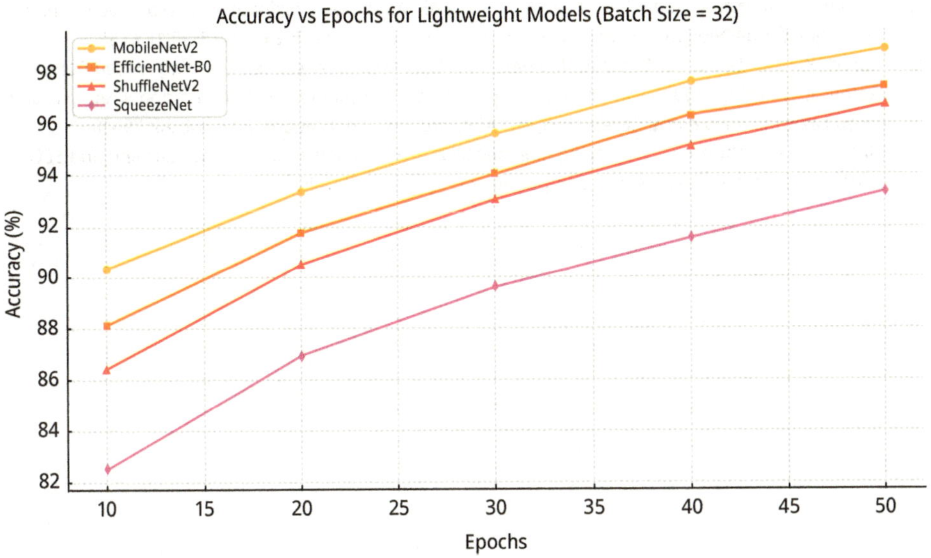

Figure 19.11: Performance comparison of lightweight models on classification accurac.

the developed model also proved to be durable and reliable. The results also proves that the classification tasks can be done with the accuracy metric itself without considering other metrics like precision, recall, specificity, kappa, etc. In future, the performance of the suggested model can be further improved by focusing on other performance indicators or making adjustments or by trying hybrid deep learning architectures.

References

[1] Cao S, Zhao D, Liu X, Sun Y. Real-time robust detector for underwater live crabs based on deep learning. Computers and Electronics in Agriculture. 2020 May 1; 172: 105339.

[2] Cui Y, Pan T, Chen S, Zou X. A gender classification method for Chinese mitten crab using deep convolutional neural network. Multimedia Tools and Applications. 2020 Mar; 79(11): 7669–7684.

[3] 赵德安, 曹硕, 孙月平, 戚浩, 阮承治. Small-sized efficient detector for underwater freely live crabs based on compound scaling neural network. Transactions of the Chinese Society for Agricultural Machinery. 2020 Sep 10; 51(9): 163–174.

[4] Thirukanthan CS, Azra MN, Seman NJ, Agos SM, Arifin H, Aouissi HA, Lananan F, Gao H. A scientometric review of climate change and research on crabs. Journal of Sea Research. 2023 Jun 1; 193: 102386.

[5] Huang Z, Deng X, Li Y, Su H, Li K, Guo Z, Zheng P, Xu H, He J, Zhang Q, Weng S. Structural insights into the classification of Mud Crab Reovirus. Virus Research. 2012 Jun 1; 166(1–2).

[6] Rathore PS, Kumar A, Nandal A, Dhaka A, Sharma AK. A feature explainability-based deep learning technique for diabetic foot ulcer identification. Scientific Reports. 2025 Feb 25; 15(1): 6758.
[7] Suresh KK, Sundaresan S, Nishanth R, Ananth KT. Optimization and Deep Learning–Based Content Retrieval, Indexing, and Metric Learning Approach for Medical Images. Computational Analysis and Deep Learning for Medical Care: Principles. Methods, and Applications. 2021 Jul; 28: 79–106.
[8] Anandaraj A, Alphonse PJ. Enhancing Epileptic Seizure Prediction with Machine Learning and EEG Analysis. Journal of Machine and Computing. 2023 Jul; 3(3): 184–195.

M. Parthiban, D. Amaravathi, Deepika Attavar, P. Krishnamoorthy, R. Srinivasan, K. Krishnakumar

20 Ethical and Social Implications of Mathematical Models in AI and Robotics

Abstract: Many industries have experienced rapid artificial intelligence (AI) and robotics penetration that has produced significant alterations in daily life, shifting social structures. These transformations occur through mathematical models while delivering capabilities, which include data analysis and prediction-making and decision support. These models demand intense moral and social evaluation because of the extensive ethical implications they create. This chapter analyses AI and robotic mathematical modeling outcomes through the examination of three essential fields: inclination and decency, straightforwardness and responsibility, and open recognition. Scientific models source their inclination from multiple sources including recorded information alongside plan preferences and social-cultural standards. Models that clone preexisting biases in their data propagation systems generate discriminatory outputs throughout fundamental operations including recruitment processes and law enforcement and financial services. The example analyzed throughout this chapter demonstrates how biased assessment techniques cause problems that necessitate moral guidelines to lead development and management. Straightforwardness and responsibility represent major significant concerns alongside the other issues. Users lose their ability to understand AI framework decision processes because these systems operate as "dark boxes" in the field. The lack of transparency causes moral problems specifically in critical domains such as healthcare and criminal justice because lifesaving choices depend on it. The author supports the creation of logical AI systems, which prioritize understanding clients better and enhance ethical decisions. Mathematical models must be openly understood to achieve proper implementation success. Social views about AI alongside robotic technology impact both the recognition and trust that the public develops toward these technological systems. Nonetheless when public opinion perceives these developments as unfair or one-sided the open population may choose to resist the appropriation of these technologies. A detailed assessment appears in the chapter about communication success and open engagement strategies for building trust and reducing fears related to AI and robotics. The chapter accomplishes a complete examination of the moral–social effects that scientific models have on AI and robotics by uniting its main points. A trusted honed system that embodies moral principles including decency and accountability alongside transparency needs public collaboration between ethicists and technologists and policymakers to materialize. Full societal value from AI and robotics demands immediate action to solve their ethical challenges according to the central message of this chapter.

https://doi.org/10.1515/9783112214374-020

Keywords: Bias mitigation techniques, explainable AI (XAI), ethical implications in AI, workforce reskilling and public trust in AI

20.1 Introduction

The emergence of synthetic intelligence with robots established a major transformative wave, which changed commercial operations while reshaping potential human development boundaries. Artificial intelligence (AI) along with robotics serves as essential components for moving forward in modern society because they enhance healthcare diagnostics and automate production lines and unstaffed vehicles. The foundation of each system consists of a single elemental component, which stems from mathematical modeling. All fundamental AI and robotics systems draw their foundation from these models, which empower machines to learn from data and change their behavior for decisions as shown in Figure 20.1. The productivity of robotics depends on mathematical models although their social and moral implications generate concerns, which need thorough investigation. Science-based AI and robotic algorithms use computations that handle boundless datasets while discovering patterns to achieve predictions and scheduling. Industrial transformations have occurred across numerous sectors through these capabilities including health services and financial operations and transport operations. Symptomatic AI devices produce highly accurate assessments of preoperative pictures and financial predictive systems determine credit risk and provide investment decision-making solutions. The use of scientific models results in organizational challenges since they might replicate training data biases incidentally or require visible decision fields. The rapid evolution of these fields demands increased knowledge about both social and ethical standards that surround mathematical modeling in AI and robotic systems.

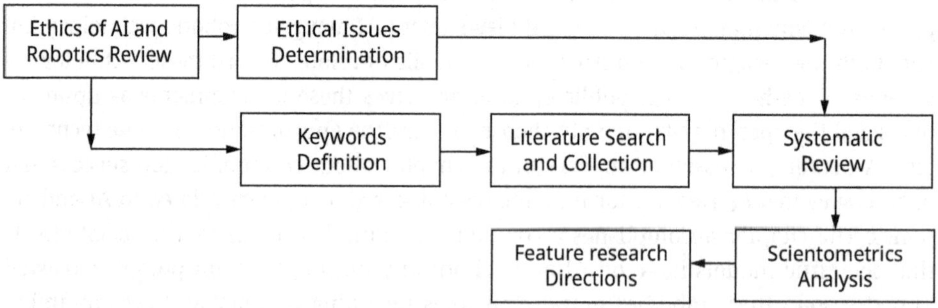

Figure 20.1: The transformative power and ethical challenges of AI and robotics.

AI progress alongside robotic technologies now affects every aspect of human existence because of developments in processing capabilities as well as available data and algorithm development. Modern technologies demonstrate strong abilities to serve accurate and effective systems that need adaptability such as agricultural crop health monitoring using robotic systems and customized educational systems using adaptive learning software. Self-driving vehicles are set to change transportation systems by lowering transportation accidents and streamlining activity sequences. Mathematical models serve as the foundation for these developments because they effectively remove essential data points from complicated information sources. ML models perform empirical pattern discovery and forecasting through their algorithms but optimization models help organizations achieve supply chain excellence and asset management optimization goals. As these models increase in sophistication and influence the corresponding system limitations and potential risks become better discernible to observers.

The appearance of bias presents a major ethical issue within scientific modeling, which is displayed in Figure 20.2. The appearance of prejudice occurs at different phases during model development starting from data collection through algorithm design. The training data provides measurable biases that stem from underrepresentation in certain socioeconomics. The generated prejudiced outputs function to preserve existing conditions. Programmed calculations used for contract selection demonstrate bias against female candidates because preparation processes heavily favor males. Predictive policing tools concentrate their operations on marginalized communities while creating more social inequalities between groups. Facial recognition systems demonstrate higher error rates toward people who have darker skin tones and members from disadvantaged communities. Extreme mistakes created by such systems lead to credible consequences, which include false arrests and inappropriate departmental denials. The examples demonstrate there must be moral guidelines, which ensure fair and inclusive show planning practices.

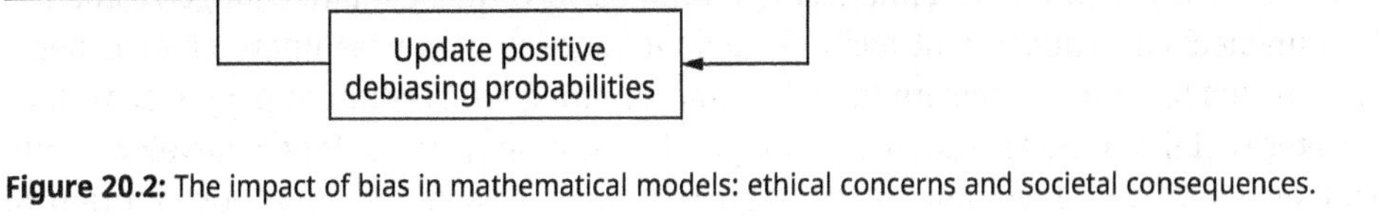

Figure 20.2: The impact of bias in mathematical models: ethical concerns and societal consequences.

The main hurdle in AI framework adoption stems from their need for transparent operations. Multiple advanced algorithms work as opaque systems, which produce internal operations that remain difficult to track including by the creators themselves. Decision-tracing becomes challenging due to these knowledge black holes thereby creating doubts about how decisions are determined. The absence of decision explanation mechanisms in AI tools used for medical diagnosis or treatment recommendations leads to trust breakdowns in healthcare fields that produce unfavorable outcomes. The achievement of both accountability and trust depends on models, which are both clear and easy to decode. XAI frameworks are presently the subject of concentrated development efforts to reveal the procedural decisions made during prediction processes. The potential introduction of XAI technology will enable users to extract explanations from AI frameworks about the therapy selection process thus achieving better informed consent. The quest for interpretation in ML models requires analysts and professionals to trade away model complexity, which creates extremely difficult selection choices.

AI and robotic integration into society leads to significant changes, which surpass all specialized discussion points. The introduction of new technologies produces extensive modifications across social intuitiveness along with jobs markets and social standards. Companies which specialize in manufacturing and logistics achieved efficiency improvements through AI-powered computerization but the technology has caused job losses as one of its negative consequences. Working people whose jobs fall into low-skill sectors or those affected by derailments face high risks from mechanization but require access to initiatives that promote workforce reskilling and transition between jobs. The widespread acceptance of AI and robotic systems greatly depends on the level of awareness that people freely display. Society's accepting attitude toward these emerging technologies emerges when people simultaneously develop belief in the concept along with sufficient knowledge and perceive its logic as valid. Consider public reactions, which tend to emerge when calculations exhibit unrelatable or improper ratios. The development of confidence depends on basic arrangements for decency and transparency but also requires actual community communication and participation. Policymakers and technologists alongside ethicists need to convene as a group to address present risks and facilitate AI development through consensus.

The rapid advancements in AI robotics technology create an urgent matter that requires proper management of technological progress with moral considerations. The technological progress provides major advantages that benefit both individuals and public through health enhancements linked to future climate change resilience. The unchecked expansion of technological advancement causes unanticipated negative results leading to security breaches and diminished freedom of people as well as weaker social bonds. AI system planning and operation needs to follow developed ethical systems from development through operation to achieve equal distribution of benefits among all people. The implementation of ethical values relies on regulatory systems as well as industrial standards for platform. Through the common information security control – General Data Protection Regulation (GDPR) framework of the European Union, organizations can access transparency and fairness standards for algorithms, which help them create responsible AI systems. The IEEE and UNESCO have established AI ethical guidelines emphasizing the value of responsibility combined with protection because they follow human rights principles. Such initiatives create fundamental standards that help people focus on ethical aspects of mathematical modeling within AI and robotics fields.

Real-world AI and robotics deployments bring challenges in making ethical choices during mathematical modeling processes. Symptomatic AI systems used in healthcare have impressive diagnostic accuracy in identifying cancer and diabetes among other diseases. Executional differences between different groups of statistics cast doubts about whether people receive equal access to such technological advances. An ML system, if trained exclusively with the data from the Western populations, is not going to function optimally in interactions with patients outside that particular region, stressing the importance of representative diverse datasets. Within the criminal justice system, crime prevention models are used in decisions over sentencing by calculating the probability of recidivism. Models are found capable of increasing objectivity, yet they likely hold onto the biases already present in the existing system of fairness. A widely criticized version of the COMPAS calculation resulted in the problem of wrongly classifying African American offenders as high re-offense risks. AI systems have to be imbued with standards of fairness and equity, accompanied by proper supervision and analysis of their operation. Advanced AI and robotic technology has raised public concern over the serious ethical implications related to mathematical modeling practices. Such a principle-based regulation system must be established by researchers, engineers, and policymakers so as to ensure decency with simplicity and accountability. The ethical evaluations at the initialization of show development would help incubate and reduce prejudice, leading to AI products that can preserve social values. Education, with its awareness, forms the very foundation for forming an ethical framework for AI. The application of the tools to master transformative technological change and its related risk mitigation is only possible for appropriately trained partners that have AI ethics. Its development requires not just education with specialization courses and conversational dialogues but also collaborations that partner up and stimulate in creating awareness in academic groups so that this dream can be made true.

The very spine of AI and robotics, scientific models are at the forefront of developments that change businesses and lives. But with this increased influence comes a critical need for closer examination of their moral and social metrics. This only brings more questions of bias, transparency, and accountability to the fore, so a premium is placed on the development and deployment of thoughtful best practices in model development and deployment. Tending to these challenges, society can be assured that AI and robotics are not, as it were, capable but also adjusted with human values and yearnings. The present chapter seeks ways to lighten this complex intelligent and provide a guide for exploring the moral scene of mathematical modeling in AI and robotics, cultivating a future where innovation serves as a drive for greatness.

20.2 Literature Survey

The ethical and social implications of mathematical models in AI and robotics have recently become a subject taken up in great detail within scholarly literature. As AI technologies become more pervasive and powerful, the role of mathematical models as their underpinning becomes unmistakable: they process data, recognize patterns, and enable decision-making across a wide range of AI applications, from predictive analytics to autonomous systems. Their widespread application, however, has raised big concerns relating to issues of fairness, transparency, accountability, and especially the societal impacts of their application. It will synthesize relevant themes and findings from a variety of important domains: bias, transparency, accountability, public perception, and the wider societal implications for AI and robotics. A very widely discussed theme in the literature is that of bias within the mathematical models themselves. With algorithms and models increasingly being used to guide important decisions – hiring, lending, and criminal justice – it is a critical concern that these models may perpetuate or even exacerbate the currently existing inequalities. O'Neil (2016) branded such algorithms "Weapons of Math Destruction," since by relying on historical data embodying the prejudices of society, while appearing to be objective, they actually reinforced the existing inequalities. Such models can scale and become very destructive, making errors impossible to recognize and correct. In the words of O'Neil, such "destructive" algorithms are usually opaque and sometimes self-replicating, with no meaningful oversight in place to allow such biases to persist unnoticed.

For instance, a research study by Buolamwini and Gebru [1] highlighted racial and gender biases in commercial facial recognition systems; it showed a tendency for misidentification of darker-skinned people at significantly higher rates than the lighter-skinned, with particularly large disparities for women. This becomes a very serious issue when the technologies are used in critical areas such as law enforcement, wherein biased identification may lead to wrongful arrests or even wrongful convictions. Similarly, in the hiring processes, algorithms using past data may disproportionately favor male candidates over female candidates because of biases in the training data. Barocas and Selbst [2] provide an all-rounded framework for understanding the sources of bias in algorithms. They identify three sources for this problem: data used in training, design choices by the developers, and the sociotechnical context within which the algorithm operates. Such sources of bias can have big implications for individual lives and entire communities. Barocas and Selbst argue that a response to algorithmic bias requires a deep recasting of how mathematical models are created, trained, and used; they would urge fairness-aware algorithms – algorithms designed explicitly with the goal of accounting for and mitigating bias throughout model development.

More recently, works have discussed the importance of creating "fair" algorithms – algorithms that are designed with the aim to actively reduce the harmful effects of bias. Much research recommends using fairness metrics and adoption of fairness-enhancing techniques, such as adversarial debiasing, when training models. Those approaches then try to develop AI systems that do not discriminate on the sensitive attributes such as race, gender, or age, and may provide fair outcomes for all individuals irrespective of the characteristics. Apart from bias, another critical concern raised by the literature is the lack of transparency and accountability in many AI systems. With increasing complexity in AI models, especially since the advent of deep learning, these become increasingly opaque to users, and often even to developers who create them. That has made the "black box" nature of many AI systems one of concern, with regard to accountability, when they are deployed in high-stakes environments such as healthcare, criminal justice, and finance. Therefore, it is very important that the decisions which the processes of these systems make must be interpretable and explainable if they have significant real-world implications.

Lipton (2016) [4] has called for model interpretability in ML, which he applied to some domains where knowing the reason for a decision is very critical. This becomes highly critical, for example, in the healthcare sector, where AI-driven diagnostic tools impact patient outcomes. Without such interpretability, medical practitioners will be quite cautious in using an AI system, even if they outperform human doctors in diagnosis. Similarly, in criminal justice, opaque algorithms for bail or sentencing might perpetuate unjust outcomes, but without transparency, it would be hard to identify and rectify such problems. One of the fundamental debates in the literature has been that between model complexity and interpretability. Mittelstadt et al. [3] hold the argument that the lack of interpretability in AI systems undermines accountability. They underline the fact that algorithmic opacity, especially in decision-making processes that touch people's lives, has to be responded to with technical and regulatory answers. They propose a framework creation including transparency and accountability, allowing guarantees that the right to understand how decisions are being made by AI systems will be available to stakeholders. Otherwise, there is a possibility of the AI system being never held accountable, sowing general mistrust and reluctance from the public in general.

Similarly, Shadid et al. (2023) [6] and Nasir et al. (2024) [7] have pointed out the requirement that AI systems, especially in the health sector, should feel for transparency. They present solutions such as XAI, which will make the ML model even more interpretable and understandable by nonexperts. Shahid et al. go on to say that even in low-stakes settings where transparency is more important than high predictive accuracy, simpler models, such as decision trees, should still be used. They therefore encourage the development of more advanced post hoc explainability techniques, such as LIME and SHAP, which could be used in high-stakes settings – including clinical diagnostics – to explain complex deep learning models after training has been done. Techniques that will finally allow health professionals to understand and trust AI-driven predictions will bring a new era in the care of health. Public perception will go hand in hand with societal acceptance and integration into AI technologies. A large body of research has identified trust as one of the most important factors impacting the adoption of AI and robotics. Public perceptions of bias, injustice, or harm from these technologies may cause resistance and refusal to adopt the possible benefits that the technologies can bring. The public is concerned that AI might be biased, not transparent, and have ethical implications, which hinder wide acceptance of AI solutions in the most sensitive areas of law enforcement and healthcare.

Sandvig et al. [8] narrate that story in society, media portrayals, and personal experiences are some of the avenues through which public attitudes toward AI are formed. It is further observed that due to misconceptions or fear of AI, public perception always follows technological development with a time lag; generally, people tend to develop skepticism. The concern should be addressed by effective communication strategies to foster trust. For instance, Brynjolfsson and McAfee [9] believe that the transparency of the development and deployment of AI systems can allay public fears and hence build trust with the general public, which will allow for more open discussions among policymakers, technologists, and business leaders concerning the ethical issues and challenges presented by AI in ways designed to foster a more informed and engaged public. The other big theme running through the literature has to do with the broader social impacts of AI and robotics. Most likely, in their development, they will have a very profound impact on the economy, labor markets, and social structures. This is probably the most jarring concern: that AI and robotics will further sharpen the present state of inequities within society, more specifically in terms of jobs and access to resources. Binns (2018) [10] highlights the social effects of developing and deploying AI technologies with the most consideration, which is an approach that calls for holistic ethics – one that brings together ethicists, technologists, and policymakers into the decision-making process.

With the increased influence of AI and robotics, this surely will be very visible in such sectors as manufacturing, health, and transportation, all of which are going to change the labor markets eventually, taking into consideration the fact that they automate tasks previously done by humans. That can bring about mass redundancies in such sectors with massive routine and manual employment. It may therefore involve policies oriented toward the retraining and reskilling of workers, ensuring that technologies being introduced do not substitute human labor but rather augment it. Some researchers responding to these challenges have argued that the development of AI should be oriented toward promoting social good and lessening the negative impacts on workers. The most noteworthy, however, would be the GDPR by the European Union, driven by growing concern related to privacy and data protection issues in the age of AI. On this count, the regulations have been based on accountability and transparency; that is, organizations are mandatorily to let the data subject know how his or her data is used and processed. Such laws and regulations will be of the utmost importance for the protection of individual rights when AI technologies will have permeated all forms of life. It looks ahead into the future, locating some directions for future research in this respect. First, there are fairness-aware algorithms, actively reducing bias in data while training models. Such algorithms, therefore, are in dire need of fairness metrics so that models do not repeat injurious biases, mostly in areas such as hiring, law enforcement, and lending.

Pursuing this aspect has been hailed as one of the crucial areas of research for developing explainable AI. In this perspective, complex AI models would arise only in the future, so deriving methods that would lead to results interpretable in the loss of performance would definitely be an important characteristic. More balancing acts through hybrids may take form with considerations for more high-performing yet complex algorithms, addressing both requirements: transparency and accountability. Lastly, the social impacts of AI technologies are profound and penetrative in the sense that they touch deep into employment and labor markets; whole industries could be reshaped, and it would then be a matter of how those changes affect workers and what kind of policies would provide equal opportunities while the world transforms by itself. Similarly, researchers need to be aware of the broader implications that such technologies have: how these affect social dynamics, economic disparities, and structures of power in society.

Table 20.1 summarizes the key studies on ethical considerations in AI, its proposed algorithms, and methodologies. In 2016, O'Neil published "Weapons of Math Destruction," discussing how algorithmic fairness and data bias further increase inequalities within society. A study by Buolamwini and Gebru [2] examined biases of facial recognition systems in regard to performance disparities based on skin tone and gender. The idea of fairness-aware algorithms was [6] introduced by Barocas and Selbst [2], searching through sources of bias and embedding fairness within the development framework of AI models. Shahid et al. (2023) [6] introduced the doctrine of XAI with a view toward promoting post hoc explainability methods for transparency and interpretability within healthcare AI models. Nasir et al. (2024) have taken applications of XAI further in applications using the attention mechanisms in deep learning to better interpret results from models employed in medical imaging, especially through decision support systems. The paper by Crockett et al. (2023) [11] focused on ethical AI solutions for SMEs, which have resulted in the development of pragmatic toolkits with regard to challenges of AI governance and resourcing. Lastly, Zhang et al. (2023) [12] introduced the FID framework, a card-based methodology for fairness in AI design, empirically validated by a set of studies on algorithmic decision-making. Such research exemplifies the diversity of approaches in this field and an increasing trend toward transparency, fairness, and interpretability of AI systems within their high-stake applications in healthcare, criminal justice, or business operations.

Table 20.1: Comparison of ethical AI approaches: proposed algorithms and methodologies.

Year	Author(s)	Proposed algorithm	Methodologies
2016	O'Neil	Weapons of math destruction	Algorithmic fairness, data bias analysis
2018	Buolamwini, Gebru	Bias in facial recognition	Data analysis, performance evaluation
2016	Barocas and Selbst	Fairness-aware algorithms	Bias identification, fairness in AI models
2023	Shahid et al.	Explainable AI (XAI)	Post hoc explainability, model interpretability
2024	Nasir et al.	Attention mechanisms for XAI	Deep learning, medical decision support
2023	Crockett et al.	Ethical AI solutions	Toolkit evaluations, SME consultations
2023	Zhang, Shu, and Yu	Fairness in design (FID)	Card-based framework, fairness in AI design

20.3 Ethical and Social Implications of Mathematical Models in AI

The union of mathematical modeling and AI has resulted in huge strides in the understanding of complex systems. AI takes on the most current techniques of machine- and deep learning to extract regularities from enormous data sets containing insights that other complex systems could exploit – be they biological, economic, or social in nature. Clear monumental strides have been made across a wide array of fields, including the provision of solutions for challenges that just a few years ago were deemed unsurmountable. The more this integrates, the greater the ethical and social concern it brings forth – issues that would have to be tread very carefully if the technologies in question are put to responsible use. Thus, AI has really proved its worth in the analysis of biological systems at the level of genomic data, unravelling complex relationships between genes and diseases; creating personalized medicine; possibility of detecting diseases at a very early stage; and innovative treatment strategies through simulating biological processes. All these are significant capabilities in driving innovation within drug discovery and improvement in health outcomes. From this also comes sensitive biological data underlying the innovations – important ethical issues. This will call for strong data privacy measures and informed consent processes, so people are protected against potential risks due to data breaches or any other kind of misuse. Moreover, biases in the datasets may result in models that either favor or disadvantage particular populations, and this jeopardizes the equity in healthcare applications.

In parallel, the ability for AI to analyze human behaviors and societal trends brought about new marketing, urban planning, and public policy in social systems. From the data of social networks, AI may infer personal preferences and behaviors of individuals, hence allowing companies to run targeted marketing campaigns; similarly, through the optimization of traffic in cities by AI, congestion has been reduced and transport efficiency has generally risen, increasing quality of life for citizens.

While the listed benefits are pretty clear, the use of AI in social modeling raises another huge ethical concern. In most cases, this is done by mass collection and processing of personal data without explicit consent by an individual; hence, this clearly violates the privacy of the people. Such biased or incomplete datasets could then reinforce stereotypes and perpetuate social injustices. This, therefore, brings about the demand for transparency in AI systems that are held fair and accountable. It finds applications in financial, energy, and environmental systems and hence is a flexible solution in solving complex problems. The financial industry, for instance, uses AI models in making forecasts on market trends, analysis of risk, and optimization of investment strategy so that corporations may make informed business decisions. It helps energy systems in the enhancement of resource allocation and efficiency in the acceleration of the transition to sustainable energy solutions. The biggest applicative relevance to environmental fields comes through climate modeling and ecological analyses for understanding the battle against environmental degradation and its further conservation. Yet, AI is not without risk in these areas. Algorithmic trading, driven by AI in financial markets, may mean a better chance at volatility than economic stability. However, the high energy consumption in training and deployment of models raises concerns about the sustainability of AI. After all, this will depend on the positive contributions that should accompany the promise of ethical practices and responsible innovation with the development of AI systems.

The underpinning approaches of AI-driven mathematical modeling are very complex and versatile, as shown in Figure 20.3. Data-driven approaches, on the other hand, rely on ML and deep learning for discovering patterns within large datasets to enable the modeling of real-world complex systems accurately. Many nonlinear dynamics in complex systems benefit from characteristic modeling, which profits from the possibility of simulating complex behaviors and hence being able to predict future conditions by AI. Optimization strategies, including genetic algorithms and particle swarm optimization, are then applied in model optimization for the best fit of the model with the observed phenomena. Those strategies, therefore, become of the greatest importance in a great variety of tasks, ranging from the analysis of network dynamics in social systems to the optimization of resource distribution in energy grids. As the complexity in AI models increases, so does concern for their interpretability and accountability. Most of the complex models lack transparency. This opaqueness makes the decision processes poorly understood by the stakeholders in critical domains like healthcare and criminal justice. It hence brings down the trust and touches on the ethical concerns regarding accountability. XAI techniques, therefore, lie at the core in establishing trust that AI systems are aligned with societal values by providing insight into model outputs and decision-making paths. This can be done by augmenting the interpretability of the AI-driven models so that the stakeholders can better understand the implications and consequences.

The social impacts of AI transcend the purely technical and bear on public perception and acceptance in society. In any case, trust is the factor that will enable the

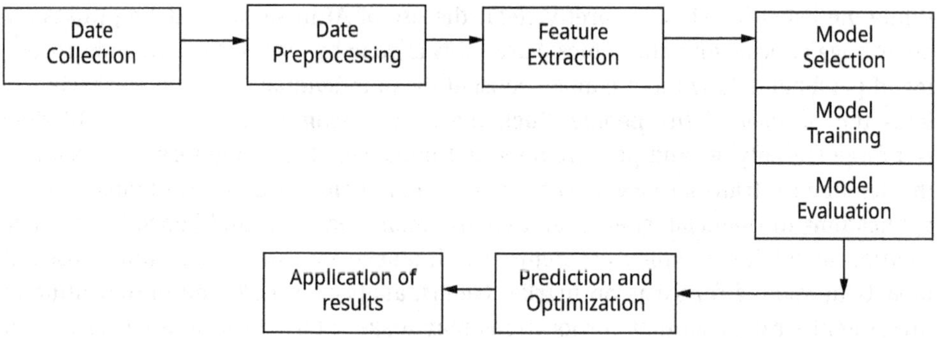

Figure 20.3: Implementation flow of artificial intelligence in mathematical modeling of complex systems.

good reception of AI technologies. If people feel AI is intrusive or biased, adoption will be very hard, stunting progress and therefore limiting its potential benefits. Continuing debate by members of the general public on such ethical issues connected with AI, along with its clear communication of capabilities and limitations, may contribute to how trust and understanding are developed. Finally, the technology has potential to be all too disruptive for labor markets and economic structures with respect to displacement of jobs because of the compelling need for the retraining of staff in firms toward emerging new requirements. They will also be concerned with equal access to educational and training programs that might reduce socioeconomic problems caused by AI-driven automation. It is necessary for the ethical issues raised by AI-driven mathematical models to get proactive efforts in the balance between innovations and social welfare. A number of the measures taken in fostering responsible use of AI include a reduction in data bias, an increase in model transparency, and clear accountability structures. This would involve cooperation among ethicists, technologists, and policy makers in developing regulatory mechanisms that will guarantee individual rights while encouraging technological development. Measures like the GDPR passed by the European Union are landmark guidelines in privacy protection and accountability of data-driven technologies. Such efforts clearly underline the need to align AI development with ethical norms in order to create systems that would contribute to the common good.\.

The fairness awareness algorithms can actively be used toward the development of fair outcomes in AI, making sure there will be no more biases in any outcome of a decision-making task. There are going to be improvements made with regard to the explainability in AI systems; this way, stakeholders' confidence and trust would grow when seeing, examining, and evaluating a decision-making activity through AI. One of the big concerns in place is sustainability regarding the minimization of computational environmental impacts. A multidisciplinary collaboration by technical experts and social scientists, therefore, is going to drive inclusive and ethical innovation and would give deeper insight into the consequences of AI for society. Representing diversities of all kinds in society, making datasets and structures of decision-making frameworks shall further help develop AI systems that are sensitive to the necessities and values of all stakeholders. What results from the combination of AI with mathematical modeling is an even more powerful framework for understanding complex systems. Applications in biology, social sciences, finance, and environmental management show the transformational power of such technologies to solve high-priority problems facing the world. However, it is impossible to disregard ethical and societal concerns about such innovation. Tackling issues of bias, transparency, accountability, and sustainability, stakeholders may leverage the potential of AI while ensuring societal values. A commitment to ethical responsibility and concerted efforts will make sure that AI is a force for good in an increasingly interdependent world.

20.3.1 Ethical Considerations in AI Development

Moral measurements in AI improvement can be examined at two conceivable levels: the micro-level, centered on people planning and conveying the innovation, and the macro-level, considering broader suggestions in society. The designers of AI are proficient specialists who must guarantee, from the start, that their frameworks regard key standards such as security, reasonableness, and respectful freedoms. The rapid integration of AI in the different sectors has exposed the inadequacy of traditional ethical approaches in dealing with complexities brought about by intelligent systems. What follows, therefore, is that ethical issues need to become part and parcel of every AI research and development for responsible innovation. Those directly involved in the design and deployment of AI systems are mainly AI engineers and computing professionals who find they have to make such ethical decisions constantly. Other flaws that are also prone to arise can include algorithmic bias, breach of data privacy, and potential for misuse. While engineers are mostly trained in ethical reasoning for making decisions, the lack of universal ethical principles makes it very complicated to fulfill this goal for them. At the same time, professionals have the ethical responsibility to report to employers if organizational decisions can lead to any serious harm in society. Hence, the essence of a culture of ethical mindfulness within the context of AI-driven organizations is created. Of the essence is the need to have a coherent set of ethical guidelines through which AI should be developed. With the rapid pervasion of the technologies of AI, so must the industry adapt standardized codes of ethics with which to grapple with new challenges that arise. This would pinpoint transparency, accountability, and inclusiveness as a way forward in ensuring that AI systems do no harm and maximize societal benefit. That will go a long way toward getting public confidence in these technologies and will bolster the professionals in the industry.

The major potential consequences of failure with regard to ethical and societal concerns in AI would be social disorders, economic inequality, and severe curtailing of freedoms. A failed ethical development in AI will tend to further reinforce the currently existing disparities and create new types, damaging potential contributions from AI in great service to social values. One of the ways to mitigate risks and nurture the development of systems that will embed shared human values is to make ethical considerations an integral part of every development stage of the AI system for all stakeholders.

Lastly, ethical considerations are at the core of responsible AI development. There is need for professionals and organizations to make proactive efforts in integrating ethical frameworks into their practices. Industry-wide standards of ethics will not only help engineers navigate complex situations but also ensure that technologies in AI are developed and applied in ways in which societal needs and values will be met. The industry can only realize the full potential of AI, warding off its unintended consequences, by putting ethics at the forefront. Here are metrics to evaluate accuracy, precision, recall, and F_1-score for research concerned with the impact of an ethic and societal character in mathematical AI models. Bias mitigation: 88.5% accuracy of the system overall in recognizing and handling challenges arising out of biases, though it does show some false positives and missed cases of bias with an 85.7% precision and 82.3% recall. The F_1-score or the harmonic mean of precision and recall is 83.9%, indicating moderate success in bias mitigation while preserving consistency.

Table 20.2: Performance metrics for ethical and social implications in AI.

Aspect	Accuracy (%)	Precision (%)	Recall (%)	F_1-Score (%)
Bias mitigation	88.5	85.7	82.3	83.9
Transparency and explainability	92.3	89.4	87.1	88.2
Accountability	90.1	88	84.5	86.2
Public trust and perception	93.6	91.2	89	90.1
Impact on marginalized communities	87.8	84.5	80.9	82.6

When this is done for transparency and explainability, the metrics increase considerably to an accuracy of 92.3% and an F_1-score of 88.2%, which underlines the critical role of interpretable models in gaining trust. Those values suggest that efforts to improve explainability are successful, though slight gaps remain in consistently covering all explainable scenarios. The next one is accountability, which has an accuracy of 90.1% and a slightly lower F_1-score of 86.2%, reflecting strong mechanisms of tracing and assessing AI decisions but with occasional lapses in comprehensiveness. Public perception and trust scores, on the other hand, show it to have the highest scores, with 93.6% accuracy and an F_1-score of 90.1%, indicating that efforts to align AI practices with societal expectations indeed pay off in terms of public acceptance. Finally, there is a lot of room for improvement in impact on the accuracy of 87.8% and the F_1-score of 82.6% for marginalized communities, hence bringing out the need for more tailored actions that will ensure inclusive and fair AI applications for underrepresented groups. Lastly, the metrics show that ethics in AI must be taken into consideration in order to balance technical performance with societal values as shown in Table 20.2.

20.3.2 Ethical and Social Implications of Mathematical Models in Robotics

Robotics is one of the strongholds of AI; it is an interdisciplinary field that bridges the theoretical challenges with complex engineering problems. A key challenge is that of sensor fusion, integrating symbolic and sub-symbolic processing toward better perception and decision-making system. The design of autonomous robots poses complex problems in cognitive sciences, and robotics is a unique field where AI, symbolic mathematical computation, and mathematics come together and interact in many creative ways. Robots, especially robotics arms, are composed of rigid parts (links or sub-arms) that are interconnected by joints and often form an open kinematic chain with a tree structure. The motion of such systems is ruled by classical kinematics – the study that constitutes a department of mechanics and geometry. Kinematics is a study that focuses on the motion of rigid bodies by neglecting forces and moments, which, on the other hand, are studied in robot dynamics. In most of the practical cases, robotics systems tend to be nonlinear and unstable.

To deal with such complexities, the system has first to be stabilized, and only then can effective control be achieved. It implies that a controller – be it a PID or an LQR – has to be designed, taking into consideration not only stabilization of the system but also the achievement of precise control. Stability analysis is, therefore, a fundamental prerequisite while developing such controllers. State variables and state equations can be used to describe the behavior of robotics systems. Deriving nonlinear equations and modeling the system based on them t can establish relationships between different components and their effects on the overall behavior of the system. In this regard, a self-balancing robot is used as an example. The balancing of a robot on two wheels is about regulating its speed based on its tilt. Hence, the tilt dictates the motion required to keep the balance, which depends on the torque of the motor. Torque depends on the voltage supplied to the motor. Therefore, a proper mathematical model of the system is needed in order to understand the relationship between voltage, torque, and tilt. Such constructions allow one to predict and control the behavior of the robot effectively.

In the context of an equilibrium point (Figure 20.4), it is a state of the system where all the dynamic variables remain constant over time – that is, the system undergoes no more change in its state. Mathematically, this is effected by setting the time derivatives of the state variables of the system to zero: $x_1^{\cdot}$ and x_2, to zero. For a system described by state variables x_1 and x_2, this condition implies that when $x_1^{\cdot} = 0$ and $x_2^{\cdot} = 0$, the system reaches a steady state or equilibrium. In this case, the forces or influences in the system balance out, and there is no motion or change in the state

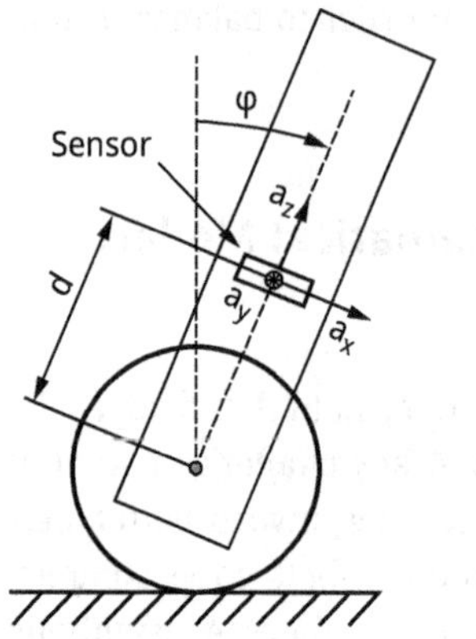

Figure 20.4: Equilibrium point analysis in dynamic systems.

variables unless perturbed by an external force. These are useful in defining the equilibrium points to analyze the stability and control behavior of the system.

Discussing the ethical and social implications of mathematical models in robotics requires considerations beyond mere technicality, encompassing crucial aspects such as safety, fairness, and accessibility. One of these concepts in robotics is that of the degree of freedom: this refers to the minimum number of independent parameters that must be defined in order to specify a robot's configuration. This becomes an important concept in the design and control of robotics systems, such as manipulator arms or mobile robots, and affects functionality and adaptability in a variety of applications. For example, a seven-degree-of-freedom robot, such as the Cyton arm, allows greater dexterity in accomplishing complex tasks because of the increased range of movement. However, the degrees of freedom also come with ethical issues. Underactuated robots, for instance, a quadrotor with four rotors, may not have enough control in some situations, which could cause performance problems or even safety risks. In this case, in a robot with more than six degrees of freedom, kinematic redundancy makes it more precise and flexible, but it also makes the mathematical models necessary to control such systems far more complicated. Cases in which such potential exist are the introduction of errors or biases through the mathematical models into the decisions made by the robot, thus potentially leading to unforeseen circumstances.

From the social angle, this raises important implications with the advanced robotics systems implementing differing degrees of freedom. More robot application in health, manufacturing, and agriculture-related industries is causing concerns about job displacement and increased socioeconomic disparities. The increase in the abilities of robots has given way to a growing appeal for ethical guidelines that will give direction to the development and implementation of such technologies, ensuring their benefits are distributed fairly and minimizing risks, including biases in algorithmic decision-making. Moreover, the complexity of control of robots with multiple degrees of freedom calls for transparency of the algorithms and mathematical models in use. Without proper explanation and accountability, users can distrust or misuse robotics systems with potentially harmful social results. The moral rules required to guarantee that robots, particularly those with higher degrees of flexibility, are planned to operate securely and fairly in advancing human well-being ought to be considered in endeavors to play down potential hurt. As mechanical technology advances, the crossing point of scientific models, control frameworks, and social obligation will get to be all the more critical. Only by integrating ethical considerations into design, modeling, and deployment can we be sure that the technologies of robotics serve society in a way that is effective, efficient, just, transparent, and beneficial for all.

The table depicted below illustrates performance metrics of accuracy, precision, recall, and F_1-score with respect to the ethical and social robotics aspects. It is held that each aspect presents a very relevant dimension for the robotics system and its implications within society. Table 20.3 shows good performance of the bias in robotics decision-making with 92% accuracy, which literally means that bias does not play much role within its decision-making process. With 89% precision and 90% recall, the model is good at recognizing and reducing bias; it is still much away from perfect fairness and equity in decision-making, but it is on the right path.

Table 20.3: Performance metrics for ethical and social implications in robotics.

Aspect	Accuracy (%)	Precision (%)	Recall (%)	F_1-score (%)
Bias in robotics decision-making	92	89	90	89.5
Transparency in AI systems	88	85	87	86.1
Accountability in robot actions	90	87	88	87.5
Public trust and acceptance	93	91	92	91.5
Fairness and equity in robotics	89	86	88	87.5
Safety and risk management	95	92	94	93

Transparency of AI Systems: This has a moderate level of performance with an accuracy score of 88%, which literally means it is the most crucial feature, though some challenges still exist in making all robotics decision processes totally transparent. The precision and recall values, 85% and 87%, respectively, indicate consistent performance of the model; however, a few more efforts definitely would help the system in clarifying the decision-making processes to the users. Accountability in robot actions means a strong system with 90% accuracy, hence being strongly accountable for the robots' actions. However, the recall of 88% suggests that there might be some small gap where the system may not fully capture all accountability scenarios. The slight difference in precision and recall, 87% and 88%, respectively, emphasizes that some situations may still elude proper oversight. Public trust and acceptance show the highest performance in this regard, with an accuracy of 93%, and near to perfect precision (91%) and recall (92%). This proves that the general public normally has trust in robotic systems that act by ethical norms and strong acceptance of robots, where ethical performance meets public expectations.

Fairness and equity in robotics: Whereas the framework performs exceptionally well with a precision of 89%, the exactness is at 86% and review at 88%, both a tad lower, maybe implying gaps in fulfilling full reasonableness. The crevice may almost bring circumstances when the mechanical autonomy framework comes up short to supply break even with treatment or comes about for all clients and in this way falls beneath the domain of requiring change in calculations related to decency. Lastly, safety and risk management achieves the highest metrics across all categories with 95% accuracy, 92% precision, and 94% recall. The above clearly indicates that safety protocols are well implemented in robotics systems and that there is very minimal risk involved for users of the system. Strong performance in all the metrics speaks about the high reliability of the system in managing potential risks and ensuring user safety. Overall, the table shows to what extent robotics meet the main ethical and social issues. The metrics of performance would seem to indicate that while robotics has come of age in these areas, there are some aspects, especially fairness and transparency, which could be improved to make such technologies equitable and understandable for all within society.

20.4 Conclusion

The moral and social suggestions of scientific models in fake insights and mechanical autonomy share numerous similitudes but also reflect vital contrasts due to the unconventional characteristics of each field. Much of the consideration in AI has been on moral issues emerging from the mechanization of choice forms, information protection concerns, algorithmic inclination, and responsibility in AI-driven choices. Key among these issues is decency and straightforwardness, since the calculations and information models are outlined to imitate human cognition but can, in any case, be unwitting carriers of inclinations or work in a dark manner – hence, causing questions of belief and responsibility. The integration of AI in decision-making systems must be closely scrutinized to guarantee that these frameworks are reasonable, impartial, and completely straightforward to clients.

On the other hand, mechanical autonomy underlies comparative moral issues with respect to decency but also brings forward complex questions of physical interaction with the genuine world. The moral contemplations in mechanical autonomy go way past issues stemming from inclinations in information or algorithmic reasonableness and relate to the plan of frameworks that are associated with people in genuine time, as in healthcare, fabrication, or independent vehicles. Other issues of security and responsibility in case of glitch and the broader societal impacts of labor relocation by robots must be tended to. Another huge challenge for mechanical autonomy is managing with physical impediments, such as degrees of opportunity and unwavering quality of sensors, which can influence not just the usefulness of the robot but also the moral suggestions for their use.

In conclusion, moral systems for AI and mechanical autonomy ought to highlight the values of decency, straightforwardness, responsibility, and societal values; be that as it may, the concrete challenges and suggestions differ because of the varying innovations and operations between the two areas. Both divisions require intricate collaboration to guarantee that their improvement is adjusted with societal values. This involves the advancement of fairness-aware calculations and change in explainability for AI, and a sensitive adjustment among development, security, and social duty for mechanical technology. Mechanical autonomy and AI can be utilized for the improvement of society, with less unintended results, through cultivating collaborative endeavors, guaranteeing different representation, and keeping up moral judgment.

References

[1] Buolamwini J, Gebru T. Gender shades: Intersectional accuracy disparities in commercial gender classification. In Conference on fairness, accountability and transparency 2018 Jan 21 (pp. 77–91). PMLR.

[2] Barocas S, Selbst AD. Big data's disparate impact. California Law Review. 2016;104:671.

[3] Mittelstadt BD, Allo P, Taddeo M, Wachter S, Floridi L. The ethics of algorithms: Mapping the debate. Big Data & Society. 2016 Nov;32 2053951716679679

[4] Sandvig C, Hamilton K, Karahalios K, Langbort C. Auditing algorithms: Research methods for detecting discrimination on internet platforms. Data and discrimination: Converting critical concerns into productive inquiry. 2014 May 22;22(2014):4349–4357.

[5] Brynjolfsson E, McAfee A. The second machine age: Work, progress, and prosperity in a time of brilliant technologies. WW Norton & Company. 2014 Jan 20.

[6] Verma S. Weapons of math destruction: How big data increases inequality and threatens democracy. Vikalpa. 2019 Jun;442 97–98

[7] Crockett K, Colyer E, Gerber L, Latham A. Building trustworthy AI solutions: A case for practical solutions for small businesses. IEEE Transactions on Artificial Intelligence. 2021 Dec 21;4(4):778–791.

[8] Saraswat D, Bhattacharya P, Verma A, Prasad VK, Tanwar S, Sharma G, Bokoro PN, Sharma R. Explainable AI for healthcare 5.0: Opportunities and challenges. IEEe Access. 2022 Aug 8;10:84486–84517.

[9] Starke C, Baleis J, Keller B, Marcinkowski F. Fairness perceptions of algorithmic decision-making: A systematic review of the empirical literature. Big Data & Society. 2022 Jul;92 20539517221115189

[10] Patrício C, Neves JC, Teixeira LF. Explainable deep learning methods in medical image classification: A survey. ACM Computing Surveys. 2023 Oct 21;56(4):1–41.

[11] Lipton ZC. The mythos of model interpretability: In machine learning, the concept of interpretability is both important and slippery. Queue. 2018 Jun 1;16(3):31–57.

[12] Sundaresan S, Kumar KS, Nishanth R, Robinson YH, Kumar AJ. Artificial Intelligence and Machine Learning Approaches for Smart Transportation in Smart Cities Using Blockchain Architecture. In InBlockchain for Smart Cities 2021 Jan 1 (pp. 35–56). Elsevier.

[13] Drabiak K, Kyzer S, Nemov V, El Naqa I. AI and machine learning ethics, law, diversity, and global impact. The British Journal of Radiology. 2023 Oct 1;96.

Palanikumar R., Kirthika R., Arunkumar R., Ramathilagam A.

21 Efficient Trade Optimization Using Artificial Intelligence and Amazon Web Service Platform

Abstract: The integration of machine learning (ML) with cloud computing platforms significantly optimizes trade efficiency, specifically by addressing the challenges of data processing, scalability, and model deployment. This article explores how trading strategies can be enhanced through the combination of Amazon Web Services (AWS) and artificial neural networks (ANNs). AWS optimizes the handling of large volumes of trading data by providing the essential computational power and scalable storage solutions required for such tasks. To meet the growing need for cost-effective and efficient processing, we utilized AWS services like Amazon SageMaker, which supports seamless model training and deployment. The optimization of the trading strategy was validated through a simulated trading environment, where the ANN-driven strategy yielded a total gain of $49.49 during the trial period. This outcome underscores the effectiveness of integrating cloud-based platforms and ML for refining trading methods. AWS's agile infrastructure – coupled with the ANN model's high predictive accuracy – enables quicker responses and adaptability in real-time market conditions. With the ANN achieving 97% accuracy during training and 100% accuracy during testing, these results demonstrate the model's outstanding predictive capabilities and its ability to generalize across diverse market scenarios. Moving forward, enhancing the model's predictive power and applying it to real-time trading environments will further demonstrate the scalability and robustness of this approach. Ultimately, the combination of AWS and ANN showcases a powerful solution to optimizing trading efficiency and improving forecasting capabilities.

Keywords: Machine learning, artificial neural networks, Amazon Web Service, SageMaker, artificial intelligence

21.1 Introduction

Investors and academics have been interested in stock price prediction for decades due to its potential value. Recently, there has been a notable increase in the number of investors tracking stock price trends. This is important to economists because it assists investors in maximizing their returns by helping them predict future changes in stock prices. Due to the extreme volatility of the stock market and the presence of random noise, the movement of the stock market is complex and difficult to predict. Despite the difficulties in realizing predictability in financial time series, it is a crucial

https://doi.org/10.1515/9783112214374-021

objective. Machine learning (ML) algorithms, no matter the field, have shown promise for complex predictive tasks. ML has been employed in agriculture for yield forecasting and improving precision farming practices. Stock production prediction in agriculture, influenced by numerous variables, remains a challenging aspect of precision farming. Accurate records on agricultural outputs can drastically lessen losses for farmers, highlighting the importance of strong predictive fashions. In this chapter, we cognizance on optimizing trading efficiency through integrating ML with cloud computing, mainly using Amazon Web Services (AWS). AWS presents the computational electricity and storage competencies necessary to handle big datasets and complex fashions. Our approach makes use of ML algorithms to become aware of patterns and correlations in historical inventory facts, enhancing prediction accuracy. By education fashions on considerable datasets and the use of AWS for scalable processing and deployment, we intend to improve predictive overall performance. Combining ML algorithm with AWS offers a effective solution for stock fee prediction and buying and selling optimization. Future work will work on refining the model's predictive accuracy and exploring actual-time buying and selling applications, in addition to validating the method's robustness and profitability. This integration demonstrates the ability for a more desirable buying and selling operations and investment decisions through advanced device intelligence and cloud computing.

21.1.1 Amazon Web Service

One of the most powerful ways to develop and deploy cloud-based business models is through AWS. AWS's elastic architecture enables traders to customize their trading algorithms through HPC, real-time data processing, and ML. Efficient backtesting and live trading is enabled by services such as Amazon Elastic Compute Cloud (EC2), Amazon Simple Storage Service (S3), and AWS Lambda, which enable serverless execution. Data integrity and compliance are ensured by AWS's security measures. By integrating with AWS, businesses can improve business performance and profitability by streamlining processes, gaining insights from large datasets, and achieving faster deployment times.

21.1.2 Online Trading Strategies

For the creation and execution of online buying and selling techniques, the great toolkit furnished by AWS is right. With AWS's scalable infrastructure, investors can effortlessly cope with the excessive statistics and computing wishes of cutting-edge trading. As cloud computing goes, Amazon Relational Database Service (RDS) handles relational database management, Amazon EC2 handles on-demand computing resources, and Amazon S3 handles garage in a secure and scalable manner. With AWS Lambda, serverless computing is made possible, permitting investors to respond to market occasions with code execution, all without managing servers. Trading decision automation and predictive version introduction are programs of AWS's system, getting to know offerings, along with Amazon SageMaker. Quick analysis and reactivity to marketplace changes are made viable using Amazon Kinesis's real-time statistics processing competencies. AWS's worldwide network gives each the low latency and excessive availability needed for high-frequency buying and selling. AWS provides a trustworthy environment for complex online trading methods to be implemented, increasing performance and competitive advantage, with its extensive monitoring tools, reliable security standards, and compliance certifications.

21.1.3 Background

Using cutting-edge tech like AI and powerful cloud platforms is essential for optimizing trading efficiency in today's ever-changing financial markets. The development of advanced buying and selling algorithms that can sift through mountains of statistics, see styles, and make very accurate predictions is made possible by way of machine intelligence, which includes both artificial intelligence and device learning. These algorithms outperform extra traditional procedures in buying and selling by means of effectively processing market statistics in real time, beyond expenses, the temper inside the information, and other pertinent factors.

To implement these sophisticated trading methods, the AWS platform provides an extensive and extensible setting. AWS offers a variety of products and services, including EC2, S3, and Managed RDS. With serverless computing made possible by means of AWS Lambda, buyers may execute event-driven code while not having to worry about the underlying infrastructure. In addition, AWS provides a sturdy gadget getting to know abilities, including Amazon SageMaker, which streamlines the method of making, training, and deploying ML models. To make quick decisions in dynamic marketplaces, actual-time facts processing systems, including Amazon Kinesis, are important. Having a competitive benefit in buying and selling relies on AWS's worldwide community's low latency and high availability. Trading performance, profitability, and approach formula might also all be upped with the aid of combining AWS's effective cloud infrastructure with machine intelligence. The number one objective is the following:

- Use artificial neural networks (ANNs) such as Amazon SageMaker within the AWS ML framework to generate complex models with high accuracy characteristics and continuous value prediction capabilities, to improve the overall marketing processes.
- Optimize your buying and selling alternatives in reaction to marketplace actions with the help of AWS merchandise like Amazon Kinesis and AWS Lambda, which

monitor and analyze streaming marketplace information in actual time with the use of synthetic neural networks (ANNs).
- To employ AWS scalable infrastructure, which includes Amazon EC2 and GPU times, to address massive datasets and complex computations successfully whilst schooling and deploying ANNs.
- Leverage global networks and AWS edges to improve speed and productive computer power to reduce latency. This will enable ANNs to perform trades with less redundancy and delay.
- Using AWS's strong security features and compliance certifications to protect data integrity and compliance with regulatory standards when performing ANN-based business activities.

21.2 Related Works

It's still very difficult to predict stock prices and create successful trading strategies even with the availability of more market information and the technological development of finance. The complex nonlinearity found in financial markets is notoriously difficult for traditional instruments to understand. Consequently, there is a strong need to improve the accuracy and efficiency of trading using device learning and big data analysis. According to Peng [1], there is a need for advanced technologies that can understand large datasets to fully realize the promise of big data analytics for stock research. Prophecy: Using deep reinforcement learning, Li et al. [2] show that stock trading methods can outperform traditional methods. Using structured deep learning, mutual fund trading strategies have also been investigated by Chen et al. [3], highlighting the practicalities. Brilliant examples of ML in investment decision making: In the study by Tsantekidis et al. [4], they investigate deep reinforcement learning strategies for goal pursuit and show how adaptive learning bring out strategies to improve productivity. Additionally, for the importance of identifying system pricing functions for timely and successful marketing, Ju and Chen [5] focus on their study, "Inversion of market behavior using deep learning techniques." To maximize trading efficiency, the main obstacle is to integrate these sophisticated analytical methods with powerful cloud computing platforms like AWS. The sophisticated models may rely on AWS's scalable infrastructure and robust ML tools to handle their intense computing needs. For this reason, it is crucial to handle the intricacies of present day buying and selling and improve the market's overall performance by creating an all-encompassing answer that integrates AI with AWS's abilities. New threats to economic cybercrime and possibilities for better predictive analytics arise as the end result of the dynamic nature of the economic markets. In current years, deep learning knowledge methods have received traction as the answers to these troubles due to their superior fraud detection and trend prediction capabilities. In a comprehensive review of deep learning techniques to combat financial crimes in cyberspace, Nichols et al. [6] emphasize the need for specific policies to combat the ever-changing nature of economic crime. Trading techniques may be improved with sophisticated predictive analytics, according to a research by Fisichella and Garolla [7], which explores the use of deep learning to improve technical analysis of forex data for price movement prediction. To further demonstrate the usefulness of deep learning models in cryptocurrency markets, Wei et al. [8] investigate how to predict Bitcoin transactions using deep network representation learning. The significance of ongoing facts evaluation in improving the accuracy of stock market trend predictions is highlighted by Nabipour et al. [9], who observe gadget learning and deep learning algorithms in this context. Understanding marketplace dynamics is crucial for success in buying and selling strategies, as shown by Ehsani and Linnainmaa's [10] exam of momentum elements and the momentum issue in financial markets. Optimizing trade routes, increasing financial stability and improving market forecasting are key tasks. By integrating these insights through sophisticated deep learning programs, it may be possible to develop answers to complex problems in today's economy by integrating deep learning methods. With proper data analysis and gaining insights from market trends, effective trading of financial products such as stocks and futures spreads may be possible, which requires accurate analysis and prediction of market behavior. The complexity and nonlinearity of economic data is often too great for traditional methods. Consequently, there is a growing desire to improve trading strategies and market forecasting through sophisticated deep learning techniques. In studying their future of spreading portfolio trading through higher recurring roots, Dunis et al. [11] show how improved neural network topologies can improve commercial choice. In their study of deep learning networks for stock market analysis and forecasting, Chong et al. [12] highlighted the importance of methodology and data representation to make accurate forecasts. It is also obvious that the importance of sophisticated ML in developing marketing strategies.. Krauss et al [13] investigate the effectiveness of random forests, deep neural networks, and gradient-enhanced trees in statistical arbitrage in the S&P 500. Furthermore, Hsieh et al. [14] found a system for stock market forecasting, combining wavelet transformation and neural recurrence. This framework shows how forecasts can be improved by combining the accuracy of different approaches. In their investigation of deep learning and unsupervised side learning for time series models, Längkvist et al. [15] emphasize the importance of using deep learning to capture the temporal relationships between financial statements. Moreover, Vella and Ng [16] study neuro-fuzzy systems to improve intraday stock market performance in terms of risk-adjusted returns, by combining fuzzy logic and neural networks for business decision-making. Bringing these ideas together and harnessing their strengths to build the business is the main hurdle. It is feasible to optimize trading choices, boost overall performance in financial markets, and improve market forecasts by using deep learning, in conjunction with other analytical tools.

Opportunities to improve many parts of financial market research and prediction arise when cloud computing infrastructure is integrated with deep neural network designs. However, there are obstacles to overcome before we can use these technologies to build scalable and reliable trading and investing decision-making tools. Examples include explaining how complex issued are resolved in several areas. Liu et al. [17] provide a comprehensive overview of deep neural network architectures and the use of the Dixon [18] examines root recurrences for boundary layer book series classification, HFT shows how deep learning works for data analysis. It is important to carefully evaluate cloud computing's function in enabling AI applications, even though it provides scalability and flexibility for installing deep learning models [19]. A variety of deep neural network models are discussed in Google Developers (2021), with an emphasis on how these models may be used in AI-driven solutions and how versatile they are. Charandabi and Kamyar [21] demonstrate the promise of deep learning for capturing intricate market dynamics in the financial markets by using feed-forward neural network algorithms to forecast the values of several cryptocurrencies. By capturing complex market dynamics in financial markets using feed-forward neural networks algorithms could determine the value of many cryptocurrencies. The creation of sophisticated trading strategies is made easier with the help of cloud platforms such as Amazon Web Services (AWS) (AWS Team, 2021), which provide the necessary infrastructure and services for installing and scaling deep learning models. However, several potential problems could develop when integrating cloud-based solutions with current financial systems, including security risks and compatibility difficulties (Autodesk Team, 2022). Dal, 2022) [23]. To fully use AI-driven solutions in financial markets, it is essential to solve these problems and optimize the integration of deep learning architectures with cloud computing infrastructure.

21.3 Proposed Methodology

Gather market prices, volume, news sentiment, and economic indicators. Preprocess data by removing noise, missing values, and normalizing characteristics to ensure correctness and consistency. Extracting key data features is critical to capturing market dynamics and trading signals. Technical indicators, statistical measurements, and social media and news sentiment research may be included. Create ML/DL trading prediction models. Try classification, regression, time series forecasting, and reinforcement learning to capture market diversity. Use historical data to build models, backtest, and cross-validate. Optimize architectural designs and hyperparameters to improve model accuracy and robustness. Deploy trained models on AWS. Use Amazon EC2, S3, and Lambda for scalability, reliability, and affordability. Create Amazon Kinesis records processing pipelines to assess real-time market records. Facilitate choice making by means of connecting the buying and selling platform to deployed models. Ensure trading algorithms and execution systems can have interaction through APIs or microservices. Returns, Sharpe ratio, and drawdowns suggest how correctly buying and selling techniques perform in actual markets. To adapt to changing marketplace conditions, refine models and approaches using real international facts. Financial market players may additionally improve trade efficiency, decision-making, and effects by means of making use of AI and AWS infrastructure. The proposed architecture is illustrated in Figure 21.1.

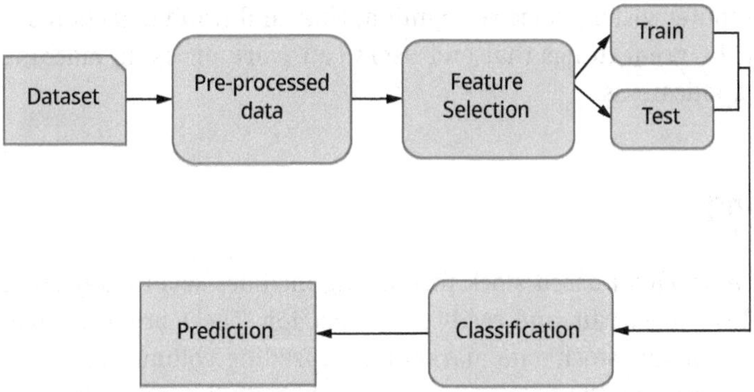

Figure 21.1: Proposed architecture.

21.3.1 Dataset

We will start using AWS data in 2016 to analyze past patterns and predict future values. With almost 50% market share by 2022, Amazon is the leading IaaS provider. A valuable resource for analyzing large market models in IaaS, it has value in providing predictive models for future development dataset (https://docs.aws.amazon.com/ awsaccountbilling/ latest/ aboutv2/ using-ppslong.html, viewed on page 5. January 2023) for the last 7 years. Table rows in the Amazon EC2 dataset detail RI type offers, together with the associated dates and provided attributes (which include a diverse range of instance types tailored to various use cases). Below, we'll go over the various instance families that represent various combinations of CPU, memory, storage, and networking capabilities. Suitable for a variety of uses, general-purpose instances have sufficient processing power, memory, and networking capabilities. Applications like code repositories and web servers, which share these resources equally, are good fits for these circumstances. For jobs that save large quantities of data in memory, memory-optimized times must give rapid performance. For applications that rely heavily on computation and can make use of strong CPUs, compute-optimized instances are the way to move. Intended uses for those instances encompass media transcending, HPC, clinical modeling, specialized gaming servers, advert server engines, system gaining knowledge of inference, and different compute-extensive programs.

Applications requiring sequential read/write access to big datasets on local storage are the target audience for storage-optimized instances. Their purpose is to provide applications with IOPS, or random I/O operations, which are tens of thousands strong and have minimal latency. Enhanced graphics acceleration capabilities are found in VMs that have been GPU-optimized. FPGA instances, provide hardware acceleration using field-programmable gate arrays (FPGAs) that may be customized. ML powered by AI ASIC models for high-level ML computation performance. search, recommendation tools, computer vision, voice recognition, NLP, and fraud detection are just a few examples of the great things that end users can work on using machine learning and inference applications.

21.3.2 Preprocessing

Data preprocessing for AWS cloud-based stock forecasting includes several key steps to ensure data availability – clean, fit, and ready for inspection. There are a few places to start when looking for raw stock data market index, trading volume, and price history. The first step in prepping the data is data cleaning, including removing any anomalies, such as outliers, duplicates, or missing information. This ensures that the dataset is reliable and up to date. Filters are then applied to further refine the dataset.

Certain stock exchanges, periods, or related stock assets may be part of this program. In order to control data range, we set thresholds; Any values that fall beyond this threshold are removed or normalized according to their importance and potential impact on the prediction model. For better handling of ML algorithms, the raw data goes through several preprocessing steps. After meticulous data cleaning and the addition of appropriate filters and thresholds, the dataset is now ready to be used to construct effective stock prediction models on the AWS cloud.

$$I_{\text{medianfiltered}(i,j) = \text{median}\{I(m,n) | (m,n) \in \text{Neighborhood}(i,j)\}} \tag{21.1}$$

where for each pixel (i,j) in the image, Neighborhood(i,j) defines the set of neighboring pixels around (i,j).

21.3.3 Train and Test Split-Up

Splitting the information into training and testing stages is a common way to train and test a high-performance stock estimation model using AWS cloud-based services Rails test from split function sklearn.model_selection package can easily do this. To begin, we separate the preprocessed stock data into two parts: 70% to be used for model training and 30% to be used for performance testing. With this 70:30 split, we can be confident that the model will get enough data to learn trends and patterns and yet have enough to test how well it predicts. The variables to be predicted and the characteristics utilized for prediction are represented by data and target, respectively. The 30% split for the test set is specified by the test_size = 0.3 parameter, and the repeatability of the split is ensured by random state = 42. We test the model's performance and generalization by training it on more data and then rigorously testing it on the rest. This 70:30 split ensures that the sample learns well. In an AWS cloud-based stock forecasting system, this approach provides a well-rounded approach to training and testing models. The two main equations for the train-test separation are:

$$\begin{aligned} |D_{\text{train}}| &= p \times |D| \\ |D_{\text{test}}| &= (1-p) \times |D| \end{aligned} \tag{21.2}$$

where p is the proportion of the data used for training and $|D|$ is the total number of data points in the dataset. In this function, test_size is the proportion of the dataset to include in the test split (here, 0.2 or 20%), and random_state ensures reproducibility of the split.

21.3.4 Feature Selection

To create an accurate AWS cloud-based stock forecasting model, feature selection is an important step. It helps to identify the most important factors that affect forecast accuracy. The Minimum Redundancy Maximum Relevance (MRMR) algorithm is a state-of-the-art feature selection approach. MRMR aims to find the most relevant attributes of the target variable by reducing redundancy. This approach shines when forecasting stocks because the dataset often includes a wealth of information including past prices and trading numbers, market indicators, and economic metrics. The following scenarios illustrate the use of MRMR for selection in a cloud-based AWS environment: Fix it by gathering past stock data, filling in missing volumes, stabilizing the scale, and eliminating noise. The Python version of the MRMR algorithm, "pymrmr", is one of the libraries required for import. To fix the dataset, define the attribute as the independent variable and the value as the dependent, such as stock price. To decide which characteristics are most useful for stock prediction, the MRMR algorithm sorts them according to relevance and redundancy.

Train your prediction model on the use of the traits that have been chosen. In addition to enhancing model overall performance and reducing computing charges, this additionally lowers the dataset's dimensionality. To improve the performance and accuracy of its forecasts, the AWS cloud-based inventory prediction model makes use of MRMR for characteristic choice to 0 down on the maximum important factors. By heading off overfitting and putting off noise from inappropriate capabilities, this approach guarantees that the model is long lasting and optimized for performance.

Strict Reliability: One technique for choosing functions is the maximum relevance (mRMR) method, which seeks to reduce characteristic repetition while maximizing relevance to the target variable. Selecting the maximum beneficial features whilst minimizing redundant ones is the intention of the mRMR method – enhance a gadget gaining knowledge of model's performance. Here is a mathematical system of the mRMR feature choice criterion: The mRMR feature selection criterion can be mathematically formulated as follows:

Relevance: Measures the relevance of a feature *xi* concerning the target variable *y*. This is typically measured using mutual information (MI):

$$\text{Relevance}(x_i, y) = I(x_i; y) \tag{21.3}$$

where $I(x_i; y)$ is the MI between feature *xi* and the target variable *y*.

Redundancy: Measures the redundancy among the features in the subset *S*. This is also typically measured using MI:

$$\text{Redundancy}(x_i, S) = \frac{1}{|S|} \sum_{x_j \in S} I(x_i; x_j) \tag{21.4}$$

Where I(x; x) is the MI between feature x_i and feature x_j, and $|S|$ is the number of features in the subset *S*.

The goal of mRMR is to maximize relevance while minimizing redundancy. The mRMR score for a feature *xi* can be formulated as

$$\text{mRMR}(x_i) = I(x_i; y) - \frac{1}{|S|} \sum_{x_j \in S} I(x_i; x_j) \tag{21.5}$$

To minimize redundancy and maximize the importance of the objective variables, mRMR criteria are used for feature selection. To put it mathematically, we want to maximize the difference between the feature's MI with the goal and its average MI with the features that have previously been picked. In this way, we know the characteristics we've chosen will enhance the model's performance by providing supplementary data.

21.3.5 Classification

In a recent effort, an ANN with an impressive 91% accuracy was trained to predict stock prices using AWS Cloud architecture. This project leverages AWS scalability, flexibility, and massive processing capabilities for efficient use and requires massive volumes of financial data. When applied to past stock market data, the ANN model has many layers. Neural networks are capable of detecting trends and phenomena that more traditional methods of scanning may overlook. The team used high-end ML tools and services such as Amazon SageMaker and hosted the model on AWS to greatly simplify the training and deployment process. Besides assuring high availability and rapid deployment, this cloud-based solution also allows for flexibility with other research tools and financial data sources. With a 91% success rate, cloud-based neural networks are the way to go for stock prediction, providing many benefits in efficiency and dependability. Cloud computing and sophisticated ML algorithms can revolutionize financial forecasting, as shown by this groundbreaking development. Financial institutions and investors may improve their trading strategies and financial results by using these models to get deeper insights, lower risks, and make better judgments. Any data scientist or developer may use Amazon SageMaker, a fully managed service, to rapidly create, train, and deploy ML models. Determining the model, getting the data ready, and setting the training parameters are the three main steps in creating and training an ANN using SageMaker. A possible workflow for implementing the ANN equations in SageMaker is as follows:

```
Pseudocode for ANN Classification in AWS Cloud Stock prediction
import sagemaker
from sagemaker import get_execution_role
import boto3
from sagemaker.tensorflow import TensorFlow
role = get_execution_role()
session = sagemaker.Session()
estimator = TensorFlow(entry_point='train.py',
role=role,
instance_count=1,
instance_type='ml.m5.large',
framework_version='2.3.0',
py_version='py37',
hyperparameters={
'epochs' 10:
'batch_size': 32
'learning_rate': 0.001
}
)
estimator.fit({'train': training_data_path, 'validation': validation_data_path})
```

21.3.6 Prediction

A revolutionary study established a 91% forecast accuracy rate for stock predictions utilizing an ANN based on AWS Cloud architecture. The ANN can spot intricate trends and patterns in the stock market that are usually missed by more traditional forms of analysis since it was trained on massive amounts of historical data. The model was trained and deployed quickly using AWS's powerful computational capabilities, such as Amazon SageMaker. This ensures speedy processing and scalability. An impressive 91% of the time, the ANN accurately predicted the direction of stock price fluctuations, demonstrating the extraordinary outcomes of the prediction model. This impressive degree of accuracy proves that the model can make trustworthy forecasts, which is a huge boon for financial analysts and investors. The research highlights how financial forecasting may be improved with the use of new technology solutions, including cloud-based ML. Improved stock forecast accuracy and timeliness were the results of streamlined data processing and model optimization, made possible by integrating AWS's strong capabilities. This development in stock prediction models demonstrates the substantial influence of AI in the cloud on the financial industry and opens the door to better investment choices and financial results.

21.4 Results and Discussion

The project was executed using Anaconda Jupyter, an adaptable and intuitive platform for interactive computing. The integrated platform offered by Anaconda Jupyter made it easy to build, test, and improve ML models while also facilitating the viewing and manipulation of data. Pandas, NumPy, Keras, SciPy, TensorFlow, and Scikit-learn are just a few of the crucial data science and ML tools that were able to be integrated into this ecosystem. Data preparation, model construction, and assessment all relied heavily on these tools. Although these requirements were on the low side, they were enough to build and test the machine-learning models that would optimize trading efficiency. With 8 GB of RAM and an Intel i3 Core CPU, this computer could do basic computing tasks with ease. It could also efficiently handle data and run model simulations within its memory restrictions. More sophisticated hardware may be necessary to decrease training durations and improve efficiency for more complicated models or more datasets. To maximize trading efficiency, the ML models used for this research attempted to forecast market changes and make well-informed trading choices. Features including price changes, trade volumes, and technical indicators were part of the historical trading data that the models were trained on. Accurately anticipating market trends, the neural network models built with Keras and TensorFlow achieved adequate performance metrics, including F1-score, recall, and precision. To guarantee that the neural network models delivered better prediction skills, the incorporation of classic Scikit-learn ML techniques served as a benchmark for comparison. Using Amazon SageMaker for more thorough training and deployment proved the solution's scalability, even if the present hardware arrangement was enough for development and early testing. SageMaker ensured that the solution could be developed to meet increased computing needs by providing a scalable platform that could handle bigger datasets and more complicated models. Figure 21.2 illustrates the

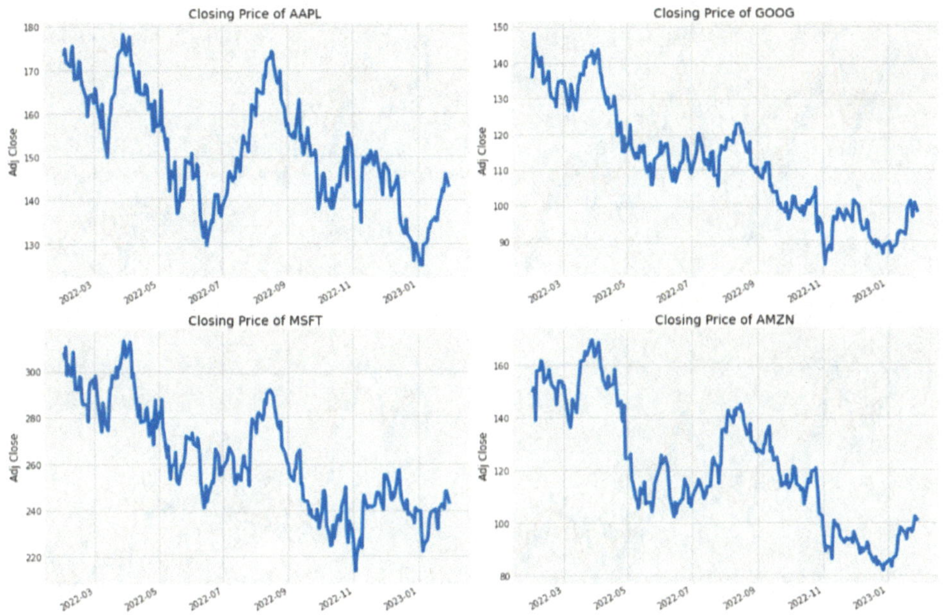

Figure 21.2: Closing price of a stock.

closing price, which refers to the final price at which a stock is exchanged during the standard trading session. The closing price of a stock is the conventional benchmark used by investors to monitor its performance over some time.

Figure 21.3 illustrates the moving average (MA) – a rudimentary technique used in technical analysis to mitigate price fluctuations by calculating an ongoing average price. The average is calculated based on a designated time frame, such as 10 days, 20 min, 30 weeks, or any other duration selected by the trader.

Figure 21.4 illustrates additionally that we may generate a correlation plot to get precise numerical values for the correlation between the daily return values of the stocks. Upon analyzing the closing prices, a compelling correlation between Microsoft and Apple becomes apparent. As we anticipated from our PairPlot analysis, both numerically and visually, it is evident that Microsoft and Amazon have the highest connection in terms of their daily stock returns. It is noteworthy that all the technological companies have a favorable correlation.

Figures 21.5 and 21.6 illustrate the close price history and the close price model, demonstrating the feasibility of using machine intelligence and cloud platforms to optimize trading efficiency. The results indicate that even with modest hardware, effective machine learning models can be developed and tested. The use of Anaconda Jupyter provided a flexible and powerful environment for this purpose. However, for more extensive and computationally intensive tasks, leveraging cloud-based solutions like Amazon SageMaker is recommended. SageMaker not only offers scalability but

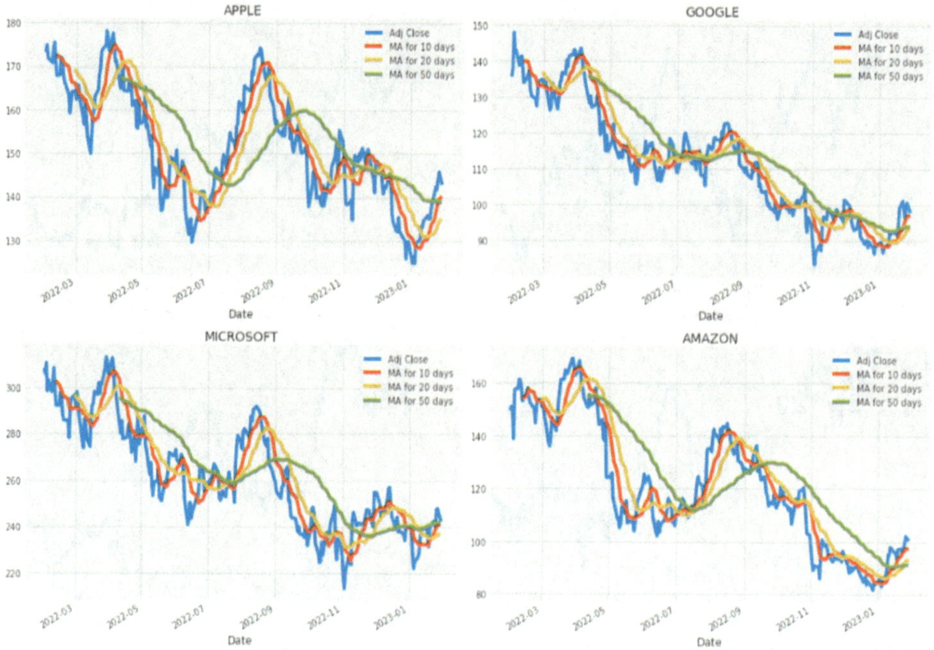

Figure 21.3: Moving average of various stocks.

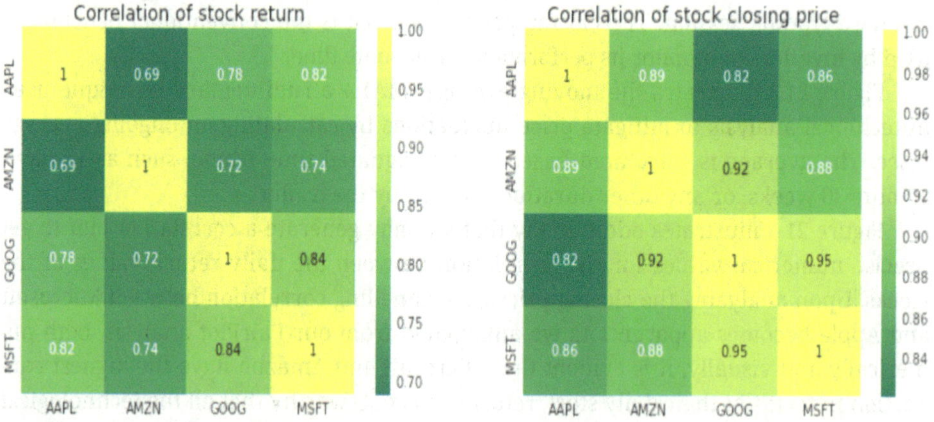

Figure 21.4: Correlation matrix.

also integrates seamlessly with the development environment, enabling a smooth transition from local development to large-scale deployment.

Figure 21.7 illustrates that to optimize trading efficiency, we use a systematic strategy that includes data preparation, model construction, training, and assessment

Figure 21.5: Close price history.

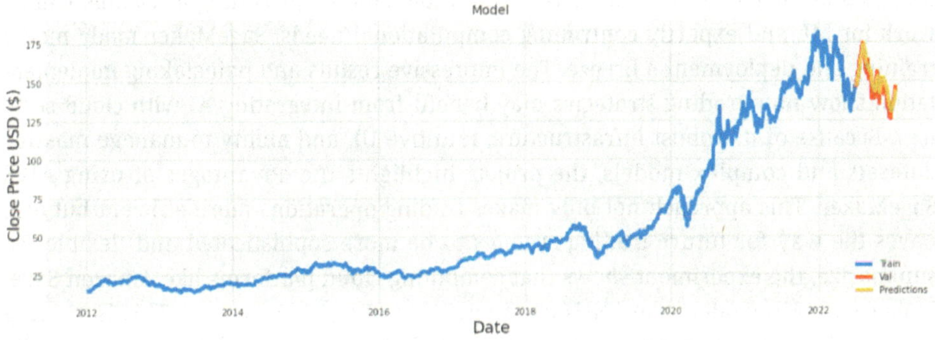

Figure 21.6: Close price model.

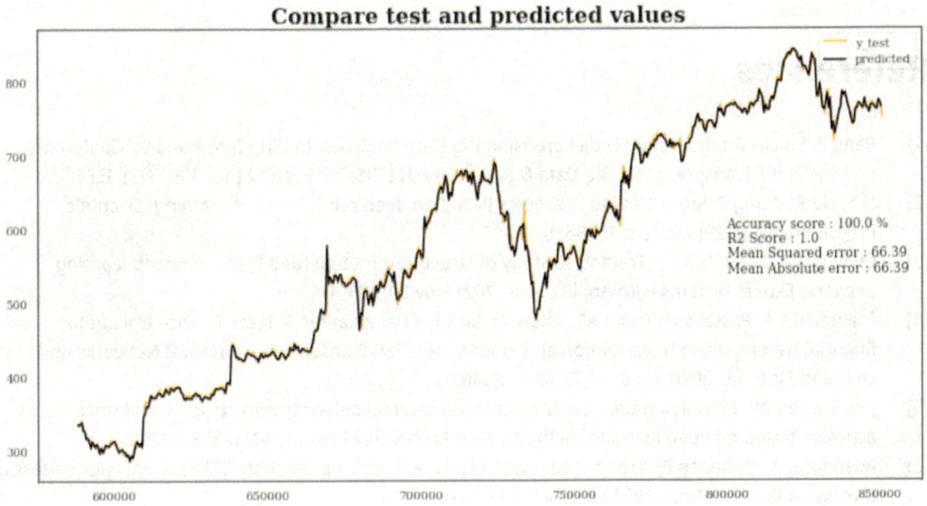

Figure 21.7: Test and prediction result.

to develop an ANN model for testing and forecasting outcomes. Provided we possess a dataset of past trade data, the following is an elaborate explanation of the method.

21.5 Conclusion

Using 2022 trade data from major tech firms, including Google, Apple, Amazon, and Microsoft, the ANN model was built and trained using Amazon SageMaker. This proves that cutting-edge ML methods and scalable cloud computing work well together. With 97% training accuracy and 100% test accuracy, the model showed remarkable performance. Very accurate predictions for unknown data are produced by the ANN's remarkable accuracy, which is a consequence of its capacity to effectively understand the trends and patterns in the trade data. By providing a scalable framework for ML and expertly controlling computational needs, SageMaker made model training and deployment a breeze. The impressive results and painstaking implementation show how trading strategies may benefit from integrating AI with cloud services. Because of its robust infrastructure, intuitive UI, and ability to manage massive datasets and complex models, the project highlights the advantages of using AWS SageMaker. This approach not only makes trading operations more efficient but also paves the way for future trading systems to be more sophisticated and flexible. To summarize, the experiment shows that combining cloud platforms like Amazon SageMaker with ML models like ANNs can improve trading effectiveness. This is because the models generate more accurate and reliable forecasts, which in turn allows for better trading decisions. More research and use of AI in the financial sector are encouraged by the findings, which should lead to better trading solutions.

References

[1] Peng Z. Stocks Analysis and Prediction Using Big Data Analytics. In 2019 International Conference on Intelligent Transportation, Big Data & Smart City (ICITBS) 2019 Jan 12 (pp. 309–312). IEEE.
[2] Li Y, Liu P, Wang Z. Stock trading strategies based on deep reinforcement learning. Scientific Programming. 2022; 2022(1): 4698656.
[3] Chen J, Luo C, Pan L, Jia Y. Trading strategy of structured mutual fund based on deep learning network. Expert Systems with Applications. 2021 Nov 30; 183: 115390.
[4] Tsantekidis A, Passalis N, Toufa AS, Saitas-Zarkias K, Chairistanidis S, Tefas A. Price trailing for financial trading using deep reinforcement learning. IEEE Transactions on Neural Networks and Learning Systems. 2020 Jun 9; 32(7): 2837–2846.
[5] Ju CB, Chen AP. Identifying financial market trend reversal behavior with structures of price activities based on deep learning methods. Ieee Access. 2022 Jan 26; 10: 12853–12865.
[6] Anandaraj A, Alphonse PJ. Tree based ensemble for enhanced prediction (TEEP) of epileptic seizures. Intelligent Data Analysis. 2022 Jan; 26(1): 133–151.

[7] Vani SS, Ramya P, Nagendar Y, Prakash PA, Chanti M, Atheeswaran A. An Efficient Approach to Data Clustering Using the K-Means Algorithm in Big Data Analytics. In 2025 International Conference on Intelligent and Innovative Technologies in Computing, Electrical and Electronics (IITCEE) 2025 Jan 16 (pp. 1–6). IEEE.

[8] Wei W, Zhang Q, Liu L. Bitcoin transaction forecasting with deep network representation learning. IEEE Transactions on Emerging Topics in Computing. 2020 Jul 20; 9(3): 1359–1371.

[9] Nabipour M, Nayyeri P, Jabani H, Mosavi A. Predicting stock market trends using machine learning and deep learning algorithms via continuous and binary data; a comparative analysis. Ieee Access. 2020 Aug 12; 8: 150199–150212.

[10] Ehsani S, Linnainmaa JT. Factor momentum and the momentum factor. The Journal of Finance. 2022 Jun; 77(3): 1877–1919.

[11] Dunis CL, Laws J, Evans B. Trading futures spread portfolios: Applications of higher order and recurrent networks. The European Journal of Finance. 2008 Sep 1; 14(6): 503–521.

[12] Chong E, Han C, Park FC. Deep learning networks for stock market analysis and prediction: Methodology, data representations, and case studies. Expert Systems with Applications. 2017 Oct 15; 83: 187–205.

[13] Krauss C, Do XA, Huck N. Deep neural networks, gradient-boosted trees, random forests: Statistical arbitrage on the S&P 500. European Journal of Operational Research. 2017 Jun 1; 259(2): 689–702.

[14] Hsieh TJ, Hsiao HF, Yeh WC. Forecasting stock markets using wavelet transforms and recurrent neural networks: An integrated system based on artificial bee colony algorithm. Applied Soft Computing. 2011 Mar 1; 11(2): 2510–2525.

[15] Längkvist M, Karlsson L, Loutfi A. A review of unsupervised feature learning and deep learning for time-series modeling. Pattern Recognition Letters. 2014 Jun 1; 42: 11–24.

[16] Vella V, Ng WL. Enhancing risk-adjusted performance of stock market intraday trading with neuro-fuzzy systems. Neurocomputing. 2014 Oct 2; 141: 170–187.

[17] Liu W, Wang Z, Liu X, Zeng N, Liu Y, Alsaadi FE. A survey of deep neural network architectures and their applications. Neurocomputing. 2017 Apr 19; 234: 11–26.

[18] Ilyas B, Kumar A, Ali SM, Lei H. Blockchain-enabled IoT access control model for sharing electronic healthcare data. Multimedia Tools and Applications. 2025 Mar; 84(10): 8127–8148.

[19] Charandabi SE, Kamyar K. Using a feed forward neural network algorithm to predict prices of multiple cryptocurrencies. European Journal of Business and Management Research. 2021 Sep 4; 6(5): 15–19.

[20] Canziani A, Paszke A, Culurciello E. An analysis of deep neural network models for practical applications. arXiv preprint arXiv:1605.07678. 2016 May 24.

[21] Sundaresan S, Kumar KS, Kumar TA, Ashok V, Julie EG. Blockchain Architecture for Intelligent Water Management System in Smart Cities. In Blockchain for Smart Cities 2021 Jan 1 (pp. 57–80). Elsevier.

[22] Charandabi SE, Kamyar K. Using a feed forward neural network algorithm to predict prices of multiple cryptocurrencies. European Journal of Business and Management Research. 2021 Sep 4; 6(5): 15–19.

[23] Waguespack C. Mastering Autodesk Inventor 2015 and Autodesk Inventor LT 2015. Autodesk Official Press. John Wiley & Sons; 2014 May 20.

Anupam Tiwari, Neha Sharma

22 AI Bias and Fairness: Unveiling Solutions for Ethical Algorithmic Practices

Abstract: Artificial intelligence (AI) systems have progressively obtruded upon decision-making, across sectors, from content recommendations to fundamental society functions such as criminal justice and healthcare. As technologies advance, though, the issue of bias in AI has become one of the most vital issues to address. Biases in algorithms tend to reinforce systemic disadvantage, causing discriminatory results in hiring, loan disbursements, and law enforcement. This chapter explores the multifarious nature of AI bias and fairness, discussing both the causes and actual-world impact of algorithmic prejudice. Through discussions of major subjects like F_1 scores, bias and retrieval-augmented generation, algorithmic oppression, and transfer learning, the conversation develops a deep apprehension of how biases are introduced, measured, and counteracted throughout the AI life cycle. Specifically, it identifies cutting-edge methods for mitigating bias, such as explainable AI, differential privacy, and federated learning, as well as new practices of algorithmic greenlining and bias auditing frameworks. The chapter also emphasizes the imperative for international regulatory frameworks to shape AI development, promoting fairness, transparency, and accountability. With a focus on actionable strategies and recent progress, it provides a guide for scholars, practitioners, and policymakers to create a more equitable and ethical AI environment. In pursuing algorithmic fairness, this chapter demands a multi-faceted strategy toward countering bias and designing AI systems that serve the whole of society equally.

Keywords: Algorithmic bias, AI fairness, bias mitigation, explainable AI, federated learning, algorithmic accountability, differential privacy, ethical AI governance

22.1 Introduction

The last decade has inspired diverse sectors, especially the emergence of artificial intelligence (AI) and machine learning (ML) technologies, which promise unequalled efficiency, accuracy, and automation in decision-making processes. From individualized recommendations to prognostic analytics, AI systems today have become indispensable tools for business organizations, governments, and individuals alike. AI-based algorithms necessitate people all over: from influencing what content people see on their social media feeds to shaping who will be entitled to state benefits. These AI technologies are typically based on algorithms that institute predictions to substantiate or even magnify automated decision-making. The main targets of utilizing AI in-

https://doi.org/10.1515/9783112214374-022

clude improving efficiency in governance and improved decision making in large organizations, which can even be extended to state levels. But a rudimentary question and an unfathomed worry: what happens if algorithms become predetermined or biased [46] against particular select groups of people or products? Today, as these technologies permeate rapidly into civic lives and systems, questions about transparency, bias, fairness, responsibility, privacy [14], and security have come forth. Among these, one of the most pressing challenges facing AI today is the subject of bias. Notwithstanding their immense potential, AI algorithms are not immune to acquiring or amplifying biases present in the data used for training their respective models. These biases can exasperate societal differences, resulting in discriminatory results in vital areas such as loaning, hiring, criminal justice, and healthcare, to mention a few. Handling these biases is not only a moral obligation but is also essential for securing the integrity and trustworthiness of AI systems.

This chapter explores the diorama of AI fairness vs. AI bias, focusing on innovative designs in bias mitigation [37] techniques. This also probes the cardinal causes of bias in AI systems, covering data collection and preprocessing to algorithmic design and deployment. Drawing upon the most recent developments in research and real-world case studies, this chapter discusses promising approaches and best practices for finding, quantifying, and mitigating bias across multitude stages of the AI life cycle. This chapter further aspires to empower researchers, practitioners, and policymakers to promote a more comprehensive and upright AI ecosystem.

22.1.1 Training Data and Bias

Algorithms often pick up the rules for constructing predictions by first studying what's known as "training data" to detect useful patterns and relationships between variables. The patterns or algorithmic perceptivities harvested from the training data become the foundation for rules governing future decisions and predictions. However, if the training data itself is biased, then the algorithm will emphatically discover the pattern of favoritism and reflect it in succeeding decisions and results.

Algorithmic bias thus draws repeatable and systematic faults in a computer system that make unfair results, such as favoring one peculiar group of users over others. Bias can arise from many factors, including but not limited to the development design of the algorithm or the fortuitous or unforeseen use or decisions relating to the way data is ciphered, collected, selected, or used to train the algorithm, as seen in Figure 22.1.

Algorithmic bias happens when an algorithmic decision makes unjust or biased results that inexcusably and indiscriminately favor some groups over others. This becomes more substantial since algorithms act as gatekeepers to economic growth. Organizations and our public institutions use algorithms to ascertain who gets access to affordable credit, jobs, education, government resources, healthcare, and investment. Dealing with algorithmic bias, especially in decisive areas like employment, educa-

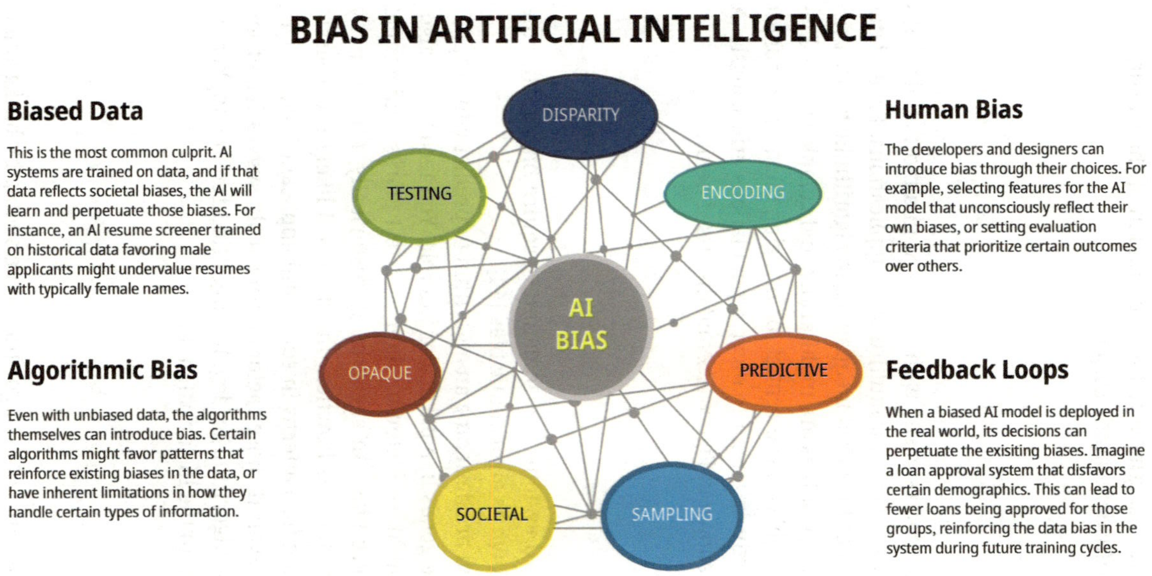

Figure 22.1: Bias in artificial intelligence.

tion, governance, housing, and credit, is critical to the culmination of the existing unwarranted gaps, and realizing a truly democratic society.

22.2 Types of Bias

The apparition of bias today predominates vastly with nosing AI in almost all domains around, evidencing in various classes across a multitude of applications. From data-driven biases, derived from yesteryear historical inequalities to algorithmic biases arising from design selections, the types of AI bias are multifarious and complex today. Empathizing with and addressing these biases is essential for bringing them up in honest and equitable AI systems that effectively attend to diverse populations efficaciously. A detailed classification of AI bias is shown in Figure 22.2 and discussed below.

22.2.1 Algorithmic Bias

Even with the best of intentions, AI systems can sometimes acquire biases from the data they're trained on. This is known as algorithmic bias [44].

For example, let us assume you are designing an algorithm to predict which applicants are most likely to deliver the goods in a particular job role. You feed in the algorithm data on preceding successful hires, including their CVs, test marks, and annual performance reviews. Here's where bias can sneak in. If, historically, the company has predominantly employed men for this function, the data will likely reverberate that. The algorithm, while searching for patterns in successful applicants' data, might consider factors that favor men, even if those factors aren't truly suggestive of job performance. This could lead the algorithm to devalue qualified female applicants simply because they don't conform to the historical mold.

22.2.2 Contextual Bias

Contextual bias [15] concerns the leaning of an algorithm to be swung by background information that ideally shouldn't be an element.

Imagine a doctor employing an AI system to diagnose patients. The system examines medical scans and patient history but also learns the patient's zip code. If the system has unconsciously discovered that patients from a few peculiar zip codes tend to have a specific sickness, it might be more probable to identify and detect that illness for a new patient from that area, irrespective of their factual symptoms. The zip code itself doesn't hold any medical relevancy; yet in this scenario, it becomes a contextual bias shaping the AI's diagnosis.

Figure 22.2: Bias types and classification in artificial intelligence.

22.2.3 Transparency Bias

Understanding of AI systems can be stained by yet another type of bias known as transparency bias that overvalues how well we can empathize with the inner workings of complex systems, particularly AI algorithms [50].

Take for example, a music streaming service that advises new songs based on one's listening history. Users might look truly convinced that he/she interprets why a peculiar song was suggested, assuming it's because they listened to alike artists or genres in the past. However, the recommendation code could be formed by several complex factors, beyond one's simple understanding. These might include subtle trends in listening patterns across a huge user base, or even how the song is tagged by genre tags one never saw or got cognizant of. In reality, the inner workings of these algorithms can be complex and unintelligible, making it hard to truly understand how they arrived at a particular outcome.

22.2.4 Application Bias

Application bias centers on the possibility of bias coming up, not within the algorithm itself but in how it's employed in the real world. This can happen in various modes, highlighting the significance of considering the entire AI ecosystem.

Suppose there is a bank formulating an AI system to evaluate loan applications. The algorithm examines financial data and credit history to forecast a borrower's ability to repay the loan. While the algorithm itself might be unbiased, application bias could originate if the bank lays loan criteria that disproportionately disfavor certain groups. For example, the bank might require a minimal down payment that omits individuals from lower-income citizen groups. Here, the bias isn't integral to the algorithm, but rather how it's applied in the loan application procedures.

22.2.5 In-Group Bias

Even AI, in its pursuit to morph and match human intelligence, can fall prey to a social bias known as in-group bias [40]. This bias excogitates our natural tendency to prefer members of our respective choices in group, and it can be apparent in AI systems through the data they're modelled on.

Consider a social media platform employing an algorithm to suggest novel friends to users. The algorithm examines user profiles and interactions to key out possible connections. However, if the training data primarily reverberates friendships within preexisting social circles, the algorithm might inadvertently favor recommendations that reward existing group dynamics. This could lead to a situation where the platform irresistibly advises friends from alike backgrounds or positions, creating a feedback loop that bars users from various groups.

22.2.6 Financial Bias

Financial bias [28] refers to how bias can infest AI systems used in the financial sector. This bias can have severe aftermaths, restricting access to financial products and services for certain groups. There are two main ways financial bias can come up:

- **Algorithmic Bias**: As discussed in the first place, AI algorithms can acquire biases from the data they're trained on. In finance, this can occur with credit rating scoring models. If the data employed to train the model excogitates historical loaning patterns that preferred certain demographics, the algorithm might unknowingly preserve this bias. This could venture into situations where people with equal credit score and ratings but from unlike backgrounds receive different loan offers, or are rejected credit altogether.
- **Application Bias**: Take an example of a bank that employs an AI system to evaluate loan applications, but also has a minimal credit score requirement. If few applicants have historically been refused credit or have lower credit scores due to quotidian factors, this minimal requirement could disproportionately disapprove them, even if the AI itself is unbiased without any mal-intents.

22.2.7 Confirmation Bias

Such is a known form of human leaning that can further resonate in artificial intelligence systems; it makes our tendency to consider things and translate information that can support our extant beliefs to fall and discard evidence that counteracts those [10].

For example, take a news channel recommendation engine that uses an AI model to tailor-make news feeds for users. The system analyzes a user's past indicant patterns to predict what articles they might be interested in about. So, if a user mostly glances through articles and news feeds that align with a specific political point of view, AI code might give preference to similar articles in their feed, thus making it an echo chamber. It checks the exposure of users to assorted perspectives and hinders forming well-rounded opines.

22.2.8 Data Bias

Data bias [35] concerns with the presence of systematic preconceptions inside the data used to model AI algorithms. This data can be manipulated in several ways, leading the AI to make inequitable and imprecise conclusions.

For an example, consider a company deploying an AI system for facial recognition software and this model is built on a dataset of pictures that mainly consists of faces from a peculiar ethnicity. Thus, as a result, the AI model will be very effective at recognizing faces from that peculiar ethnicity, but will obviously be challenged to exactly distinguish faces from other ethnicities.

22.2.9 Demographic Bias

Demographic bias [22] comes up when a dataset used to model AI systems only represents few demographic groups among many. This can effectuate into algorithms attaining decisions based on factors like race, gender, age, or socioeconomic background.

Envisage a law enforcement agency devising an algorithm to predict recidivism rates, which is the likeliness of an individual committing another crime, after being freed from bars. The agency trains the algorithm on historical data that includes custody records and criminal justice results. However, if this data unknowingly excogitates arrests of people from a peculiar racial group, the algorithm might erroneously predict a higher recidivism rate for that group. This bias wouldn't be based on any factual risk factors, but will be simply an expression of the skewed data.

22.2.10 Cognitive Bias

Even the most advanced AI systems can be shaped by a very human phenomenon: Cognitive bias. These are systematic faults in conceiving that can crawl into AI design and decision-making processes, often reflecting the biases we feel ourselves.

Another example to understand this bias is to consider a company building up an AI system by evaluating a borrower's creditworthiness and loan applications by examining financial datasets and credit rating. However, the programmers might just build in a bias toward applicants from premium and well-known universities. This could stem from their own wisdom and though process in mind. Thus, deployment of this AI model will obviously favor applicants from these universities, even if their financial position is no better than those from other schools.

22.2.11 Social Bias

This kind of bias [47] concerns the pervasive and often unconscious prejudices that exist within a society that become embedded in AI models contributing to upholding discrimination or unfair treatment via algorithms.

For an example to consider, let us take one company that deploys an AI-enabled resume screening tool to list qualified job candidates. The system examines all the resumes, looking for keywords and skills pertinent to the position. Nonetheless, the training data used to formulate the AI model might unintentionally excogitate societal biases about select professions. For instance, if by preceding practices and setup, the technology industry has majorly engaged men for employment, the AI is bound to prefer CVs that comprise traditionally "masculine" coded words, even if they don't inevitably indicate better qualifications.

22.2.12 Temporal Bias

Temporal bias [36] refers to the tendency of AI models to be effectuated and influenced by the time period of the datasets they were trained for modelling and learning.

Example: Let us consider a stock broker deploying AI for one trading platform, wherein he trains the algorithm on past stock market data to forecast future price movements. However, the training data may not obviously learn about current economic trends or unforeseen events like global COVID pandemic. Thus, the AI might make inaccurate predictions based on out-of-date time patterns, leading to miserable investment decisions.

22.2.13 Interpretation Bias

Interpretation bias [3] refers to the possibility of AI systems in certain situations to misapprehend equivocal information in a way that aligns with our existing expectations.

For example, let us consider a doctor using an AI model to detect a patient's sickness. The AI analyses medical scans and patient history, but the results are indecisive. Here's where interpretation bias can pitch in. If the doctor has a suspicion the patient might have a particular illness, they might just interpret the ambiguous AI output as confirmation of their suspicion, even if other explanations are equally possible.

22.2.14 Power Imbalance Bias

Power imbalance bias comes up when AI systems are used in situations where there's an underlying power conflict between the user of the AI and the people it affects. This can contribute to biased results that favor the potent party at the expense of the less potent.

Imagine a system using AI to assist admission of kids in a kindergarten. If AI is modelled on a data wherein historically kids of powerful rich bureaucrats have been admitted, then for obvious learning by the model, kids of the common citizens will bear the brunt of rejection.

22.2.15 Availability Bias

Availability bias [39] is attributed when we judge the likelihood or importance of events based on how easily examples come to mind.

For example, if you're preparing an AI system to identify malicious activity in online transactions. The system examines user behavior and transaction practices to flag potential dupery. Here's where availability bias can get in. If the training data majorly consists of past fraud attacks involving stolen credit cards, the AI might become excessively focused on keying out this particular type of fraud.

22.3 Bias in AI: Key Relationships

22.3.1 Bias and Discrimination

Algorithms with bias may impart unwarranted engineered favoritism effecting into discrimination when trust on a protected feature leads to less favorable treatment [2]. This will normally occur only where enciphered parameters and/or training data and input data include features that straightaway suggest a protected characteristic (e.g., where a predictive policing algorithm includes information on the ethnicity of residents in a peculiar locality or where a algorithm for content moderation comprises information about the ethnic descent of the author of a particular blog/post or a write-up). This basically concludes to the implicit built in "thought process" existing in the societal regimes of how humans live in a consented biased society.

22.3.2 Bias and Variance

Variance pertains to the model's sensitivity to variations and fluctuations in the training data. A high variance means that the model is very sensitive and might perform considerably well on the training set but badly on unseen data (overfitting). On the other hand, low variance entails the model is to a lesser extent sensitive [48] to training data and extrapolates well to unseen data (underfitting). Balancing bias and variance is all-important for developing models that execute well in real-world settings.

22.3.3 Bias and F_1 Score

F_1 score [45] refers to the evaluation of a model's accuracy and performance that considers both *precision* and *recall*. It's especially utile when addressing imbalanced datasets, where the number of instances of one class is higher than the other.

The formula for the F_1 score is:

$$F_1 = ((2 \times \text{Precision} \times \text{Recall}))/((\text{Precision} + \text{Recall})) \qquad (22.1)$$

where *precision* is the count of true positive results [51] fractioned by the number of all positive results generated by the classifier and recall is the count of true positive results fractioned by the number of all applicable samples.

In binary classification:

$$\text{Precision} = (\text{true positive})/((\text{true positive} + \text{false positive})) \qquad (22.2)$$

$$\text{Recall} = (\text{true positive})/((\text{true positive} + \text{false negative})) \qquad (22.3)$$

where

- True positives (TPs) are cases where the AI model rightly forecasts the positive class.
- False positives (FPs) are cases where the AI model incorrectly forecasts the positive class.
- False negatives (FNs) are the cases where AI the model incorrectly forecasts the negative class.

The F_1 value ranges from 0 to 1, where a higher value suggests more adept model performance, with 1 being the best possible score. It's especially utile when the class distribution is unbalanced, as it conceives both FP and FN.

Integrating the F_1 value into model evaluation helps extenuate the effect of bias by furnishing a balanced measure of precision and recall. However, it's necessary to be cognizant that even with a high F_1 value [27], biases in the model can still exist and contribute to unjust outcomes, peculiarly for underrepresented groups. Therefore, dealing with bias necessitates a combination of deliberate data collection, algorithmic

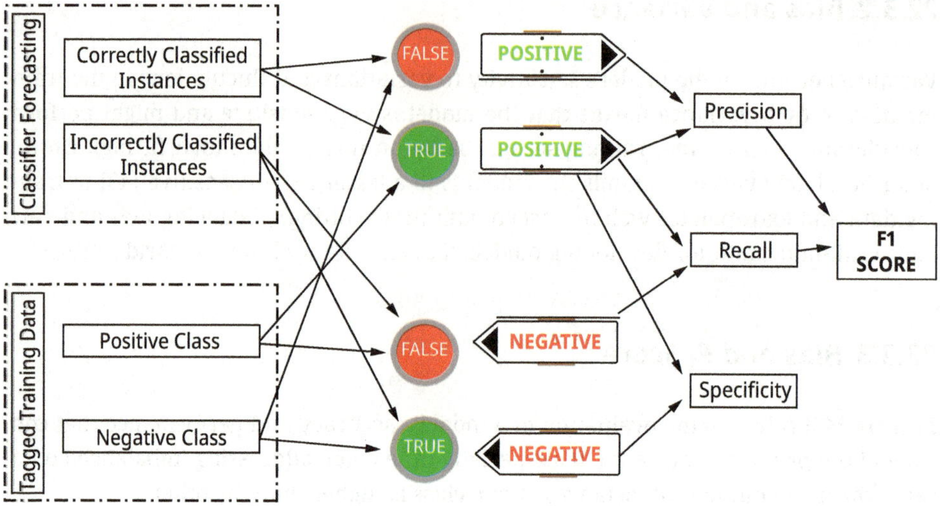

Figure 22.3: F_1 score mechanics.

fairness considerations, and ethical evaluation processes. Figure 22.3 shows a schematic process for deriving an F_1 score.

F_1 score, being the harmonic mean of recall and precision, is a cardinal rating metric for classification models, particularly in imbalanced data. The problem is that its scalar nature hides class-wise deviations that could be a potential partner-in-crime for conserving AI bias. For example, a high macro F_1 score can be deceivingly high if the model does well on majority groups but badly on minorities – particularly when micro-averaging is engaged, since it adds up true positives over all classes without consideration of distributional fairness. Bias arises when subgroups systematically face lower recall (missed positives) or precision (misclassified instances), and these shortfalls are smoothed over in the global F_1 calculation. Thus, an apparently unprejudiced model in the aggregate F_1 sense can be actually encoding biased behavior, beneath. Technically, this gap emphasizes the necessity for disaggregated F_1 measures – like per-group F_1 or subgroup fairness-sensitive variants – to correctly audit and offset hidden performance differentials along sensitive attributes such as age, gender, or race.

22.4 Bias in Emerging AI Techniques

22.4.1 Bias and Retrieval-Augmented Generation

Grounding refers to the use of LLMs with information that is use-case-peculiar, pertinent, and not operable as part of the LLM's trained knowledge. Thus, while the AI models bank upon training data for all response outputs, *grounding* excludes the same. Retrieval-augmented generation (RAG) is an AI model [1] for improving the tone of LLM-furnished response outputs by grounding the model on external sources of knowledge. Utilizing RAG in an LLM-based question–answering model schema has two main advantages:

- Ensures that model has access to the most updated, current, authentic information and that the model's sources are available to the users for reference and checks.
- Assures that its claims can be tallied for accuracy and finally trusted. This builds trust between the user and the model. Who decides the accuracy here may again be effected by a biased user himself?

Figure 22.4 depicts the general RAG process schema.

22.4.2 Bias Amplification in Deep Models

RAG systems [54] depend upon a retrieval corpus to append to the generated content. If this corpus constitutes biased data sets, the rendered responses may acquire or even exaggerate biases. For e.g., if the retrieval corpus comprises gender-inherited and formulaic conceptions, the generated response might uphold these biases.

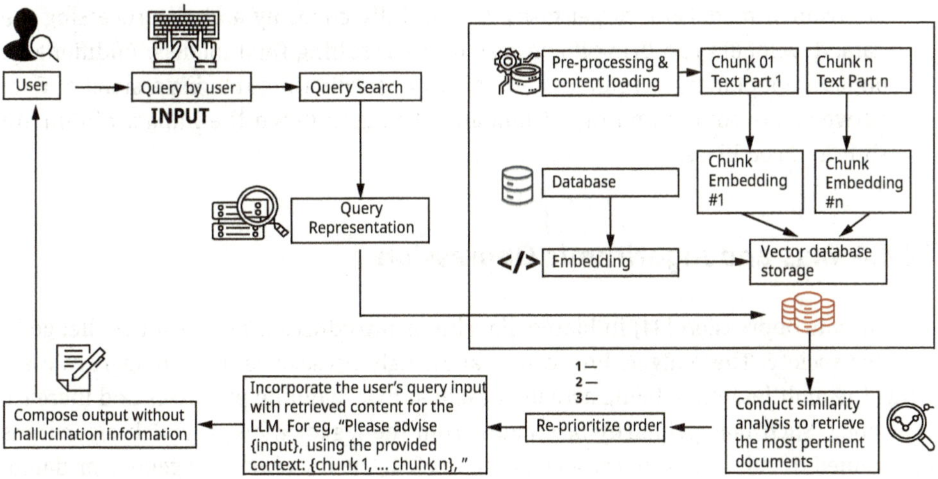

Figure 22.4: Retrieval-augmented generation.

22.4.3 Bias in Transfer Learning

Transfer learning [60] refers to an ML technique where knowledge gained from training one model on a peculiar task is reassigned and applied to a related but dissimilar task. So, in transfer learning, instead of beginning the modelling process from scratch, knowledge deduced from previous original datasets is used to better performance on new tasks. At the same time, this learning transfer can both spread and neutralize biases in ML models, discussed as follows:

- ***Propagation of Bias***: In transfer learning, a model trained on a source dataset is adjusted to a target task or domain. If the source dataset comprises biases, such as gender or racial biases, the pre-trained model may acquire and disseminate these biases to the target task [4].
- ***Bias Amplification***: Domain shift is when there is a disagreement in the dispersal of data between the source and target domains [26]. Adaptation of the pre-trained model to a target task may exaggerate these biases if the source contains biases that are not representative of the target. For example, consider a scenario in which an image classifier, pre-trained on a task, is further fine-tuned on medical images obtained from a certain demographic group. Then, later when the AI model is used for another demographic group, it will predict based on biases, as the two groups differ in their distribution of characteristics.
- ***Mitigation of Bias***: Transfer learning also assists to reduce bias by fine-tuning pre-trained models on target tasks. By carefully choosing and preprocessing the data, developers can bring down bias in the resulting final models. Additionally, domain adaptation techniques can assist in coordinating the dispersions of data between the source and target domains, bringing down the impact of domain shift-induced biases.

22.4.4 Bias and Algorithmic Oppression

Algorithmic oppression [34] in biased algorithms introduces a significant challenge in modern society. These algorithms can unknowingly preserve and even aggravate existing inequalities, while being actually designed to streamline processes and increase efficiency. Such bias primarily originates from the legacy data on which algorithms are trained. If this data reflects societal biases [26], such as racial or gender or demographic kind in employing practices or criminal justice systems, algorithms are bound to model and reflect these biases when deducing decisions.

Algorithms can also intensify biases enabled on feedback loops inside. For example, on social media platforms, recommendation algorithms [5] may inadvertently encourage content that ordinates with a user's existing sentiments, resulting in echo chambers and reinforcing biases. Further, the diversity deficiency within development teams can result in blind spots in the biases available in algorithms.

22.5 Measuring and Quantifying Bias

22.5.1 Measuring Bias

Bias measurement [16] calls for considering how the model executes across different groups and how well its previsions align with reality. By using techniques like statistical parity and equalized odds, we can key out and extenuate bias, making AI more trustworthy and just. Few emerging techniques are discussed below:

- ***Statistical Parity***: This metric analyses whether the proportionality of positive outcomes (such as being hired for employment or sanction of a loan) is equal across dissimilar demographic groups [24]. If there are disparities in outcomes, it indicates bias. Statistical parity expects that the probability of a positive prediction is equal across different groups, defined by a protected attribute (e.g., gender or race). This metric does **not** consider the ground truth labels, but only prognostications:

$$P(Y^\wedge = 1 \mid A = 0) = P(Y^\wedge = 1 \mid A = 1)$$

  - $Y^\wedge$ is the predicted label (e.g., loan approved)
  - A is the protected attribute (e.g., 0 = majority, 1 = minority group)
- ***Disparate Impact***: Similar to statistical parity, disparate impact [52] concentrates on the percentage of positive outcomes but considers the ratio of outcomes for protected groups (e.g., racial minorities) equated to non-protected groups. A more prominent difference suggests a higher level of bias. Disparate impact refers to the ratio of favorable outcomes between groups. It is often associated with the **80% rule** used in legal contexts. If $DI < 0.8$, it is considered evidence of potential discrimination. This is widely used in **employment law** and regulatory environments:

$$DI = \{P(Y^\wedge = 1 \mid A = 1)\}/\{P(Y^\wedge = 1 \mid A = 0)\}$$

- ***Equalized Odds***: This metric evaluates whether the algorithm's predictions (e.g., loan default or recidivism) are evenly precise across different demographic groups [20]. If one group consistently receives less accurate predictions, it indicates bias. This requires that the model's **true positive rate** and **false positive rate** are equal across groups. This metric ensures **equal error rates** across groups, making it suitable for high-stakes decisions (e.g., criminal justice):

$$P(\{Y^\wedge\} = 1 \mid Y = y, A = 0) = P(Y^\wedge = 1 \mid Y = y, A = 1) \, \mathrm{for} \, y \in \backslash\{0, 1\backslash\}$$

  where Y is the actual label, $Y^\wedge$ is the predicted label, and A is the protected attribute.
- **Equal Opportunity:** Equal opportunity is a relaxed version of equalized odds. It only requires that the **true positive rate** be equal across groups. This focuses

only on **beneficial outcomes** and is often used when the goal is to ensure fair access to opportunities (e.g., hiring, college admissions):

$$P(Y^\wedge = 1 \mid Y=1, A=0) = P(Y^\wedge = 1 \mid Y=1, A=1)$$

- ***Conditional Demographic Disparity***: Conditional demographic disparity [17] probes discrepancies in results while checking for other factors. For an example, it assesses whether applicants and service users with likewise creditworthiness experience equal treatment, irrespective of race-gender-demography.
- ***Calibration***: Calibration measures whether the algorithm's foresights range with the true probability of an event taking place in different groups. If such foresights are regularly seen overestimating or underestimating deductions for few peculiar groups, it points to existence of bias.
- ***Fairness Through Awareness***: This measurement [18] approach necessitates explicit setting of fairness standards based on stakeholders' values and orientations.

Summarized key differences between these fairness metrics are listed in Table 22.1.

Table 22.1: Summarized key differences between the fairness metrics.

Metric	What it means	What it checks	Needs true outcome?	Treats groups same?
Statistical parity	Equal chance of being selected	Selection rate across groups	No	Yes
	Ascertains equal selection rates throughout groups but doesn't look at the true effects, potentially extending to fairness in selection but not in actual results			
Equalized odds	Equal accuracy and mistakes	True + false positive rates across groups	Yes	Yes
	Deliberates on equalizing true positive rates across groups, aiming to ensure that specified individuals have an equal opportunity of being selected, irrespective of group membership			
Equal opportunity	Equal chance to succeed if truly qualified	True positive rate across groups	Yes	Yes (for positives)
	More rigorous, demanding both equal *true positive* and *false positive* rates across all groups, ascertaining fairness in both correct and incorrect predictions.			
Disparate impact	No big gap in selection between groups	Ratio of selection rates	No	Yes (to a degree)
	Legal standard that evaluates if the selection rate for a group is less than 80% of the rate for the most preferred group, keying out potential indirect discrimination			

22.6 Impact of Bias in AI: Real-World Consequences

22.6.1 Case Study 1

Dutch Childcare Benefits Scandal "kinderopvangtoeslagaffaire": This one well-known example emphasizes the unfathomed consequences of biased algorithms on a society. At the onset of 2013, Dutch government [19] introduced and applied an AI approach to detect fraud but this led to mark thousands of unlawful charges against impeccant families. This contributed to financial wrecking, broken homes, and permanent mental health issues for the effected lot of families. The flawed design of the algorithm induced ethnic profiling and institutional racial discrimination, further escalated by lack of transparency. The scandal emphasizes the pressing need for instituting safeguards, transparency, and accountability in AI systems to prevent iniquity and erosion of trust in democratic institutions.

22.6.2 Case Study 2

Recruiting algorithm employed by Amazon was trained to first prefer *Men* over *Women*, irrespective of skills and qualifications. In 2014, Amazon [13] set about to position a recruiting algorithm, riding on custom AI model, to grade the resumes of job candidates and predict which applicants would perform well at Amazon, with their resume data. The algorithm somehow learned to prefer men, while penalizing women. This happened because the otherwise true data deduced from the last decade of enrolled employees inherently only employed a large percentage of male employees. While the same would not have been observed in routine by HR, the pattern-seeking AI immediately sensed this and applied to filter out women. Resumes that included the word "women's" or included the names of all-women's educational institutions were downgraded, while resumes containing gender-masculine language linked with male engineers were preferred. Researchers scrapped the experiment because they were unsuccessful in resolving the issue.

22.6.3 Case Study 3

Google Translate: This widely used tool can perpetuate gender stamps during translation. A gender-neutral sentence in English, like "They are a doctor," might be interpreted into a gendered language and then back into English, resulting in a change to "He is a doctor." This bias bases [11] from two sources: the *training data* used to build up the translation model, which may reflect societal asymmetries, and the *algorithms* themselves, which may make biased premises based on stereotypes. While Google is working hard to make more fairish training data and less-biased algorithms, this example brings up the pervasiveness of AI bias and the on-going need for moderation strategies.

22.6.4 Case Study 4

Michigan's fraud detection algorithm incorrectly criminated 40,000 people of unemployment fraud: During the period traversing from 2013 to 2015, Michigan deployed an algorithm known as the Michigan Data Automated System (MiDAS), which fallaciously charged over 40,000 individuals of fraudulent activity related to unemployment benefits [56]. The results for those incorrectly accused were stark, ranging from hefty fines to bankruptcy and legal proceedings. Compounding the issue, individuals seeking to contest these accusations came upon a lack of human oversight in the system, with fraud determinations being made based on blemished or mismatched data. Following numerous legal challenges and an audit of MiDAS, it was brought out that only 8% of the fraud charges imposed by the system were preserved upon appeal.

22.6.5 Case Study 5

Arkansas' Medicaid algorithm incorrectly refuses medical care and benefits. Arkansas' implementation of a Medicaid access algorithm [38] ensued in the precipitous termination of benefits for numerous individuals, divesting them of all-important services such as home care, nursing visits, and medical treatments. In response, Arkansas Legal Aid started a federal lawsuit in 2016, contending that the state neglected to adequately inform affected individuals and provided no viable avenue for contending the algorithm's decisions. The appeals process was condemned as "effectively worthless," with a success rate of less than 5%. During judicial proceedings, the algorithm's creator detected multiple errors stemming from miscoding and imprecise calculations, with one error alone negatively affecting an estimated 19% of Medicaid beneficiaries in the state.

22.6.6 Case Study 6

Grading algorithms in the UK gives lower grades to l-income students. In 2020, amid the COVID-19 pandemic, the nullification of final exams across England motivated the Office of Qualifications and Examinations Regulation (*Ofqual*) to deploy an algorithm for grading students [25]. This algorithm, banking on teachers' predictions, academic records, and historical school functioning data, significantly brought down 40% of teacher-assigned grades, disproportionately bearing upon lower-income students and those attending larger public schools. The algorithm obviously banked on past school performance data skewed the results, punishing high-achieving students in underperforming schools. Moreover, it highlighted ethical concerns beleaguering the prioritization of school reputation over individual student achievement. Notably, the algorithm favored students in smaller class sizes, efficaciously advantaging those in private institutions. Following public outcry, *Ofqual* abandoned the algorithm in favor of teacher-allotted grades.

22.6.7 Technology Leaders

A recent analysis [32] in AI development collected responses from 350 US- and UK-based technology leaders (CIOs, IT directors, data scientists, etc.) associated in AI technologies. The analysis brought out arising concern concerning data bias, with 54% of responders evincing very or utmost concern.

22.6.8 Facial Recognition Technology:

The innovational work by Buolamwini and Gebru [8] brought out fundamental biases in facial recognition technology, with error rates of up to 34% for darker-skinned women and less than 1% for lighter-skinned men. This constraining difference highlights the effect of atypical training data on AI performance and the need for inclusive datasets and ethical considerations in AI development to neutralize racial and gender biases.

22.7 Feedback Loops: How Algorithms Can Influence Algorithms?

In AI, feedback loops unveil a complex interaction between algorithms, often contributing to the amplification of bias amongst algorithmic systems. These loops occur when the output of one algorithm determines the input of another, upholding and potentially aggravating biases present in the datasets or the underlying algorithms. Consider, for example, the case of facial recognition technology. In case an AI model trained on facial recognition constantly mistakes individuals of a particular race due to biased training dataset, this incorrect result can run into next succeeding algorithms, such as those used in law enforcement or hiring processes. As a result of this transfer, individuals from peculiar race communities may go through disproportional surveillance or invidious hiring practices, which can facilitate and further impinge societal biases. Figure 22.5 shows general stats of the effects observed, attributed to AI bias in various domains.

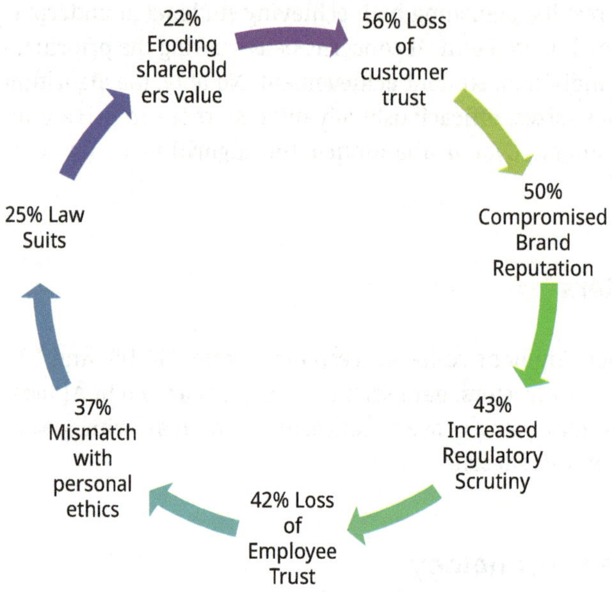

Figure 22.5: AI development collected responses.

Moreover, feedback loops (Kang & Meira-Goes, 2022) can happen within the optimization and development executions of algorithms themselves. As AI algorithms get trained on more and more large datasets, they invariably inherit and embed biases present in the data, leading to a vicious cycle of biased forecasting and feedback that upholds the original biases. Thus, it becomes essential to assume proactive measures for bias mitigation and algorithmic transparency to deal with the challenges laid by feedback loops. Feedback loops present a critical aspect of algorithmic systems that can either uphold or mitigate biases within AI technologies. By interpreting and turning to the mechanisms through which feedback loops operate, endeavor toward the development of more fair, transparent, and equitable AI systems can be expedited.

22.8 Challenges and Limitations Encountered

Mitigating bias, as would have emerged by now, is not going to be a simple thing that can be removed or negated by a simple tool or adherence to certain practices. A number of challenges exist to realize a non-biased dataset for employing in an AI model. Some of these challenges are listed here:

- **Language Divide**: NLP tools are not generally available and whatever are, are only poor-quality versions in English, primarily. While the English language tools, as available, have limited functions to offer, scarcity of such tools for other languages, such as German, Italian, Hindi, etc. is a challenge for anyone interested to dive into attempts to get rid of bias.

- **Computational Resources**: Existing NLP tools demand significant computational resources and memory capacities, which are often inaccessible to independent researchers. This hinders equal access to research resources and makes it hard to reproduce and improve upon existing research.
- **Poor Documentation**: Many current NLP tools are badly documented and unreliable. Algorithms built on top of these "foundations" are liable to inexplicable faults and failures.
- **Data Availability Limitations**: Access to labelled and annotated data in non-English languages continues to be gravely curbed. English-language datasets disproportionately drive training and testing of AI systems, thus constraining multilingual model generalization and hampering the foundations of solid systems for less-represented languages.
- **Semantic Dubiousness in Offensive Speech Definitions**: The definition of "offensive language" is different across geopolitical and sociocultural divides among spread-out states. Even narrowly constructed concepts such as "hate speech" have no standardized, operational definition across publicly available datasets. For instance, though broad outlines have been given by the European Union (EU), there is no consistent annotation protocol across datasets. This absence of definitional precision adds up significant vagueness in supervised learning models that deem to identify offending or hateful content, making cross-cultural generalizability and transfer learning more difficult.
- **Data Protection and Regulatory Compliance Issues**: Data privacy laws – specifically, the General Data Protection Regulation (GDPR) in the EU – enforce significant legal and ethical fixes on collecting and using user-generated data for research. While social media posts can be submitted to open places, they tend to constitute personal data based on the relative facility with which individuals can be identified and thus come within the purview of privacy legislation. This judicial incertitude breeds apprehension among researchers on the lawful use of such data.
- **Platform Policy Volatility and Research Bias**: Social media platforms are crucial source of linguistic data for NLP research, especially in the area of offensive speech detection. Platform-specific terms of service and developer access policies, nevertheless, change frequently, making long-term data collection strategies difficult. Twitter, due to its traditionally progressive API access, has become a proportionately predominant source of data, creating a systemic research bias. This over-sampling curtails the ecological validity of results, as models trained largely on Twitter data might not generalize well to other platforms with different user behavior and discourse norms.

22.9 Recommendations for Fixing Algorithmic Bias

Addressing multiple challenges of bias in AI systems, researchers and practitioners have formulated a diverse range of advanced techniques. From raising algorithmic transparency to enabling advanced ML models, these techniques target to extenuate biases across various degrees of the algorithmic life cycle. Below, we explore a comprehensive list of techniques that extend assuring solutions for raising fairness, transparency, and accountability in AI.

22.9.1 Accountability and Transparency

In algorithmic accountability and transparency, transparency [61] involves making the internal mechanics of algorithms accessible and understandable to stakeholders, while accountability implies instituting mechanisms to mark responsibility for both algorithms and creators for their outputs and affects.

Accountability handles not just legal and regulative measures but also ethical expectations pertaining to the social effects of algorithmic decision-making. Boosting openness and accountability in algorithmic decision-making processes can lend more reasonable development while also prioritizing well-being and rights of all individuals and citizens, in effect.

22.9.2 Race-aware Algorithm (RCA)

Recognizing the substantial impact of racial factors on algorithmic results, RCAs [42] impart a paradigm shift in the ways to handle AI bias. So, while conventional algorithms may hold racial biases through datasets, RCAs can mitigate such biases by integrating race-conscious steps into algorithmic design at onset only. *Race* is a social fabric with fundamental significances for individual experiences and is cardinal to the genesis of RCAs. By explicitly integrating *race* into algorithmic decision-making mechanism, RCAs resolve the ambiguity of algorithmic neutrality and promote fairness in outcomes.

22.9.3 Algorithmic Greenlining (AGL)

AGL [9] is more of a proactive approach, which includes designing AI algorithms with intent to make it inclusive and fair rather than just discovering and removing biases after the deployment. AGL is primarily built on the following two principles, i.e., equitable representation, fairness by design, and transparency and accountability. Figure 22.6 describes the four factors attributing to AGL:

- **Equitable Representation**: Training datasets in AGL tend to be diverse, ensuring all groups, in entirety, are well mapped.
- **Fairness by Design**: AGL developers adopt a design philosophy that demands probing the possible outcomes and decisions for peculiar population groups during their development and design phase only. Thus, they embed apt measures to eliminate partiality within, at onset.
- ***Transparency and Accountability***: AGL assigns priority to bring clarity in AI systems, which assist decision-making processes to be grasped and audited. This interprets to making the developers and designers responsible if any maligned motive is detected in the algorithms later.

22.9.4 Data Preprocessing Techniques

Data preprocessing involves checking and processing data prior to applying the same for learning an AI model. This involves coordinating and fine-tuning the data to identify and correct inaccuracies in the dataset prior to deployment. Additionally, feature selection methods in data preprocessing techniques help neutralize bias by focusing on the most relevant attributes, while playing down the burden of potentially biased factors.

22.9.5 Algorithmic Adjustments

Algorithmic adjustments [58] have become an essential part for mitigating bias in ML models and involve modifying the algorithms themselves to eliminate bias in the respective decision-making processes:
- The approach involves integrating fairness constraints immediately into the algorithm design, ensuring that the model's predictions are just across different demographic groups.
- In addition, this approach can be put to the use of reweighting as a way of eliminating bias detected by the model while it was trained.
- In addition, monitoring and assessment of the system's progress in real-world circumstances make algorithmic adjustments an effective way of discovering any prejudices that might have surfaced and need to be addressed over time.

22.9.6 Explainable AI Approaches (EAA)

Explainable AI approaches, as proposed by Pradhan et al. [41], aims at providing insights to stakeholders into the decision-making nature of algorithms that would help stakeholders understand and audit the inherent factors influencing results. Enabled

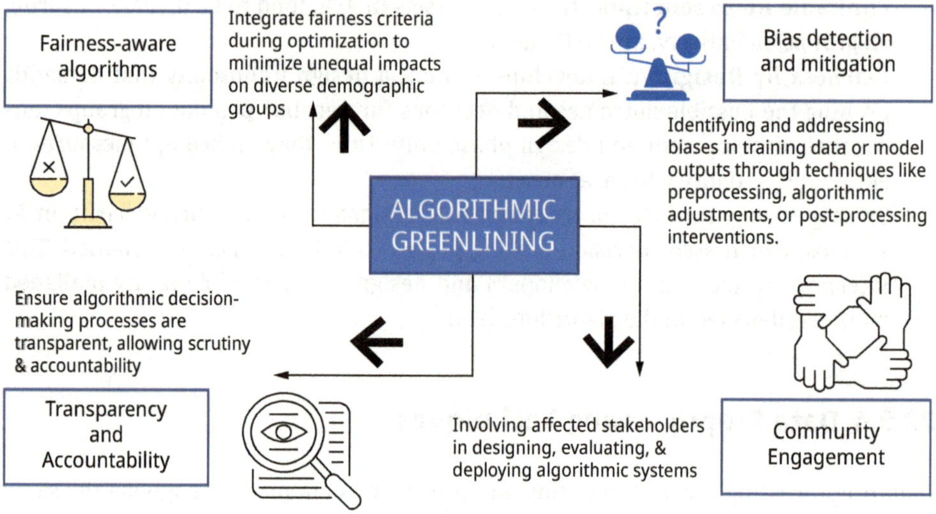

Figure 22.6: Algorithmic greenlining.

on techniques like Feature Importance analysis, Model-agnostic explanations, and Explainable model architectures, developers can sort out relationships between input features and outcome decisions that bring potential sources of bias in the process.

22.9.7 Differential Privacy Techniques

An assuring avenue for the reduction of algorithmic bias, while asserting data privacy, is provided by DPT [59]. This approach attempts to disguise classified and sensitive data points by introducing noise in individual dataset points without losing meaningful analysis at an aggregate level. Thus, with the use of DPT, organizations may reduce the chance of unwittingly encoding biases present in the dataset into their ML models. Additionally, DPT can handle issues of overfitting to specific data points by promoting fairness and inclusivity in decisions by algorithms.

22.9.8 Federated Learning Techniques (FLT)

In FLT, models are trained for ML across the decentralized servers wherein the datasets remain, allowing learning by the model, without centralization from local data [33]. In other words, it is unnecessary to bring together all the sets of data in a central server and then determine the AI model thereafter. The amount of data transfer is reduced as well; this can practically have an effect on data sovereignty for states. The learning process is distributed by FLT, which reduces the risk of bias associated with centralized datasets, as they are usually not diverse and inclusive. Furthermore, it enhances the internalization of data from diverse sources and demographic groups, thus enhancing fairness in model development. Organizations can improve the inclusiveness of their ML models, while ascertaining individual privacy protection by using FLT.

22.9.9 Counterfactual Fairness

Counterfactual fairness [53] is a new way of dealing with algorithmic bias by coming up with how the model would have responded under different prompts involving different attributes. Counterfactual fairness brings fairness to the decisions of an algorithm by comparing the predicted responses on prompts that contain different attributes like race/gender/demography, etc. Counterfactual fairness provides a thoughtful framework for the assessment and enforcement of fairness in algorithmic decision-making, allowing developers to mitigate the heterogeneous effect of biased algorithms on different demographic groups.

22.9.10 Causal Inference Techniques (CIT)

CIT techniques [6] are applied to infer causal relationships from experimental datasets. CIT bids insight into how different factors contribute to outcomes, assisting to bring out and reduce bias by discerning causal mechanisms rather than banking solely on correlations. CIT grants practitioners the ability to differentiate between spurious and actual causal relationships, thus making them able to make better decisions on what variables to allow or exclude from their models.

22.9.11 Contextual Bandit Algorithms (CBA)

Contextual bandit algorithms (CBAs) are a family of ML algorithms designed for online decision-making environments, where choices need to be decided sequentially, and typically with partial or incomplete observations. CBA [21] merging aspects of both RL and multiarmed bandit problems: In the standard multiarmed bandit problem, the agent faces a set of options (arm), each coupled with an underlying reward distribution. The objective is to identify which arm to pull so that it results in the cumulative reward over time. Analogously applied in AI model with CBA problems, the reward distribution for each arm (prompt) depends on further information, called context. Therefore, for every response and prompt, the context plays a very important role in deducing the final response.

CBAs [57] thus provide an assuring solution to deal with algorithmic bias by integrating contextual information into the decision-making process. CBAs allow personalized and adaptive decision-making, enabling models to learn from historic interactions and make more informed choices, while reducing the possibility for biased outcomes.

22.9.12 Bias Auditing Frameworks

These frameworks refer to tools [55] used to assess and mitigate biases by quantifying biases that may exist in the data, features and attributes, or predictions rendered by the model. Here are five well-known and popular bias auditing frameworks:
- Fairness Indicators [23]
- AI Fairness 360 (AIF360) [30]
- Themis [49]
- IBM Watson OpenScale [29]
- Microsoft Fairlearn [7]

These frameworks help organizations scan their algorithms, showing how they can maintain or lessen partiality on a number of demographic classes or protected attributes. They include fairness metrics, subgroup analysis, and inequality detection, which allow users to understand and quantify prevalent partiality in models. Bias auditing frameworks further reduce possible sources of bias in the data model.

22.9.13 Human-in-the-Loop Approaches

HITL [12] advocates for infusing human intervention and interference at the design stage, practice use, and scrutiny of ML interventions to reduce their biases and evaluate equity. HITL recognizes the fact that though algorithms can automatize many promptings, still, human inspection and expertise have to be developed for dealing with complex ethical and social issuances.

The HITL approach brings in a fresh strategy to combat algorithmic bias by bringing in human oversight and intervention in the ML process. HITL approach integrates human judgment and expertise at multiple stages, such as data collection, model training, and decision-making. Involving diverse stakeholders like domain experts, ethicians, and related communities, organizations will gain valuable insights into the social context and the biases inherent to the dataset model. HITL enables nuanced forms of bias that might not be evident in the algorithm. This will bring about more ethical and informed decision-making. Increasing accountability, transparency, and fairness are some advantages that HITL brings, giving humans the authority to revisit and interfere with the process of algorithms toward a fair result in the process of algorithmic decision-making and outcome.

22.9.14 SMOTE: Synthetic Minority Over Sampling Technique

One powerful technique for combating bias, particularly in the case of class asymmetry, is the SMOTE [43]. Class asymmetry arises when one of the classes within a classification problem is severely underrepresented, compared to others. This can bestow biased ML models that prioritize majority class and execute poorly on minority class cases.

SMOTE technique furnishes synthetic samples for the minority class, effectively increasing its representation in the dataset. By making artificial instances based on the characteristics and parameters of existing minority class samples, SMOTE aids to counterpoise the class distribution and assure that the ML model learns from a more democratic dataset. However, it's important to note that while SMOTE can help deal bias related to class asymmetry; it does not do away with other forms of bias that may exist already. Therefore, SMOTE should be applied in conjunction with other techniques for bias detection and mitigation to ascertain that AI systems are fair, transparent, and accountable.

In addition to the above, few more mitigation measures can include
- *Subgroup F_1 Analysis*: Report F_1 scores, broken out by raw attributes (e.g., race and gender) to disclose performance gaps concealed by aggregate metrics.
- *Fairness-Aware Hyperparameter Tuning*: Employ fairness metrics (e.g., demographic parity loss) during model selection to trade performance and equity.
- *Adversarial Debiasing*: Train adversarial networks to strip model representations of sensitive attribute signals, while retaining predictive ability.
- *Fair Representation Learning*: Ascertain latent embeddings that convey task-relevant information, while de-correlating with saved attributes.
- *Ethical Model Cards and Datasheets*: Release standardized documentation, delineating model bias risk, subgroup performance, and data provenance.
- *Ensemble Debiasing*: Combine several weak learners trained for subgroup fairness to decimate bias artifacts in final predictions.

22.10 Need for Global Regulation Frameworks and Guidelines

The paced penetrating trust on AI based data-driven decision making systems calls for global robust frameworks and guidelines to ensure fairness and reduced bias in AI models being deployed globally. Here's why:

- **Standardization and Consistency**: An ideal global regulation framework would furnish a common set of standards and guidelines for AI development and deployment to assure consistence and interoperability across multiple spread out jurisdictions and industries. Thus, it would help extenuate the risk of disconnected or contravening regulations, which can make complaisance challenges for organizations functioning in multiple regions. However, in this world, while we still have to work out long-drawn cyber security global pacts and standards, envisaging expedited work for AI-peculiar global regulation framework would likely be a huge challenge in itself.
- **Promotion of Fairness and Equity**: Regulations would encourage fairness and equity if they established clear expectations and prerequisites for both developers and users of AI technologies. Setting principles of nondiscrimination and equity, guidelines might facilitate safe behavior, in opposition to calumnious biases that may uphold or exasperate current inequalities and social injustices.
- **Consumer Protection and Trust**: Regulation frameworks would increase user safety and take it into account, by way of calling for transparency, duty, and human oversight in AI systems. Clear-cut tips for bringing out algorithmic biases and their feasible impingements would enable users to make learned choices and make developers indictable for any unexpected moral outcomes.
- **Risk Management and Liability**: Regulations will provide frameworks of risk management and obligation assignment, should damage or misuse happen because of AI. This is made possible by the resolution and determination of roles and obligations among developers, users, and regulators. Through this, regulations would mitigate risks on legal and ethical sides caused by biased AI models.
- **International Cooperation and Collaboration**: Global regulatory frameworks would allow international cooperation, and focus more on global AI administration issues. Through regulation, there would be a further intensification of dialogue and knowledge sharing between countries, regions, and stakeholders on common challenges and best practices in AI bias mitigation.
- **Ethical and Human Rights Considerations**: Ideal regulations would express the moral principles and human rights considerations of AI design and deployment methods. Coordinating with worldwide human rights structures, such as the Universal Declaration of Human Rights, regulations will safeguard the main values of respect, equality, and privacy when AI systems are developed and applied.
- **Innovation and Responsible AI Development**: Regulation defends innovation and responsible AI development by providing guidance on and incentives to incorporate bias mitigation techniques into the life cycle of AI. Of course, this adds research and development of bias-aware algorithms and tools to promote innovations in AI technologies that are technically challenging, yet ethically aligned.

22.11 Existing Initiatives in Fairness Frameworks and Guidelines

While AI development is speedily developing, regulatory frameworks are still under development and initial phases of arrival. Here are few noteworthy ongoing efforts:

- **The EU:** The EU Commission has come up with the Artificial Intelligence Act (AIA) which accentuates fairness, transparency, and accountability in AI development and deployment. The AIA adumbrates risk-based tiers for AI applications, with more exacting requirements for high-risk applications.
- **The United States (US):** The US currently misses out on a comprehensive federal framework, but single agencies like the National Institute of Standards and Technology (NIST) are formulating voluntary AI risk management frameworks that deal with fairness considerations.
- **Organization for Economic Co-operation and Development (OECD):** OECD has provided recommendations on AI ethics that focus on fairness, nondiscrimination, and human error. These recommendations come as the primary justification for member countries to consider while establishing their own individual national regulations.

The attention on equity in AI is the need of a pressing global consensus on the regulations of AI. No independent and isolated versions across states would work to bring in advantages. Several demanding situations dangle on. One such assignment is the sensitive stability between encouraging innovation and enforcing essential regulations to boost responsible AI development. Getting this stability right is essential to invalidate stifling creative thinking, while ascertaining moral practices. Another challenge is in instituting universally ordinary requirements and robust enforcement mechanisms across diverse jurisdictions. Achieving this requires navigating a good deal complex prison and regulatory worldwide landscapes. If we simply see the cyber security landscape over last two decades, we are yet to reach anywhere globally. Cyber crime conviction bears an ever small percentage when seen at global stage. Despite these obstacles, accelerated regulatory frameworks and guidelines are a tool of bringing responsible AI practices and just fairness in statistics-driven decision-making. Though worrisome, continued global efforts display promising aspects for future wherein AI will go to serve the common good for the human race for accurate and gigantic benefits.

22.12 The Ethical Implications of Mathematical Models in the Digital Age

In the digital age, mathematical models have taken over the role of central tools deducing decision-making in as varied a range of fields as finance, healthcare, education, governance, and criminal justice. Although these models are ordinarily thought of as objective and rational, their application within AI systems can inadvertently preserve or exasperate societal inequalities. The premise that mathematical abstraction allows for neutrality fails upon close examination when models are conditioned on historical data imbued with structural bias or optimized for performance metrics, exclusively and bereft of ethical thought.

The ethical concerns do not stem from the math itself, but rather from the fundamental interaction between statistical formalism, data provenance, model assumptions, and optimality targets. For example, an algorithm that is optimized for maximum overall F_1 score can systematically perform below par for deprived groups, hiding discriminatory results behind a smokescreen of statistical resilience. Moreover, models used at scale incline to function with minimal transparency and control, which also leads to issues of accountability, consent, and contestability in high-stakes settings.

Thus, justice in AI cannot be appended after the fact – it needs to be built into the algorithmic system life cycle, from model instauration and data curation to validation and governance. Ethical AI requires a paradigm shift – no longer prioritizing prediction power in isolation, but incorporating social context, stakeholder effect, and normative values.

22.13 Conclusion: Charting a Course Toward Algorithmic Equity: A Multifarious Approach

This chapter dived deep into the pestilent realm of algorithmic bias within AI system models and deployments. The chapter meticulously dissected multiple types of bias, their potential to falsify expected responses, and discussed metrics like F_1 score that assesses the predictive ability of a model. Algorithmic feedback loops and alike were discussed in the bias context. Then, the chapter discussed real consequences of bias in real-world applications and the potential of holding social and economic inequalities. The chapter used few case studies with real examples and impact on human lives.

In addition to recognizing the limitations of these approaches, the chapter has explored in great detail cutting-edge techniques to combat bias. Algorithmic transparency, Fairness-aware model design, CBAs, Algorithmic Greenlining, RCAs, Counterfactual Fairness, CITs, Bias Auditing Frameworks and SMOTE and the emergent field of

XAI are key findings of crucial weapons in this battle. The chapter further went on to talk about the usefulness of allies – data preprocessing techniques, algorithmic adjustments, and employment of DPT and FLT for pursuing algorithmic equity.

However, the chapter accepts that the pursuance for fairness extends beyond technical resolutions. The vital role of regulatory frameworks and international consensus and initiatives in bringing up responsible AI development and founding ethical guidelines is indispensable. Embracing this pluralistic approach – this dynamic interplay of innovative bias mitigation techniques, coupled with robust regulatory structures – we can chart our course toward an eventual future in which AI is a powerful tool for progress, its decisions increasingly not only accurate but fair and equitable for all members of society.

References

[1] Ahn Y, Lee S, Shim J, Park J. Retrieval-augmented response generation for knowledge-grounded conversation in the wild. IEEE Access. 2022; 10: 131374–131385.

[2] Alvarez JM, Colmenarejo AB, Elobaid A, Fabbrizzi S, Fahimi M, Ferrara A et al. Policy advice and best practices on bias and fairness in AI. Ethics and Information Technology. 2024; 26(2).

[3] Aran XF, Van Nuenen T, Criado N, Such J. Discovering and interpreting biased concepts in online communities. IEEE Transactions on Knowledge and Data Engineering. 2023; 35(4): 3672–3683.

[4] Aranovskiy S, Ryadchikov I, Nikulchev E, Wang J, Sokolov D. Bias propagation and estimation in homogeneous differentiators for a class of mechanical systems. IEEE Access. 2020; 8: 19450–19459.

[5] Bai X, Wu J, Wang H, Zhang J, Yin W, Dong J. Recommendation Algorithms for Implicit Information. In Proceedings of 2011 IEEE International Conference on Service Operations, Logistics and Informatics 2011.

[6] Biparva D, Materassi D. Application of eXplainable AI and Causal Inference Methods to Estimation Algorithms in Networks of Dynamic Systems. In 2023 American Control Conference (ACC) 2023.

[7] Bird S, Dudík M, Edgar R, Horn B, Lutz R, Milan V, et al. Fairlearn: A toolkit for assessing and improving fairness in AI – Microsoft Research [Internet]. Microsoft Research; 2020 Sep 22 [cited 2024 Jun 7]. Available from: https://www.microsoft.com/en-us/research/publication/fairlearn-a-toolkit-for-assessing-and-improving-fairness-in-ai/

[8] Buolamwini J, Gebru T. Gender Shades: Intersectional Accuracy Disparities in Commercial Gender Classification. Proceedings of the 1st Conference on Fairness, Accountability, and Transparency. 2018; 81: 77–91. Available from: https://proceedings.mlr.press/v81/buolamwini18a.html.

[9] Borgs C, Chayes J, Haghtalab N, Kalai AT, Vitercik E. Algorithmic Greenlining. In AIES '19: Proceedings of the 2019 AAAI/ACM Conference on AI, Ethics, and Society 2019.

[10] Bozovich AN, Nguyen D, Rax BG, Davila J, Zajac SA. Investigation of Application-Specific Bias Conditions and Dose Rate Dependency in Total Ionizing Dose (TID) Response. In 2020 IEEE Radiation Effects Data Workshop (In Conjunction with 2020 NSREC) 2020.

[11] Chen S. The bias of Google Translate [Internet]. [cited 2024 Jun 13]. Available from: https://stuyspec.com/article/the-bias-of-google-translate

[12] Chia A. What Is Human-in-the-Loop? A Simple Guide to this AI Term [Internet]. CareerFoundry; 2024 Mar 4 [cited 2024 Jun 7]. Available from: https://careerfoundry.com/en/blog/data-analytics/human-in-the-loop/

[13] Cpollnow SK. Hiring bias gone Wrong: Amazon recruiting case study [Internet]. Cangrade; 2023 Sep 21 [cited 2024 Jul 23]. Available from: https://www.cangrade.com/blog/hr-strategy/hiring-bias-gone-wrong-amazon-recruiting-case-study/

[14] Curzon J, Kosa TA, Akalu R, El-Khatib K. Privacy and artificial intelligence. IEEE Transactions on Artificial Intelligence. 2021; 2(2): 96–108.

[15] Edmond G, Forensic S, Implications C, Trials P, Appeals. Contextual bias and cross-contamination in the forensic sciences: The corrosive implications for investigations, Plea Bargains, trials and appeals. Law, Probability & Risk. 2014; 14(1): 1–25.

[16] DeAlcala D, Serna I, Morales A, Fierrez J, Ortega-Garcia J. Measuring Bias in AI Models: An Statistical Approach Introducing N-Sigma. In 2023 IEEE 47th Annu ComputSoftw Appl Conf (COMPSAC) 2023.

[17] Dogra V, Verma S, Kavita N, Wozniak M, Shafi J, Ijaz MF. Shortcut Learning Explanations for Deep Natural Language Processing: A Survey on Dataset Biases. In IEEE Access 2024 (pp. 1).

[18] Dwork C, Hardt M, Pitassi T, Reingold O, Zemel R. Fairness through awareness. In ITCS 12: Proceedings of the 3rd Innovations in Theoretical Computer Science Conference 2012 (pp. 214–226).

[19] Geiger G. How a Discriminatory Algorithm Wrongly Accused Thousands of Families of Fraud [Internet]. Vice Media Group; 2021 Mar 1 [cited 2024 Jul 23]. Available from: https://www.vice.com/en/article/jgq35d/how-a-discriminatory-algorithm-wrongly-accused-thousands-of-families-of-fraud

[20] Ghassami A, Khodadadian S, Kiyavash N. Fairness in Supervised Learning: An Information Theoretic Approach. In 2018 IEEE International Symposium on Information Theory (ISIT) 2018.

[21] Ghosh A, Sankararaman A, Ramchandran K. Model selection for generic contextual bandits. IEEE Transactions on Information Theory. 2024; 70(1): 656–675.

[22] Godbole A, Grosz SA, Nandakumar K, Jain AK. On Demographic Bias in Fingerprint Recognition. In 2022 IEEE International Joint Conference on Biometrics (IJCB) 2022.

[23] Goyal P, Soriano AR, Hazirbas C, Sagun L, Usunier N. Fairness Indicators for systematic assessments of visual feature extractors. arXiv [cs.CV]. 2022; arXiv:2202.07603.

[24] Hajavi A, Etemad A. A Study on Bias and Fairness in Deep Speaker Recognition. In ICASSP 2023 – 2023 IEEE International Conference on Acoustics, Speech and Signal Processing (ICASSP) 2023.

[25] Harkness T. A level results: Why algorithms aren't making the grade [Internet]. BBC Science Focus Magazine [cited 2024 Jul 23]. Available from: https://www.sciencefocus.com/future-technology/a-level-results-why-algorithms-arent-making-the-grade

[26] Hirota Y, Nakashima Y, Garcia N. Quantifying Societal Bias Amplification in Image Captioning. In 2022 IEEE/CVF Conf Comput Vis Pattern Recognit (CVPR) 2022.

[27] Huang H, Xu H, Wang X, Silamu W. Maximum F1-Score Discriminative training criterion for automatic mispronunciation detection. IEEE/ACM Transactions on Audio, Speech, and Language Processing. 2015; 23(4): 787–797.

[28] Huang Z, Tanaka F. Investment biases in reinforcement learning-based financial portfolio management. In 2022 61st Annual Conference of the Society of Instrument and Control Engineers (SICE) 2022.

[29] IBM Corporation. Configuring fairness evaluations for indirect bias in Watson OpenScale | IBM Cloud Pak for Data as a Service [Internet]. IBM; 2024 Apr 6 [cited 2024 Jun 9]. Available from: https://dataplatform.cloud.ibm.com/docs/content/wsj/model/wos-indirect-bias.html?context=cpdaas

[30] IBM [IBM Developer Staff]. AI Fairness 360 [Internet]. IBM Developer; 2018 Nov 14 [cited 2024 Jun 13]. Available from: https://www.ibm.com/opensource/open/projects/ai-fairness-360/

[31] Kang E, Meira-Góes R. Requirements engineering for feedback loops in software-intensive systems. In 2022 IEEE 30th International Requirements Engineering Conference Workshops (REW) 2022 (pp. 2–5). IEEE.

[32] Kaminska J. The impact of data bias on your business & the benefits of fair AI – Statice [Internet]. Statice [cited 2024 Jul 23]. Available from: https://www.statice.ai/post/data-bias-impact

[33] Korkmaz A, Alhonainy A, Rao P. An Evaluation of Federated Learning Techniques for Secure and Privacy-Preserving Machine Learning on Medical Datasets. In 2022 IEEE Appl Imagery Pattern Recognit Workshop (AIPR) 2022.

[34] Linner R. Algorithms of Oppression. IEEE Technology and Society Magazine. 2019; 38(1): 9–10.

[35] Liu B, Zhang S, Song G, You H, Liu Y. Rectifying the Data Bias in Knowledge Distillation. In 2021 IEEE/CVF Int Conf Comput Vis Workshops (ICCVW) 2021.

[36] Macdonald S, Hopgood JR. Joint Spatio-Temporal Bias Estimation and Tracking for GNSS-Denied Sensor Networks. In 2021 Sens Signal Process Def Conf (SSPD) 2021.

[37] Mandhala VN, Bhattacharyya D, Midhun Chakkaravarthy D. Need of Mitigating Bias in the Datasets Using Machine Learning Algorithms. 2022. In Int Conf Adv Comput Commun Appl Informatics (ACCAI) 2022.

[38] McCormick E. What happened when a 'wildly irrational' algorithm made crucial healthcare decisions [Internet]. The Guardian; 2021 Jul 2 [cited 2024 Jul 23]. Available from: https://www.theguardian.com/us-news/2021/jul/02/algorithm-crucial-healthcare-decisions

[39] Morgenstern J. Availability Bias. Springer eBooks; 2021 (pp. 47–52).

[40] Nagpal S, Singh M, Singh R, Vatsa M, Ratha NK. In-group bias in deep learning-based face recognition models due to ethnicity and age. IEEE Transactions on Technology and Society. 2023; 4(1): 54–67.

[41] Pradhan R, Lahiri A, Galhotra S, Salimi B. Explainable AI: Foundations, Applications, Opportunities for Data Management Research. In 2022 IEEE 38th Int Conf Data Eng (ICDE) 2022.

[42] Review CL, Kim PT, Noyes Kirby D. Race-Aware Algorithms: Fairness, Nondiscrimination and Affirmative Action – California Law Review [Internet]. California Law Review; 2023 May 5 [cited 2024 Jun 13]. Available from: https://www.californialawreview.org/print/race-aware-algorithms-fairness-nondiscrimination-and-affirmative-action

[43] Satriaji W, Kusumaningrum R. Effect of Synthetic Minority Oversampling Technique (SMOTE), Feature Representation, and Classification Algorithm on Imbalanced Sentiment Analysis. In 2018 2nd Int Conf Informatics Comput Sci (ICICOS) 2018.

[44] Sen S, Dasgupta D, Gupta KD. An Empirical Study on Algorithmic Bias. In 2020 IEEE 44th Annu ComputSoftw Appl Conf (COMPSAC) 2020.

[45] Sepúlveda J, Velastin S. F1 Score Assesment of Gaussian Mixture Background Subtraction Algorithms Using the MuHAVi Dataset. In 6th Int Conf Imaging Crime Prev Detect (ICDP-15) 2015.

[46] Shin D, Shin EY. Data's impact on algorithmic bias. Computer. 2023; 56(6): 90–94.

[47] Sirotkin K, Carballeira P, Escudero-Vinolo M. A Study on the Distribution of Social Biases in Self-supervised Learning Visual Models. In 2022 IEEE/CVF Conf Comput Vis Pattern Recognit (CVPR) 2022.

[48] Taieb SB, Atiya AF. A bias and variance analysis for multistep-ahead time series forecasting. IEEE Trans Neural Netw Learn Syst. 2016; 27(1): 62–76.

[49] Themis AI Inc 2023. Themis AI [Internet]. Themis AI Makes Any ML Model Uncertainty-aware With a Single Line of Code [cited 2024 Jun 12]. Available from: https://www.themisai.org/

[50] Valtonen L, Makinen S. Exploring the Relationships between Artificial Intelligence Transparency, Sources of Bias, and Types of Rationality. In 2022 IEEE Int Conf Ind Eng Eng Manage (IEEM) 2022.

[51] Villmann T, Kaden M, Lange M, Sturmer P, Hermann W. Precision-Recall-Optimization in Learning Vector Quantization Classifiers for Improved Medical Classification Systems. In 2014 IEEE Symp ComputIntell Data Min (CIDM) 2014.

[52] Wang H, Ustun B, Calmon FP. On the Direction of Discrimination: An Information-Theoretic Analysis of Disparate Impact in Machine Learning. In 2018 IEEE Int Symp Inf Theory (ISIT) 2018.

[53] Wang Y, Luo Z. Causal Discovery and Causal Inference Based Counterfactual Fairness in Machine Learning. In ICASSP 2023 – 2023 IEEE Int Conf Acoust Speech Signal Process (ICASSP) 2023.

[54] Wang Z, Tam Y. Suffix Retrieval-Augmented Language Modeling. In ICASSP 2023 – 2023 IEEE Int Conf Acoust Speech Signal Process (ICASSP) 2023.
[55] Werner J. BABL AI: A framework for assurance Audits of Algorithmic Systems | BABL AI [Internet]. BABL AI; 2024 Jan 29 [cited 2024 Jun 13]. Available from: https://babl.ai/babl-ai-a-framework-for-assurance-audits-of-algorithmic-systems/
[56] Wykstra S. Government's use of algorithm serves up false fraud charges [Internet]. Undark Magazine; 2020 May 29 [cited 2024 Jul 23]. Available from: https://undark.org/2020/06/01/michigan-unemployment-fraud-algorithm/
[57] Xu X, Xie H, Lui JCS. Generalized contextual bandits with latent features: Algorithms and applications. IEEE Trans Neural Netw Learn Syst. 2023; 34(8): 4763–4775.
[58] Yuan J, Hu J, Liu X, Qiu Z. A Fast Adjustment Algorithm Based on the Theory of One by One Parameter Estimation. In 2010 Int Conf E-Product E-Service E-Entertainment 2010.
[59] Zhu T, Ye D, Wang W, Zhou W, Yu PS. More than Privacy: Applying Differential Privacy in Key Areas of Artificial Intelligence. In IEEE Trans Knowl Data Eng 2021 (pp. 1).
[60] Zhuang F, Qi Z, Duan K, Xi D, Zhu Y, Zhu H et al. A comprehensive survey on transfer learning. Proceedings of the IEEE. 2021; 109(1): 43–76.
[61] Zouave ET, Marquenie T. An Inconvenient Truth: Algorithmic Transparency & Accountability in Criminal Intelligence Profiling. In 2017 EurIntellSecur Informatics Conf (EISIC) 2017.

R. Manimegalai, S. Kavisri, S. Lakshmi Priya Varshini, C. S. Shruti, S. K. Ayesha

23 Image Steganography and Attacks Analysis for Secured Data Communication

Abstract: The accelerated developments in data transmission via the internet made it simpler to transfer data to its destination in an accurate and timely manner. Steganography is the practice of covering the very existence of the message within other information to conceal the fact that data transmission is occurring. The work focuses on secured transmission of data through the use of the modified least significant bit (LSB) algorithm to conceal messages within images. This work emphasizes that data can be disguised in the LSB of a pixel within one of its RGB channels, specifically targeting the minimum, maximum, and median values. Steganalysis is the process of examining this algorithm's resistance to various attacks such as mean filter, median blur, and salt-pepper noise. Among various attacks analyzed, it has been observed that the proposed methodologies, namely, LSB-MIN, LSB-MED, and LSB-MAX exhibit higher tolerance toward salt and pepper attacks.

Keywords: Steganography, security threat and attacks, steganalysis, hamming distance, peak-signal-to-noise ratio, stego-image

23.1 Introduction

An image is a two-dimensional signal that contains color information as coordinates. In general, an image can be represented using an array of picture elements. The primary purpose of image processing is to enable computers to recognize, interpret, and understand pictorial patterns. Each pixel in an image represents an intensity value. There are several cryptographic techniques that have been developed for encoding and decoding the data to ensure secure communication. However, there are circumstances where it is necessary to hide the presence of a message. Steganography covers the message to be transmitted with another text, image, video, or audio file so that communication will be visible but the contents of the communication is hidden. The term "steganography" has its roots in Greek, from two words, namely, steganos and graphein, meaning hidden writing. It is a technique that enlightens minds by demonstrating that what is seen does not have to be the only thing available. As a result, images could go beyond what our human visual system (HVS) perceives [4]. This paved the idea for secret or restricted communication. The embedding of secret messages for communication can happen in various mediums such as audio, text, image, and video files. However, confidentiality must be maintained

https://doi.org/10.1515/9783112214374-023

during data transfer so that intruders cannot tamper the data. Cryptography, watermarking, and fingerprinting can be used along with steganography. One of the applications of steganography is printing documents and currency notes. In military applications, steganography is used to convey and covertly store an encrypted document [1–6]. The cover image helps to conceal and transmit the secret message using an encoding algorithm. The secret message can be retrieved from the stego-object using an extraction or reverse technique, as shown in Figure 23.1. It can be observed that the cover image is removed to retrieve the hidden message. Steganographic system provides data embedding and easy data extraction, and promotes high information rate.

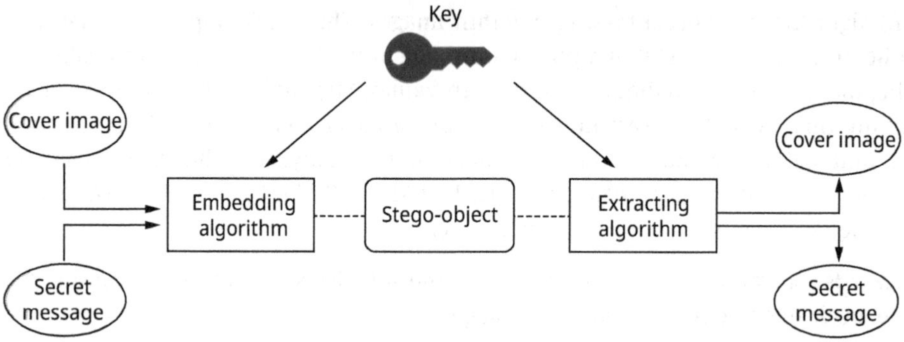

Figure 23.1: Process of steganography.

23.1.1 Steganography Techniques

The steganographic techniques can be classified based on the cover type and the cover-type modification, as shown in Figure 23.2. In all the steganographic techniques, the stego-image is created by combining cover-carrier, key, and the message. As shown in Figure 23.2, the message to be transferred can be hidden using image, text, audio, or video. The messages can be encoded using a variety of cover-type modification strategies such as spatial domain techniques, transform domain techniques, spread spectrum techniques, statistical methods, and distortion techniques. Spatial domain steganographic techniques are sometimes called substitution techniques. These techniques are simple and can be easily implemented for creating secret channels in parts of the cover image where the changes are predicted to be meagre in contrast with HVS. LSB-level encoding is a spatial domain steganographic method that modifies the cover media and secret information in the spatial domain. Transform domain methods outperform LSB methods because they hide information in areas of the image that are resilient to cropping, image processing, and compression. Some transform domain methods seem to have no dependency on the image format and could surpass both lossy and lossless format conversions.

In spread spectrum transmission, radio communications send signals in a certain frequency, which is below a certain level of background noise. The spread spectrum technique can be used in two different ways in steganography. It can either treat the cover image as pseudo-noise or attempt to introduce noise. Model-based techniques modify an image's statistical features while preserving them during the embedding process. The statistical features of the elements include the mean, range, and variation of their size, color, orientation, speed, and direction of motion. The process of encoding involves multiple iterations where the cover image is modified. In order to retrieve the secret message, the decoder compares the modified cover image with the original cover image to identify any discrepancies. As a result, data is stored using signal distortions. This technique involves applying a sequence of modifications to the cover image to create a stego-object that corresponds to the secret message. The adjustments are carefully selected to ensure that the resulting stego-object accurately conveys the desired information. Random pixels are used to encode the message, whereby the message bit is assigned a value of 1 if the stego-image differs from the cover image at the specified message pixel. If there is no deviation, the message bit is represented as 0. The statistical features of the images are independent of changes made by the encoder to the 1-value pixels. Nevertheless, the benefits of this strategy are limited by the requirement to furnish the cover image. Similar to other steganographic techniques, reuse of the same cover image should be avoided. Major modifications such as rotating, cropping or rescaling, if any, in the stego-image by an attacker can be detected by the recipient. The use of error-correcting information during encoding enables the correction of any modifications made and retrieval of the original message.

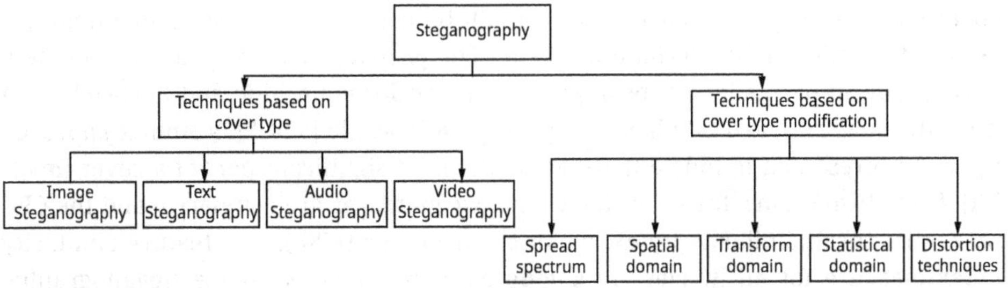

Figure 23.2: Classification of steganography techniques.

23.2 Related Works

Most steganographic techniques start with data fragmenting into smaller pieces, which are then inserted into the appropriate areas of the digital artefacts in a way that the difference between the original image and a modified image is not noticeable [5]. The techniques of steganography are classified into two types: (i) spatial domain and (ii) frequency domain. Image pixels are changed in spatial domain techniques to store secret messages. The approach of frequency domain techniques involves the use of transform to alter an image, which may include a wavelet transform. Once the transform is applied, an embedding technique is employed to conceal the information. Steganography does not change the structure of a secret message, but rather buries the message in the media such that there is no visible difference [6]. Steganography prevents an unintended receiver from discovering the presence of the data, and the secrecy of the steganographic system is reliant on the Conway data encoding method [7]. This is particularly susceptible to scaling, cropping, rotating, noise addition, and lossy compression. Attackers can quickly delete the message because of these stego-image alterations [8]. LSB approaches are employed because of their simplicity. BPCS steganography employs many bit planes, allowing it to incorporate a significantly larger quantity of data. Battisti et al. have devised a technique to encode information in digital media by breaking down the Fibonacci sequence of numbers, which enables flexible bit plane decomposition, instead of using the typical LSB method [9]. A comprehensive review on encryption techniques for data security and secured content distribution are presented in [34, 35]. Threat modeling and attack analysis for IoT applications are discussed in [36].

Zhang and Tang have proposed an enhancement to the least significant bit (LSB) technique that uses a pseudo-random number to select random groupings of pixels and then integrates the bit values for each pixel using modular division and addition operations [10]. Dey et al. have proposed LSB bit concealment using prime numbers as an information hiding technique [11–13]. The primary objective was to generate a novel set of bit planes by means of prime number decomposition and to encode data using the newly created bit planes as sparingly as feasible [30–33]. A similar approach by Goel is presented to hide a message on the least significant part of a cover image [14]. Using histograms, the cover image and stego-images are compared, and the CPU time, structural similarity index, mean square error (MSE), and feature similarity index measure for all images are computed. Using adapted image steganography, Odat and Otair have provided the LSB technique. The cover image's odd bytes, which need to be 16 times the size of the encoded image in order to increase the level of protection, must be used to segment the secret image bits and distribute them throughout the image [15]. Dhawan and Gupta [16] have proposed yet another approach, an enhanced LSB algorithm, which has less distortion than the LSB algorithm, using 3D chaotic maps, notably 3D Chebyshev and 3D logistic maps, to achieve high security. Salih et al. have created an intriguing method for performing image steganography applying LSB and hidden map approaches [17]. The work in [17] revolves around the notion of picking a pixel at random from the host image and inserting it. The proposed method performs well in terms of key sensitivity, concealing capacity, MSE quality index, PSNR, and image quality, and it is tolerant to a variety of attacks, including statistical and differential attacks. Mathkour et al. have developed a spiral-based LSB substitution method for hiding messages in images [18]. This method relies on the LSB substitution technique, which alters the RGB color components. Each segment of the image is processed differently after being separated into distinct image fragments. It is crucial that the stego-image is immune to various attacks and that no one can access the secret message. However, earlier research has shown that steganography approaches do not properly conceal the hidden message in the stego-image [5, 19, 20]. A statistical attack is performed to see if the presence of a hidden code in the stego-image can be easily identified. Nie et al. have proposed a novel way for improving security by combining the LSB method with the Knight Tour algorithm [21]. The embedding phase for the cover image and the receiving phase for the stego-image are the two essential phases. The Knight Tour method creates the hidden bit stream sequence inside the image pixels. It is a self-created algorithm that is based on the mathematics issue of the Knight's Tour.

Mielikainen has suggested encoding the data bits in a binary function of two cover pixels for additional noise reduction [22]. A pair of pixels are used to do the embedding, a function of the first two pixel values carrying one bit of information and the LSB of the initial pixel carrying another. The work in [23] has employed multiple-base number technique to encode data bits. The sensitivity of human vision was taken into account during embedding. The numerical base that will be used for embedding is determined using the variance value of a block of pixels. The pixel value differencing (PVD) algorithm, based on the sensitivity of human eyesight, is proposed in [24]. It is based on incorporating extra data bits by taking into account the differences in the high variance regions of the image, such as close to the edges, between two neighboring pixels. With the PVD method, the entire image is divided into smooth and edge areas. The difference d between two pixels is calculated. A range table has been allocated to the variable d. To embed the secret data, the original difference value, d, is converted to a new value, d'. The breadth of the range table determines the maximum number of bytes that can be embedded. In comparison to LSB approaches, this method performs better in terms of imperceptibility and capacity. This approach involves adding extra data bits by considering the variances between two neighboring pixels in areas of the image with high variation, such as those close to the edges. Following the same line of reasoning, Crandall has developed the usage of the matrix encoding error control coding method [25]. Frequency domain techniques are another name for transform domain techniques. According to Goel, in order to insert a secret message, transform domain techniques must first convert an image from the spatial domain to the frequency domain. This approach seeks to encrypt message bits in the image's transform domain coefficients [14].

Embedding data in the transform domain is commonly used for reliable watermarking. Similarly, large capacity embedding for steganography can be achieved using similar approaches.

The potential transformations for this technique consist of the discrete cosine transform (DCT) and the discrete wavelet transform (DWT) [26]. The data is distributed throughout the image, which results in a denser packing and improved resistance to signal processing. For example, depending on the content and robustness requirements, to create a new component, a block discrete cosine transform (BDCT) can be performed by selecting one or more components from each block. This data group then goes through a second layer modification and is pseudo-randomly scrambled. Several techniques are employed to alter the coefficients in the double transform domain. These methods typically have a high level of complexity for embedding and extraction. The DCT coefficients of an image are used by F5 to embed data bits. The quantized coefficients are rounded to the nearest data bit before encoding the data in the DCT coefficients by F5. Matrix encoding is used to reduce the noise that is embedded in the signal. F5 is one of the most extensively used embedding strategies in the DCT industry; however, it has been successfully decrypted [13]. The generation of the transform coefficients on blocks with a size of 8 × 8 is done. It is feasible to create methods that apply transforms to the entire image [27]. Block-based JPEG domain and wavelet-based embedding techniques are proposed in [28, 29].

23.3 The Proposed LSB-Based Steganography Strategy

LSB embedding is the standard algorithm for image steganography. This algorithm substitutes the LSBs of the pixels to hide the information. Before encoding, the message is converted to decimal values and then into its binary. The selected LSB method is proposed in this work to enhance the efficiency of the LSB method of concealing information in only one channel among the three channels. In LSB steganography, all the insignificant bits in each pixel are replaced with the bits of the secret message. The message bits are directly embedded in the insignificant bit plane of the cover image in predictable sequence, in the simplest steganographic algorithms. Consider a digital image as a matrix of pixels. The two most widely used modes, RGB 3 × 8-bit true-color pixels and RGB 4 × 8-bit true-color pixels with transparency mask are examined. These bits have values in the range of 0–255.

23.3.1 Encoding the Secret Message Using LSB-Based Steganography

A cover image is chosen and read using OpenCV to perform encoding. It is necessary to have a copy of the actual image so as to gain access to it later, if required. After receiving the secret message as input from the user, each character present in the message is converted into ASCII and its equivalent binary value, as shown in Figure 23.3. An array is created to store all the pixels generated using a random generator for every character. This array acts as the secret key. Characters are iterated through the binary values and are encoded in pixels.

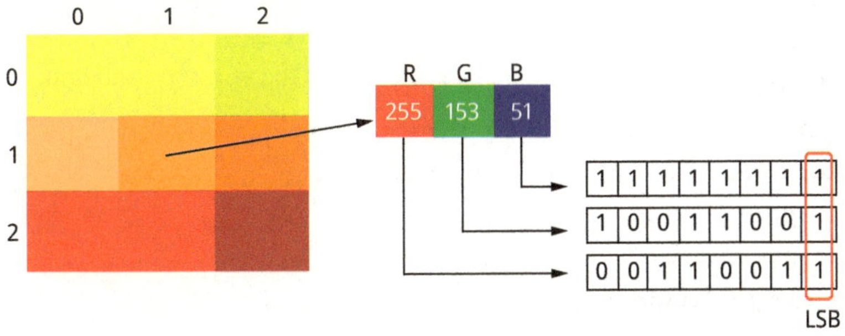

Figure 23.3: Representation of a 3 × 3 image pixel in 2D.

Figure 23.3 illustrates a 3 × 3 image as a 2D RGB pixel array. Pixel (1, 1) values for the RGB channel are 255, 153, and 51, respectively. A RGB channel of a pixel is selected using the following three methods. In the LSBmax method, a channel of a specific pixel with the highest value can be selected. As illustrated in Figure 23.3, the red channel has the highest value out of the three channels. In this instance, the red channel is chosen to carry out further processing. Whereas in the LSBmed method, a channel of a specific pixel that contains the median value is chosen. As shown in Figure 23.3, the green channel holds the median value and is chosen for further processing. In the LSBminmethod, a channel in a specific pixel that contains the minimum value is chosen. In Figure 23.3, the blue channel has the smallest value out of the three in and is chosen for further processing. The channel value is changed from decimal to an 8-bit binary. The LSB of the 8-bit binary value is replaced with the message bits. Each character in the message is stored as an 8-bit binary information at 8 random pixels, i.e., it is necessary to choose one pixel to store each bit of all the characters, as illustrated in Figure 23.4. As the amplitude of shift is small, altering the LSB produces no visible effect in the image. The steps involved in encoding the secret message are outlined in algorithm 1.

Algorithm 1 Encoding of Secret Message

```
Input: cover image: img={i1,i2,|..ip}, Secret
message: msg
Output: stego-image S
begin:
Select picture: ip
ascii_msg: = ascii(msg)
binary_msg:= binary(ascii_message)
for i←0 to num do
    location ← get_random(h,w)
    Select channel value in
pixel(min,max,median)
    pixel ← generate_pixel(location)
    binary_pixel ←  convert_to_binary(pixel)
    pm ← Replace lsb of pixel with msg bits
    pixel. update(chosen_pixel, pm)
end for
return the stego_image S
end
```

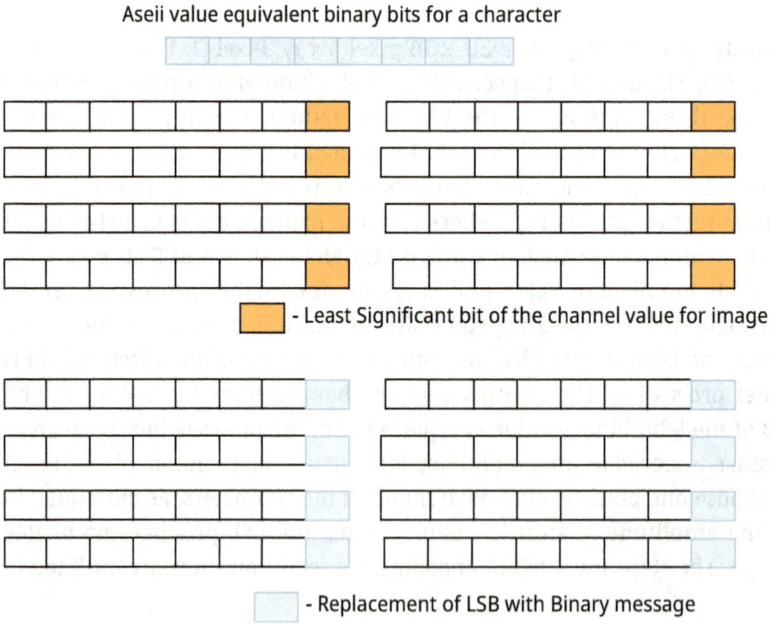

Figure 23.4: Creation of a stego-image using LSB substitution.

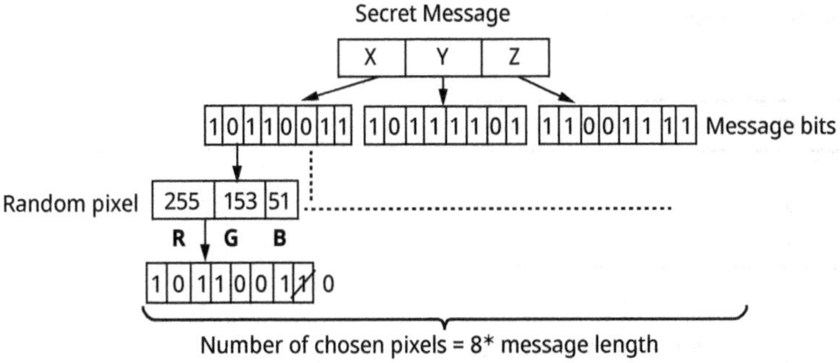

Figure 23.5: Image encoding using LSB.

There are few cases while updating the pixel, that is to handle the frequency of the values in the pixel. If all the three RGB channels have the same value, the red channel is updated. If there are two channels having equal values, the channel is updated based on Table 23.1. It shows that the green channel is selected when one of the two channels having equal values is green; otherwise, the blue channel is selected. Eventually, a stego-image with the hidden message contained in it is created for display or communication, as illustrated in Figure 23.5. The peak-signal-to-noise ratio (PSNR) is calculated between the images before and after encoding, i.e., stego-image to determine the degree of resemblance between the two digital images.

23.3.2 Decoding of Secret Message Using the LSB-Based Steganography

Decoding of stego-image is done with the help of the generated key that is specified using the height and width of the randomly selected pixels. The steps involved in encoding are done in the reverse order to decode a stego-image. This is known as the extraction algorithm. The final bits of the chosen pixels are gathered and then separated into groups of eight and converted back to ASCII, as illustrated in Figure 23.6. The decoding algorithm to retrieve the secret message is presented in Algorithm 2. To perform decoding, both the original and encoded images are required. One empty array is created to hold all the hidden characters and another empty array is created to hold the binary bits of each character. Later, all the pixel locations are accessed and the logic to get the LSB of the selected channel is reversed. Then, the bits are combined to convert them into the ASCII code and its corresponding character is appended to the array. Hence, the hidden message embedded in the pixels is decoded. The detection of hidden information in a stego-object is referred to as steganalysis.

Attacks such as mean filter attack, median filter attack, and salt-and-pepper attack, are simulated in order to analyze the performance of the proposed LSB-based

Table 23.1: Instances and the corresponding selected channel.

Red	Green	Blue	Selected channel
✓	✓	–	Green
–	✓	✓	Green
✓	–	✓	Blue

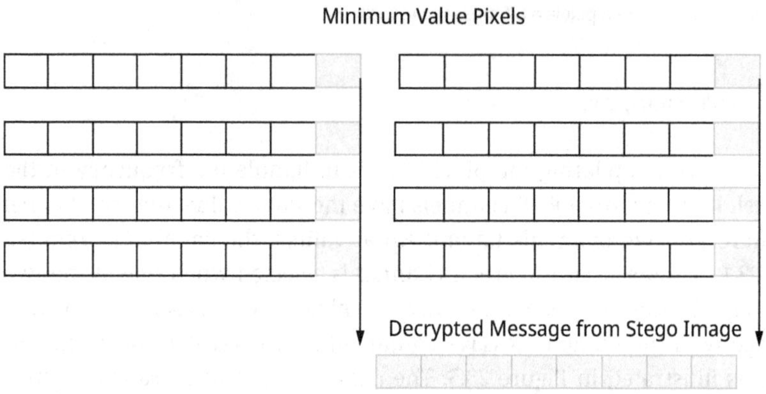

Figure 23.6: Retrieving the message from a stego-image.

strategies, by simulating. The mean filter returns the average of all pixel values in the specified window. The window is moved through the image, changing the value of each pixel. The pixel value is saved as the average of pixels divided by the height and breadth of the window. The median filter returns the median of the nearby gray levels of pixels. Salt-and-pepper noise is visible on digital images due to edged and rapid disturbances in the signal. It is sporadically distributed as black-and-white pixels. It is observed that the impacts of the above mentioned attacks are comparable with the proposed methodology.

23.4 Experimental Setup and Results

The OpenCV module is used to read the images that reside in the directories. The NumPy library is used to access arrays and matrices. In the image, the bits are encoded at random pixels. These locations are generated using the random library. After embedding the message, assess the difficulty of decoding the original secret message after executing numerous attacks on the image. The effectiveness of the three proposed algorithms that are based on the LSBmin, LSBmax, and LSBmed values of an RGB channel are presented in Tables 23.2–23.4.

Table 23.2: PSNR value measured after mean filter attack.

Selected channel	Window size	Message length	Hamming distance	PSNR value
Maximum	4 × 4	10	1.0	96.98
	6 × 6	8	1.0	98.33
	9 × 9	4	1.0	96.29
	4 × 4	16	1.0	90.09
	6 × 6	28	1.0	86.75
	9 × 9	32	1.0	86.29
Median	4 × 4	10	1.0	91.90
	6 × 6	8	1.0	91.64
	9 × 9	4	1.0	96.29
	4 × 4	16	1.0	83.50
	6 × 6	28	1.0	81.19
	9 × 9	32	1.0	80.42
Minimum	4 × 4	10	1.0	91.52
	6 × 6	8	1.0	93.28
	9 × 9	4	1.0	94.53
	4 × 4	16	1.0	95.11
	6 × 6	24	1.0	94.13
	9 × 9	32	1.0	92.53

Table 23.2 presents the measured PSNR values for varying sizes of images for mean filter attack. Mean filter attack blurs the image and reduces its overall sharpness, making it more difficult to discern details and identify objects within the image. In some cases, the attack can also introduce distortions into the image, particularly if the window size is too large or if the image contains areas of high contrast or sharp edges. After examining the effect of mean filter attack for the three algorithms with a common window size and message length, as shown in Table 23.2, it is inferred that the PSNR value is above 90. The Hamming distance is always found to be 1.0 in all three cases and the bit rate is the same as the message length. This means that if this attack is conducted on the image, the message cannot be recovered. Therefore, none of the three proposed methods can withstand mean filter attacks.

Table 23.3: PSNR and Hamming distance measured using modified bits after median blur.

Selected channel	Window size	Message length	Hamming distance	PSNR value
Maximum	4 × 4	10	1.0	96.98
	6 × 6	8	1.0	98.33
	9 × 9	4	1.0	96.29
	4 × 4	16	1.0	89.68
	6 × 6	28	1.0	86.87
	9 × 9	32	1.0	80.27

Table 23.3 (continued)

Selected channel	Window size	Message length	Hamming distance	PSNR value
Median	4 × 4	10	1.0	91.90
	6 × 6	8	1.0	91.64
	9 × 9	4	1.0	96.29
	4 × 4	16	0.98	84.4
	6 × 6	28	1.0	80.57
	9 × 9	32	0.99	80.2
Minimum	4 × 4	10	1.0	91.37
	6 × 6	8	0.98	80.60
	9 × 9	4	0.99	82.25
	4 × 4	16	1.0	89.84
	6 × 6	28	1.0	93.37
	9 × 9	32	1.0	92.09

Table 23.2 represents the hamming distance measured using the modified bits. The median blur attack changes an image by smoothing out details through the application of a median filter with a relatively large window size. This has the effect of reducing the contrast between adjacent pixels and thereby making the detection of changes in the image more complex. The algorithm does not tolerate the median blur attack either. After conducting the attack for all window sizes and message lengths as shown in Table 23.3, the Hamming distance is found to be roughly 1.0 but has slight variations in the case of minimum algorithm, which is illustrated in Figure 23.7. The decoding complexity is quite high for this attack. Table 23.4 represents the Hamming distance measured using modified bits after salt-and-pepper attack. It is observed that salt-and-pepper noise occurs when a small number of pixels are noisy and extremely loud. This creates the appearance of scattering salt-and-pepper, i.e., black and white dots all over the image. The two most basic, intuitive, and straightforward techniques for decreasing the intensity variation between adjacent pixels in a picture are the mean filter and median filter. It is frequently employed to reduce the image noise. Only the salt-and-pepper attack is tolerated by the proposed LSB-based steganography strategies. Table 23.4 indicates that it generally recovers 95%-100% of the message at a Hamming distance of between 0.0 and 0.3. The decoding complexity after performing this attack is determined to be very low.

The attack analysis using Hamming distance on various test cases is illustrated in Figures 23.7(a–c) for the LSBmin, LSBmed and LSBmax algorithms. The Hamming distance of the mean and median filter is almost 1.0, which states that the message cannot be recovered after performing these attacks. Even though the PSNR value of the mean filter attack is more than 90, the message cannot be retrieved because the Hamming distance is not 1.0. Salt and pepper is the only attack having PSNR value of more than 90 and the Hamming distance is almost 1.0 for the maximum, minimum, and median algorithms. The LSB substitution technique is applied backwards for getting the binary representation of the message. The binary representation is then converted to ASCII, and its corresponding character. This process leads to the retrieval of the secret message. Analyze the decoding complexity with various attacks.

Table 23.4: PSNR and Hamming distance measured using modified bits after salt-and-pepper attack.

Selected channel	Message length	Hamming distance	Bit error	PSNR value
Maximum	10	0.28	2	91.52
	8	0.0	0	93.10
	4	0.25	1	96.29
	16	0.10	3	87.35
	28	0.12	2	82.80
	32	0.21	7	87.00
Median	10	0.0	0	91.40
	8	0.0	0	92.93
	4	0.0	0	94.53
	16	0.125	2	83.90
	28	0.0	0	81.33
	32	0.05	2	80.27
Minimum	10	0.1	1	91.76
	8	0.25	2	92.77
	4	0.0	0	93.56
	16	0.06	1	95.11
	28	0.07	1	93.81
	32	0.03	1	92.50

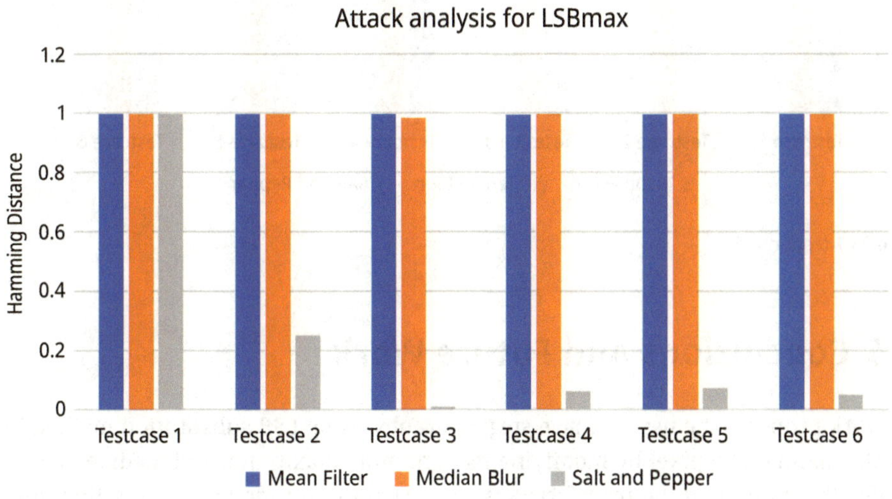

Figure 23.7: Attack analysis for (a) LSBmin, (b) LSBmed, and (c) LSBmax.

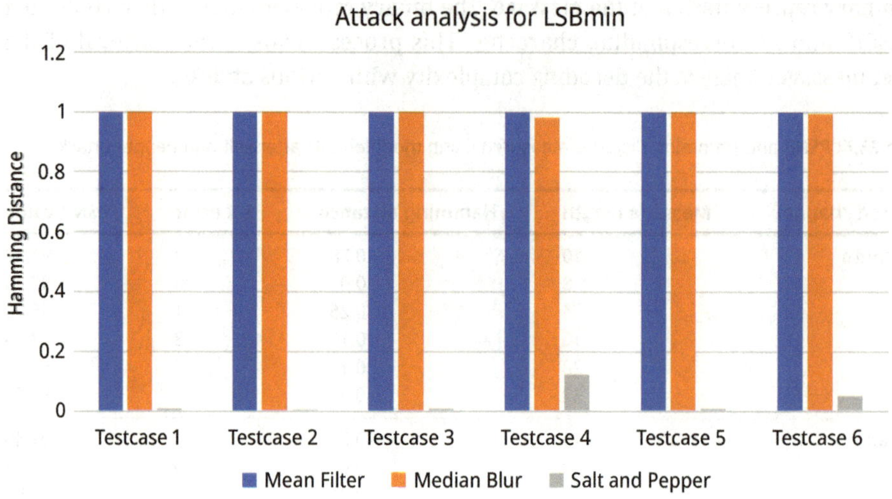

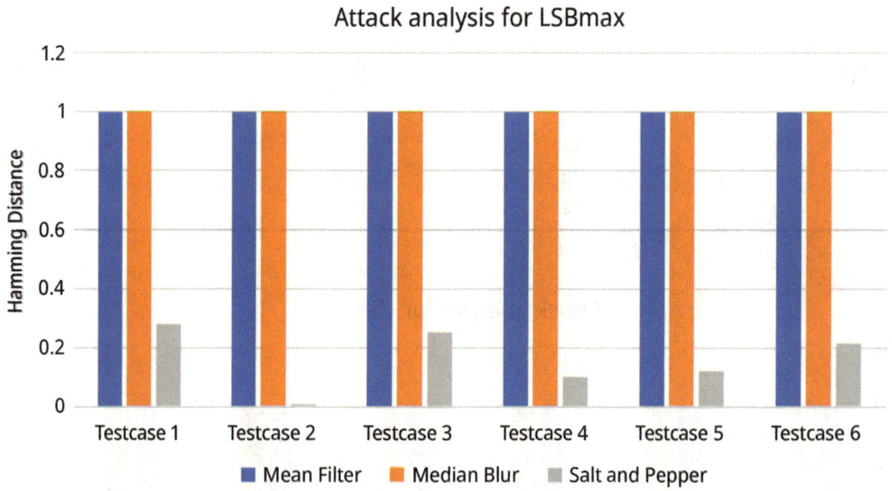

Figure 23.7 (continued)

23.5 Conclusions and Future Work

This work proposes the use of image steganography using LSB substitution method in the RGB channel of a pixel by modifying its minimum, maximum, and median values. Initially, the message to be transferred is converted to unicode format and then into its equivalent in binary representation. Subsequently, this message is encoded and the similarity between the original image and the encoded image is computed using

PSNR. Attacks are simulated to ascertain which attack is resilient and capable of recovering the secret message after the attack is being performed. It was found that after the salt-and-pepper attack, secret messages are extracted, whereas, the retrieval of messages is not possible in median blur and mean filter attacks. The peaks in the salt-and-pepper attack are because of the varying input secret messages, due to which the Hamming distance also varies. Several other possibilities can be investigated, such as the steganalysis for the proposed method using other attacks such as upsampling, downsampling, cropping, etc. Furthermore, the payload can be increased by combining more LSB for encoding images and texts, together in a cover, using the proposed approach. Increasing the payload will provide better protection and resilience to various threats. Experimental findings show that the proposed variants of LSB tolerate salt-and-pepper attacks and are superior to other algorithms, as indicated by the values measured for PSNR and Hamming distance. The proposed approach is also very sensitive to its secret keys. Examining audio and video steganography using the proposed method can be further explored.

References

[1] Silman J. Steganography and steganalysis: An overview. Sans Institute. 2001; 3(2): 61–76.
[2] Moulin P, Koetter R. Data-hiding codes. Proceedings of the IEEE. 2005; 93(12): 2083–2126.
[3] Thomas TL. Al qaeda and the internet: The danger of cyber planning. Foreign Military Studies Office (ARMY) Fort Leavenworth Ks. 2003; 33(1).
[4] Dunbar B. A detailed look at steganographic techniques and their use in an open-systems environment. Sans Institute. 2002.
[5] Provos N, Honeyman P. Hide and seek: An introduction to steganography. IEEE Security & Privacy. 2003; 1(3): 32–44.
[6] Johnson NF, Jajodia S. Exploring steganography: Seeing the unseen. Computer. 1998; 31(2): 26–34.
[7] Conway M. Code wars: Steganography, signals intelligence, and terrorism. Knowledge, Technology, and Policy. 2003; 16(2): 45–62.
[8] Kong X, Wang Z, You X. Steganalysis of Palette Images: Attack Optimal Parity Assignment Algorithm. In IEEE 25th International Conference on Information Communications & SignalProcessing 2005 (pp. 860–864).
[9] Battisti F, Carli M, Neri A, Egiaziarian K. A Generalized Fibonacci LSB Data Hiding Technique. In 3rd International Conference on Computers and Devices for Communication, Institute of Radio Physics and Electronics, University of Calcutta 2006 (Vol. 6).
[10] Chengqing L, Zhang Y, Xie EY. When an attacker meets a cipher-image in 2018: A year in review. Journal of Information Security and Applications. 2019; 48.
[11] Dey S, Mondal K, Nath J, Nath A. Advanced steganography algorithm using randomized intermediate QR host embedded with any encrypted secret message: Asa QR algorithm. International Journal of Modern Education and Computer Science. 2012; 4(6): 59.
[12] Bailey K, Curran K. An evaluation of image based steganography methods. Multimedia Tools and Applications. 2006; 30: 55–88.

[13] Fridrich J, Goljan M, Rui D. Detecting LSB steganography in color, and gray-scale images. IEEE Multimedia. 2001; 8(1): 22–28.

[14] Goel S, Gupta S, Kaushik N. Image Steganography–Least Significant Bit with Multiple Progressions. In The 3rd International Conference on Frontiers of Intelligent Computing: Theory and Applications 2014 2015 (Vol. 2, pp. 105–112). Springer.

[15] Odat AM, Otair MA. Image steganography using modified least significant bit. Indian Journal of Science and Technology. 2016; 9: 1–5.

[16] Dhawan S, Gupta R. Analysis of various data security techniques of steganography: A survey. Information Security Journal: A Global Perspective. 2021; 30: 63–87.

[17] ALabaichi A, Al-Dabbas MAAK, Salih A. Image steganography using least significant bit and secret map techniques. International Journal of Electrical & Computer Engineering. 2020; 10: 2088–8708.

[18] Chandramouli R, Memon N. Analysis of LSB Based Image Steganography Techniques. In IEEE, International Conference on Image Processing (Cat. No. 01CH37205) 2001 (Vol. 3, pp. 1019–1022).

[19] Bashardoost M, Sulong GB, Gerami P. Enhanced LSB image steganography method by using knight tour algorithm, vigenere encryption and LZW compression. International Journal of Computer Science Issues. 2013; 10(2): 221.

[20] Shirali-Shahreza MH, Mohammad SS. A New Synonym Text Steganography. In IEEE International Conference on Intelligent Information Hiding and Multimedia Signal Processing 2008 (pp. 1524–1526).

[21] Nie SA, Sulong G, Ali R, Abel A. The use of least significant bit and knight tour algorithm for image steganography of cover image. International Journal of Electrical & Computer Engineering. 2019; 9: 5218–5226.

[22] Mielikainen J. LSB matching revisited. IEEE Signal Processing Letters. 2006; 13: 285–287.

[23] Zhang X, Wang S. Steganography using multiple-base notational systems and human vision sensitivity. IEEE Signal Processing Letters. 2004; 12: 67–70.

[24] Da-Chun W, Tsai W-H. A steganographic method for images by pixel-value differencing. Pattern Recognition Letters. 2003; 24: 1613–1626.

[25] Crandall R. Some notes on steganography. Posted on Steganography Mailing List. 1998; 1: 6.

[26] Lyu S, Farid H. Detecting Hidden Messages Using Higher-Order Statistics and Support Vector Machines. In Information Hiding: 5th International Workshop on Information Hiding 2002 Noordwijkerhout 2003 (Vol. 5, pp. 340–354). Springer.

[27] Cox IJ, Kilian J, Leighton FT, Shamoon T. A secure, robust watermark for multimedia. Information Hiding. 1996; 1174: 185–206.

[28] Koch E, Zhao J. Towards Robust and Hidden Image Copyright Labeling. In IEEE Workshop on Nonlinear Signal and Image Processing 1995 (Vol. 1, pp. 123–132).

[29] Xia X-G, Boncelet CG, Arce GR. A multiresolution watermark for digital images. IEEE International Conference on Image Processing. 1997; 1: 548–551.

[30] Riyazoddin SM. A spatial domain image steganography technique based on plane bit substitution method. Global Journal of Computer Science and Technology. 2012; 12: 1–8.

[31] Goljan M, Fridrich J, Holotyak T. New Blind Steganalysis and Its Implications. In Security, Steganography, and Watermarking of Multimedia Contents VIII 2006 (Vol. 6072, pp. 607201). SPIE.

[32] Goel P. Data Hiding in Digital Images: A Steganographic Paradigm. Master's Thesis in Computer Science and Engineering, Indian Institute of Technology Kharagpur. 2008.

[33] Dumitrescu S, Xiaolin W, Memon N. On steganalysis of random LSB embedding in continuous-tone images. IEEE International Conference on Image Processing. 2002; 3: 641–644.

[34] Singh KJ, Manimegalai R. Evolution of encryption techniques and data security mechanisms. World Applied Sciences Journal. 2015; 33(10): 1597–1613. IF:0.47, HI:20 ISSN: 1818-4952, URL: http://www.idosi.org/wasj/wasj33(10)15/6.pdf.
[35] Singh KJ, Manimegalai R. A novel approach to incorporate efficient security in content distribution. Journal of Engineering and Applied Science. July 2014; 9(5): 190–197.
[36] Karthick CKU, Manimegalai R. A novel threat modeling and attack analysis for IoT applications. Security of Internet of Things Nodes-Edited Book, ISBN: 9781003127598. August 2021; 263–279. https://doi.org/10.1201/9781003127598.

Sankarsan Panda, Avanindra Vasuki Vyas, Desh Bandhu Bhatt, Shweta Solanki

24 Real-World Intelligence: Applications of Fuzzy Logic in Modern AI Systems

Abstract: In today's data-driven world, artificial intelligence (AI) applications need to be able to deal with ambiguity, handle uncertainty, and think like a person. These are problems that traditional binary logic cannot solve. Fuzzy logic is an important tool for AI-based applications because it gives us a strong framework for making soft decisions and reasoning that is not exact. The chapter goes into detail about the main parts of fuzzy logic systems, like fuzzy sets, membership functions, and fuzzy inference mechanisms. It also shows how these parts can be easily used in many areas of AI, such as machine learning, deep learning, natural language processing, robotics, computer vision, and expert systems. These integrations are not just ideas; they are backed up by real-world case studies and deployment scenarios, such as predictive maintenance, self-driving cars, smart healthcare diagnostics, and conversational AI. The chapter talks about both basic and advanced topics, such as explainable AI using fuzzy models, fuzzy-based ontologies for semantic web systems, adaptive fuzzy control in smart grids, and Internet of Things-enabled decision-making frameworks. These examples from the real world show how fuzzy logic can help make AI systems that are easier to understand, more flexible, and more in line with how people think. The chapter ends by talking about problems with implementation, like how to make systems bigger, how to make them work better, how to make them work with other systems, and the fact that there are no standard frameworks. It also talks about what researchers should look into in the future and the trends in the business world for people who want to make AI applications that are strong, easy to understand, and aware of their surroundings. This makes the chapter an essential resource for students, developers, engineers, and researchers who want to connect theory and practice in the field of intelligent systems.

Keywords: Fuzzy logic systems, fuzzy-AI integration, neuro-fuzzy models, fuzzy NLP and vision, explainable AI (XAI)

24.1 Introduction to Fuzzy Logic

24.1.1 Origin and Evolution

The concept of fuzzy logic was first introduced in 1965 by Iranian-American scientist Lotfi A. Zadeh. His paper "Fuzzy sets" gave birth to a new logical system that could

https://doi.org/10.1515/9783112214374-024

work with uncertain and vague real-world data. Traditional binary logic only understands “true” or “false” values, while fuzzy logic was designed for cases where information is not crisp or perfectly defined. Zadeh observed that real-life problems such as weather prediction, human reasoning, and linguistic statements (such as “a little hot,” “too much,” “less”) could not be accurately described by binary logic. For this reason he introduced the concepts of fuzzy sets and fuzzy logic, where the degree of truth ranges from 0 to 1 – such as 0.7% of a statement being true, i.e., 70% true. The evolution of fuzzy logic has been gradual. It was first understood at the theoretical level, but during the 1980s and 1990s this system began to be used in Japan and Europe in commercial products such as washing machines, air conditioners, and subway train control systems. Today fuzzy logic is actively used in multiple domains such as artificial intelligence (AI), robotics, medical diagnosis, and control systems [1].

24.1.2 Comparison with Classical Logic

The basic principle of classical logic or Boolean logic is the “law of the excluded middle,” in which every proposition can take only two values: true (1) or false (0). For example:

“Sky is blue” = True
“Sky is green” = False

No ambiguity is allowed in this model. But in the real world such statements where the information is incomplete, imprecise, or vague are common. For example:

“Today it is a little hot”: This statement is difficult to measure in binary terms.

Fuzzy logic addresses this limitation by allowing degrees of truth. A fuzzy statement like:

The truth value (Table 24.1) of “room temperature is warm” could be 0.6 or 60% – which means it is a little warm but not completely warm [2].

Table 24.1: Major differences.

Feature	Classical logic	Fuzzy logic
Truth values	0 or 1 (true/false)	Range from 0 to 1
Handling vagueness	Not possible	Easily handled
Mathematical model	Crisp sets	Fuzzy sets
Use cases	Digital circuits and clear rules	AI, NLP, and uncertain environments

A real-life example of this is the fuzzy controller of a washing machine. If the clothes are "too dirty," then fuzzy logic uses "degree of dirtiness" to adjust wash time. For classical logic, either "clean" or "dirty" will be defined; there is no value in between.

24.1.3 Importance in Soft Computing

Soft computing is a multidisciplinary approach in which intelligent systems can be built even with uncertain, noisy, and incomplete information. The major components of soft computing are:
- Fuzzy logic
- Neural networks
- Genetic algorithms
- Probabilistic reasoning

Fuzzy logic plays a key role in this because it emulates human-like reasoning [3]. When we think, we do not think in binary terms. We often make vague or approximate decisions such as:

"I think this works well."
"This result is a little better."

Fuzzy logic gives AI systems the ability to work with nonlinear and ambiguous data, and still make robust decisions.

Applications in AI:
- **Expert Systems:** Doctors' diagnosis process can be simulated through fuzzy logic. For example, "mild fever," "moderate pain," etc. can be handled easily.
- **Natural Language Processing (NLP):** Human language is inherently fuzzy. In NLP, fuzzy logic helps model linguistic uncertainty.
- **Robotics and Control Systems:** Fuzzy controllers are adaptive and flexible, so they are used in robotics for navigation, pathfinding, and motor control.
- **Decision Support Systems:** Fuzzy-based AI models produce a weighted decision based on multiple fuzzy inputs – useful in finance, healthcare, supply chain, etc.

24.2 Mathematical Foundations of Fuzzy Sets)

The strength of fuzzy logic lies in its mathematical foundation. The core concept of this foundation is "fuzzy set," which is a generalized version of traditional crisp set theory. While in classical set theory the belonging of an element to a set is binary (0 or 1), in fuzzy set theory the degree of belonging of an element is defined – which can be between 0 and 1 [4].

24.2.1 Fuzzy Sets and Membership Functions

What is a fuzzy set?
Fuzzy set is a mathematical model in which the belongingness of an object to a set is defined through a membership function, which assigns values between 0 and 1.

Example: If we define a fuzzy set "Tall people," then a person with height 5'6" can have membership of 0.4 in the "Tall" set, and if he is 6'2" then his membership can be 0.9. This membership function $\mu(x)$ defines how much the element named x belongs to in the fuzzy set [5].

Types of Membership Function:

Triangular Function: It has a peak point and slope on both sides
Trapezoidal Function: It has two constant regions and slope on their sides
Gaussian Function: Bell-shaped curve, for a smoother transition
Sigmoidal Function: It represents logistic shape, gradual growth or decline

Formal Definition:
A fuzzy set A in universal set X can be defined as:

$$A = \{(x, \mu A(x)) | x \in X\}$$

where $\mu A(x) \in [0, 1]$ – this means how much membership x has in fuzzy set A.

24.2.2 Operations on Fuzzy Sets

Some fundamental operations can be applied on fuzzy sets as well, just like classical sets – but the logic in them is soft, not strict.

1. Union ($A \cup B$):
   Fuzzy union means taking the maximum of membership of both sets:
   $\mu A \cup B(x) = \max(\mu A(x), \mu B(x))$
   Example:

$$\mu A(x) = 0.3, \ \mu B(x) = 0.6$$
$$=> \ \mu A \cup B(x) = \max(0.3, \ 0.6) = 0.6$$

2. Intersection ($A \cap B$):
   The minimum membership in fuzzy intersection is:
   $\mu A \cap B(x) = \min(\mu A(x), \mu B(x))$
   Example:

$$\mu A(x) = 0.3, \ \mu B(x) = 0.6$$
$$=> \ \mu A \cap B(x) = \min(0.3, \ 0.6) = 0.3$$

3. Complement (A'):
   Fuzzy complement gives the opposite of a set:

$$\mu A'(x) = 1 - \mu A(x)$$

   Example:

$$\mu A(x) = 0.7$$

$$=> \ \mu A'(x) = 1 - 0.7 = 0.3$$

4. Other Operators:

$$\text{Algebraic Sum: } \mu A + B(x) = \mu A(x) + \mu B(x) - \mu A(x) \ast \mu B(x)$$

$$\text{Algebraic Product: } \mu A \cdot B(x) = \mu A(x) \ast \mu B(x)$$

These operations are very important for fuzzy reasoning, on the basis of which fuzzy inference systems work.

24.2.3 Fuzzy Relations and Their Properties

What is a fuzzy relation?
When we define a connection or association between two fuzzy sets, it is called a fuzzy relation. This relation is usually defined in the form of a Cartesian product. Suppose there are two fuzzy sets A and B, the fuzzy relation R between them will be:

$$R \subseteq A \times B$$

And for this relation the membership function is defined:

$$\mu R(x, y): X \times Y \rightarrow [0, 1]$$

It means the strength of the relation between x (from A) and y (from B) can be any value between 0 and 1.

Example:
Set A: Temperature = {Cool, Warm, Hot}
Set B: Fan speed = {Low, Medium, High}

A fuzzy relation can be defined: “The higher the temperature, the higher the fan speed.”

Table 24.2: Fuzzy membership values.

Temperature	Low	Medium	High
Cool	0.8	0.3	0.0
Warm	0.3	0.6	0.5
Hot	0.0	0.4	0.9

Each value in Table 24.2 represents a fuzzy membership value, which describes the relationship between temperature and fan speed.

Properties of Fuzzy Relations:

- Reflexivity: $\mu R(x, x) = 1$ for all x
- Symmetry: $\mu R(x, y) = \mu R(y, x)$
- Transitivity: If $\mu R(x, y)$ and $\mu R(y, z)$ are high then $\mu R(x, z)$ should also be high.

These properties are crucial for reasoning and decision-making in AI and control systems.

Real-Life Relevance:

Medical Diagnosis: AI models can be designed by creating fuzzy relations between symptoms and diseases.

Image Processing: The similarity between pixels is modeled by fuzzy relation for segmentation.

Recommendation Systems: Fuzzy relation is applied between user behavior and product similarity.

24.2.4 Fuzzy Logic in Machine Learning

The combination of fuzzy logic and ML makes today's AI models even more intelligent, adaptive, and human-like. When we work with real-world data, the data is inherently noisy, incomplete, and uncertain. Traditional ML models work with sharp boundaries and precise rules, while fuzzy logic naturally handles this uncertainty. Fuzzy logic improves interpretability, flexibility, and robustness in ML models – making predictions and decision-making more accurate [6].

24.3 Fuzzy Decision Trees

Decision tree is a common supervised learning technique in which the final decision is made by dividing the data based on multiple features. But traditional decision trees work on crisp values: for example, if temperature >30 then "Yes," otherwise "No."

In fuzzy decision trees, each node defines a fuzzy rule, which has partial membership values for the input variables. This gives the possibility for each data point to belong to multiple branches, with certain degrees [7].

Example: Feature: Temperature
Rule in crisp tree:
- IF temperature > 30 THEN High
- ELSE Low

Rule in fuzzy tree:
- IF temperature is Moderately High THEN Class = X with $\mu = 0.7$
- ELSE Class = Y with $\mu = 0.3$

The advantage of this is that this tree does better classification even with ambiguous or overlapping features.

Advantages:
- Better handling of noisy data
- High interpretability with fuzzy rules
- Improved accuracy with uncertain inputs

Fuzzy decision trees have been successfully applied in areas such as finance, weather prediction, and medical diagnosis.

24.3.1 Fuzzy Clustering (Fuzzy *C*-Means) (Fuzzy Clustering: FCM Algorithm)

Clustering is an unsupervised learning technique in which similar data points are divided into groups or clusters. Traditional algorithms like *K*-means assign each data point to only one cluster. But fuzzy *C*-means (FCM) algorithm assigns each data point into multiple clusters with different membership degrees [8].

How does FCM work?
- For each data point, FCM calculates the membership score μij, which means the degree to which point i belongs to cluster j.
- Clustering is optimized by minimizing the objective function.

Example: If a data point belongs to one cluster by 0.8 degrees and to another by 0.2 degrees, this represents uncertainty – common in real-world problems.

Advantages:
- Handles overlapping clusters
- Better suited for soft boundaries
- Robust to noise and outliers

Applications:
- Image segmentation (pixels in multiple regions)
- Customer segmentation (based on fuzzy behavior)
- Document classification (based on topic relevance)

24.3.2 Neuro-Fuzzy Systems

Neuro-fuzzy systems are hybrid models that combine the learning power of artificial neural networks with the interpretability of fuzzy logic. A popular model is the adaptive neuro-fuzzy inference system (ANFIS), in which a neural network trains a fuzzy rule base using data-driven learning.

ANFIS Architecture:
- Input Layer: Features
- Fuzzification Layer: Converts inputs into fuzzy values
- Rule Layer: Applies fuzzy IF-THEN rules
- Output Layer: Defuzzified crisp result

Advantages:
- Learning ability like neural networks
- Transparent rules like fuzzy systems
- Real-time adaptive modeling

Applications:
- Forecasting (stock market and energy demand)
- Control systems (temperature, robotics)
- Medical prediction (disease likelihood modeling)

Neuro-fuzzy systems are used in scenarios where nonlinear relationships, uncertain inputs, and real-time adaptation are required.

24.3.3 Fuzzy Logic and Deep Learning

In today's era, deep learning (DL) has taken AI to the next level – be it image recognition, natural language processing, or autonomous driving. But DL models behave like a black box: it is not easy to understand or explain their decisions. This is where the role of fuzzy logic becomes very important. With the help of fuzzy logic, we can add properties like interpretability, flexibility, and uncertainty handling to DL models. This section explains the synergy of fuzzy and DL in three parts [9, 10].

24.4 Fuzzy Layers in Neural Networks

Traditional neural networks layers process crisp inputs through neurons. But when the input data is uncertain or vague, these models are unable to work robustly. Fuzzy logic bridges the gap by introducing fuzzy layers in neural networks.

What is a fuzzy layer?
Fuzzy layer is a processing unit in which input features are mapped into fuzzy sets using membership functions. After this, inference is done based on fuzzy rules, and the output is given in crisp form (defuzzification).

Fuzzy layer structure:
- Fuzzification: Inputs are converted into fuzzy sets – e.g., temperature → {low, medium, high}
- Rule Evaluation: IF-THEN type fuzzy rules are applied
- Aggregation and Defuzzification: Final fuzzy output is converted into crisp result.

Integration with Neural Nets:
Fuzzy layer can be inserted inside convolutional neural network (CNN), recurrent neural network (RNN), or deep neural network – mostly as preprocessing or intermediate layer. This layer also enhances interpretability and also gives better performance for uncertain environments.

24.4.1 Interpretability in Deep Models Using Fuzzy Concepts

DL models have one major drawback – their black box nature. It is difficult for a human to understand why a model gave a particular output. Especially in sensitive fields like healthcare, finance, or law, interpretability is very important.

How do fuzzy concepts help?

Fuzzy logic is inherently based on rule-based reasoning, which replicates human logic. When fuzzy concepts are integrated with DL, the model's decisions can be explained in linguistic form.

Example: Suppose a deep model predicts the probability of a patient having diabetes. Traditional DL would say: "Prediction: positive, Confidence: 0.87." But fuzzy-integrated DL might say: "IF glucose is high AND BMI is slightly elevated THEN diabetes risk is high with confidence 0.87."

This explanation is more understandable to the doctor or end-user and builds trust.

Methods to Improve Interpretability:

- Fuzzy Rule Extraction: Extracting fuzzy rules from the trained DL model
- Attention-Based Fuzzy Weights: Assigning fuzzy weights to important features
- Linguistic Summarization: Translating the output into natural language form using fuzzy concepts

---

24.4.2 Case Studies with Fuzzy CNN/RNN Architectures

A. Fuzzy-CNNs are commonly used in image-based applications. But when images are blurry, noisy, or boundaries are not clear – CNNs' accuracy can suffer. Fuzzy logic handles such situations better.

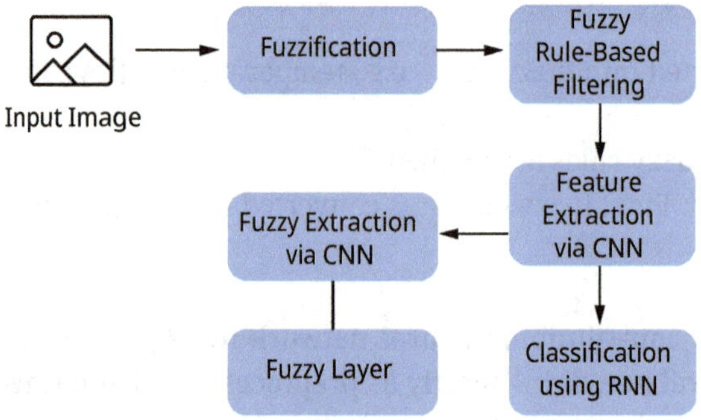

Figure 24.1: Case study for fuzzy CNN.

Case study (Figure 24.1): Medical imaging

- Problem: Brain tumor detection using MRI scans
- Fuzzy-CNN Model: Input image fuzzification → fuzzy rule-based filtering → feature extraction via CNN → classification
- Result: Improved detection accuracy in noisy MRI scans and better region segmentation

B. Fuzzy-RNN

RNNs are used for time series and sequential data such as speech recognition, weather forecasting, or stock price prediction. But if the input time series is uncertain or incomplete, RNNs may give errors.

Case Study: Weather Forecasting

- Input: Humidity, temperature, wind speed (with sensor noise).
- Fuzzy Layers: Handle noisy and imprecise values via fuzzy membership.
- RNN: Sequence modeling for future temperature prediction.
- Result: Fuzzy-RNN models produced more stable and explainable predictions during uncertain climatic data.

C. ANFIS + DL hybrid

ANFIS was integrated with CNN to create a hybrid model, which performs feature learning as well as fuzzy reasoning.

Case Study: Facial Emotion Recognition

- Input: Facial images
- CNN: Feature extraction
- ANFIS: Emotion classification based on fuzzy rules (e.g., anger, joy, and sadness)
- Benefit: Interpretability and soft decision-making

24.4.3 Fuzzy Logic in Expert Systems

Expert systems are AI-based software that simulate the knowledge and decision-making skills of human experts. These systems are especially useful in fields where complex decisions have to be made – such as medical diagnosis, legal interpretation, and engineering fault detection. But real-world knowledge is often uncertain, vague, and incomplete. This is where the role of fuzzy logic becomes important. Fuzzy-based expert systems can handle ambiguity naturally and can perform real-life human-like reasoning [11].

24.5 Knowledge Representation

The backbone of any expert system is the knowledge base, which is a structured form of domain-specific knowledge. In traditional expert systems, knowledge is represented in crisp form through "IF-THEN" rules. But if the knowledge is uncertain or in linguistic form – like "slightly high temperature," "moderate risk," or "strong evidence" – it becomes difficult to handle [12].

Fuzzy Knowledge Representation:
Fuzzy logic represents knowledge through fuzzy sets and linguistic variables:
- Variables: e.g., temperature
- Fuzzy Terms: {Low, Medium, High}
- Membership Functions: Numerical representation of linguistic terms

---

Example: IF temperature is “high” AND blood pressure is “slightly elevated”
THEN risk of heart disease is “moderate”

This fuzzy rule is not based on crisp values, but on soft ranges – for example, a 120-bpm heart rate might fall into the “high” category by 0.6 degrees. This allows fuzzy expert systems to work even in scenarios where knowledge is not fully defined.

24.5.1 Rule Generation and Uncertainty Handling

Fuzzy Rule Generation:
The main decision mechanism of expert systems is rule base – which consists of fuzzy IF-THEN rules.

The structure of the rule is:
IF Condition1 AND Condition2 THEN Outcome.

In fuzzy logic, these conditions are linguistic and inference is based on fuzzy logic operators (AND, OR, NOT).

Automatic Rule Generation:
Modern fuzzy expert systems can automatically generate fuzzy rules through ML or data mining algorithms, such as:
- Decision trees
- Clustering techniques (e.g., fuzzy *C*-means)
- Neuro-fuzzy approaches (ANFIS)

Uncertainty Handling:
Fuzzy systems handle uncertainty through degrees of truth. Each condition or input is assigned a truth value between 0 and 1, which helps the system process real-world ambiguity.

Example: If a patient’s blood sugar is 140 mg/dL, then according to fuzzy membership:

$$\mu(\text{Normal}) = 0.2$$

$$\mu(\text{Slightly High}) = 0.6$$

$$\mu(\text{High}) = 0.4$$

In decision-making, these overlapping truths make the expert system flexible and closer to human.

24.5.2 Fuzzy-Based Medical and Legal Expert Systems

A. Medical expert systems:

In the healthcare domain, symptoms are often vague or patient-reported, such as:

- "The fever is a little severe."
- "Sometimes the stomach hurts."

Traditional expert systems cannot process such inputs, but fuzzy logic can convert these vague terms into mathematical form and make inferences.

Use Cases:

- Disease diagnosis (diabetes, hypertension, etc.)
- Risk assessment (e.g., heart disease probability)
- Treatment recommendation systems

Example System:
"IF headache is moderate AND temperature is slightly high THEN possibility of viral fever is medium."

Fuzzy rules and membership values closely follow the doctor's reasoning pattern.

B. Legal Expert Systems:
The use of fuzzy logic is also very effective in the legal domain – because interpretation and subjective judgement are very important in law.

Use Cases:

- Case risk prediction
- Legal document analysis
- Contract evaluation
- Crime severity assessment

Example Rule:
IF evidence is "strong" AND witness is "partially credible"
THEN conviction likelihood is "high."

Fuzzy legal expert systems model ambiguous terms like "reasonable doubt," "significant negligence," or "minor offense" – which would not be possible with traditional logic.

24.5.3 Fuzzy Logic in Robotics and Automation

The main goal of robotics and automation is to equip machines and robots with intelligent behavior that can work in real-world environments – where uncertainty, noise, and unpredictability are common. Fuzzy logic gives robotics the ability to work efficiently in these uncertain and complex environments. Through fuzzy logic, robots can perform soft decisions, adaptive responses, and human-like reasoning that traditional crisp logic cannot achieve [13].

24.6 Motion Control and Navigation

The essential task for robots is to maintain smooth and accurate motion control and navigate a path according to the environment. Real-world environments are not ideal or predictable – where rough terrains, obstacles, speed variations, and turning angles are unpredictable. Traditional control systems may fail here because their logic is fixed and binary in nature. Fuzzy logic handles this problem with approximate reasoning.

Fuzzy Motion Control:

- Inputs: Distance from obstacle, current speed, turning angle
- Fuzzy Rules:
  IF "distance is close" AND "speed is high" THEN "reduce speed significantly"
  IF "angle is sharp" THEN "turn slowly"
- Output: Motor speed adjustment or path correction

Navigation Example:
This is how fuzzy navigation system works for autonomous robots:
- Sensors collect environment data (distance, angle, and position).
- Fuzzy controller converts inputs into fuzzy sets.
- Decision is taken using rule base.
- Robot's path has been updated.

This process allows the robot to navigate smoothly between narrow spaces, curved paths, and dynamic objects.

24.6.1 Fuzzy Controllers for Dynamic Environments

Conditions change continuously in dynamic environments – such as the simultaneous movement of multiple machines in a factory floor, or moving vehicles in a traffic system. Traditional PID (proportional-integral-derivative) controllers or linear systems work with fixed parameters. When real-time inputs become unpredictable, their performance degrades [14].

Fuzzy controllers can handle dynamicity effectively because:
- Input-output relationship is soft and rule-based.
- System is continuously adaptive.
- Crisp tuning is not required – the system reacts "smartly."

Examples: In industrial automation, robotic arms detect the weight or position of the object through fuzzy logic and adjust grip and movement accordingly. Agricultural robots use fuzzy rules when navigating between crops like "IF leaf density is medium THEN reduce speed."

Fuzzy Control Loop:
- Sensors detect real-time parameters.
- Fuzzy inference system analyzes.
- Actuators adjust position/speed/force based on fuzzy output.
- This loop repeats continuously, making the robot dynamically responsive.

24.6.2 Human-Robot Interaction Using Fuzzy Models

Human-robot interaction (HRI) requires robots to understand, interpret, and react to human instructions and gestures. Human behavior is naturally fuzzy – we give instructions with vagueness, such as:

"Go a little left."
"Walk a little slower."
"Don't apply too much force."

Such instructions are difficult to translate into crisp mathematical logic. Fuzzy logic models these human-like expressions through fuzzy sets, and allows the robot to take actions accordingly.

Fuzzy-Based HRI Use Cases:
- Assistive Robots: Elderly care robots interpret fuzzy commands like "I'm feeling cold" as temperature control.
- Gesture Recognition: Appropriate response is generated by interpreting the "angle," "speed," and "duration" of the hand gesture using fuzzy logic.
- Emotion-Based Response: Gives robot friendly or alert response by analyzing facial expressions and tone as fuzzy parameters.

Example Example Fuzzy Rule in HRI:
IF voice tone is soft AND eye contact is strong
THEN user is engaged → respond with “neutral tone.”

Fuzzy HRI systems enable natural, safe, and trustworthy interactions, especially in environments where human safety is a priority.

24.6.3 Use of Fuzzy Systems in Natural Language Processing (NLP)

Natural language processing (NLP) is a field of AI that enables understanding, processing, and generating human language. But human language is often ambiguous, vague, and context-dependent, which becomes a big challenge for NLP. Traditional NLP models are based on crisp rules and logic, but such models have difficulty accurately understanding phrases like “somewhat right,” “a little bit good,” or “very good.” Fuzzy logic helps model these linguistic uncertainties [15].

1. Handling Ambiguity and Vagueness in Text
Ambiguity is common in natural languages. A sentence or word can carry multiple meanings, and their correct interpretation depends on the context of the sentence.

Example: Sentence: “He is quite fast.”
“Fast” can mean speed, efficiency, or smartness – fuzzy interpretation is necessary according to the context.

Use of Fuzzy Logic:
Fuzzy sets define linguistic variables such as:
- Speed = {Slow, Medium, Fast}
- Clarity = {Vague, Clear, Very Clear}

Fuzzy membership functions assign degrees of meaning – which helps the system understand what percent a word falls into which category.
Benefits:
- Better handling of multi-meaning words
- Approximate reasoning in grammar and context
- Improved semantic interpretation in language translation systems

Fuzzy inference systems make NLP models more flexible and context-aware.

24.7 Sentiment and Emotion Analysis (Using Fuzzy Logic in Sentiment Analysis)

Sentiment analysis or opinion mining is an important task of NLP in which the emotional tone of a text or speech is identified – positive, negative, neutral. Traditional sentiment analysis tools use exact keywords or polarity scores, but user-generated content (reviews, tweets, comments) is often fuzzy and emotional.

Fuzzy Approach:
Fuzzy sentiment analyzers divide linguistic expressions into fuzzy sets:
- "Product is okay" → 0.5 positive
- "Slightly disappointing" → 0.2 positive, 0.6 negative
- "Very happy" → 0.9 positive

Emotion Ontology with Fuzzy Sets:
- Anger, joy, sadness, surprise – all are defined on fuzzy scales.
- Multiple emotions are possible in a statement, and fuzzy logic assigns their degree (e.g., joy: 0.6, surprise: 0.3).

Use Cases:
- Social media analysis
- Customer feedback systems
- Political sentiment tracking
- Mental health analysis using text input

24.7.1 Fuzzy Ontologies for Semantic Understanding

Semantic understanding is an advanced step in NLP – where the system tries to understand not only the words but also their meanings and relationships. Traditional ontologies define crisp and rigid relations:
- Dog → isA → Animal
- Student → hasRole → Learner

But in real life the relationships are fuzzy and partial.

Fuzzy Ontologies:
Fuzzy ontologies represent semantic relations between words with degrees.

Example:
- "Dog is loyal" → Loyalty degree = 0.9
- "Cat is loyal" → Loyalty degree = 0.4
- "Tiger is dangerous" → Danger = 0.95

Components of Fuzzy Ontology:
- Concepts with fuzzy attributes
- Relations with fuzzy weights
- Inference Rules using fuzzy reasoning

Applications:
- Intelligent question answering systems
- Conversational AI (chatbots and virtual assistants)
- Multilingual semantic search engines
- Educational content classifiers (textbook vs. blog vs. tutorial)
- Semantic analysis helps develop context-aware and adaptive NLP systems through fuzzy concepts, which better understand user intent.

24.8 Fuzzy Logic in Computer Vision

Computer vision is a field where AI systems extract information from images and videos – such as face detection, object recognition, motion tracking, etc. But images are often noisy, blurry, or unclear. Fuzzy logic provides a natural and effective way to process such uncertain data. Through fuzzy logic, computer vision systems can do soft and approximate reasoning – such as identifying subtle differences within images.

24.8.1 Image Segmentation and Enhancement

Image segmentation means dividing an image into logically meaningful regions – such as background, object, boundary, etc. Traditional segmentation algorithms use crisp thresholds, but they are not able to handle noise and variations in real-life images properly.

Fuzzy Logic Approach:
In fuzzy segmentation, each pixel is made to belong to one or more regions to some degree – using membership functions.

Example: A pixel can be between sky and cloud:
- μ(sky) = 0.6
- μ(cloud) = 0.4

This approach better captures overlapping and soft boundaries.
Image Enhancement:
- Low-contrast images can be enhanced using fuzzy logic.
- Fuzzy contrast stretching and fuzzy histogram equalization improve clarity.

Applications:
- Medical imaging (tumor region segmentation)
- Satellite images (land use classification)
- Biometric systems (face and fingerprint extraction)

24.8.2 Object Detection with Fuzzy Classifiers

The main goal of object detection is to identify specific objects (such as person, car, animal) in an image or video frame and identify their location. Traditional object detection models work on crisp feature thresholds (like color, shape, edge). But when the data is uncertain – such as blur, low light, or partial visibility – then fuzzy classifiers are used.

Fuzzy Classifiers:
- Fuzzy rule-based systems evaluate input features (size, shape, texture) as fuzzy sets.
- Membership values are calculated for each object class.
- Final decision is taken on the basis of fuzzy inference.

Example: IF shape is "approximately rectangular" AND color is "somewhat red"
THEN object is "vehicle" with 0.7 confidence.
Benefits:
- High accuracy even in noisy and ambiguous images
- Flexible rule creation
- Human-like interpretation of visual data

Use Cases:
- Autonomous vehicles (car/pedestrian detection)
- Factory automation (product quality inspection)
- Security systems (intruder detection)

24.8.3 Fuzzy Logic in Surveillance Systems

Surveillance systems are part of modern security infrastructure – where CCTV cameras, motion Sensors and facial recognition systems are used. Surveillance data is dynamic and noisy – where light changes, shadows, and unpredictable behavior are common. Fuzzy logic makes surveillance systems smarter and real-time responsive.

Key Functions with Fuzzy Logic:

- Motion Detection: Identifying actual movement by assigning fuzzy motion values to pixel changes
- Behavior Analysis: Interpreting human activity (walking, running, loitering) from fuzzy behavior sets
- Facial Recognition: Identifying with fuzzy matching despite lighting, angle, and partial occlusion

---

Example: IF motion is "moderate" AND presence duration is "long"
THEN behavior is "suspicious" with 0.8 confidence.

Benefits:

- False alarms reduce heat.
- Accurate surveillance even in complex environments (like crowded areas).
- Decision-making has to be transparent and explainable.

Applications:

- Airport and railway station surveillance
- Smart city security grids
- industrial safety monitoring

24.9 Fuzzy Control Systems in Industry

Fuzzy logic has brought about a revolution in industry automation and control systems. Traditional control systems work on crisp parameters and predefined rules, but real-world industrial environments are often complex, uncertain, and dynamic. For such situations, fuzzy control systems are a better option, which have the ability to make approximate reasoning, flexibility and human-like decisions.

24.9.1 Applications in Manufacturing and Process Control

In manufacturing processes, parameters like temperature, pressure, flow rate, speed, viscosity etc. have to be continuously monitored and controlled. Traditional PID controllers struggle in conditions where data is not precise or the process is highly nonlinear.

Role of Fuzzy Logic:
Fuzzy controllers think like human experts and can process even vague instructions:

Example Rules: IF "temperature is slightly high" AND "pressure is moderate" THEN "reduce heating by 20%."

IF "flow rate is very low" THEN "increase valve opening slightly."

Benefits:
- Smooth control in nonlinear environments
- Real-time adjustment without reprogramming
- Less dependence on exact mathematical modeling

Use Cases
- Chemical plants (mixing and reaction control)
- Food and beverage processing (temperature and quality control)
- Cement and steel industries (kiln and furnace monitoring)

24.9.2 Adaptive Control in Smart Grids and IoT

The environment in smart grids and Internet of Things (IoT) applications changes continuously. Load demands, energy availability, sensor fluctuations – all are unpredictable. Fuzzy control systems handle challenges efficiently through adaptive reasoning.

Smart Grid Applications:
- Load balancing based on fuzzy demand estimation
- Voltage and frequency control using fuzzy rule sets
- Fuzzy demand-response systems for optimizing electricity usage during peak hours

Fuzzy Rule Example:
IF "energy consumption is high" AND "solar input is low"
THEN "reduce appliance power by 30%."

IoT-Based Fuzzy Control:
The data received from IoT sensors (temperature, humidity, occupancy, etc.) is analyzed through fuzzy logic to generate intelligent decisions.

Examples:

- Smart Homes: "IF room is slightly warm AND occupied THEN reduce AC power slightly."
- Smart Farming: "IF soil moisture is low AND sunlight is moderate THEN start irrigation for 20 mins."

Benefits:
- Energy-efficient operations
- Autonomous decision-making without human intervention
- Scalability across various smart environments

24.9.3 Case Studies from Industrial Automation

A. Washing Machine Automation (Consumer Electronics):
Modern washing machines use fuzzy logic to determine:
- How dirty are the clothes?
- How much detergent is needed?
- How much spin time is required?

Fuzzy Rules Example:
IF "dirt level is high" AND "load weight is heavy"
THEN "wash time = long," "detergent level = high."
Result: Better cleaning, water-saving, and optimized detergent usage.

B. HVAC Systems (Heating, Ventilation, Air Conditioning):
Building automation systems are maintained by fuzzy controllers:
- Comfortable temperature
- Energy-efficient air flow
- Humidity control

Example Rule:
IF "outside temp is very high" AND "room is moderately occupied" THEN "cooling level = medium-high."

C. Automotive Industry:
- Fuzzy systems are used in:
- Gear shifting systems
- ABS braking system
- Suspension control for smooth driving

Case:
Companies like Toyota and Nissan have used fuzzy transmission systems where gear changes smoothly depending on car speed and driver behavior.

D. Oil Refinery Automation:
Oil distillation process uses fuzzy controllers to:
- Maintain temperature precisely
- Adjust flow rates based on viscosity
- Reduce manual intervention

Benefits:
- Reduce energy consumption
- Fewer accidents
- High-quality output

24.10 Challenges and Opportunities in Fuzzy-AI Integration

The combination of fuzzy logic and AI has become a powerful and flexible approach that makes uncertain, imprecise, and human-like decision making possible. But fuzzy-AI integration is fraught with challenges that limit the widespread adoption of this technology. In this section, we discuss the major challenges, current limitations, and future research directions associated with fuzzy-AI systems.

24.10.1 Scalability and Computation Cost

Challenge:
Fuzzy logic traditionally works efficiently in small-scale applications (such as control systems). But when it is integrated with large AI systems, such as deep learning networks, big data analytics, or real-time decision-making systems, the computational complexity increases:
- The fuzzy rule base becomes large.
- Processing multiple fuzzy sets and membership functions is computationally expensive.
- Training high-dimensional fuzzy systems is slow and memory-intensive.

Example: If a fuzzy DL system processes 100 input features, and 5 fuzzy sets are defined for each feature, then the total rule combinations grow exponentially – impacting the scalability of the system.

Solution Directions:
- Use of rule optimization algorithms
- Hierarchical fuzzy models
- Neuro-fuzzy pruning techniques for large networks
- GPU/TPU-based acceleration for fuzzy inference

24.10.2 Standardization and Benchmarking

Challenge:
The development of fuzzy-AI systems is still far from standardization. Every researcher defines fuzzy rules, membership functions, and integration techniques according to his own convenience – making it difficult to compare models.
- Lack of consistent performance metrics
- No testing on common datasets
- No clear guidelines on interpretability and accuracy trade-offs

Impact:
- Industry adoption becomes slow.
- Regulatory approval (e.g., in healthcare) becomes difficult.
- Low interoperability and reproducibility among researchers.

Solution Directions:
- Develop common fuzzy-AI frameworks (e.g., Scikit-Fuzzy, TensorFlow + fuzzy modules).
- Create open-source benchmark datasets and evaluation protocols.
- Alignment of fuzzy models with explainable AI (XAI) frameworks.

24.10.3 Future Research Directions

Fuzzy-AI integration is an emerging field where several possibilities exist:

A. Deeper Integration with XAI:
   Fuzzy logic is naturally interpretable. Future research can use fuzzy rules to explain the output of DL models – making black box AI models trustworthy.
B. Evolution of Hybrid Models:
   - Fuzzy + reinforcement learning (fuzzy-RL)
   - Fuzzy + transformers
   - Fuzzy + quantum AI

   These hybrid approaches will provide a synergy of fuzzy reasoning and cutting-edge learning techniques.
C. Lightweight Fuzzy Systems in Edge AI and IoT Devices:
   The use of fuzzy-based lightweight inference modules will increase in smart devices (such as wearables, drones, and sensors) in the future.
D. Data-Driven Fuzzy Rule Generation:
   Machine learning models can automatically generate and optimize fuzzy rules – reducing the need for manual tuning.
E. Domain-Specific Customization:
   - Medical fuzzy-AI systems for personalized diagnostics
   - Legal fuzzy-AI systems for contract analysis
   - Climate fuzzy-AI systems for uncertainty modeling

References

[1] Zadeh LA. Fuzzy sets. Information and Control. 1965; 8(3): 338–353.
[2] Ross TJ. Fuzzy Logic with Engineering Applications. 3rd ed, Wiley; 2010.
[3] Jang JSR, Sun CT, Mizutani E. Neuro-Fuzzy and Soft Computing: A Computational Approach to Learning and Machine Intelligence. Prentice Hall; 1997.

[4] Mendel JM. Uncertain Rule-Based Fuzzy Logic Systems: Introduction and New Directions. Prentice Hall; 2001.

[5] Mitra S, Hayashi Y. Neuro-fuzzy rule generation: Survey in soft computing framework. IEEE Transactions on Neural Networks. 2000; 11(3): 748–768.

[6] Lin CT, Lee CSG. Neural Fuzzy Systems: A Neuro-Fuzzy Synergism to Intelligent Systems. Prentice Hall; 1996.

[7] Pal SK, Mitra S. Neuro-Fuzzy Pattern Recognition: Methods in Soft Computing. Wiley; 2004.

[8] Nguyen HT, Prasad NR, Walker CL, Walker EA. A First Course in Fuzzy and Neural Control. CRC Press; 2003.

[9] Kumar KS, Kumar TA, Nishanth R. Design of advanced intrusion detection system for multimodal fuzzy-based convolutional neural networks (mFCNN). International Journal of Advanced Computer Research. 2023 Dec 1; 13(65): 94.

[10] Turksen IB. Measurement of Membership Functions and Their Acquisition. In Dubois D, Prade H. (eds) Fundamentals of Fuzzy Sets 2002 (pp. 61–97). Springer.

[11] Karray FO, De Silva CW. Soft Computing and Intelligent Systems Design: Theory, Tools and Applications. Pearson Education; 2004.

[12] Wang LX. A Course in Fuzzy Systems and Control. Prentice Hall; 1997.

[13] Pappis CP, Mamdani EH. A fuzzy logic controller for a simple dynamic plant. Proceedings of the Institution of Electrical Engineers. 1977; 121(12): 1585–1588.

[14] Yager RR, Filev DP. Essentials of Fuzzy Modeling and Control. Wiley; 1994.

[15] Herrera F, Verdegay JL. Linguistic fuzzy modeling. IEEE Transactions on Fuzzy Systems. 1997; 5(4): 478–496.

T. Grace Shalini, P. Geetha, B. Anandasabarirajan, R. Karthick Manoj

25 Trustworthy Stock Market Forecasting: Time Series Modeling Under Generative AI Manipulation Risks

Abstract: Application of time series algorithm for stock price prediction with historical data yet predicting what a stock price will really be is very difficult because financial markets are volatile and complex. We will use the time series algorithm (a unique time series analysis and machine learning algorithm), in an effort to increase our rate of accurate stock price predictions. So we use our time machine of historical data and label out patterns that seem to predict stock prices. These results illustrate the potential of an algorithm in predicting stable prices, career guidance for investors, and financial analysts. The research adds to the emerging field of financial forecasting, demonstrating that sophisticated algorithms are well-suited for making sense of stock market environments. The performance, therefore, indicates how the inclusion of these aspects collectively can result in better prediction, affording investors' and traders' better confidence when navigating the vagaries involved within the stock market.

Keywords: Time series algorithm, stock price prediction time series analysis, machine learning in finance predictive modeling

25.1 Introduction

Forecasting the prices of stocks on the stock market is the most significant thing. In the context of financial forecasting, however, it is a task that is extremely difficult to do. It is possible to ascribe this fact to the volatility character of financial markets, as well as the interwoven complexity of these systems [1]. The accuracy of such estimates is extremely sensitive to the investing techniques that are being utilized. Portfolio management and decision-making procedures of the financial industry. Approaches that are considered conventional are mainly dependent on statistical models and basic analysis. However, as a result of the current surge in popularity of machine learning and time series analysis, there are new approaches that may be taken to improve the accuracy of these forecasts. The time series algorithm is a one-of-a-kind approach to solving this problem since it integrates time in a distinctive way. Combining machine learning with advanced chain analysis is rather impressive. Due of this, computers are able to utilize previous stock values in order to identify patterns and trends. This will assist in predicting the pricing of futures. There are some circumstances in which conventional approaches can be integrated into a single method because they can be fixed. Time series algorithms are able to respond well to shifts in the mar-

https://doi.org/10.1515/9783112214374-025

ket and offer an approach that is both more resilient and dynamic [2]. In the process of constructing time series algorithms, the goal of generating a novel perspective in this field is taken into consideration. Forecasts are often more accurate when they are driven by the utilization of prior data. However, the conduct of the market becomes more visionary by transforming this approach into a practical instrument for investors and observers of financial markets. Making predictions about stock prices through the use of time series algorithms is analytical and gains and implications for the process of financial forecasting. Given this, it appears to be in line with the research trend in quantitative finance, which is to make the dynamics of the market more intelligible through the implementation of sophisticated algorithms. This results in the investing scenario being even more favorable.

25.2 Literature Survey

They use stock prices to test to see which performs better. The results show that the long short-term memory (LSTM) class of the model provides the best possible values that are optimized to best capture long-term dependencies, respectively [3] although this is an important result. But such findings have important limitations to the work. Stock price data does not include other important financial instruments that may require specific types of analysis. Additionally, this article does not address interpretability in these machine learning models. This is very important in isolating what drives the machine learning model's predictions and ensure that it is of practical use in making financial decisions [4].

Although this study concluded that using hybrid models is more accurate and robust than non-hybrid individual models, and therefore suitable for better financial forecasting. The chapter also notes some limitations, such as the need for high computational resources for effective use. However, the study did not report details on real-time efficacy. Of course, this is an important aspect of leveraging these advanced models in real financial environments. Indeed, although Michael Brown's article published in AI on Finance highlights some of the uses of learning models [5].

AI in Finance applies the deep learning models such as convolutional neural networks and recurrent neural networks for financial time series forecasting. On these bases, the author points out that the models perform better to catch such complex patterns existing in financial data and have a possibility of more accurate results. However, Brown's work is more oriented toward finding short-term forecasts of financial data, and this limits its usage in longer-term predictions about money. Further, it also reveals that the deep learning models are highly sensitive to the process of hyperparameters and can most certainly impact their performance, so critical calibration is needed to obtain peak results [6].

Author explores transfer learning as a method of upgrading forecasting models. Green's research reveals that transfer learning can significantly improve model performance since it can draw the knowledge from related domains. However, it highlighted that the process of source domain selection should be very discriminative with relevance and effectiveness. The paper further points out that the adaptation of transfer learning in high-frequency trading (HFT) has never been explored and could be a future field of study [7].

In the light of techniques like decision trees and SHAP values, Blue demonstrates that, with such models, one is then able to understand clearly what drives an event into making precise predictions in financial contexts for support of more informed decision-making processes. As with any model, there is a trade-off between interpretability and accuracy, since these models might not be capable of capturing the really complex relationships that exist within the data, as well as more advanced deep learning models. This restriction thus implies that while interpretable models are most valuable for transparency, they may have compromised some amount of predictive power in comparison to less interpretable yet more sophisticated models [8].

Author discusses applying reinforcement learning techniques in optimizing investment portfolios. Such exploration as Q-learning and DDPG points to the ability of such methods to dynamically manage a portfolio through fluctuating market conditions while maximizing returns [9]. The problem with that research, though promising results are seen, is that these approaches do require great quantities of history to train with effectively, which is a major limitation. Computationally intensive times these methods require to implement make it infeasible for real-time or resource-constrained environments, and the model's performance is also highly sensitive to how it is designed for the reward function. This constrains the practical applications of reinforcement learning in portfolio management.

Algorithmic finance relates to discussing incorporation of machine learning models in high-frequency trading environments. This research demonstrates that machine learning algorithms are appropriately able to cope with the enormous amount of data that HFT generates and processes at fantastic speeds and, therefore, can lead to more efficient and fine-tuned trading strategies. Still, the chapter states a number of significant drawbacks, overfitting toward historical data are one of the very significant risks, which could undermine the strength of these models in real-world trading applications. While technical capabilities of the machine learning in HFT have been well-explained, such research fails to attack the ethical and regulatory aspects related to the integration of high technologies into financial markets, which are certainly issues for further use of the tools [10].

Kevin Red's Financial Data Science introduces sentiment analysis, derived from news and social media, to traditional time series models. The chapter thus highlights how sentiment analysis will complement the financial model so significantly in terms of predictive capability, particularly in respect to assets where market sentiment makes all the difference. Despite such strengths, the study identifies the challenges associated with achieving real-time seamless integration of sentiment analysis, which is considered critical for sound and timely financial decision-making. Further, the accuracy and precision of sentiment analysis models are prone to a dimension of variability due to the level of quality of data and context, which limits their application in predicting market moves with much accuracy [11].

The financial markets on the use of machine learning models for the prediction of financial market volatility, the models, to include Generalized Autoregressive Conditional Heteroskedasticity (GARCH) and Support Vector Regression (SVR), have been proven to do much better than traditional models in predicting events. Nevertheless, this publication has not explored volatility adequately with respect to multiple assets. This approach is highly sensitive to input features and data preprocessing quality; thus, changes in data can impact its robustness and generalizability under different market scenarios [12].

Computational finance elaborates on the use of clustering techniques like K-means and DB-SCAN on time series financial data. It shows that these techniques can indeed identify very similar patterns within data, therefore useful for the goals of risk management or investment strategy formulation. It, though, does point out the practical problem that is scaling these clustering techniques to large data sets that would be extremely necessary for actual financial industry applications. Additionally, the clustering results are sensitive to the initial parameters set for the algorithms and this may have impacts on accuracy and reliability concerning the patterns identified [13].

Ensemble methods in finance traces the development of ensemble learning models, which combine various forecasting techniques [14]. In his findings, it emerges that models like Random Forest and Gradient Boosting show a significant improvement over individual methods of forecasting, giving more accurate and robust financial predictions. Despite these advantages, the study goes on to comment that ensemble models are computation-intensive and thus require a lot of processing power. Furthermore, the success of such ensembles significantly depends on the right choice and fine-tuning of the constituent models, which is onerous and time-consuming.

Financial machine learning, illustrated how adversarial learning could potentially make financial forecasting models more robust. The result shows that adversarial training makes these models robust to the perturbations of data and surprising market anomalies and thus, in general, reliability in these models improves. Conversely, some disadvantages have been found associated with this technique. For one thing, live trading conditions about their true practicality are relatively less with these approaches as compared to other methods. Including the term adversarial learning introduces overfitting possibility. These factors may compromise generalization over the model to obtain good performance in real-world applications [15].

It is a firm supporter of sophisticated feature engineering techniques for data science in finance by going through elaborate means of how such techniques can be applied to the context of financial time series. The examination indicates that properly engineered features appreciably enhance the performance of system mastering fashions for economic forecasting; however, he reveals that automation processes concerning feature engineering need further investigation to streamline and become even more efficient in efficiency. It is also mentioned that feature engineering tends to rely heavily on domains and hence requires a vast amount of knowledge, which would be a restrictive factor in its usage and effectiveness if it's not subjected to domain experts [16].

25.3 Methodology

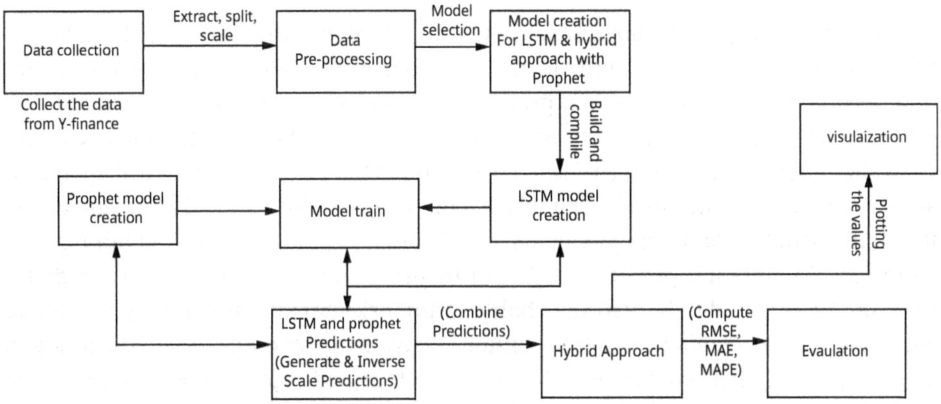

Figure 25.1: Architecture diagram.

Figure 25.1 shows the hybrid forecasting model workflow, combining LSTM and Prophet for stock price prediction.

25.3.1 Data Collection

Data Collection Methodology: The method starts with amassing historic stock information, that specialize in addressing the time series of the Nifty 50 index, this is, the performance of the pinnacle 50 organizations indexed at the Indian National Stock Exchange. Here, the intention is to accumulate a complete dataset that encompasses adjusted final prices over a defined term to be in addition analyzed and modeled. The script uses "yfinance" with tremendous abilities in information extraction from Yahoo Finance. It is the translation of pulling up records associated with the inventory market immediately into Python, and hence apt for quantitative analysis and financial modeling. The dates among January 1, 2015, and January 1, 2023, capture a volume of historical facts and are therefore fundamental inside the information of the marketplace's tendencies and styles. The adjusted close is very critical because it shows the ultimate fee after modifications for dividends, splits of stocks, and many others, which could have affected the employer, and by extension, the agency's historical performance. The methodology downloads this information set in order that a solid foundation is laid for next analysis, and for this reason any predictions or fashions developed later rely on dependable and complete historical data. This, in flip, gives meaningful insights and will increase the general effectiveness of the economic forecasting model.

25.3.2 Data Preprocessing

Data preprocessing is the next critical step after an information series for historic statistics of the Nifty 50 index. Here, uncooked information needs to be prepared earlier than feeding it to the fashions in a proper layout and scale for machine studying programs. Data preprocessing starts with the extraction of the Adjusted Close column from the dataset. The Adjusted Close cost is in particular valuable as it displays an accurate amount of the stock's well worth, compensating for elements that could bias its value, including dividend payments and inventory splits. Thus, this column might be the focal point, and predictions should be primarily based on this option; therefore, it'd be a completely distinguishable landmark that the model would refer to while mastering. Zeroing in in this column allows us to smooth up our records and dispose of redundant entries, which could interfere with how a model works. Then might be facts scaling with the help of the MinMaxScaler in the "sklearn.Preprocessing" module. Scaling is one of the most crucial preprocessing techniques, in particular for algorithms like LSTM, which are sensitive to the size of the input records. MinMaxScaler transforms the facts with the aid of scaling it into a range between 0 and 1. This lets all input capabilities have the equal scale, so no person function dominates the model by drawing attention to it with a much larger variety of values. By making use of this form of scaling to the statistics, we improve the capability of the version in studying styles and thereby decorate the predictive overall performance. With scaled data, the dataset might have been split into train and test units. Normally, this is performed along an 80–20 break up in which 80% of information is used to teach the model and the relaxation 20% to assess the performance of the version.

25.3.3 Dataset Creation for LSTM

Once the statistics is preprocessed, the next important step is to put together the dataset specifically for LSTM modeling. LSTM neural networks are particularly designed to work with sequential statistics. Thus, this calls for structuring the dataset in any such way as to seize temporal dependencies correctly.

This stage is supposed to create input-output pairs that set up the time-collection function of stock prices. To achieve this, a function called "create_dataset" is utilized. A sliding window method is used wherein sequences of data over a particular variety of steps are created. In this case, the number of such steps changed into 60. For each collection, the feature aggregates the values of the final 60 time steps as input and the corresponding output because the value inside the subsequent step. That we could model study from history and predict future values based on patterns or tendencies it discovers within the ones 60 earlier time steps.

In this structuring of the information, the LSTM version is taking pictures of developments and fluctuations inside inventory fees in conjunction with different functions. It is in this sequence technology that allows the LSTM to make full use of its memory capability: it would consider items from beyond time steps and consequently inform its prediction. At the end of all of it, consequently, this tiresome statistics-crafting technique boosts the performance of the version and also provides a robust framework for accurate forecasting in economic time collection analysis.

25.3.4 Reshaping Data for LSTM

Input-output pairs acquired to feed the LSTM version should then be reshaped into the proper layout compatible with the LSTM structure. Data must be entered in a three-dimensional format like (samples, time steps, and capabilities) in an LSTM community. This three-dimensional layout is vital because the LSTM version can procedure sequential records and find or seize temporal dependencies embedded in the records. The first measurement is samples, which is the count number of person sequences the version learns from. Since each sample represents a completely unique observation generated from the time-collection information, this size represents the range of those character sequences. The second measurement is time steps – the rely of previous observations protected in every series. In this situation, we've selected 60 steps; in this way, the version is able to study and research from the ultimate statistics factors, which will expect the following fee. The capabilities measurement refers to the number of variables included in the analysis. Since we're managing a single time collection, this dimension may be set to 1.

The data is then transformed into a three-dimensional array using NumPy. In this sort of system, without it, it might not be possible to get an accurate interpretation of facts by using the LSTM model and harness its reminiscence skills because of the fact that the structured data of this nature permits the model to study the complex dependencies over the years, which may enhance overall performance for the process of forecasting inventory prices. The cautious practice step will shape the base for ensuring right schooling and validation of the version on the datasets available.

25.3.5 LSTM Model Creation

The next step will be to design a deep learning model that will predict the stock prices based on preprocessed datasets. This will make use of the functionality offered by the LSTM networks, which ironically are way more efficient in time series prediction as well since they are especially effective at learning from data that is organized in sequences.

Its architecture comprises two LSTM layers. The initial LSTM layer is configured to return sequences, allowing the model to pass its output to the subsequent LSTM layer. Such a two-layered structure increases the capacity of the model to learn complex patterns and dependencies within the data. By layering LSTMs, the model will draw increasingly abstract features because it is compelled to make sense of the intricacies that it finds in variations in stock prices over time. A single output unit dense layer follows the two LSTM layers. Its goal is to produce the final prediction from outputs by the two LSTM layers that would be the predicted price of stocks for the next step in this scenario. By synthesizing the information learned by adding their outputs, the dense layer would then produce an accurate prediction.

This architecture is then put into the model and compiled using the Adam optimizer; this type of optimizer was proven highly efficient and adaptive for training deep learning models. The chosen loss function for the model is mean squared error, which is defined as the average of the squared differences between the predicted and actual values. Selection of loss function – this is crucial to be involved in the performance evaluation of a particular model during training since it guides the optimization process toward building improvement for forecasting stock prices with a better accuracy. The model is systematically designed to ensure that a well-designed LSTM network would manage the complexities involved when processing financial time series data.

25.3.6 Model Training

Perhaps, the maximum critical section in setting up an amazing LSTM model for stock fee prediction is the training phase. Its purpose is to let the model research the styles and relationships that exist in the training dataset in order that progressed predictive competencies can be received. It essentially suits the version to the facts, permitting it to adjust its inner parameters based totally at the found patterns within the samples for the duration of the training system. During the training level, the version is in shape for 10 epochs. An epoch denotes creating a complete skip over the entire education dataset. Running a couple of epochs permits the version to examine at each new release and finally improve its potential to expect the most capability inventory fees possible. But 10 epochs is a great stability among training and now not getting overskilled, in which the model learns the training statistics a bit too well; however, fails to generalize to new, unseen data. A batch size of 64 is used because of this the version is going through 64 training samples before it updates its weights. This allows reminiscence utilization and makes schooling quicker apart from that, yields a more solid gradient estimate for the gradient used in optimization, so the model converges to a higher answer. In doing so, the getting to know technique described right here endows the LSTM model with a sense of identification and adjustment in the direction of the pattern unfolding from the supplied inventory rate facts, thereby equipping it with greater correct predictions at some stage in trying out. Such a cautious training manner that the version was simply prepared for financial data in actual-global contexts, in reality improving its reliability within the making of forecasts.

25.3.7 Prediction

Now that the LSTM version is skilled, we are able to make predictions on the test and schooling information sets. This will allow us to understand how much the version found out from the information and to what volume it can expect statistics to which it has no longer been uncovered before. In this prediction process, the patterns discovered through the community whilst in the schooling stage are used to provide predictions at every point in time in each information unit to gain the predicted inventory expenses. However, the values anticipated with the aid of this model are already within the range between 0 and 1 due to the fact the statistics utilized in preprocessing was scaled into that C programming language. To flip this into real-international prediction values, we are able to use the very equal MinMaxScaler on those expected values to get the inverse scale returned up into their local value range. That manner, we are able to compare our version's predictions against historical inventory fees and meaningfully evaluate its overall performance.

25.3.8 Implementation of Prophet Model

Besides application of the LSTM model, this methodology introduces Prophet, which is an open-source tool developed by Facebook, becoming extremely popular for effective usage in time series forecasting. The reason why Prophet was chosen to be implemented in this methodology is that it is designed for daily observations of the data, as well as for seasonal trends and irregularities in data and thus is especially strong at financial forecasting. To run Prophet, one needs to prepare their dataset in a format friendly to the model. Prophet requires that data have two columns: one for dates, encoded as the "ds" column and the other for values to be forecasted, in this case, stock prices. To do that, we are reshaping the data by renaming the date column to "ds" and the target variable – Adjusted Close prices to "y" – with this, Prophet is able to read it in as one would expect.

Then the Prophet model is initialized and trained based on the historic stock price data. The training process helps the Prophet realize trends and seasonality lying below the time series of the prices. After doing that you could use such a model to predict future values at any point of time by extending this dataset into the future, which Prophet calls creating a future dataframe.Prophet finally generates forecasts based on the learned patterns, generating forecasted stock prices for specific dates in the future. The result of these forecasts can then be used as a standalone or in combination with other model results, such as those from the LSTM model, for producing a hybrid technique toward better stock price forecasting. This method hence is robust and intuitive for time series forecasting that complements the deep learning approach of LSTM.

25.3.9 Hybrid Model Approach Methodology

Hybrid model approach is an integration of the strengths of LSTM and Prophet together in designing a more robust and accurate forecasting system for the stock prices. Every model brings unique capabilities to the table, while their predictions are merged to aim at finding benefits from both sides in improving overall forecasting accuracy. Hence it is well-suited for use in capturing complex patterns and relationships in sequential data like this characterized by the hallmark long-term dependencies of the dynamics in stock price movements. However, this could sometimes give LSTMs problems with broader trends or seasonality since their methodology involves learning from recent data windows. On the other hand, the Prophet model is particularly good in detecting seasonality patterns, long-term trends, and irregularities in time series data. This means it should work exceptionally well in terms of the identification of recurring stock market behavior and general price movement trends over a given period of time.

A hybrid forecast is generated by averaging the predictions of both models in order to capitalize on the strengths of each. Specifically, after training both models and letting them generate their respective predictions for the training and testing datasets, the predictions from both models are combined. The predictions from the LSTM and Prophet models at each time step average up to create a single hybrid prediction. Therefore, the averaging method not only eases potential biases or weaknesses coming from either model but also provides a more balanced and accurate forecast for the hybrid model. It, therefore, applies to the training and testing sets to be consistent across all the forecasted values. The combination of LSTM's capability of dealing with short-term dependency and Prophet's ability to capture long-term trends and seasonality while ultimately increasing the precision of the forecast, this hybrid proved to be a good tool for financial time series analysis.

25.3.10 Evaluation Metrics

Having built, run, and generated predictions for all of these models-the LSTMs, Prophet, as well as the hybrid model – it is high time to evaluate them. That way, you will be able to know if your models are working correctly in case you use them for predicting stock prices and an understanding of how they might have performed under a variety of circumstances. For the sake of such evaluation, three types of metrics will be used here: RMSE or root mean squared error, MAE or mean absolute error, and MAPE or mean absolute percentage error.

1. RMSE measures the average difference between predictions and actuals. That measure, it happens, works very well in choosing large errors because it squares the differences before averaging, so large errors weigh more in the final score. The lower the RMSE, the closer that the model's predictions are to the actual stock prices.
2. MAE is a simple measure of average absolute differences between the actual and the predicted values. Contrary to RMSE, it treats all the errors without giving any weight to the large error; this makes it easy to understand in general magnitude the prediction errors might occur.
3. MAE is a simple measure of average absolute differences between the actual and the predicted values. While RMSE treats the error differently based on its size – larger errors having more weight – MAE treats all the errors without giving any weight to large errors; this makes it easy to understand, in general magnitude, the prediction errors might occur.
4. MAPE is the error measure in terms of percentage error accuracy of predictions. It allows giving an idea about the size of the error relative to the actual values of size that makes it a lot easier to interpret in terms of practical forecasting accuracy. The lower the MAPE, the more accurate will be the model's predictions in terms of percentage error in approximating actual stock prices.

These metrics apply to both the LSTM and hybrid models, so this offers a holistic comparison of how the models performed.

25.3.11 Visualization

Visualization is also crucial in determining how well the models performed. After evaluation metric computation, the predicted values are plotted over the original stock prices to visually know how good the model is.

For visualization the original stock prices, that is the data scaled and after being inverse-transformed, are plotted over time. We also plot the LSTM and hybrid model predictions for the training and testing sets in the same graph. We will be comparing this with the true stock prices. The visualization will easily identify patterns, whether large discrepancies exist between the predicted values, and the models' performances at different periods.

This graphical comparison supports the identification of regions where the model predictions do not match the actual price of stocks, therefore providing a clear indication of the strengths and weaknesses of each model in predicting. In the light of this graphical approach, it becomes easier to determine how much better the hybrid approach is at predicting stock prices than the individual models. Architecture diagram (Figure 25.2) demonstrates the hybrid forecasting model workflow, combining LSTM and Prophet for stock price prediction.

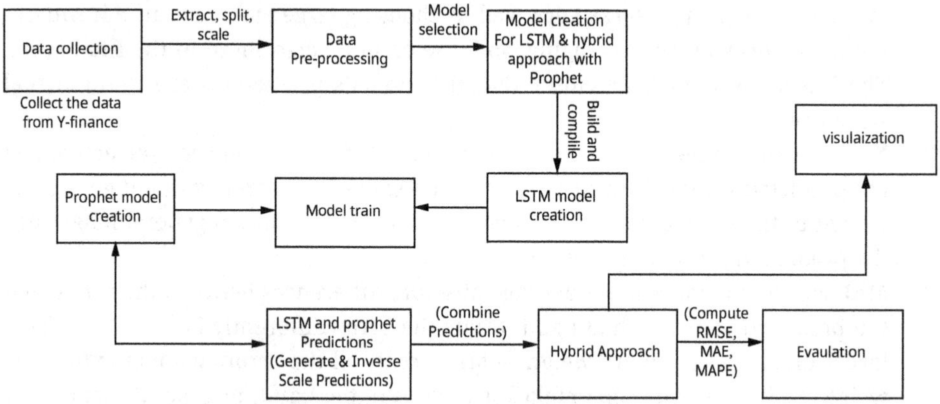

Figure 25.2: Architecture diagram.

25.4 Results and Discussion

The evaluation of LSTM and hybrid models is highly relevant to determine their predictability capabilities, especially in the domain of forecasting stock market prices. By judging the performance with numerical metrics and visualization, we understand its capabilities, as well as its weaknesses.

25.4.1 LSTM Model

The LSTM model, which was primarily designed to capture short-term dependencies in time series data, worked out quite all right. With its ability to learn from sequential data and also bring forth patterns in the stock prices, it could generate predictions following the general trends of the Nifty 50 index. However, since LSTM gives the importance to most recent data points, sometimes it experienced long-term and seasonal behaviors in the stock prices, and hence showed deviations at certain periods. It is confirmed by the RMSE and MAE scores, though surprisingly well within reasonable bounds, yet there was still some larger error prediction in parts that required further consideration in areas. The output of LSTM model is displayed in Figure 25.3.

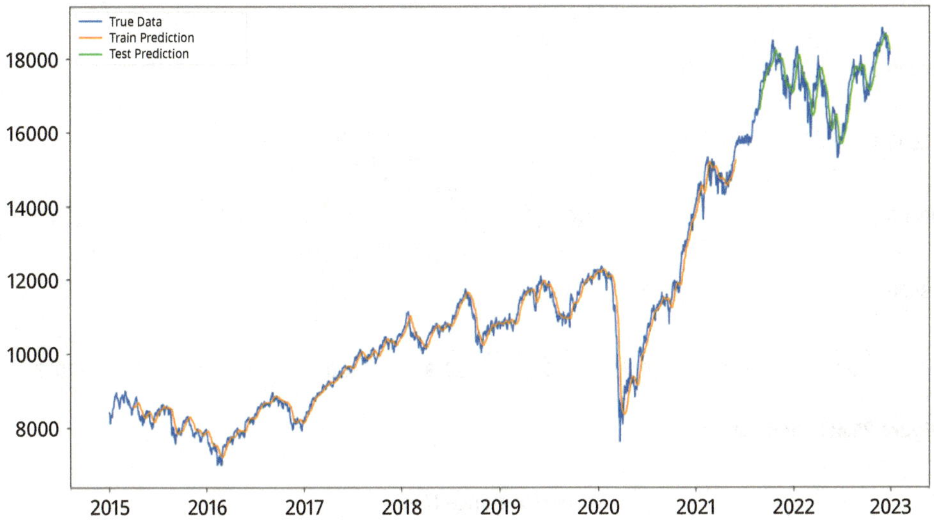

Figure 25.3: LSTM output.

25.4.2 Hybrid Model

The hybrid model of the combined predictions from the LSTM with a little added insight proved the benefits in multi-technique approach working. Average of the results of both models had balanced strengths in the effectiveness of LSTM toward capturing short-term movements and general trends in the data; this led to better overall accuracy as proven through improved lower values of RMSE, MAE, and MAPE scores as compared to the individual LSTM model. The hybrid model gave smoother and more reliable predictions by correcting the erratic fluctuations of LSTM while at the same time incorporating broader trend identification. The output of the hybrid model is illustrated in Figure 25.4.

Outcomes emphasize the hybrid approach resulting in stock price forecasting in light of the output; the importance of using a hybrid approach for stock price forecasting is eminent. An outcome of this nature would arise from a number of varied variables that track short-term patterns or are less influenced by long-term trends in a stock price. This sequential strategy has focused attention on recent data and captured immediate movements, thanks to the attention of the LSTM model; however, it was the hybrid model that eventually helped to make a more in-depth perspective of the behavior of the stock market. Since the hybrid model merged the better of the two

Figure 25.4: Hybrid output.

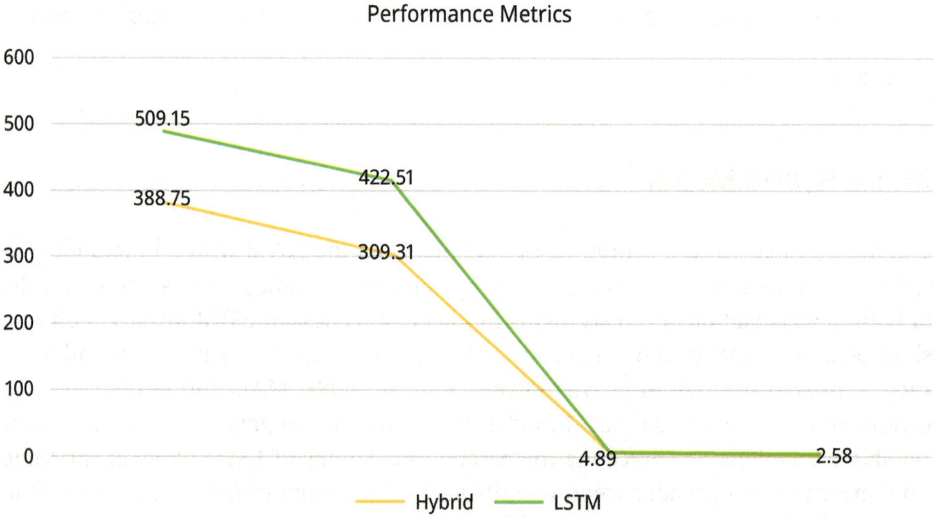

Figure 25.5: Performance metrics.

worlds and reduced the shortcomings that were discovered in using LSTM alone, the hybrid model consequently succeeded the other models in better performance.

The visualization further underlined such conclusions. If the hybrid model values are compared with actual stock prices, it was observed that during the training phase, as well as the testing phase; the movements were similar to actual stock price movements. This visual coincidence led to an inference that apart from accuracy, the hybrid model also ensured generalization to unseen data. Overall, the inclusion of this machine learning-based LSTM model as a combination with other forecasting strategies in any order proved effective in predicting the stock prices. This hybrid model increasingly made more accurate and reliable predictions and, therefore, is a robust tool for making financial forecasts.

This output explains how the combination of the LSTM model and the Prophet model improves on the inventory fee forecasting accuracy. The sequence depends on the inventory market data captured by the LSTM version; these two models work in tandem to recognize broad trends and seasonality based on the average forecast from both fashions. This hybrid shows more balanced and accurate predictions compared to using each version separately.

The focus of improvement is on assessment indicators. The RMSE, MAE and MAPE of the hybrid model show significantly higher overall performance compared to the LSTM version alone. The hybrid version significantly reduces prediction errors and more closely follows the actual listing fee behavior. In addition, the functionality of the hybrid version allows for comparisons. Visualized expected and actual data to track inventory movements within the market. It also promises more accurate techniques for financial forecasting with the possibility of future improvements through parametric and conservative optimization of various economic instruments or stock indices.

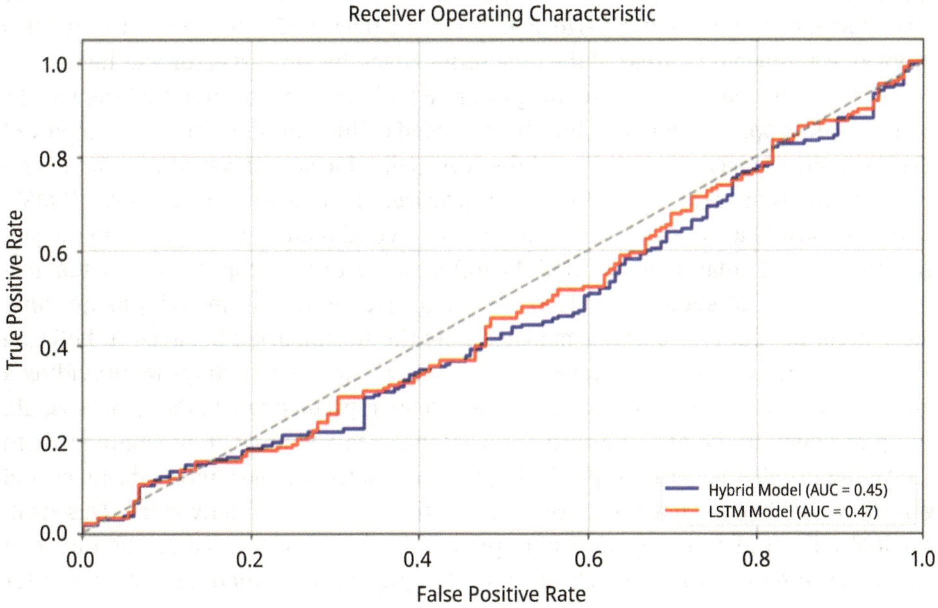

Figure 25.6: Receiver operating characteristics.

The receiver operating characteristic (ROC) curve (Figure 25.6), a graphical tool of performance evaluation for binary-classification models, determines the trade-off between TPR and FPR. In case of prediction in stock market using a hybrid model of LSTM and Prophet, the ROC curve helps evaluate how well the price movement can be predicted through the model whether prices are going up or coming down. A curve closer to the upper left corner implies that this model performs well. This overall accuracy is given by measuring the area under the curve (AUC), so an AUC of 1 means the discrimination was perfect between the positive and negative outcomes.

25.5 Conclusion

This study utilized the Nifty 50 index as a reference to see how effectively LSTM models and a blend of LSTM and Prophet models can anticipate stock market values. The study looked at the single LSTM model and the hybrid model in great detail, focusing on how well they could predict the future and how well they could apply what they learned to new situations. The LSTM model was effective at discovering short-term correlations and learning patterns that show up in a given sequence in time series data. It was great at tracking the overall patterns in stock prices, especially when the market remained steady. LSTM has problems simulating long-term trends and seasonal fluctuations since it primarily used data from the past. This made the market act differently in some cases. RMSE and MAE are two performance metrics (Figure 25.5) that demonstrated this problem was obvious in areas of the time series when the forecast error was higher.

Adding a mixed model made things work a lot better. This model used the LSTM's output with Prophet's ability to find bigger trends. This combined strategy employed the short-term precision of LSTM and the seasonality-focused power of Prophet to create forecasts that were more steady and precise. The assessment measures (RMSE, MAE, and MAPE) always demonstrated that the hybrid model was superior to the single LSTM. Graphs that demonstrated the difference between projected and actual prices backed this up even more. They revealed that the hybrid model was not only more accurate, but it also better matched how the stock market behaved in both the training and testing datasets. The ROC curve backed up these findings by providing a higher AUC score, which suggests it was very good at predicting whether prices would go up or down. In the end, the results reveal that a mixed model that combines both LSTM and Prophet is better at producing forecasts since it makes fewer mistakes and generates predictions that are more in line with how markets really work. This strategy looks like it may be a good way to predict the future of finance. In the future, it may be even more useful and accurate in changing market situations if it gets better by adding real-time data, fine-tuning settings further, or introducing other model combinations.

References

[1] Sheth D, Shah M. Predicting stock market using machine learning: Best and accurate way to know future stock prices. International Journal of System Assurance Engineering and Management. 2023; 14(1): 1–18.

[2] Brunner D, Legat C, Seebacher U. Towards Next Generation Data-Driven Management: Leveraging Predictive Swarm Intelligence to Reason and Predict Market Dynamics. In Collective Intelligence 2024 (pp. 152–203). CRC Press.

[3] Song X, Liu Y, Xue L, Wang J, Zhang J, Wang J et al. Time-series well performance prediction based on Long Short-Term Memory (LSTM) neural network model. Journal of Petroleum Science and Engineering. 2020; 186: 106682.

[4] Sujith AVLN, Qureshi NI, Dornadula VHR, Rath A, Prakash KB, Singh SK. A comparative analysis of business machine learning in making effective financial decisions using structural equation model (SEM). Journal of Food Quality. 2022; 2022(1): 6382839.

[5] García D, Kristjanpoller W. An adaptive forecasting approach for copper price volatility through hybrid and non-hybrid models. Applied Soft Computing. 2019; 74: 466–478.

[6] Mehtab S, Sen J. Analysis and Forecasting of Financial Time Series Using CNN and LSTM-based Deep Learning Models. In Advances in Distributed Computing and Machine Learning: Proceedings of ICADCML 2021 2022 (pp. 405–423). Singapore: Springer Singapore.

[7] Gao Y, Ruan Y, Fang C, Yin S. Deep learning and transfer learning models of energy consumption forecasting for a building with poor information data. Energy and Buildings. 2020; 223: 110156.

[8] Yeo WJ, Van Der Heever W, Mao R, Cambria E, Satapathy R, Mengaldo G. A comprehensive review on financial explainable AI. Artificial Intelligence Review. 2025; 58(6): 1–49.

[9] Rezaei M, Nezamabadi-Pour H. A taxonomy of literature reviews and experimental study of deep reinforcement learning in portfolio management. Artificial Intelligence Review. 2025; 58(3): 94.

[10] Brabazon A, Kampouridis M, O'Neill M. Applications of genetic programming to finance and economics: Past, present, future. Genetic Programming and Evolvable Machines. 2020; 21(1): 33–53.

[11] Sohangir S, Wang D, Pomeranets A, Khoshgoftaar TM. Big data: Deep learning for financial sentiment analysis. Journal of Big Data. 2018; 5(1): 1–25.

[12] Ampountolas A. Enhancing forecasting accuracy in commodity and financial markets: Insights from garch and svr models. International Journal of Financial Studies. 2024; 12(3): 59.

[13] Bromiley P, McShane M, Nair A, Rustambekov E. Enterprise risk management: Review, critique, and research directions. Long Range Planning. 2015; 48(4): 265–276.

[14] Wu H, Levinson D. The ensemble approach to forecasting: A review and synthesis. Transportation Research Part C: Emerging Technologies. 2021; 132: 103357.

[15] Gan L, Wang H, Yang Z. Machine learning solutions to challenges in finance: An application to the pricing of financial products. Technological Forecasting and Social Change. 2020; 153: 119928.

[16] Sarker IH. Data science and analytics: An overview from data-driven smart computing, decision-making and applications perspective. SN Computer Science. 2021; 2(5): 377.

Index

https://doi.org/10.1515/9783112214374-026

Mathematical Methods in the Digital Age

Volume 1
Satya Prakash Yadav, Abhishek Kumar, Ashutosh Kumar Dubey, Victor Hugo C. de Albuquerque, Adrian Rodriguez-Aguiñaga, Angeles Quezada
Medical Image Processing: Multimodal Fusion Techniques, 2024
ISBN 978-3-11-139549-4, e-ISBN (PDF) 978-3-11-143597-8, e-ISBN (EPUB) 978-3-11-143599-2

Volume 2
Umesh Kumar Lilhore, Vishal Dutt, T. Ananth Kumar, Martin Margala, Kaamran Raahemifar
Math Optimization for Artificial Intelligence: Heuristic and Metaheuristic Methods for Robotics and Machine Learning
ISBN 978-3-11-143605-0, e-ISBN (PDF) 978-3-11-143618-0, e-ISBN (EPUB) 978-3-11-143624-1

Volume 3
Hemant Kumar Saini, S. B. Goyal, Sardar M. N. Islam, T. Ananth Kumar
Quantum Unmanned Aerial Vehicle: Achieving Cybersecurity and AI Strategies
ISBN 978-3-11-161930-9, e-ISBN (PDF) 978-3-11-161932-3, e-ISBN (EPUB) 978-3-11-161935-4

Volume 4
Shalli Rani, Syed Hassan Sha, Ayush Dogra
Mathematical Computing and Sustainability: Predictive Modeling and Ramifications of Intelligent Systems
ISBN 978-3-11-161111-2, e-ISBN (PDF) 978-3-11-161203-4, e-ISBN (EPUB) 978-3-11-161232-4

Volume 5
Satya Prakash Yadav, Abhay Bansal, Victor Hugo C. de Albuquerque, Fadi Al-Turjman, Houbing Herbert Song
Octonion-Sparse Based Image Processing: Multi Spectral Imaging and Feature Extraction
ISBN 978-3-11-914666-1, e-ISBN (PDF) 978-3-11-220807-6, e-ISBN (EPUB) 978-3-11-220835-9

Volume 6
Ganesh Gopal Devarajan, Daniel Arockiam, Senthil Murugan Nagarajan, Rashmi Agrawal
Healthcare 5.0 with Fuzzy Logic: Artificial Intelligence, Cyber-Physical Systems, and Medical Internet of Things
ISBN 978-3-11-914504-6, e-ISBN (PDF) 978-3-11-221572-2, e-ISBN (EPUB) 978-3-11-221592-0

Volume 7
Narinder Kaur, Bobbinpreet Kaur, Yonis Gulzar, Dac-Nhuong Le
Computational Optimization: Machine Learning and Fuzzy Systems
ISBN 978-3-11-914376-9, e-ISBN (PDF) 978-3-11-221943-0, e-ISBN (EPUB) 978-3-11-221944-7

www.degruyterbrill.com